D0741974

$13.95

The Universe

For Rose

The Universe

Donald Goldsmith
Drawings by Lawrence Anderson

W. A. Benjamin, Inc.
Menlo Park, California • Reading, Massachusetts • London
Amsterdam • Don Mills, Ontario • Sydney

The Universe

Cover and book design by Joseph di Chiarro, Palo Alto

Cover photograph: The Whirlpool Galaxy

Printed in the United States of America.

Published simultaneously in Canada.

Library of Congress Catalog Card Number 75-28643

ISBN 0-8053-3324-X
ABCDEFGHIJKL-HA-79876

W. A. Benjamin, Inc., 2725 Sand Hill Road, Menlo Park, California 94025

Acknowledgments

I would like to thank the astronomers who helped me most in preparing this text: Carl Heiles, Tobias Owen, Lawrence Anderson, Robert Wagoner, and Andrew Fraknoi. Howard Preston gave me invaluable assistance in obtaining photographs, and many others were kind enough to provide copies upon request. I also would like to salute the tireless and competent efforts of James Hall and John Nelson of Author–Publisher Services, without whom this book could not have appeared in its present form.

The publisher and I thank the following individuals and institutions for granting permission to reproduce illustrations and quotations in this book.

Ron Allen: Plates 2 and 3.

American Science and Engineering, Inc.: Figure 9–18.

Arkham House Publishers, Inc., Sauk City, Wisconsin: Opening quotation to Chapter 4, from the fragment *Azathoth* by H. P. Lovecraft (from *Dagon*, copyright © 1965 by August Derleth).

Halton Arp: Figures 5–10 and 5–12.

Cambridge University Press: Opening quotation to Chapter 2, from A. N. Whitehead's *The Concept of Nature*.

Dell Publishing Co., Inc.: Opening quotation to Chapter 10, from Kilgore Trout's *Venus on the Half-Shell*, copyright © 1974–1975 by Scott Meredith Literary Agency, Inc.

Fermi National Accelerator Laboratory: Figure 12–10.

Hale Observatories: Plates 1, 4, 5, 6, 7, and 8; Figures 1–7, 2–2, 4–1, 4–3, 4–4, 4–10, 4–13, 4–15, 4–16, 4–18, 4–24, 5–1, 5–4, 5–5, 5–7, 5–8, 5–10, 5–12, 6–1, 6–3, 6–6, 6–11, 6–20, 7–5, 7–16, 8–13, 9–19, 9–21, 9–22, 10–31, and 10–37. Copyright © by California Institute of Technology and Carnegie Institution of Washington, D.C.

Harvard College Library: Figure 10–14.

Dr. V. P. Hessler, Professor of Geophysics, Emeritus (University of Alaska), and Consultant to the Geophysical Institute of the University of Alaska: Figure 9–16.

International Creative Management, Inc.: Opening quotation to Chapter 9, from *A Study in Scarlet* by Arthur Conan Doyle in *The Complete Sherlock Holmes*, (copyright © 1927 by Doubleday & Company, Inc.).

Leonard Kuhi: Figure 7–3.

Lick Observatory of the University of California: Figures 1–9, 2–17, 3–7, 3–10, 4–2, 4–9, 4–11, 4–12, 4–27, 5–6, 6–2, 6–10, 6–13, 7–2, 8–11, 8–12, 8–14, 9–17, 10–8, 10–32, 10–33, 10–34, 10–35, 12–13, and the cover photograph.

Lund Observatory: Figure 1–6.

Macmillan Publishing Co., Inc.: Opening quotation to Chapter 1, "He Wishes for the Cloths of Heaven," from *Collected Poems* of William Butler Yeats (copyright © 1906 by Macmillan Publishing Co., Inc.; renewed 1934 by William Butler Yeats).

Meteor Crater Enterprises, Inc.: Figure 9–13.

William C. Miller: Figure 8–15.

William Morrow and Company, Inc.: Opening quotation to the book, from *Zen and the Art of Motorcycle Maintenance*, copyright © by Robert M. Pirsig, 1974.

National Aeronautics and Space Administration: Figures 2–21, 9–18, 10–1, 10–3, 10–19, 10–21, 10–22, 10–23, 10–24, 10–25, 10–27, and 11–6.

National Astronomy and Ionosphere Center: Figure 11–5.

National Radio Astronomy Observatory: Plates 2 and 3; Figure 11–4.

Physics Today: Figure 12–15.

Dr. Arno Penzias: Figure 3–8.

Dr. P. Buford Price: Figure 8–17.

Princeton University Press: Opening quotation to Chapter 3, from *Myths and Symbols in Indian Art and Civilization* by Heinrich Zimmer (copyright © 1946 by Bollingen Foundation).

Arnold Rots: Plates 2 and 3.

Prof. Martin Ryle and Dr. P. J. Hargrave: Figure 5–2.

W. W. Shane: Plates 2 and 3.

Hyron Spinrad: Figures 4–22 and 6–27.

Stanford Linear Accelerator Center: Figure 2–6.

Alar Toomre: Figure 6–31.

University of Chicago Press: Figure 3–2.

Viking Press, Inc.: Opening quotation to Chapter 11, from "Humming-Bird." From *The Complete Poems of D. H. Lawrence*, edited by Vivian de Sola Pinto and F. Warren Roberts. Copyright © 1964, 1971 by Angelo Ravagli and C. M. Weekley, Executors of the Estate of Frieda Lawrence Ravagli. All rights reserved. Reprinted by permission of the Viking Press.

Prof. Joseph Weber: Figure 12–14.

Yerkes Observatory: Figure 9–12.

Zuiver Wetanschappelijk Onderzoek: Plates 2 and 3.

Preface

Astronomy, which is as old as any science, has been bursting with discoveries during the past two decades. These new results come from our observations of celestial objects, and from detailed calculations of what may be going on in regions of space quite unlike our own. In this book I have attempted to explain what astronomers now know.

The text follows the history of the universe from its apparent origin in the "big bang," through the formation and evolution of galaxies and of stars, and into the structure and exploration of our solar system. Two final chapters deal with the origin and development of life, and with black holes, pulsars, and the future of the universe. This ordering of topics, although not the usual one for astronomy textbooks, can provide an introduction to the astronomy of today in one-quarter or one-semester courses. Longer courses that also include topics such as coordinate systems, astronomical instrumentation, and the history of astronomy can use this textbook to cover modern astronomical research.

For several thousand years, astronomers have sought to explain what is happening in the heavens. Any set of facts can have many explanations, and everyone tends to choose the explanation that appears best for a given set of facts. Sometimes the "facts" themselves are not generally agreed upon, which makes for lively debate, persecution, confusion, and progress.

The word "astronomy" itself reflects the interaction of human beings with the universe. It comes from the Greek words "astron" (star) and "nomos" (human law). The noun "nomos" springs from the verb "nemo" (to select or to apportion) and the "law" in "nomos" refers specifically to the individual sets of laws that various Greek communities created. This type of law stands in contrast to "logos" (divine law), which appears in the root of the word "astrology," ancient sister of astronomy. The aim of astrology has been to find the universal laws behind the human-made laws for celestial objects. Astronomers content themselves—or they try to—with an understanding of the physical laws that *summarize observations in a coherent way.* Astrologers seek a more basic framework in which a few key facts can provide a master-plan interpretation of life. This search for a permanent set of rules makes it hard for astrology to react to any new piece of information by changing its premises, and this reluctance to change is what makes astrology "unscientific" in astronomers' eyes. All people are fairly stubborn, but it probably would be more difficult to convince an astrologer that the interpretation of Pisces personalities must be changed than it would be to convince an astronomer that the interpretation of quasar distances must be changed.

I myself prefer astronomy, with its strange names, weird personalities, and shifting interpretations of reality. Of course, like anyone else, I am prejudiced: I was trained as an astronomer, and used to teach the subject at the State University of New York at Stony Brook. I have written this book because I feel that people ought to know more about astronomy, and that they will provide a fair return to the person who can explain astronomy in an enjoyable way. Like many students who passed through an introductory astronomy course, I believe that I can improve on the textbooks in use, and I have enjoyed facing this challenge in writing this book.

I have not used algebraic formulas in the text (except for a few footnotes) because, although scientists find mathematical expressions a convenient shorthand, most people do not enjoy symbols and equations. I have employed metric units (grams, centimeters, meters, kilometers), because the metric system makes much more sense than the "Anglo's honest pound." Only the United States now clings to the clumsier system of weights and measures, and attempts at conversion have begun. Some rules for using the metric system are given in the Appendix (Page 404), but the only part of astronomy where this system may cause confusion is in measuring distances within the solar system. For this it is enough to remember that a kilometer is about half a mile.

One final note: At many points in the text, the phrases "as far as we know," "seems to," "apparently," and "appears as though" show up as reminders of the uncertain state of our knowledge. Boring as these uncertainties may be, the plain fact is that I should have inserted more of these phrases, not fewer. I always have enjoyed the realization that what we know about certain areas of astronomy may prove to be all wrong; to separate what we know for sure (almost) from what we know probably (maybe) has a deep and permanent attraction once you let yourself get into it.

January 1976

Donald Goldsmith
Interstellar Media
1655 Twelfth Avenue
San Francisco, CA 94122

Contents

"Phaedrus remembered a line from Thoreau: 'You never gain something but that you lose something.' And now he began to see for the first time the unbelievable magnitude of what man, when he gained power to understand and rule the world in terms of dialectic truths, had lost. He had built empires of scientific capability to manipulate the phenomena of nature into enormous manifestations of his own dreams of power and wealth—but for this he had exchanged an empire of equal magnitude: an understanding of what it is to be a part of the world, and not an enemy of it."

Robert M. Pirsig,
Zen and the Art of Motorcycle Maintenance

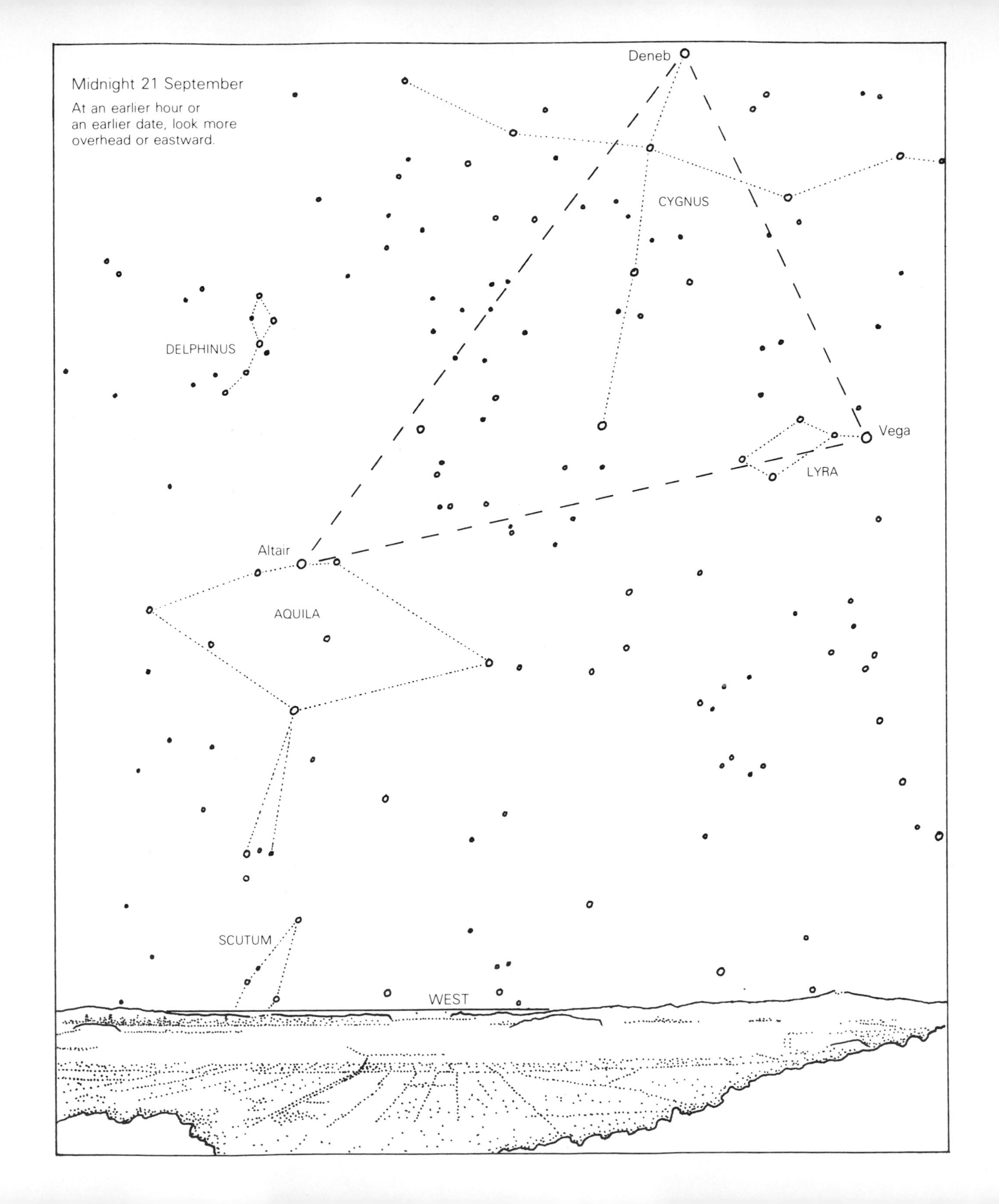
Midnight 21 September
At an earlier hour or
an earlier date, look more
overhead or eastward.
Deneb
CYGNUS
DELPHINUS
Vega
LYRA
Altair
AQUILA
SCUTUM
WEST

Introduction: A Look at the Night Sky

Chapter **1**

"Had I the heavens' embroidered cloths,
Enwrought with golden and silver light,
The blue and the dim and the dark cloths
Of night and light and the half-light,
I would spread the cloths under your feet:
But I, being poor, have only my dreams,
I have spread my dreams under your feet,
Tread softly because you tread on my dreams."

W. B. Yeats

If you stand outside on a clear evening in the fall and look toward the west, you will notice three bright stars that form a large isosceles triangle, called the "Summer Triangle" because it is seen most easily in late summer and early fall (Figure 1–1). Each of these three stars—Vega, Altair, and Deneb—ranks among the twenty brightest stars in the sky, and they are called stars "of the first magnitude," using a naming system that originated two millennia ago.

Vega, Altair, and Deneb seem bright because they are close to us, compared with the distances to most other stars. The next sixty brightest stars in the sky are "of the second magnitude," and the faintest stars visible to ordinary eyesight are some five thousand "sixth-magnitude" stars.[1] From inside a city, even a clear night will reveal only fourth-magnitude stars, because the city lights and haze dim our vision. When astronomers managed to establish the distances to some of the nearest stars, they found that by human standards all stars are incredibly far away. The distance from the Earth to the moon (400,000 kilometers) seemed a mighty journey for humankind to the first men who traveled there, but this journey covered one four-hundredth of the distance from the Earth to the sun, the nearest star. The Earth orbits around the sun at a distance of 150 million kilometers, basking in the heat and light from its giant overlord. Yet, if we compare the distance from the Earth to the sun with the distance to a nearby star such as Vega, the former distance shrinks to nearly nothing, because Vega is one *million* times farther from us than the sun is.

[1]Each successive magnitude is about 2½ times fainter than the preceding one; thus a sixth-magnitude star is one hundred times fainter than a first-magnitude star.

◀ **Figure 1–1.** The western sky at about midnight in late September. The Summer Triangle consists of Vega (the brightest star in Lyra, the Lyre), Altair (in Aquila, the Eagle), and Deneb (in Cygnus, the Swan).

The tremendous leap in distances to other stars from the distance to the sun means that the Earth's orbit around the sun covers no distance at all, cosmically speaking. Every year the Earth makes a journey of 940 million kilometers in its solar orbit. Thus in three hundred thousand years the Earth will cover a distance equal to Vega's distance from us. We can compare our orbit around the sun to a seat on a seashore merry-go-round, looking at ships that are ten thousand kilometers away, in Naples or Yokohama.

Figure 1–2. If we are riding on a lakeside merry-go-round and look at boats on a lake, we will see the nearer boats appear to shift their positions against the background of more distant boats as we move. This apparent shift is called the "parallax effect."

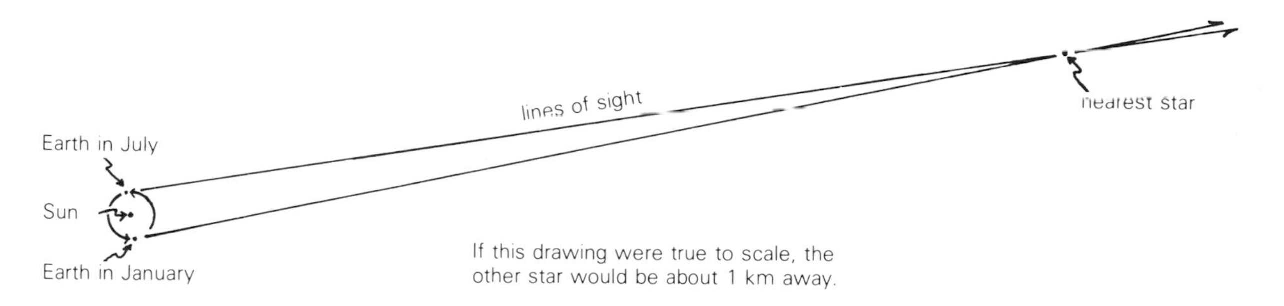

Figure 1–3. The parallax effect in the apparent positions of stars on the sky that arises from the Earth's motion around the sun is extremely small. The reason for the small parallax shift is that the Earth's orbit spans such a tiny distance in comparison with the distances to even the nearest stars.

When people realized that the Earth orbits around the sun once each year they asked: Why don't we see an apparent shift in the position of stars when we look at them from different points in our orbit, say in January and July? Riders on a lakeside merry-go-round looking at rowboats on the lake would see such an apparent displacement of the boats' apparent positions relative to one another (Figure 1–2). This apparent shift in position, called the *parallax* effect, in fact does occur (Figure 1–3), but the amount of this shift is very small because the stellar boats are so far away—thousands of kilometers and more in our merry-go-round model. Even for the stars nearest to us, the apparent shift in position from the parallax effect never equals as much as one "second of arc"—less than the angular size of a dime two kilometers away (Figure 1–4). For comparison, the sun and the moon each have an apparent angular size of half a degree. There are 360 degrees in a complete circle, so half a degree is 1/720 of a full circle around the sky. Each degree contains 60 "minutes of arc," and each minute of arc contains 60 "seconds of arc."

Figure 1–4. The angular size of a dime two kilometers away is one second of arc (1/3600 of a degree). Even this small angle, which is sixty times less than the smallest angular change that the human eye can see, is greater than the parallax shift of Alpha Centauri, the nearest star to the sun.

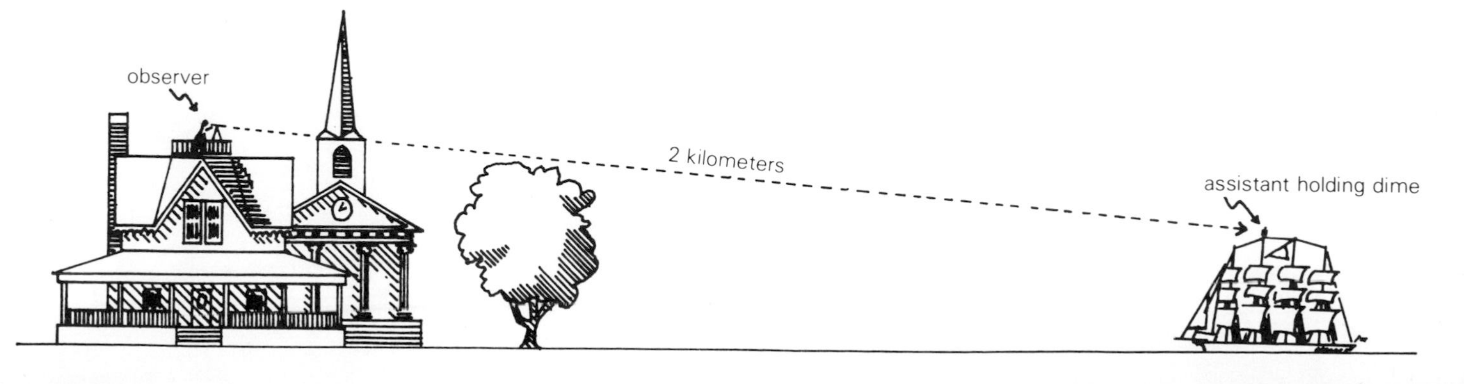

Thus half a degree can be expressed as thirty "minutes of arc" or 1800 "seconds of arc." When astronomers finally succeeded in measuring the parallax effect caused by the Earth's orbital motion, they invented a unit of distance called the "parsec" (from *par*allax plus *sec*ond of arc). If a star's apparent position appears to shift back and forth by one second of arc as the Earth moves around the sun, the star has a distance of one parsec (Figure 1–5). The farther away a star is, the *less* will be the amount of the parallax shift. If the parallax effect shifts the star back and forth by half a second of arc, the star is two parsecs away; if the shift is one tenth of a second of arc, the star is ten parsecs away. That is, any star's distance from us, measured in parsecs, equals one over the parallax shift, measured in seconds of arc. Using trigonometry, we can calculate that one parsec equals about 200,000 times (in actuality, 206,265 times) the distance from the Earth to the sun, so one parsec is about thirty trillion kilometers. Light rays, which travel at a speed of 300,000 kilometers per second, take 3¼ years to cover one parsec of distance, so one parsec equals 3¼ "light years" of distance.

The sun is the only star that lies within one parsec of the Earth. The nearest star to the sun, Alpha Centauri, has a parallax shift of three quarters of a second of arc, so its distance is four thirds of a parsec, and the light from Alpha Centauri takes 4½ years to reach us. Vega has a parallax effect of about one seventh of a second of arc, so Vega lies seven

Figure 1–5. If a star's apparent angular shift back and forth is one second of arc, then the star's distance from us, by definition, would be one "parsec." That is, the long side of a right triangle with the short side equal to the distance from the Earth to the sun, and the vertex angle equal to one second of arc, is one parsec, which is equal to 3¼ light years or about thirty trillion kilometers.

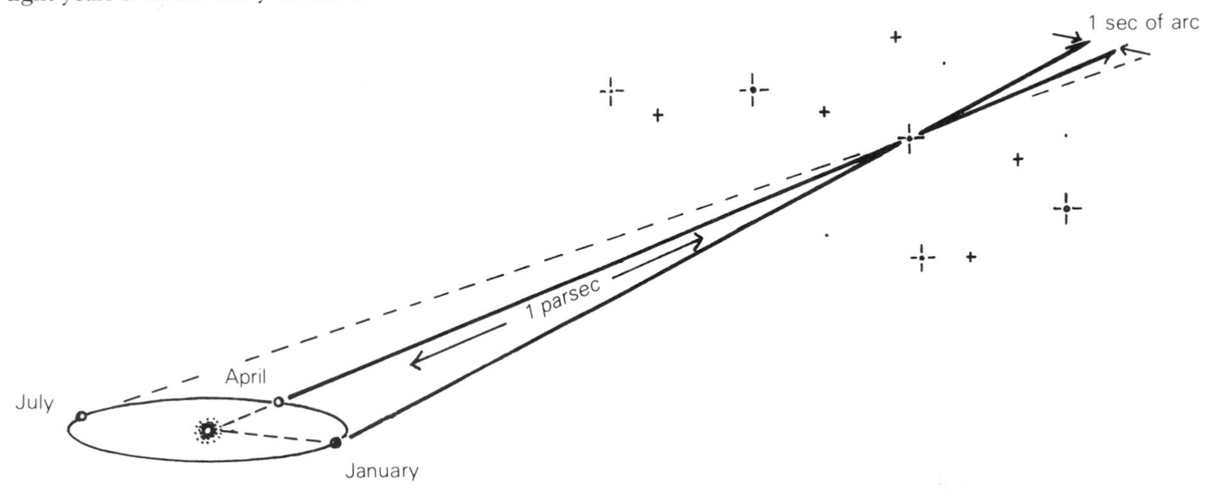

Figure 1–6. When we look toward the central regions of our galaxy, we see stars crowded together to form a "milky way" across the sky. This mosaic drawing made from many photographs shows this band of light, as well as the presence of dark, obscuring matter in front of some parts of the central regions of our galaxy.

parsecs away and its light takes 23 years to cover this distance. Altair, which is slightly closer to us, has a parallax of one fifth of a second of arc, a distance of five parsecs, and a travel time of 18 years for its light to reach us. Deneb is considerably farther from the Earth—far enough, in fact, that the amount of its parallax shift cannot be measured accurately, because we cannot determine changes in position to better than about one thirtieth of a second of arc. The best estimate of Deneb's distance from us is 450 parsecs, so its distance is ninety times the distance to Altair, and Deneb's light takes fifteen hundred years to reach us. Thus today we can determine only how things on Deneb were at the time of the fall of the Roman empire.

The light that reaches us from the stars carries messages from the past. These messages arrive continuously in the form of tiny "elementary" particles called photons, and these messages allow us to study the universe. At any moment, the photons' messages—light waves, radio waves, x rays, and infrared radiation—are out of date, because the photons take time to travel at the speed of light from the source that produced them to the person who decodes their information. Moving at a speed of 300,000 kilometers per second, light waves from the sun take more than eight minutes to reach the Earth—eight minutes of grace should the sun suddenly snuff out. The same speed through space allows the photons from Polaris, the pole star, to reach us in 650 years, and the light from Rigel, the giant blue star in Orion's foot, takes 900 years to cover the 275 parsecs from there to here.

All these stars, with their distances from Earth of 4/3 to 450 parsecs (400 trillion to fourteen thousand trillion kilometers), lie basically next door to us. Our sun belongs to a giant group of a hundred billion stars called the Milky Way galaxy, which has a diameter of forty thousand parsecs. The sun, Alpha Centauri, Vega, Altair, Deneb, Polaris, and Rigel

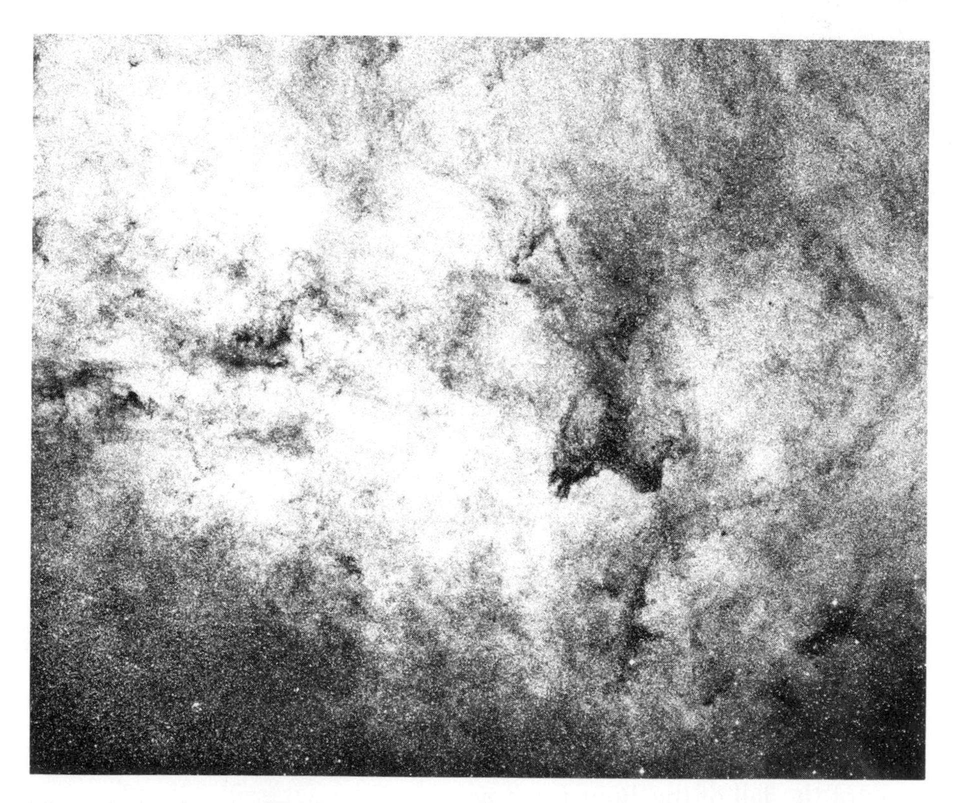

Figure 1–7. A powerful telescope can resolve many of the stars in the Milky Way into individual points of light. This photograph shows the central part of the Milky Way, located in the constellation Sagittarius.

all are within a small volume far out from the central regions of the Milky Way galaxy. In the summer and fall we see these central regions forming a diffuse band or "milky way" across the sky (Figure 1–6). This band of light comes from millions of stars that are about eight thousand parsecs away, near the center of our galaxy. The stars' great distances from us—a thousand times the distance to Vega—prevent us from seeing them as individuals, although a telescope lets us separate them into points of light (Figure 1–7).

Is the Milky Way the farthest that we can see without a telescope? Not at all, if you know where to look. If you examine the sky directly to the south and about two thirds of the way up from the horizon to the zenith (directly overhead), around eight o'clock on a November night. you will see four second-magnitude stars that form a rectangle with its long axis parallel to the horizon (Figure 1–8). This is the "great square of Pegasus," which is supposed to represent the flying horse of Greek mythology. From the top left-hand star in the rectangle, two parallel lines of three stars each, one bright (second-magnitude stars) and one faint (fourth-magnitude stars), stream upward and to the left, forming the maiden Andromeda (whom Perseus, riding on Pegasus, rescued from Cetus, the sea-monster, when her parents, King Cepheus and Queen Cassiopea, had left her as a propitiatory sacrifice). If you find the second star in each of these lines, and look upward from the lower through the upper of these middle stars, going the same distance as the separation of these stars (Figure 1–8), you can see a fuzzy object (best observed by averting your vision slightly), the great Andromeda galaxy, which is much like our own Milky Way galaxy, but *half a million parsecs away* (Figure 1–9). The light from the Andromeda galaxy has taken more than 1½ million years to reach us: It left at a time when human beings were combing the Olduvai gorge for edible roots and berries.

The Andromeda galaxy has the greatest distance of any objects readily visible to the unaided (naked) eye, although in the right circumstances, say on a clear night in Mexico when the proper precautions have been taken, you can see glimpses of galaxies ten times farther away than the Andromeda galaxy, shining by light that has taken fifteen million years to reach us. Telescopes have revealed more and more distant galaxies, farther and farther away from us—as far away as three *billion* parsecs. The light from a galaxy three billion parsecs away has taken ten billion years to reach us; thus it left before the Earth and the sun had condensed from the primeval blob that became the solar system.

Is this the most distant source of light; are these the oldest photons? Not at all, once again. The oldest photons that reach us come from a time when the universe was completely different from the way it is now, a time when no galaxies, no stars, no planets existed, a time in fact before any grouping of matter into separate clumps had occurred.

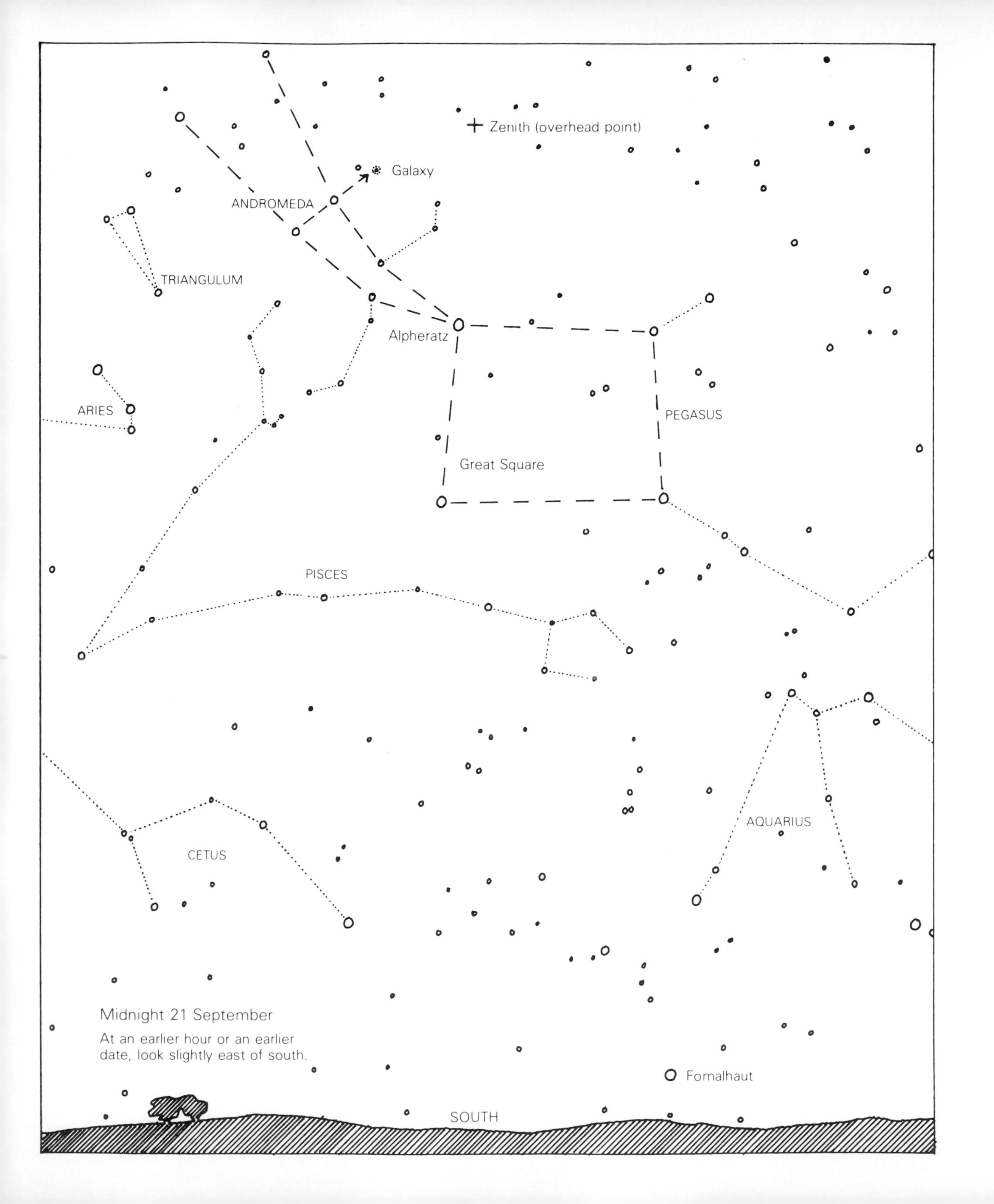
Zenith (overhead point)
Galaxy
ANDROMEDA
TRIANGULUM
Alpheratz
ARIES
PEGASUS
Great Square
PISCES
AQUARIUS
CETUS
Midnight 21 September
At an earlier hour or an earlier date, look slightly east of south.
Fomalhaut
SOUTH

◀ **Figure 1–8.** At midnight about September 21, or at 8 p.m. about November 21, the constellations of Pegasus and Andromeda lie almost directly to the south. We can use our identification of the Great Square of Pegasus to find the two chains of stars in Andromeda, and thus to find the Andromeda galaxy.

These oldest of all photons fill the universe with "microwave radiation" (Chapter 3) that has been traveling for fifteen to twenty billion years. The photons in this microwave radiation carry a message at least fifteen billion years old, which was detected first in 1965 A.D. on our calendar. The entire Christian epoch has lasted one ten millionth or so of the time that these photons have been traveling; surely it is worth trying to find out what they have to tell us. One clear message, so often frightening to humanity, contrasts our brief lives to the mighty roll of the universe. If we compare the flow of time in the universe since the microwave photons were formed to the story of a railroad journey from San Francisco to New York, then the photons from the farthest galaxy that we can see got on in Omaha. The Earth formed when we passed by Detroit, life on Earth appeared around Cleveland, mammals got on in Newark, and humans have been riding for about half a block. The record of human "civilization" would occupy at most one millionth of the trip, or about five meters, and an average human lifetime about two centimeters.

Within these figurative two centimeters lies our chance to connect with the rest of the universe. We can decipher the messages that photons bring us, can reconstruct the history and development of the universe, and even can start to understand how it all works. The plan of this book is to ride that train from the edge of darkness through the past toward the unknown future, as we examine how the universe began, how it has changed, and how it will continue to change in ways only partly known to us now.

Summary

The apparent brightness of a star depends on two quantities: the star's actual brightness and its distance from us. If we can determine the distance to a particular star, then we can calculate the star's true brightness from its apparent brightness. This determination depends on the fact that the star's apparent brightness varies in proportion to one over the square of its distance from us. If two stars have the same true brightness, but one is three times more distant, then the farther star will *appear* nine times fainter than the nearer star.

Astronomers measure distances in light years (the distance that light travels in one year) and in parsecs (one parsec = 3¼ light years). Par-

◀ **Figure 1–9.** The Andromeda galaxy is a giant spiral much like our own Milky Way galaxy. Two smaller galaxies, satellites of the Andromeda galaxy, appear as elliptical blobs above and below the main galaxy.

secs are natural units for measuring the distances to the nearer stars because they reflect our measurements of the "parallax effect." The Earth's motion around the sun would make a star one parsec away appear to shift its position back and forth by one second of arc against the background of much more distant stars. A star five parsecs away would have a proportionately smaller parallax shift of one fifth of a second of arc. We can measure the distances of stars up to about 30 parsecs (100 light years) with the parallax effect; for greater distances, the parallax shift becomes too small for accurate measurement.

Because light takes time to travel immense distances, whatever we see on the sky represents the past history of the universe—four years ago for the star nearest the sun, 1½ million years ago for the Andromeda galaxy. The light from the most distant galaxy that we can see has taken ten billion years to reach us, a time that is at least half the age of the universe.

Questions

1. As we go from first-magnitude to sixth-magnitude stars, are we going from brighter to fainter stars, or from fainter to brighter stars?
2. The Earth's motion around the sun causes nearby stars to appear to shift their position in the sky back and forth across the background of much more distant stars. Why does the amount of this "parallax effect" *decrease* for the more distant stars?
3. The unaided human eye can detect changes in position as small as one minute of arc, which is equal to sixty seconds of arc. If an object shows a parallax shift (back and forth) of one minute of arc, what would the distance to the object be, measured in parsecs?
4. Light takes 3¼ years to travel one parsec of distance. How many years does light take to travel from the star Arcturus (10 parsecs away) to reach us? How many years does it take for the light from the galaxy M 87 (10 million parsecs away) to reach us?
5. One year equals about 32 million seconds. If we were to compress the entire history of the universe (about 16 billion years) into one year, how many seconds would the history of the human race (about 1.6 million years) occupy?

6. The magnitude scale of brightness is constructed such that a difference of *five* magnitudes implies a brightness ratio of one hundred. That is, one magnitude difference means 2.512 times brighter, two magnitudes means $(2.512)^2 = 6.310$ times brighter, and five magnitudes means $(2.512)^5 = 100$ times brighter. What is the ratio of brightnesses of a first-magnitude and an eleventh-magnitude star? What is the ratio of brightnesses of a star of the third magnitude and a star of the twenty-third magnitude, which is about as faint as can be seen with the largest telescopes?

Further Reading

Charles Whitney, *Whitney's Star Finder* (Alfred Knopf, New York, 1974).

H. J. Bernhard, D. A. Bennett, and H. S. Rice, *New Handbook of the Heavens* (Mentor Books, New York, 1948).

W. Ley, *Watchers of the Skies* (Viking Compass Books, New York, 1969).

The Physics of Astronomy

Chapter 2

"The aim of science is to seek the simplest explanation of complex facts. We are apt to fall into the error of thinking that the facts are simple because simplicity is the goal of our quest. The guiding motto in the life of every natural philosopher should be 'Seek simplicity and distrust it.'"

Alfred North Whitehead

Photons: Energy, Frequency, and Wavelength

What we know about astronomy we know from photons, which are the elementary particles that form light waves, radio waves, and all the other kinds of "electromagnetic radiation." If we see the stars of Orion or Gemini, we see them because photons that the stars produced have crossed trillions of kilometers of space, entered our eyes, and pressed on the rods and cones of our retinas to cause chemical changes that triggered messages down our optic nerves to our brains, which note that we are seeing stars. In a similar way, radio signals from the sun, from other stars, and even from exploding galaxies can be detected when the photons that form these waves reach and excite a properly designed receiver mounted at the focus of a radio telescope. Although some elementary particles that are not photons do reach the Earth from outer space (these particles, called "cosmic rays," will be discussed in Chapter 8), the facts about astronomy that we have gleaned so far consist essentially of a summary of observations of photons. Thus photons ought to begin our look at the particles that comprise the universe.

What are these special carriers of astronomical information that we call photons? They are tiny, "elementary" particles, one variety of the basic building blocks of the universe as we know it. Photons compose and carry all forms of electromagnetic radiation (light waves, radio waves, x rays, gamma rays, ultraviolet radiation, infrared radiation, and microwave radiation). Furthermore, photons communicate electromagnetic forces, such as electrical attraction and repulsion, or magnetic fields of force, from one region of space to another. Photons have properties that are highly unusual compared to the world of large particles. These unusual properties remind us that the elementary building blocks of nature are not simply smaller versions of what they can form. Instead, in the world of elementary particles, we can glimpse the basic mysteries of nature, which run counter to our intuitions that arise in a world of larger objects.

Most notable among the properties of photons are the following: First, photons have no mass, and thus weigh nothing whatsoever. Second, photons always travel through unobstructed regions of space at the same speed, 299,793 kilometers per second, which we shall round off to 300,000 kilometers per second and call "c." Third, each photon carries some energy of motion, which ranges from nearly zero to an energy greater than that of a speeding bullet. Fourth, each photon has a characteristic *frequency*, which varies in direct proportion to the photon's energy, and an equally characteristic *wavelength*, which varies inversely with the photon's energy. The product of the frequency times the wavelength for *any* photon equals the speed of light, c. We shall consider these facts a bit more carefully.

1. Photons have no mass. The "mass" of an object gives a characteristic measure of the object's resistance to motion. We cannot drive a football three hundred meters with a golf club because a football has more mass than a golf ball. Everything that we familiarly call an "object" has some mass, but photons do not.[1]

2. If photons can travel freely through unobstructed space, they will move in straight lines at the speed (c) of 300,000 kilometers per second. If some kind of matter blocks a photon's path, the photon may be deflected or even destroyed by its encounter with this matter, but in the free space of a vacuum, all photons travel at the same speed.

3. Each photon carries a certain amount of energy of motion (also called "kinetic energy") that characterizes the photon. Photons all travel at the same speed, but they differ from one another in the amount of energy carried along at the speed of light. High-energy x-ray photons can pass through our bodies, whereas lower-energy light waves are stopped by even a layer of cloth. When a photon interacts with matter, the photon can deposit some, or all, of its energy in the matter. If a photon loses some of its energy, it becomes a different photon; if it loses all of its energy, it disappears. For example, ultraviolet photons from the sun can deposit their energy of motion in the outer layers of our skin, thereby causing sunburn in white people. Energy from the sun's photons allows life on Earth to exist and to develop, mainly through the chemical changes that occur when photons interact with chlorophyll molecules in plants.

   If large numbers of photons strike an object, the energy they deposit produces a pressure on the object called "radiation pres-

[1]As Einstein's equation $E = mc^2$ implies, mass can turn into energy and energy into mass (Page 25). Since energy equals mass times the speed of light squared, mass equals energy *divided* by the speed of light squared. In this sense, a photon's energy of motion gives it some mass, but not the kind of mass we usually think about when we think of an object such as a golf ball. However, the photon's mass-out-of-energy does allow the photon to feel the gravitational effect of massive objects (Page 143).

sure." The sun's photons produce a small amount of radiation pressure on objects here on Earth, although human beings cannot sense this pressure directly. However, the radiation pressure inside stars can be quite important because of the great number of photons there. The basic reason that stars much more massive than the sun cannot exist is that the increasing radiation pressure would push the star apart.

4. Each photon has a characteristic frequency and wavelength that are directly related to the photon's energy. We can think of a photon as a tiny bundle of energy that travels through space in a straight line, wiggling as it goes (Figure 2–1). The number of the photon's wiggles per second that pass a given point in space gives the photon's frequency, measured in wiggles per second (or "cycles per second"). Scientists call one wiggle per second one "hertz," in honor of their deceased colleague Heinrich Hertz, and they abbreviate this quantity as "Hz." Photons in light waves have frequencies of trillions of Hz; that is, they wiggle trillions of times per second. Radio-wave photons have frequencies of millions of Hz ("megahertz," abbreviated MHz, denotes one million wiggles per second). The frequency of a photon varies in direct proportion to its energy, so a photon with a frequency of six trillion Hz has an energy three times the energy of a photon with a frequency of two trillion Hz.

In addition to its frequency, each photon also has a characteristic wavelength, which measures the distance the photon advances every time it wiggles once. Light waves have wavelengths measured in one hundred-thousandths (10^{-5}) of one centimeter, so we may imagine that light waves consist of photons that advance a few hundred-thousandths of a centimeter every time they wiggle. For any photon, the frequency (wiggles per second) times the wavelength (distance advanced per wiggle) equals the speed of light (c = 300,000 kilometers per second). This means that the larger the frequency, the smaller the wavelength, and vice versa, since frequency times wavelength always gives the same

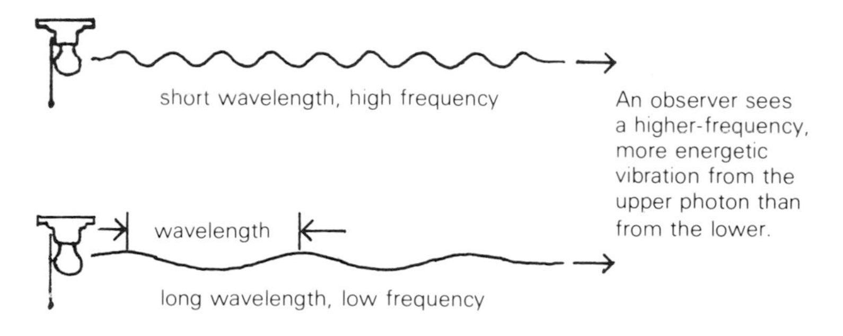

Figure 2–1. We can picture a photon as a bundle of energy, wiggling its way through space with a characteristic frequency of vibration and an equally characteristic wavelength of oscillation. The energy of each photon varies in direct proportion to the photon's frequency, and as one over the photon's wavelength.

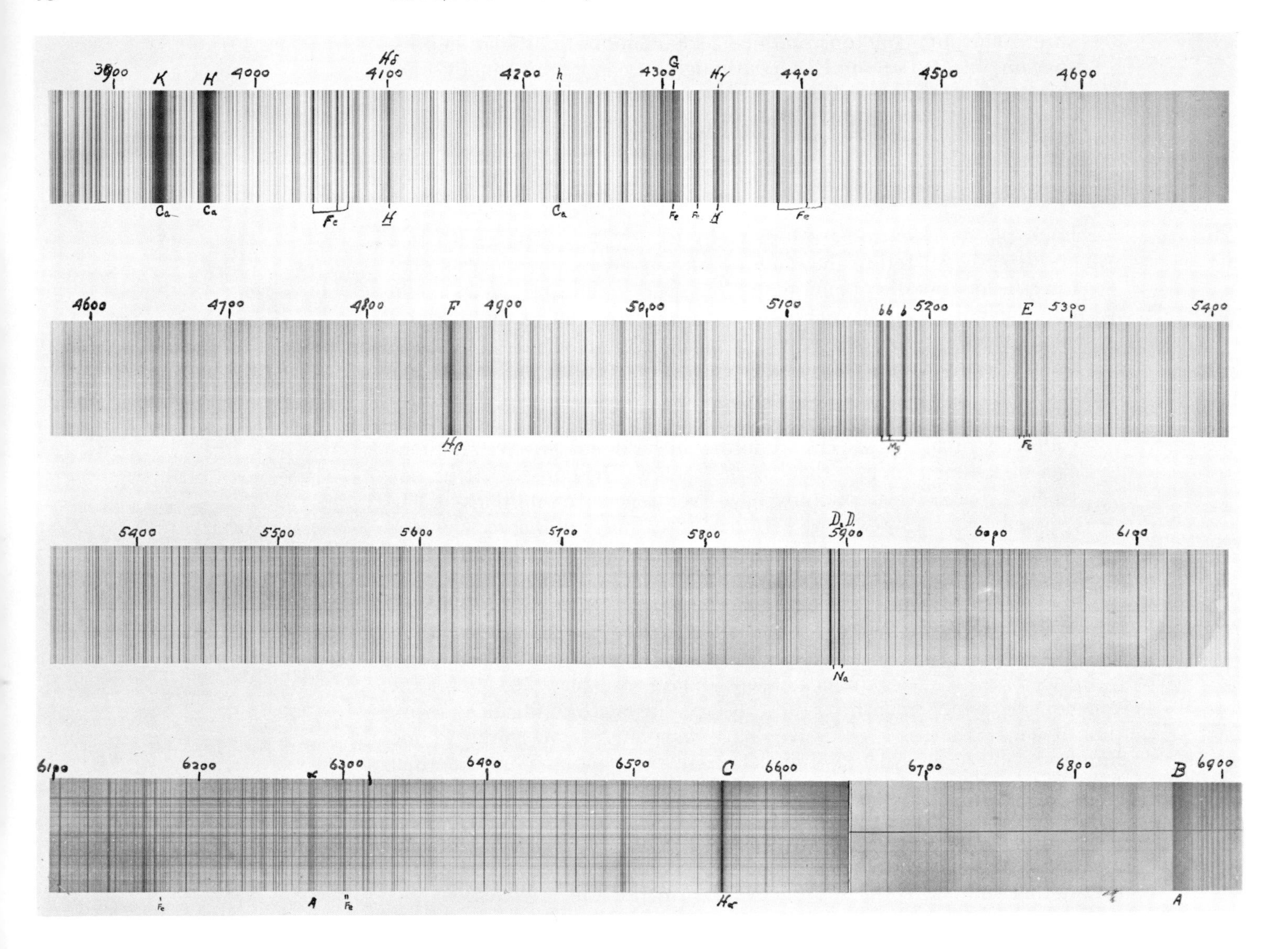

Figure 2–2. A photograph of the sun's light separated into its spectrum of photon energies, frequencies, and wavelengths shows dark lines at those frequencies or wavelengths where few photons emerge from the sun. Here the spectrum is cut into strips placed one above the other for easier viewing. The numbers above the spectrum give the photon wavelengths in angstroms (one angstrom equals one ten-billionth of a meter). The darkest "absorption lines" show the removal of photons by calcium, sodium, hydrogen, and iron atoms and ions at the frequencies or wavelengths at which these elements can absorb photons.

product. Therefore, a photon's wavelength *decreases* if the photon's energy increases, and the wavelength increases if the photon's energy decreases. Since the energy and frequency are directly proportional, it follows that if we know *one* of these three quantities—energy, frequency, and wavelength—we can calculate the other two. The essence of any photon consists of its energy, its frequency, or its wavelength. We usually will deal with the photon's energy as its primary characteristic, but all three quantities always are linked together.

Our eyes respond to a small band of energies (or of frequencies or wavelengths) called "visible light." Photons with energies within this range interact with the molecules in our eyes that react to light, and our eyes thus can discriminate within this energy range among various energies or "colors" of the photons that arrive. Among visible-light photons, red light has the least energy per photon (hence the longest wavelength) and blue light has the largest energy per photon; other colors fall in between.

Astronomers regard detecting photons as their particular province of science, and they bend all their subtle powers toward the goal of more accurate photon detection. Since photons differ from one another in the amount of energy they carry (or, equivalently, in their frequencies and wavelengths), astronomers count the number of photons of different energies that arrive from a photon source such as our sun. The most convenient way to describe the results of this counting is to show the "spectrum" of photons. Such a spectrum presents, for each energy that a photon can have, the number of photons with that particular energy that have been detected. Astronomers have developed "spectroscopes" and "spectrographs" to spread light into its spectrum of energies and to measure the number of photons with each energy. If we spread visible light waves into the spectrum of photons with various energies, we can use the photons themselves to show the number of photons of each energy. We do this simply by photographing the light once we have separated it into its different energies (or colors) by passing the light through a prism or bouncing it off a finely ruled grating: More photons of a particular energy will leave more light at that energy's spot in the picture. Figure 2–2 shows the spectrum of light from the sun, arranged by decreasing energy (from left to right and from top to bottom).

How large is a photon? If a photon can be thought of as having a "size," that size is the photon's wavelength. Table 2–1 shows the energies, frequencies, and wavelengths of various kinds of photons. The photons' energies are measured in "ergs," a unit of energy much loved by physicists. (One erg is the amount of energy needed to give an object that has a mass of two grams a velocity of one centimeter per second. One erg also approximately equals the energy that an underweight housefly expends in taking off as the fly swatter descends, or the amount of energy needed to power a light bulb for one billionth of a second.)

Table 2–1
Typical Energies, Frequencies, and Wavelengths of Various Types of Photons

	Gamma rays	X rays	Ultraviolet	Visible light	Infrared	Microwaves	Radio and television
Energy per photon (in ergs)	10^{-6}	10^{-8}	10^{-10}	5×10^{-12}	10^{-13}	10^{-15}	10^{-17}
Frequency of photon (in Hertz)	10^{20}	10^{18}	10^{16}	5×10^{14}	10^{13}	10^{11}	10^{9}
Wavelength of photon (in centimeters)	3×10^{-10}	3×10^{-8}	3×10^{-6}	6×10^{-5}	3×10^{-3}	0.3	30

We can see from Table 2–1 that radio-wave photons have *wavelengths* of many centimeters or even meters. Does this mean that these photons are really this large? In a practical sense, the answer is yes: If we want to interfere with such photons, or capture them, we must make a trap larger than they are—larger, that is, than the photons' wavelength. Photons interact most with objects equal to their own size, that is, their own wavelength. Therefore, we build a television antenna about one meter long to capture the photons with one-meter television wavelengths. These photons pass through thin walls as if they didn't exist. Our eyes have special cells (the rods and cones) a few times 10^{-5} centimeters wide that interact with light waves that have wavelengths of a few times 10^{-5} centimeters.

Photons, then, each carry a certain energy, which in turn indicates their characteristic frequency and wavelength. They all have no mass and travel at the speed of light, diffusing outward from the sources that produce them to spread radiation throughout the universe.

Distances and Total Energy Output

How do we measure the total photon output of a source of photons such as our sun or another star? There are two things to determine: the *number* of photons produced each second, and the total amount of photon *energy* produced each second. Since various kinds of photons have different energies, these two quantities need not be in direct proportion for all stars. Some stars produce fewer high-energy photons each second than other stars, which produce many low-energy photons per second. Our sun, which is a rather average star, emits about 10^{45} photons each second. The average energy per photon is about 4×10^{-12} erg, so the total solar energy output is 4×10^{33} ergs per second. Thus in each second the sun produces enough energy to satisfy human needs for almost a million years at our present rate of energy consumption.

Of course, most of this energy never reaches the Earth. If we imagine the photons as fiery tadpoles shot out at the speed of light from the sun,

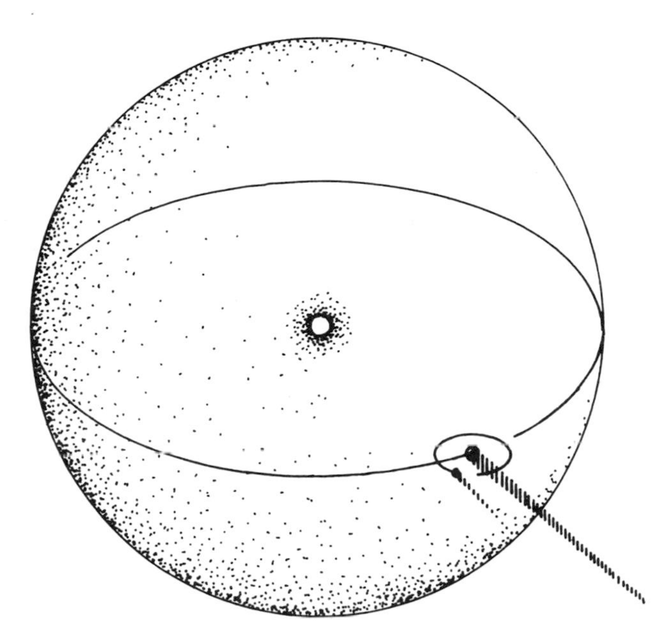

Figure 2–3. The Earth intercepts a fraction of the sun's total photon output that equals the target area of the Earth divided by the total area of a sphere drawn around the sun at the Earth's distance from the sun. This fraction is less than one part in a billion.

we can see that most of them miss the Earth entirely and shoot by us out into space. In Fact, the Earth is so far from the sun that less than one billionth of the sun's photons strike the Earth's surface.[2] If the Earth were farther from the sun, a still smaller fraction of the sun's photons would hit the Earth. From geometry we know that one sphere that is twice as large as another has four times the surface area of the smaller sphere (the area increases as the square of the sphere's radius).

Picture the Earth as a dot on the surface of a sphere with the sun at the center (Figure 2–3). Photons from the sun pass outwards through the sphere's surface in all directions, and a small fraction of them hit the Earth. If the Earth were twice as far from the sun, one quarter as many photons would strike it each second. The area of the spherical shell through which the photons pass would be four times larger, while the Earth's size remains the same. If the Earth were three times farther from the sun, one ninth as many photons would arrive from the sun each second. This relationship of the number of arriving photons with the distance from Earth is at the heart of observational astronomy. On Earth we can measure only the number of photons—from the sun, from a star, from a galaxy—that *reach* the Earth. To determine how many photons a source *emits* per second, we must know the fraction of the surface of the sphere around the source that the Earth occupies, as

[2]The solar energy that does reach the Earth each second exceeds our present energy consumption by several thousand times.

drawn in Figure 2–3. To find this fraction we must know the distance from ourselves to the source of photons.

If we can measure the distance to a source of photons, we can calculate the number of photons it emits each second by measuring the number we receive each second and allowing for the fact that the number of photons that reach us decreases in proportion to one over the square of the distance to the source. The stars Vega and Castor emit about the same number of photons and the same amount of energy each second, but Vega is twice as close to us as Castor and *appears* to be about four times brighter. If we somehow could determine the actual number of photons emitted each second from a particular source (for

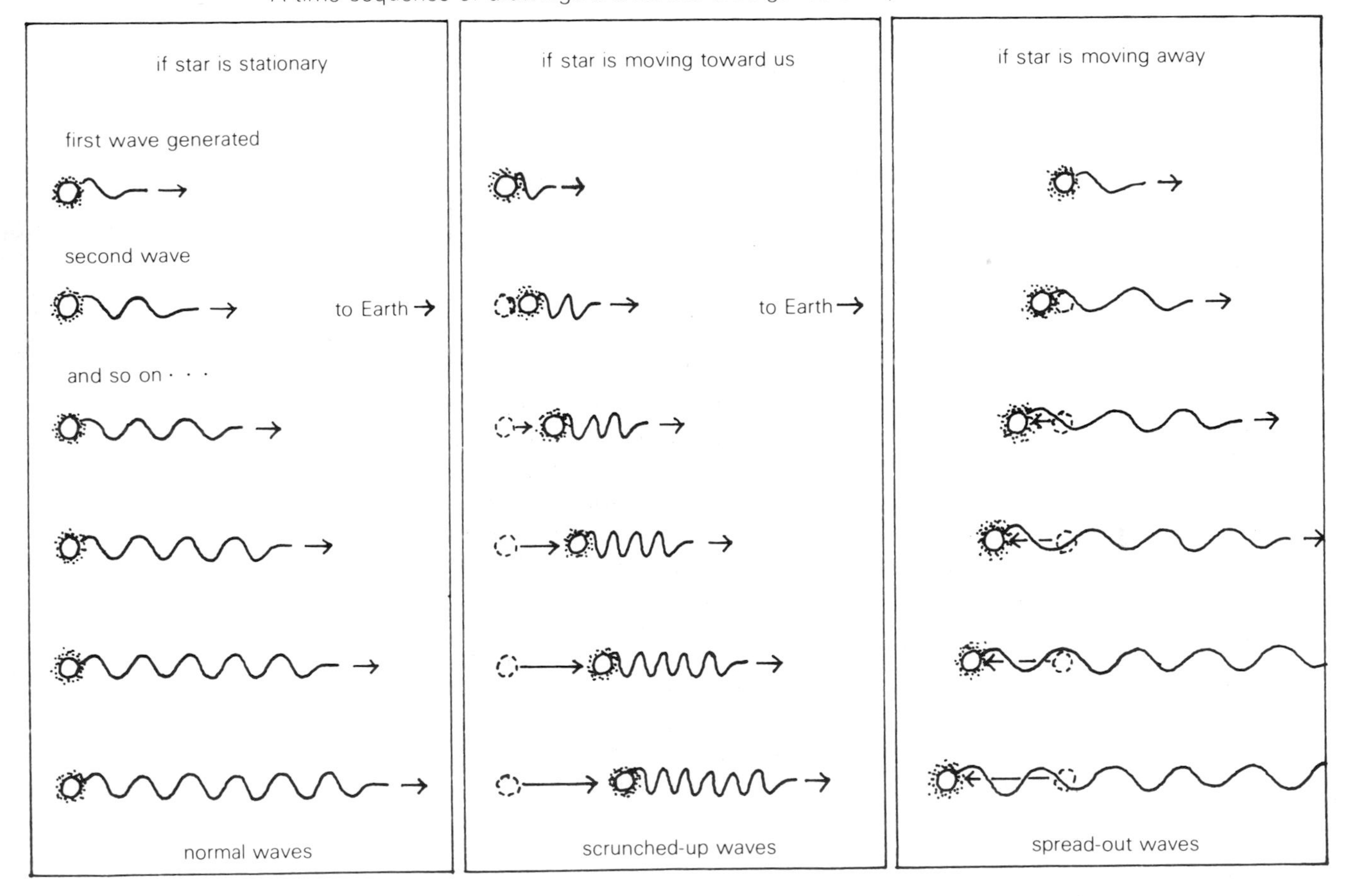

In each case, the waves are emitted once per time step for six equal time steps. Notice that in each case the first wave moves the same distance from the old star position during the six time steps. However, in the second sequence, each succeeding wave is emitted from a position slightly closer to us, thereby producing a shorter wavelength.

example, by recognizing it as almost identical to a source we already have measured), then we can turn the relationship between distance and apparent brightness around, and find the distance by comparing the actual number of photons the source emits with the number that we observe. Thus if we knew that Vega and Castor each emit the same number of photons and the same amount of energy each second, we could deduce that Castor must be twice as far from us as Vega is, from the fact that Castor appears to be only one fourth as bright as Vega. Both lines of reasoning—either working from the distance to the number of photons a source emits, or from the number it emits to finding its distance—arise from the geometrical fact that *the observed number of photons decreases with the square of the distance.*

Astronomers call the number of photons observed each second from a source the "apparent luminosity" of the source (measured in photons per second), and the actual number of photons emitted each second the "absolute luminosity" of the source (in photons per second). Similarly, if we measure the total photon *energy* reaching us, or the total photon energy emitted by the source, we have the apparent luminosity or the absolute luminosity measured in ergs per second (energy per second). Vega and Castor have the same absolute luminosity, but Vega's apparent luminosity is four times Castor's. Many astronomers constantly work at measuring the apparent luminosities of photon sources and then try to determine the sources' absolute luminosities. This is the first step toward understanding what the sources themselves consist of.

The Doppler Effect

One property of photons that we have not discussed yet has been a godsend to astronomers. Photons always are observed traveling at the speed of light, *no matter how the source of photons may be moving.* This fact has been checked repeatedly by experiment, although it contradicts all our inner, human intuition about how speeds add or subtract. A photon arriving from a star that is moving toward us at 100 kilometers per second arrives at the speed c, *not* $c + 100$ kilometers per second. Photons from stars moving away from us also arrive with speed c, as do photons from stars that are moving neither toward us nor away from us. Although the source's speed toward us or away from us does not affect the photons' velocities, this motion does affect the photons' energies, and hence their frequencies and wavelengths.

If a photon source is moving toward an observer (us), we observe the photons to arrive with greater energies than they would have if the source were not moving toward us. Conversely, motion away from an observer decreases the energies of the arriving photons (Figure 2–4).

◀ **Figure 2–4.** The relative motion of a photon source toward us *increases* the energies of the photons that we observe, in comparison to the energies we would see if the source were not moving with respect to us. Conversely, relative motion away from us *decreases* the energy per photon that we observe.

Only the *relative* motion is important: Whether we are moving toward the source or the source is moving toward us, in either case we would observe that the photons from the source have greater energies than they would if we were not moving in relation to the source. The same is true for a source's motion away from us: The observed photon energies decrease whether we are moving, the source is moving, or both are moving. The amount of the change in the energies of the photons from a moving source, whether the change is an increase (approach) or a decrease (recession), depends on the *relative* velocity of the source with respect to us. Large velocities produce large changes, small velocities produce small changes. If the velocities involved are much smaller than the speed of light, the photon gains or loses a fraction of its original energy that is equal to the source's relative velocity (approaching or receding) divided by the speed of light.[3]

This relationship between velocity of approach or recession and the change in a photon's energy was discovered first by C. J. Doppler and therefore is called the "Doppler effect." This change in photon energy allows astronomers to determine whether stars and galaxies are moving toward us or away from us. From long years of observation astronomers concluded that stars preferentially produce photons with certain energies. If they then observed the photons from a new star and found an especially large number of photons with energies a little bit larger than the usual ones, and if the energies were all larger than the usual ones by the same proportion, then astronomers concluded that the star must be approaching us (or we must be approaching the star), and that the Doppler effect must be responsible for the change in the observed photon energies. From the measured amount of the energy change, called the "Doppler shift," astronomers can determine directly the relative velocity, since the fractional change in the energy equals the relative velocity divided by the speed of light.

Using the Doppler effect, astronomers looking along the line of sight toward Vega have found that Vega is receding from the solar system (or

[3]The mathematical relationship between the observed photon energy and the original photon energy is

$$\frac{\text{observed energy}}{\text{original energy}} = \sqrt{\frac{1 - (v/c)}{1 + (v/c)}}$$

in which the velocity v is negative for velocities of approach and positive for velocities of recession; c is the speed of light. If v is very much less than c, this relationship is nearly the same as

$$\frac{\text{observed energy}}{\text{original energy}} = 1 - \frac{v}{c}$$

which means that the *change* in energy divided by the original energy equals v/c (an increase for velocities of approach, a decrease for velocities of recession).

we from Vega) at 14 kilometers per second, while Rigel, the bright star in Orion's foot, is approaching us at 24 kilometers per second. By now, astronomers have measured the relative velocities of recession or of approach for more than 10,000 of the brightest stars in the sky.

Another important aspect of the Doppler effect is that it measures only the velocity along our line of sight (toward us or away from us). The true velocity of an object through space can be considered as the sum of two "velocity components," one along our line of sight and the other perpendicular to our line of sight. If we want to know the full story of an object's velocity, we must measure both velocity components, just as a football quarterback must know how fast his pass receiver is running down the field as well as how fast he is running across the field before he can throw the ball accurately. The Doppler effect provides us with only one of the two velocity components, the one *along* our line of sight. We cannot determine the velocities of stars *across* our field of view from the amount of the Doppler shift. To say that Rigel has an approach velocity, measured by the Doppler effect, of 24 kilometers per second does not mean that Rigel is heading straight at us (Figure 2–5). Instead, it means

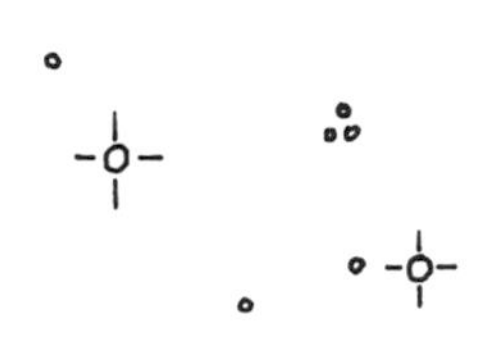

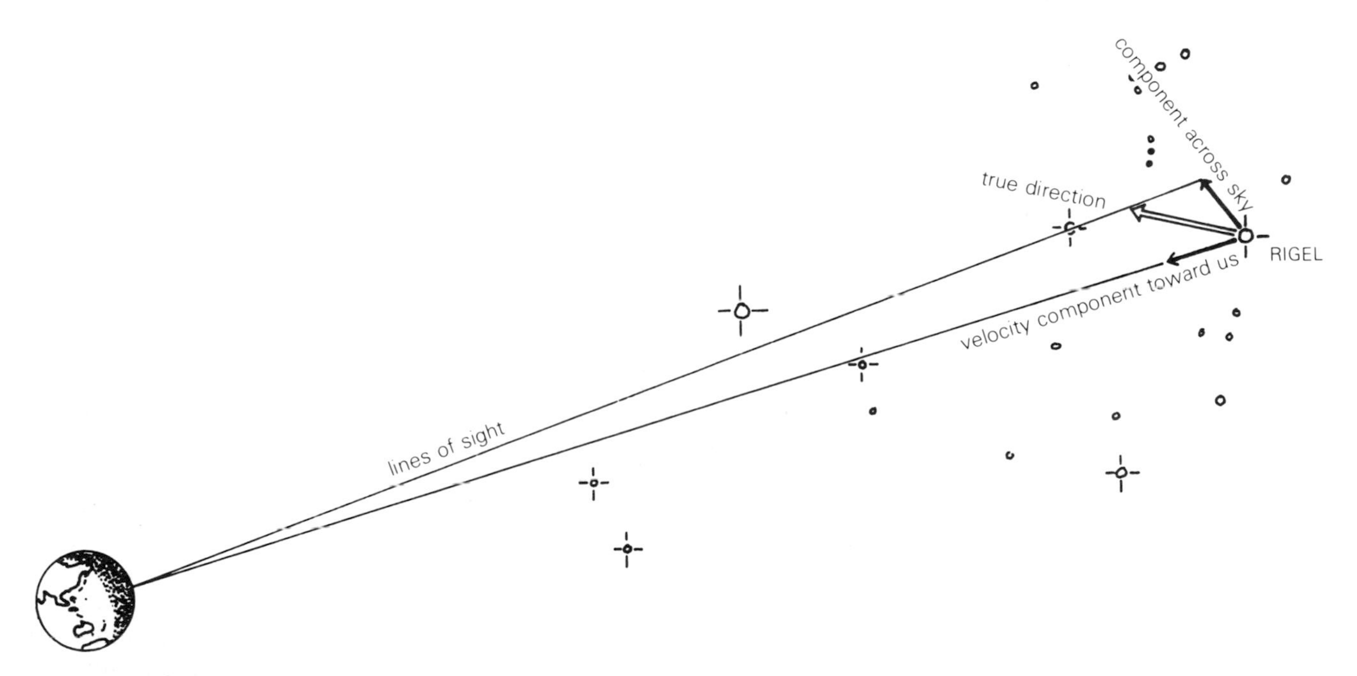

Figure 2–5. If we measure the Doppler shift of light from the star Rigel, we find that Rigel has a relative velocity of approach equal to 24 kilometers per second. However, the Doppler shift measures only the component of Rigel's velocity along the line of sight, that is, toward us or away from us. The velocity component across our line of sight cannot be determined from the Doppler effect, so we cannot tell whether Rigel is coming straight at us or (more likely) is merely headed more toward us than away from us.

that Rigel's motion is more toward us than away from us, but to determine the true orientation of Rigel's velocity relative to ourselves we also must measure the velocity across the line of sight, and this cannot be done using the Doppler effect. So we can see that valuable as the Doppler effect can be—and astronomers consider it a pearl beyond prize—we cannot use it to find the complete motion of a star or a galaxy.

Other Elementary Particles

Everything that we call "matter" is made from a few kinds of elementary particles, which combine in various ways to form atoms, molecules (groups of atoms), and large groups of molecules such as people and planets. The infinite variety of the world springs from the different ways in which elementary particles can group together, but we can understand easily the structure that lies at the heart of the matter once we know a few types of elementary particles and how they interact with each other.

The matter familiar to us consists almost entirely of protons, neutrons, and electrons. For example, a human body that weighs 70 kilograms contains about 2×10^{28} protons, 2.1×10^{28} neutrons, and 2×10^{28} electrons. Each of these protons, as far as we know, is like every other proton, each electron is like any other electron, and each neutron is the same as every other neutron. Protons and electrons each carry one unit of electric charge, denoted "e," but each proton has a *positive* charge ($+e$) and each electron has a *negative* charge ($-e$). The total number of electrons in our bodies equals the total number of protons; thus the *total* electric charge in our bodies is zero, because the neutrons have *no* electric charge. The mass of an electron is much smaller than a proton's mass (1836 times smaller, in fact); thus they dart about more easily than protons do. Each neutron has a mass equal to 1839 electron masses, which is a little bit larger than a proton's mass.

Neutrons and protons tend to group together, under the influence of an attractive force called the "strong force" by physicists. The strong force does *not* affect electrons, but it does hold protons and neutrons together in "nuclei." Atoms, the most elementary unit that can be made from these three particles, all have a central "nucleus" of protons and neutrons, around which electrons orbit rapidly, filling a volume much larger than the space occupied by the nucleus. In any atom the number of electrons always equals the number of protons, so the atom's total electric charge is zero. The number of neutrons that join the protons in the nucleus varies from one element (type of atom) to another, and even from one atom of the same element to another. An element such as hydrogen, oxygen, or carbon may have different atomic *isotopes*, each of which has a different number of neutrons in the atomic nucleus. What distinguishes one kind of atom—hydrogen, helium, nitrogen, iron—from another is the number of protons in the atom's nucleus, which always equals the number of electrons in orbit around the nucleus.

Although an enormous variety of elementary particles exist, only the three mentioned above—protons, neutrons, and electrons—compose all of the objects familiar to us. These kinds of elementary particles tend to combine into atoms, and atoms can link together to form molecules, each of which may contain anywhere from two to millions of atoms. Complicated molecules can transmit information by dividing in two and forming duplicates of the original molecules, as in cell replication during biological reproduction (Chapter 11). But at the smallest level everything depends on protons, neutrons, and electrons, the nearest thing to pointlike particles that we know (an electron has a size of less than 10^{-13} centimeter, and a mass of only 9.1×10^{-28} gram).

The elementary particles we have just described are more like "ordinary" particles than photons are, but some strange things happen when elementary particles combine. Photons, the elementary particles that carry electromagnetic radiation, represent pure energy: Each photon carries some energy of motion, and this energy defines the photon, which has no mass. Particles that do have mass possess an *energy of mass* (measured in ergs) which, as Einstein ably demonstrated, equals the product of the particle's mass (in grams) times the square of the speed of light, c ($c = 3 \times 10^{10}$ centimeters per second). Thus a two-gram penny has an energy of mass of 1.8×10^{21} ergs, which is ten times the energy used each year in the United States. This energy of mass resides in the penny or in any other object with mass, so long as the mass exists.

Einstein's great contribution to our understanding of energy was the demonstration that energy of mass in fact can be converted into energy of motion (kinetic energy), not an easy process on Earth, but possible all the same. The interior of our sun converts about 4.5×10^{12} grams of matter (about the mass of a small mountain) into 4×10^{33} ergs of energy of motion each second. Inside the sun, matter (protons) actually disappears, and "pure energy" (photons) appears in its place. Such photons eventually carry energy away from the sun into space, where the Earth intercepts a tiny fraction that allows life to continue. Human beings have made particle accelerators in which elementary particles such as protons and electrons are accelerated to nearly the speed of light. When these particles collide with each other, sometimes as a result of their collision mass turns into the amount of energy of motion given by Einstein's formula. Conversely, inside these machines, mass can be made from energy: It is possible for photons to come together and turn into particles with mass, such as electrons. "Mass" and "energy" indeed are interchangeable, although in an isolated system of particles, the total amount of energy, which is given by the usual form of energy (energy of motion) *plus* the energy of mass that resides in every particle with mass, remains a constant. Thus by converting energy of motion into mass, or mass into energy of motion, we merely exchange one kind of energy for another. In the quiet of the Earth, matter usually remains matter, and we must look to violent locations such as particle accelerators (Figure 2–6), or outward to the stars, for evidence of the interconvertibility of matter and energy.

Figure 2–6. The Stanford Linear Accelerator in Menlo Park, California passes beneath Interstate Route 280. Antimatter particles are produced here by the millions every working day; they quickly meet particles of ordinary matter and end in annihilation. The headquarters of the firm that published this book can be seen at the extreme right.

Table 2–2
Types of Particles and Their Corresponding Antiparticles

Particle name	Mass of particle and antiparticle (in grams)	Electric charge on particle or antiparticle
Photon (identical with antiphoton)	0	0
Neutrino Antineutrino	0	0 0
Electron Antielectron (positron)	0.91×10^{-27}	$-e$ $+e$
Proton Antiproton	1.6724×10^{-24}	$+e$ $-e$
Neutron Antineutron	1.6747×10^{-24}	0 0

The most spectacular possibility for this conversion process is the meeting of matter and antimatter. Each kind of elementary particle, so far as we know, has a corresponding kind of antiparticle, which has the same mass as the particle but the opposite electric charge. Thus an antiproton has one unit of negative charge, an antielectron (or positron) has one unit of positive charge, and an antineutron has no electric charge (Table 2–2). If a particle collides with its corresponding antiparticle, *all* the energy of mass of both particles disappears and only energy of motion remains. This energy of motion appears in particles with *no* mass, and no energy of mass: photons, neutrinos, and antineutrinos. Photons are indistinguishable from antiphotons; neither has any mass or electric charge, and we call them all photons. Neutrinos and antineutrinos also have neither mass nor electric charge, yet neutrinos and antineutrinos are different from photons and different from one another. These differences appear in the way that these particles interact with other particles, such as protons or electrons. Like photons, neutrinos and antineutrinos travel at the speed of light. Particles whose mass is not zero can never move quite as fast as light, though their speed can approach the speed of light as their energy increases without quite getting there.

Because particles and antiparticles annihilate one another, they cannot coexist for long or they will turn into bunches of photons, neutrinos, and antineutrinos. A beam of photons does not have this problem, because photons and "antiphotons" are already the same particle type.

We can imagine large regions of the universe made of antiprotons, antielectrons, and antineutrons, forming antiatoms and antimolecules, all with the same mass as the particles that we know but with opposite electric charges. However, a meeting with an emissary from such an antimatter part of the universe, even in empty space, would be a brief and hazardous affair, for the embrace of a person and an antiperson would take them off with more energy of motion than all the weapons on Earth could produce.

The atoms from which our universe is made each have the same kind of structure: Electrons orbit around a nucleus of protons and neutrons, with the number of electrons always equal to the number of protons in an atom. The reason for this equality is that unlike electrical charges attract each other, so the protons in the atom's nucleus keep attracting electrons until they have enough negative charges to balance their own positive charge. At this point, the atom's total electric charge is zero, and the atom no longer will attract distant electrons through the electromagnetic forces between opposite charges.

Because electrons have much less mass than protons and neutrons, they dance around the atom's nucleus, using their constant motion to avoid being pulled into the nucleus by the attractive electromagnetic forces from the protons. Detailed study of the behavior of atoms shows that the electrons do not dart about in a totally random way. Instead, on the average, the electrons tend to follow certain orbits around the nucleus. Each orbit has a definite average distance of the electron from the nucleus. Part of the strange behavior of elementary particles appears in the fact that *only certain orbits* with certain average distances are possible for the electrons in atoms, as it is only in these orbits that the electrons can establish a proper resonance with the nucleus that will allow the atom to exist.

We might expect that all the electrons in an atom would be found in the orbit with the smallest average distance from the nucleus, since the nucleus attracts all the electrons toward it. However, electrons have the strange property that they cannot all get together in one place. Electrons tend naturally to repel one another, because they all have negative electric charges, and like charges repel each other through electromagnetic forces. This mutual repulsion, though, could be overcome by a strong enough force, but there is a more important, *additional* unwillingness of electrons to get together, called the "exclusion principle." The exclusion principle places an absolute upper limit on the number of electrons that can be in a given atomic orbit. Calculations and experiments both show that the smallest orbit can have at most two electrons in it, the next orbit can have at most eight electrons in it, the third orbit can have at most eighteen electrons, and so on. And so on? The facts of life for electron orbits are that if we call the smallest orbit number one,

Table 2–3
Atomic Numbers and Atomic Weights of Some Common Nuclei

Name of nucleus	Atomic number (number of protons)	Number of neutrons	Atomic weight (number of protons plus number of neutrons)
Proton	1	0	1
Deuteron	1	1	2
Helium-3	2	1	3
Helium-4	2	2	4
Lithium-7	3	4	7
Beryllium-9	4	5	9
Boron-11	5	6	11
Carbon-12	6	6	12
Carbon-13	6	7	13
Nitrogen-14	7	7	14
Nitrogen-15	7	8	15
Oxygen-15	8	7	15
Oxygen-16	8	8	16
Fluorine-19	9	10	19
Neon-20	10	10	20
Sodium-23	11	12	23
Magnesium-24	12	12	24
Sulfur-32	16	16	32
Argon-36	18	18	36
Iron-56	26	30	56
Nickel-58	28	30	58
Copper-63	29	34	63
Silver-107	47	60	107
Gold-197	79	118	197
Lead-208	82	126	208
Uranium-235	92	143	235
Uranium-238	92	146	238

the second smallest number two, and so on, then the maximum possible number of electrons that can occupy a given orbit equals twice the square of the orbit's number. This means that if an atom has, say six electrons, then only two of them (at most) can be in the smallest orbit. Since the atom's electrons, under the influence of the attraction from the nucleus, seek the smallest possible orbits, an atom with six electrons probably will have two electrons in the smallest orbit and four in the second smallest orbit.

Atoms and Molecules

Atoms make contact with one another through their various electrons, which fill the atoms' outsides while the atoms' nuclei stay in the center. The behavior of atoms with regard to one another thus depends on the number of electrons they contain, but since the number of electrons always equals the number of protons, the behavior of atoms can be said to reflect the number of protons instead of the (equal) number of electrons. Either of these numbers gives the "atomic number." The neutrons in an atom's nucleus do not affect the number of electrons or their orbits around the nucleus, but the neutrons do affect the structure of the nucleus itself. An atom's total number of "nucleons," protons and neutrons, is the "atomic weight." Table 2–3 shows the atomic number and atomic weight of some common atoms. Hydrogen, the simplest of all atoms, has one proton and no neutrons, with one electron in orbit around the proton nucleus. Deuterium, an isotope of hydrogen, has one proton and one neutron in its nucleus, with one electron in orbit around it. Helium, the simplest atom with more than one proton in its nucleus, has two protons and two neutrons in its nucleus, around which orbit two electrons. Two electrons are the maximum number allowed in the smallest orbit. Atoms with the maximum number of electrons in each

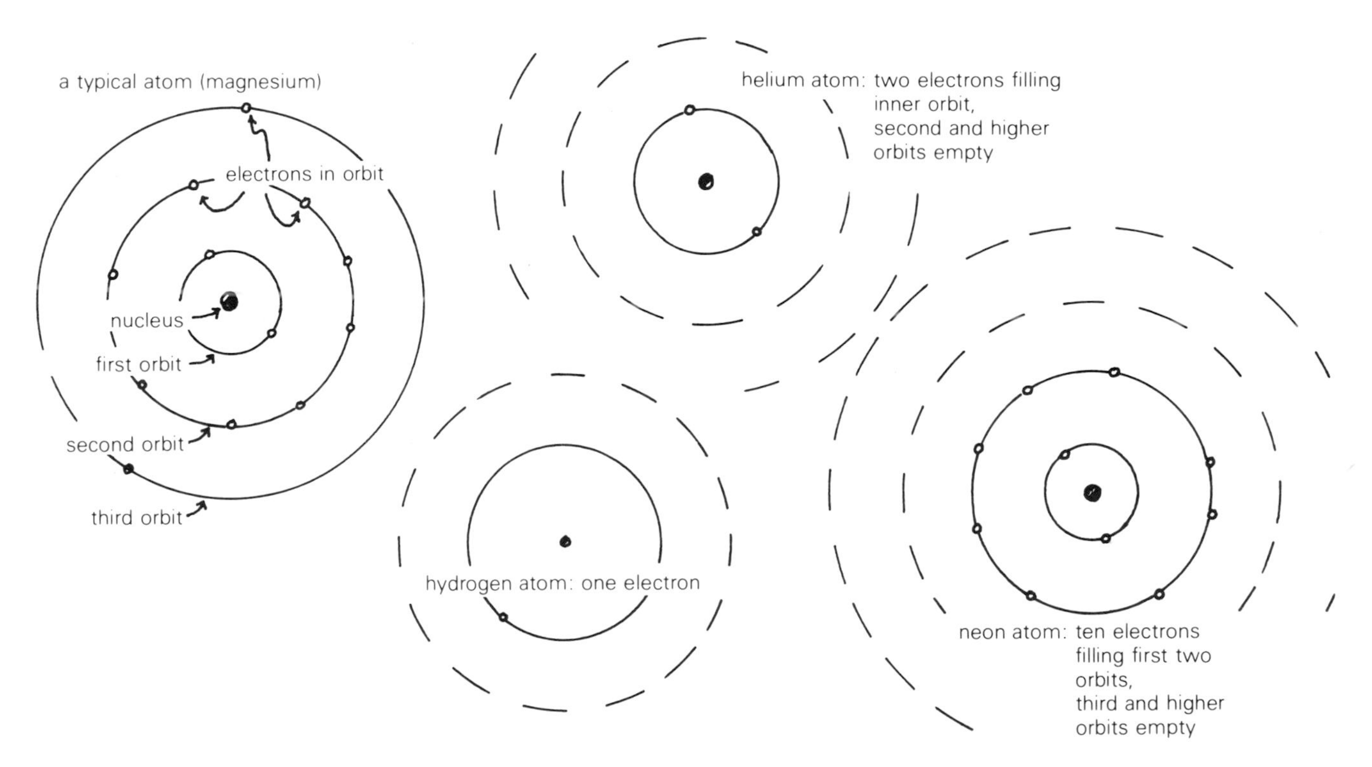

Figure 2–7. Because of the exclusion principle, no atom can have more than two electrons in the smallest possible orbit, eight electrons in the next smallest, and so on. An atom such as helium (two electrons) or neon (ten electrons) that has all of its electrons in orbits that contain the maximum allowed number of electrons is said to be "inert."

orbit (two in the smallest, eight in the next smallest, eighteen in the third) are said to have "closed shells" of electrons. Examples of such atoms are helium (two electrons) and neon (ten electrons: two in the smallest orbit and eight in the next smallest, as shown in Figure 2–7). Atoms with all their electrons orbiting in closed shells do not interact with other atoms and therefore are called "inert" atoms.

When atoms do not have all their electrons in closed-shell orbits, they tend to combine readily with one another to form molecules. In forming molecules, either the atoms share one or more of the electrons in their outer orbits ("covalent bonding"), or else the electrons stay relatively close to either one nucleus or the other ("ionic bonding"). In either case, atoms are most likely to combine if the molecule that they form has a total number of electrons equal to the number that will "close" a given electron orbit with the maximum allowable number of electrons. Figure 2–8 represents the manner in which an atom of lithium can combine with an atom of fluorine to form (at temperatures of several thousand degrees or more) the gas lithium fluoride. The lithium fluoride molecules are held together mainly by ionic bonding. Each lithium atom has three electrons (two in the smallest possible orbit and one in the next smallest), while each fluorine atom has nine electrons (two in the smallest orbit and seven in the next smallest). The outer electron in the lithium atom tends to fill the second shell in the fluorine atom, leaving a closed shell of two atoms behind. The two nuclei in a lithium fluoride molecule remain apart and do not coalesce into a single nucleus like that of a lithium atom or of a fluorine atom.

An example of covalent bonding comes from the simplest molecule, made of the simplest atom, hydrogen. Hydrogen atoms like to form pairs, thereby making hydrogen molecules (H_2) that cannot be separated easily into individual atoms. Here again the total number of electrons (two) is the right number to close an electron shell, in this case the smallest one.

In water molecules (H_2O), we find two hydrogen atoms, each with one electron, combined with an oxygen atom (two electrons in the smallest orbit, six in the next smallest) to provide a total of ten electrons, again the right number to fill both the smallest and the next smallest orbits.

The Interaction of Light and Matter

Our knowledge of atomic structure has led us to picture atoms as massive nuclei orbited by much less massive electrons, each of which seeks the smallest possible orbit but must obey the rules that set a maximum allowable number of electrons in any given orbit. Isolated from outside disturbances, atoms soon will reach a state where the electrons settle into an orbital arrangement as close to the atom's nucleus as is consistent with the "laws" of nature, that is, with the observed regularities in the electron orbits that we have outlined above. If one or more of an atom's

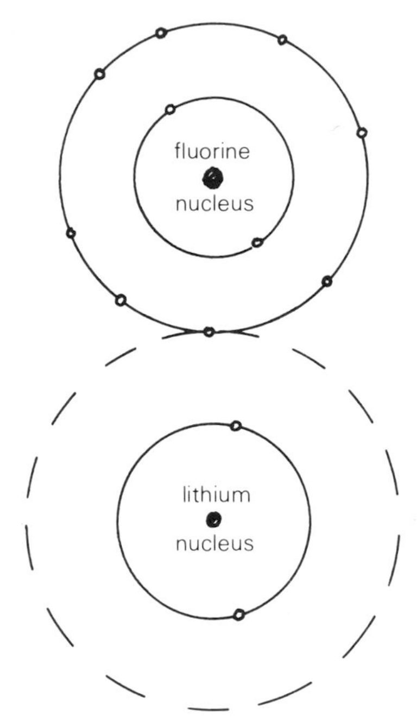

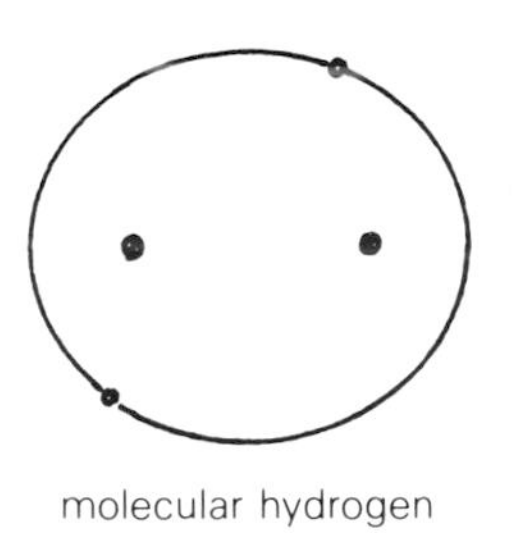

Figure 2–8. When atoms combine to form molecules, often one electron jumps back and forth from orbiting one nucleus to orbiting the other, or else an electron can manage to orbit around both nuclei simultaneously.

electrons moves in an orbit larger than is necessary to fulfill the maximum-number rules, we say that the atom is "excited." For example, a hydrogen atom with its single electron in the second smallest possible orbit, not in the smallest orbit, is an "excited" atom. When the electron jumps from the second smallest to the smallest orbit—and this jump will occur rapidly, in about 10^{-8} second after the electron has been excited to the second smallest orbit—the atom produces a photon that carries away some energy from the atom (Figure 2–9). This photon energy must appear because the atom has more energy when it is excited than when it is not excited. The fact that the atom has a smaller amount of energy when the electron moves in an orbit closer to the nucleus reflects the fact that atoms are held together by the mutual attraction of the nucleus and the electrons. Some energy must be added if we want to pry the electrons away from the nucleus, and the closer the electrons are to the nucleus, the more energy we must apply.

When an atom's electrons jump from larger to smaller orbits, the atom's energy *decreases*, since it would now take more energy to separate the electrons from the nucleus. This decrease in the atom's energy appears in the photon that accompanies the change in orbit. The photon's energy of motion (greater than zero) exactly equals the change in energy (less than zero) of the atom. Thus an electron's jump from a larger to a smaller orbit produces a photon with a specific energy, equal to the change in the atom's energy that the jump makes. For instance, when the electron in a hydrogen atom jumps from the third smallest to the second smallest orbit, the atom emits a photon with an energy of 0.3×10^{-11} erg, which is a photon of red light. The jump from the second smallest to the smallest orbit produces a photon with a larger energy, 1.6×10^{-11} erg, which is a photon of ultraviolet light. In the sun's outer layers, hydrogen atoms undergoing such electron jumps are extremely numerous, and a part of the sun's light consists of a big bunch of photons with energies of 0.3×10^{-11} erg and another big bunch of photons with energies of 1.6×10^{-11} erg. A third, and equally fascinating bunch of photons arises from electrons in hydrogen atoms that jump from the third smallest orbit directly to the smallest orbit. These jumps produce photons with energies of 1.9×10^{-11} erg, which is the sum of the energies of the third-smallest to second-smallest orbital leap (0.3×10^{-11} erg) and the second-smallest to smallest leap (1.6×10^{-11} erg). See Figure 2–10.

The most important fact about these electron jumps is that photons that result from them do not have just any energies; instead, each photon has a definite energy that is related to the number of electrons in the atom and the orbits that these electrons jump from and jump into. Each specific decrease in the atom's energy produces a photon that carries away this exact amount of energy.

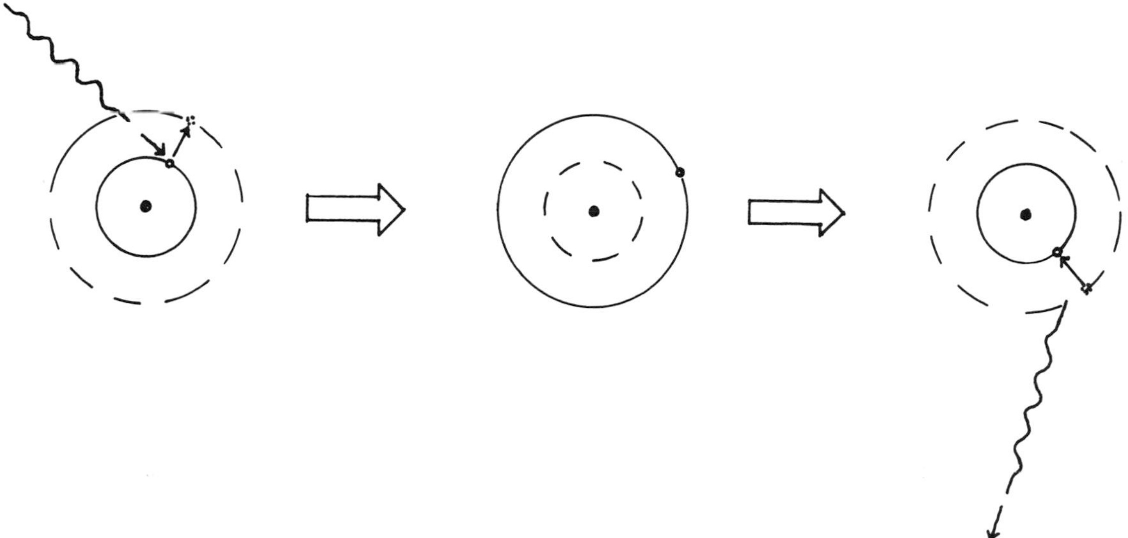

Figure 2–9. If a photon strikes a hydrogen atom with the electron in the smallest allowed orbit, the photon can "excite" the atom's electron into a larger orbit, provided that the photon carries just the right amount of energy to produce this "excitation." Afterwards, the electron will jump back to the smallest allowed orbit, emitting a photon as it does so to carry away an energy that corresponds to the atom's decrease in energy.

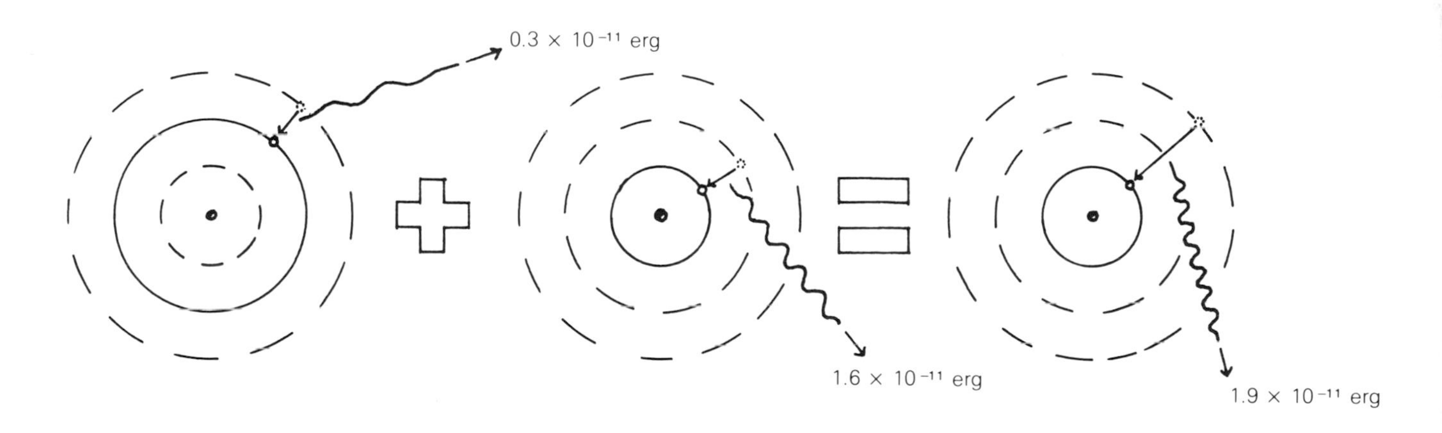

Figure 2–10. When the electron in a hydrogen atom jumps from the third smallest to the second smallest orbit, the atom emits a photon with an energy of 0.3×10^{-11} erg. When the electron jumps from the second smallest into the smallest orbit, the atom emits a photon with an energy of 1.6×10^{-11} erg. The sum of these energies, 1.9×10^{-11} erg, equals the amount of energy carried off by the photon that a hydrogen atom emits when the electron jumps directly from the third smallest into the smallest orbit.

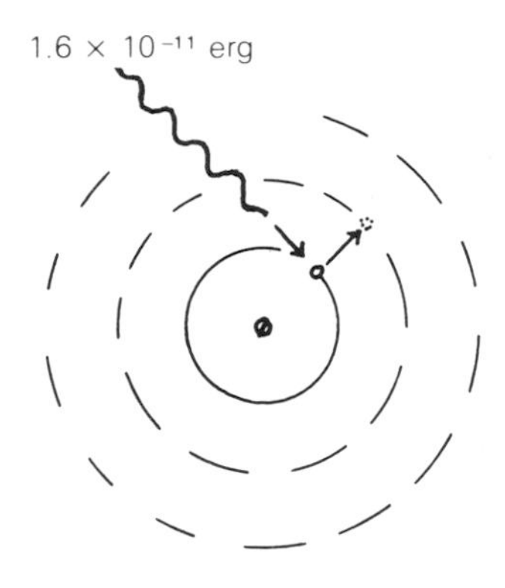

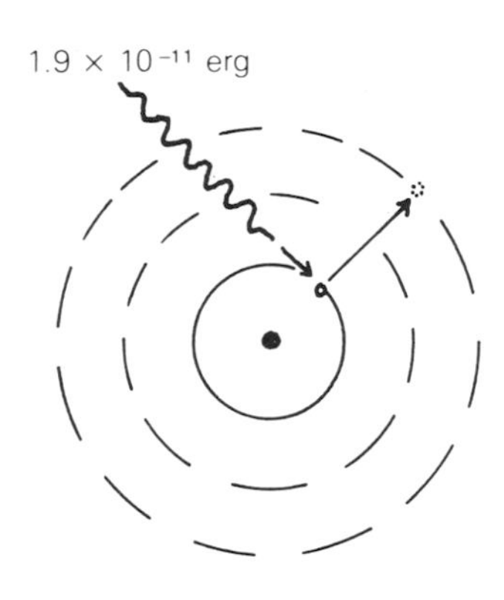

Figure 2–11. If a photon with an energy of 1.6×10^{-11} erg strikes a hydrogen atom whose electron is in the smallest orbit, the photon can make the electron jump into the second smallest orbit. If the photon carries an energy of 1.9×10^{-11} erg, it can make the electron jump into the third smallest orbit. Each jump requires a specific amount of energy that the photon must have if it is to produce this jump.

The reverse process, outward jumps of electrons orbiting an atom's nucleus, also occurs frequently. If a photon strikes an atom, the photon can cause one of the electrons in the atom to jump into a larger orbit. In this event the photon will disappear, devoting its energy to the electron's leap outward. This will occur only if the photon's energy exactly equals the energy needed to make the particular electron leap into a larger orbit (Figure 2–11). Photons almost never can give *part* of their energy to an atom: Only when the photon's energy is just the precise amount needed will the photon disappear and the electron jump to a larger orbit.

The fact that a photon will interact with an atom only if its energy matches one of the atom's possible changes in energy means that if photons of various energies encounter some type of atom, most of the photons will pass by the atoms, while only those photons with definite energies will cause a jump in electrons' orbits and thus disappear. For example, photons leaving the sun pass through its outer layers, which contain many sodium atoms. Each sodium atom has eleven electrons, two in the smallest possible orbit, eight in the next smallest, and one in the third smallest. Photons with 6.0×10^{-12} erg of energy can cause the outermost electron in a sodium atom to jump from the third smallest to the fourth smallest orbit. Photons with 6.1×10^{-12} erg or 5.9×10^{-12} erg of energy will not do this, but photons with 6.0×10^{-12} erg of energy each are likely to encounter a sodium atom, produce an outward jump of the outermost electron, and disappear as they do so, thereby donating their energy to the atom. These photons therefore are *absorbed by* the sodium atoms, and if we measure the energies of photons from the sun, we find that this solar spectrum (Figure 2–2) has a noticeable lack of photons with energies of 6.0×10^{-12} erg, that is, of a certain shade of yellow light. The removal of photons of one particular energy produces an "absorption line," which is a dark region across the spectrum that indicates an absence of photons at the particular energy of the absorption line.

The opposite process, the *emission* of photons when an atom's electron passes to a smaller orbit, produces an "emission line" in the energy spectrum of a source where such leaps to smaller orbits occur. Figure 2–12 shows the ultraviolet part of the solar spectrum. Bunches of photons are emitted by "excited" hydrogen atoms whose electrons jump from a larger to a smaller orbit. Each jump produces a photon with an energy equal to the decrease in the energy of the atom as the electron makes the leap.

If a photon with a great amount of energy strikes an atom, the photon may knock one (or more) of the atom's electrons completely free from the atom. This process requires more energy than it takes simply to move an electron into a larger orbit, because it gives the electron

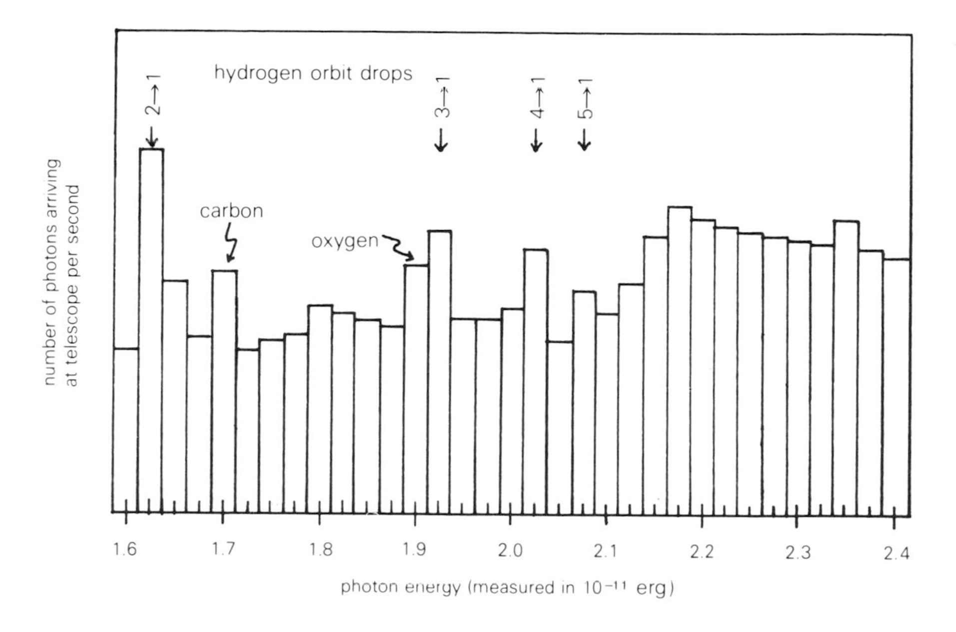

Figure 2–12. When we measure the spectrum of the sun's light (that is, the number of photons of different energies) for ultraviolet photons, we find peaks of photon emission at certain energies. These peaks arise from electrons jumping to smaller orbits in hydrogen atoms and in other atoms and ions as well.

enough energy to get all the way away from the atomic nucleus and even to have some energy of motion left over for itself once it is free of the nucleus. Atoms with one or more electrons missing are called *ions*, and the process of knocking electrons loose is called *ionization*; if we ionize atoms with photons, it is photo-ionization. We also can knock electrons loose from atoms by making them collide with one another at high speeds; this is collisional ionization. Ions have a net positive charge, so they constantly attract electrons. If any "free" electrons are near the ions they eventually will be captured by the ions to re-establish the equality of charges between an atom's protons and electrons.

Although we need a photon with a definite amount of energy for the *excitation* of an atom (making the electron jump to a larger orbit), for the *ionization* of an atom we require only that a photon have at least a certain minimum energy. Any additional energy probably will be transferred to the newly liberated electron, but we also may find that after the ionization we still have a photon, whose energy has been reduced in comparison to the original photon by the amount needed to knock the electron away from the atom plus the amount of energy of motion given to the free electron.

When ions capture free electrons to form atoms once again, this process of *recombination* produces photons. Just as a photon can lose its energy in liberating an electron from an atom, in the reverse process, when an electron joins an ion, a photon appears and carries away some energy, equal to the decrease in energy of the atom.

In summary, both the excitation and ionization of atoms require the *absorption* of energy from photons, whereas both de-excitation (leaps to smaller orbits) and recombination lead to the *emission* of photons.

These are not the only ways that photons can appear or disappear. We already have mentioned that a particle and its antiparticle will annihilate each other and produce photons (plus neutrinos and antineutrinos), and that photons themselves can turn into particle-antiparticle pairs. We shall discuss two more ways in which photons can be produced. Although all these processes are important for astronomy, the most significant processes that involve photons are the interactions with atoms that we have described above: excitation and de-excitation, ionization and recombination.

Hyperfine Structure and the 21-Centimeter Emission of Hydrogen

Throughout the universe, hydrogen atoms occasionally and spontaneously emit photons with an energy of 9.5×10^{-18} erg, hence with a frequency of 1420 MHz (1.42 billion cycles per second) and a wavelength of 21.1 centimeters. Hydrogen atoms, which are by far the most common type of atom in the universe, do this because of a subtle interaction between the electron and the proton in each atom. We have described the mass and the electric charge of protons and electrons, but in addition each of these particles has a certain amount of magnetic "spin." We can think of a proton or an electron as something like a spinning bar magnet. The amount of "spin" is the same for any proton or any electron, and a proton's spin is equal to an electron's. When an electron orbits a proton in a hydrogen atom, the electron's spin direction can line up either parallel (spins in the same direction and with the same orientation), or antiparallel (spins in opposite directions though in the same orientation), as shown in Figure 2–13. These are the only two possibilities. Unlike ordinary bar magnets, the spins of the proton and the elec-

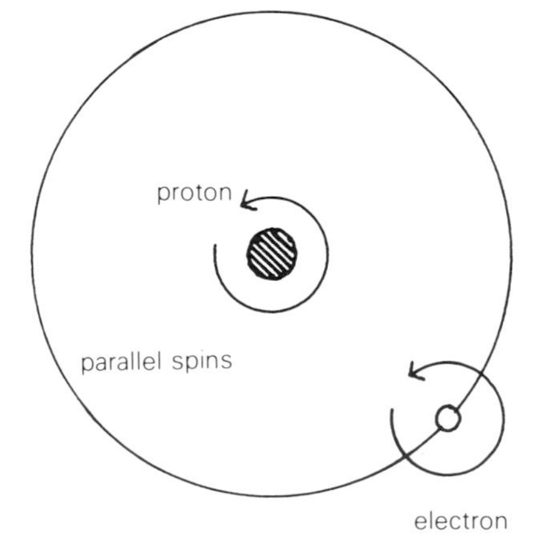

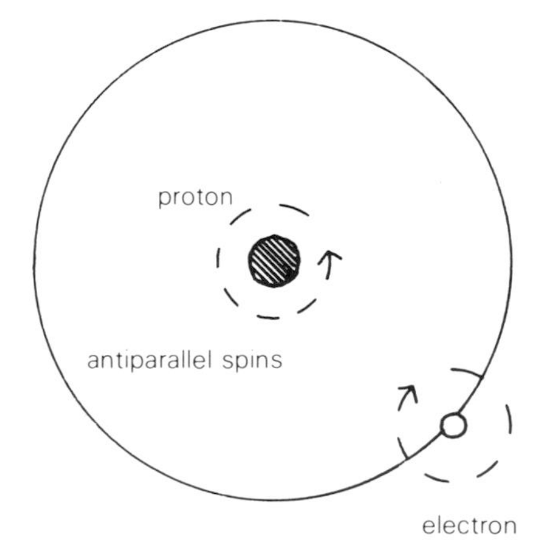

Figure 2–13. An electron orbiting around a proton in a hydrogen atom can have its spin either parallel or antiparallel to the proton's spin. If the spins are parallel, they are in the same direction, whereas antiparallel spins are in opposite directions though along the same axis.

tron must point in the same direction; only the directions of the two spins can vary. The two possibilities define the "hyperfine structure" of hydrogen atoms: either the spins are parallel, or they are antiparallel.

If a hydrogen atom whose spins are antiparallel is hit with a photon of just the right energy (9.5×10^{-18} erg), the electron's spin can flip over and thus become parallel to the proton's spin (Figure 2–14). The hydrogen atom has this tiny amount more energy (9.5×10^{-18} erg) when the spins are parallel than when they are antiparallel, because the magnetic forces between the spinning electron and the spinning proton tend to attract when the spins are in the opposite directions and to repel when the spins are in the same direction. In a similar way, bar magnets will attract one another when they are pointed in opposite directions, but will repel when they point in the same direction. To make the spins in a hydrogen atom change from antiparallel to parallel, we must add energy to overcome the fact that the spins now provide a repulsive magnetic force instead of the attractive magnetic force that arose in the antiparallel configuration.

Because a hydrogen atom has a bit more energy when the proton and electron spins are parallel, any hydrogen atom with its spin parallel to the proton's spin eventually will flip its spin over to the antiparallel position and emit a photon that carries away just the energy difference between the parallel-spin and antiparallel-spin positions (Figure 2–15). For an average hydrogen atom it takes ten million years before such a spin flip occurs, but there are so many hydrogen atoms in space that in fact some spin flips always are happening. Our own galaxy contains about 10^{69} atoms of diffuse hydrogen gas, of which about three quarters

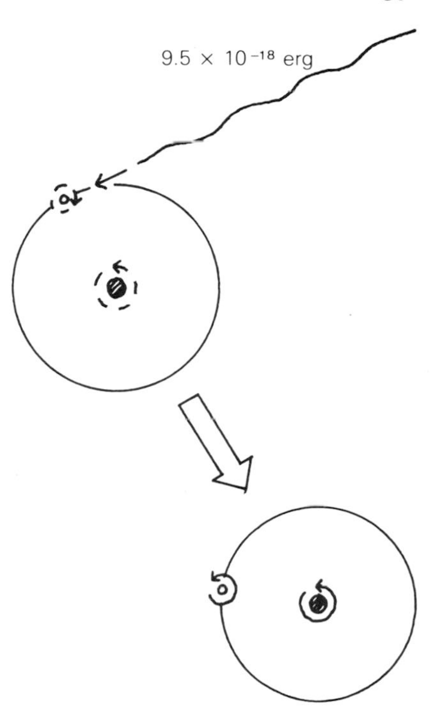

Figure 2–14. To make a hydrogen atom's electron change its spin from the antiparallel to the parallel position, we can hit the atom with a photon of energy equal to 9.5×10^{-18} erg, the energy difference between the two spin possibilities.

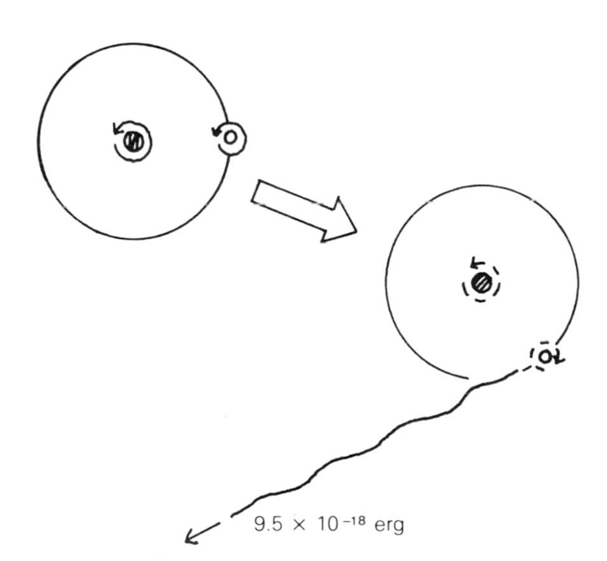

Figure 2–15. A hydrogen atom in which the electron and proton spins are parallel can flip its electron spin and emit a photon with an energy equal to 9.5×10^{-18} erg. In an isolated hydrogen atom, this will occur after a few million years.

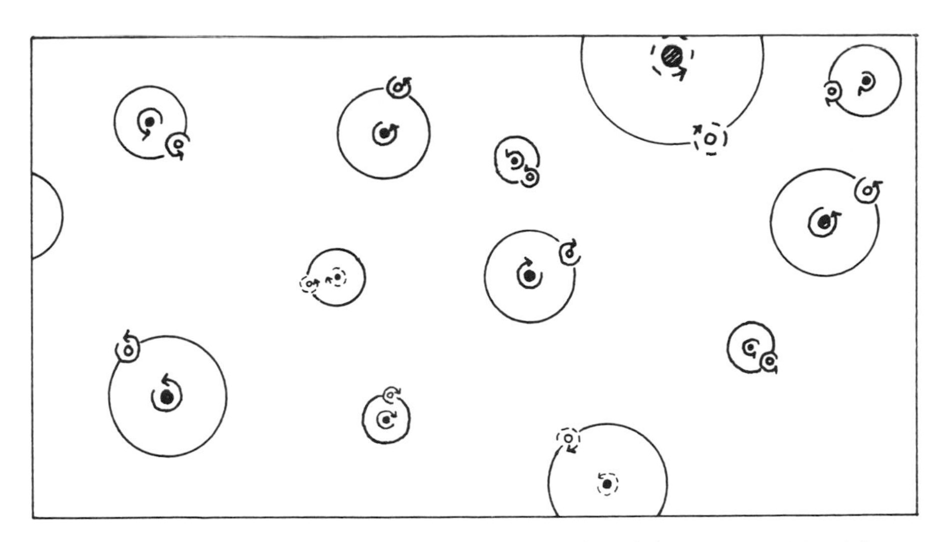

Figure 2–16. Because of collisions among the hydrogen atoms, the electron spins often are flipped from being parallel to being antiparallel to the proton spins, or conversely. At any moment, about three quarters of the hydrogen atoms will have their electron and proton spins parallel, while the other one quarter will have the spins antiparallel.

have their spins lined up parallel to each other[4] (Figure 2–16). Almost all of these atoms have the electron in the smallest possible orbit, because the atoms only rarely are excited by a passing photon. Of the $\frac{3}{4} \times 10^{69}$ atoms with parallel electron and proton spins, every second about 10^{54} such atoms have the electron spin flip over into the antiparallel configuration. These flips produce 10^{54} photons every second, each of which has an energy of 9.5×10^{-18} erg and a wavelength of 21 centimeters. Specially designed radio telescopes can capture some of these photons and thus enable astronomers to measure the amount of hydrogen gas in our galaxy. By counting the numbers of these photons with 21-centimeter wavelength, radio astronomers can even determine how much hydrogen other galaxies contain. Furthermore, if we find a "Doppler shift" of these photons from hydrogen gas clouds, we can measure the velocity toward us or away from us of the atoms that produced them, just as we do for the photons of visible light when we measure those energies or wavelengths characteristic of visible-light photons.

Hydrogen atoms are not the only atoms with "hyperfine structure"; any atom with an odd number of protons and electrons and an even number of neutrons can emit photons through the electron spin flip process. Sodium atoms (eleven protons, eleven electrons, twelve neutrons) provide another example of an atom with an electron spin flip, and astronomers now are trying to detect the photons of 17-centimeter wavelength that are emitted when the outermost electron in sodium flips its spin from one orientation to the other.

---

[4]Collisions among the various hydrogen atoms keep bumping some atoms with antiparallel spins into the parallel-spin position. As a result, at any given time we always have about three quarters of the atoms with parallel spins and one quarter with antiparallel spins.

Synchrotron Radiation

We have seen that atoms can produce photons when their electron leaps from a larger to a smaller orbit, or (in some atoms) when the electron's spin flips over. But to make photons it is not necessary to have atoms. One way that photons can appear has the name "synchrotron radiation," from the particle accelerators called "synchrotrons" in which these photon emission processes first seemed important. In the larger universe, of course, synchrotron photons from outside the Earth dominate by far the few such photons we make in our own particle accelerators.

Synchrotron radiation consists of photons that appear when a charged particle accelerates or decelerates because of an electromagnetic force. Suppose that such a force, which acts on particles with some electric charge (such as protons or electrons), makes a charged particle change its speed or the direction of its velocity. Then photons will be produced that carry away an amount of energy equal to the change in energy necessary to produce the new velocity of the particle. This photon emission from particles accelerated or decelerated by electromagnetic forces becomes extremely important if the charged particles are moving at almost the speed of light. "Synchrotron" particle accelerators use large electromagnets to make electrons and protons move in circles to higher and higher velocities. Scientists working with these "synchrotrons" noticed the glow of the photons that appeared as a result of the charged particles' velocity changes, and named the photons "synchrotron radiation." Some astronomical objects—especially the neutron stars discussed in Chapter 12—possess strong magnetic fields that affect the motions of charged particles near them and thus produce photons by the "synchrotron" process. No atoms are involved: the change in each particle's velocity caused by the electromagnetic forces are enough by themselves to produce photons, if the particles are moving at almost the speed of light. An astronomical source of "synchrotron" photons is the Crab Nebula (Figure 2–17). This filamentary object is the remnant of a star that exploded, shooting out many charged particles at almost the speed of light.

The Effect of Our Atmosphere on Photons

Whether photons arise from electron leaps in atoms, from atoms' spin flips, or from the synchrotron emission process, they all must reach the Earth by passing through our life-giving, protective, mistreated atmosphere. The air around us consists mainly of nitrogen molecules (N_2), oxygen molecules (O_2), a few water molecules (H_2O), some argon, and trace amounts of carbon dioxide, sulfur dioxide, and various other pollutants, both natural and human-made. If we could construct a vertical column of air one centimeter square at the Earth's surface and rising through the entire atmosphere, the column would contain about 3×10^{19} molecules.

We use the oxygen molecules in the atmosphere as we breathe, some plants use the nitrogen molecules to aid their growth, and almost all living creatures use the water vapor when it condenses into rain and renews our lakes and oceans. But the atmosphere also provides us with a protective shield against the photons that would destroy us if we had no atmosphere, or if we had a slightly different atmosphere, such as the one that we could produce if we tamper too successfully with the one to which we are accustomed. The sun emits not only photons of visible

Figure 2–17. The Crab Nebula, the remnant of a supernova explosion that appeared in 1054 A.D., produces visible light partly by means of the sychrotron radiation process, in which charged particles accelerate or decelerate in magnetic fields.

light, which easily pass through the atmosphere, but also many ultraviolet photons with energies greater than 6.6×10^{-12} erg. Each of these ultraviolet photons can destroy a molecule in our bodies by breaking the molecule apart, whereas visible-light photons do not have enough energy to perform such a "dissociation" of our molecular structure. Luckily for us, these ultraviolet photons almost never reach the surface of the Earth, because a layer of *ozone* molecules high in the atmosphere absorbs them (Figure 2–18). Ozone molecules (O_3) contain three oxygen atoms; they are rarer than the ordinary oxygen molecules (O_2) that we breathe, and almost never appear in the lower layers of the atmosphere. Ozone molecules are held together more delicately than oxy-

Figure 2–18. Ozone molecules ten to fifty kilometers above the Earth's surface absorb photons with ultraviolet energies and wavelengths. Any photon with more than 6.6×10^{-12} erg of energy is capable of dissociating an ozone molecule; if the photon does so, it will lose energy and not reach the Earth's surface as harmful ultraviolet radiation.

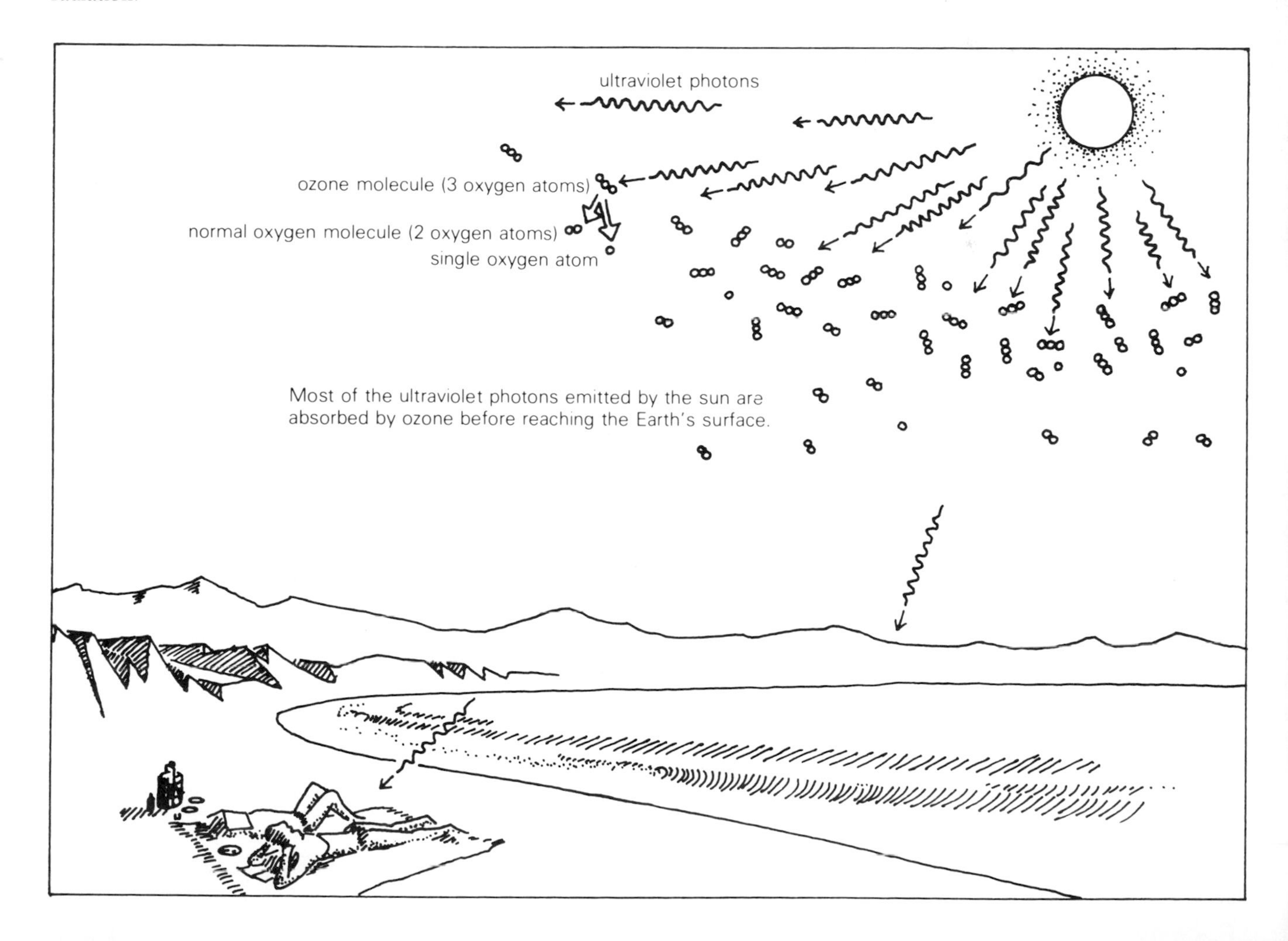

gen molecules: A photon must have at least 8.2×10^{-12} erg of energy to dissociate an oxygen molecule, but any photon with an energy greater than 6.6×10^{-12} erg will dissociate an ozone molecule. Such a molecular dissociation causes the photon to disappear; thus the ozone molecules twenty kilometers above us, rather than ourselves, feel the impact of the sun's ultraviolet light. Oxygen molecules do their share in stopping photons with energies greater than 8.2×10^{-12} erg, but if it were not for the ozone molecules we would die. Even a small reduction in the number of ozone molecules above us would lead to an increase in the amount of skin cancer among humans. The number of ozone molecules in the atmosphere now represents a balance between the destruction of molecules (by ultraviolet photons) and their creation from oxygen atoms freed when other ultraviolet photons break apart oxygen molecules. When we release chemicals that can combine with ozone or with oxygen atoms, we tend to deplete the amount of ozone in our atmospheric shield, at considerable risk to ourselves.

In addition to ultraviolet photons, many other kinds of photons never reach the Earth's surface, while others come through the atmosphere in reduced quantities. Figure 2–19 shows the effect of the Earth's atmosphere on photons with different energies. Only visible-light and some radio-wave photons penetrate the atmosphere easily. Since the sun produces many visible-light and relatively few radio-wave photons, our eyes have evolved a sensitivity to the photons that are abundant at the Earth's surface.[5] Infrared photons are absorbed in the lower atmosphere (at altitudes below 10 kilometers), while ultraviolet, x-ray, and gamma-ray photons are absorbed at higher altitudes. Furthermore, water vapor in the Earth's atmosphere absorbs not only many infrared photons, but also some of the shorter-wavelength (microwave) radio photons. We can see clearly out of our atmosphere only through two "windows" in the spectrum of photon energies: visible light and radio.

If we want to observe photons with energies that would be absorbed by the atoms and molecules in our atmosphere, the thing to do is to go above the atmosphere, by launching a rocket, a satellite, or even a balloon that can rise above the water vapor in the lower atmosphere. Astronomers are as fond of the air as other people, but their admiration for our atmosphere is tinged with ambiguity because it prevents them from seeing much that they deeply want to observe. In addition to the absorption of photons of certain energies, the atmosphere has another effect on the passage of light waves. This effect, called refraction, bends the light rays slightly from the straight-line paths that they would follow

---

[5]Because radio-wave photons have far less energy than visible-light photons, the small number of radio photons provides a total energy output in radio photons far, far less than the sun's output of visible-light energy.

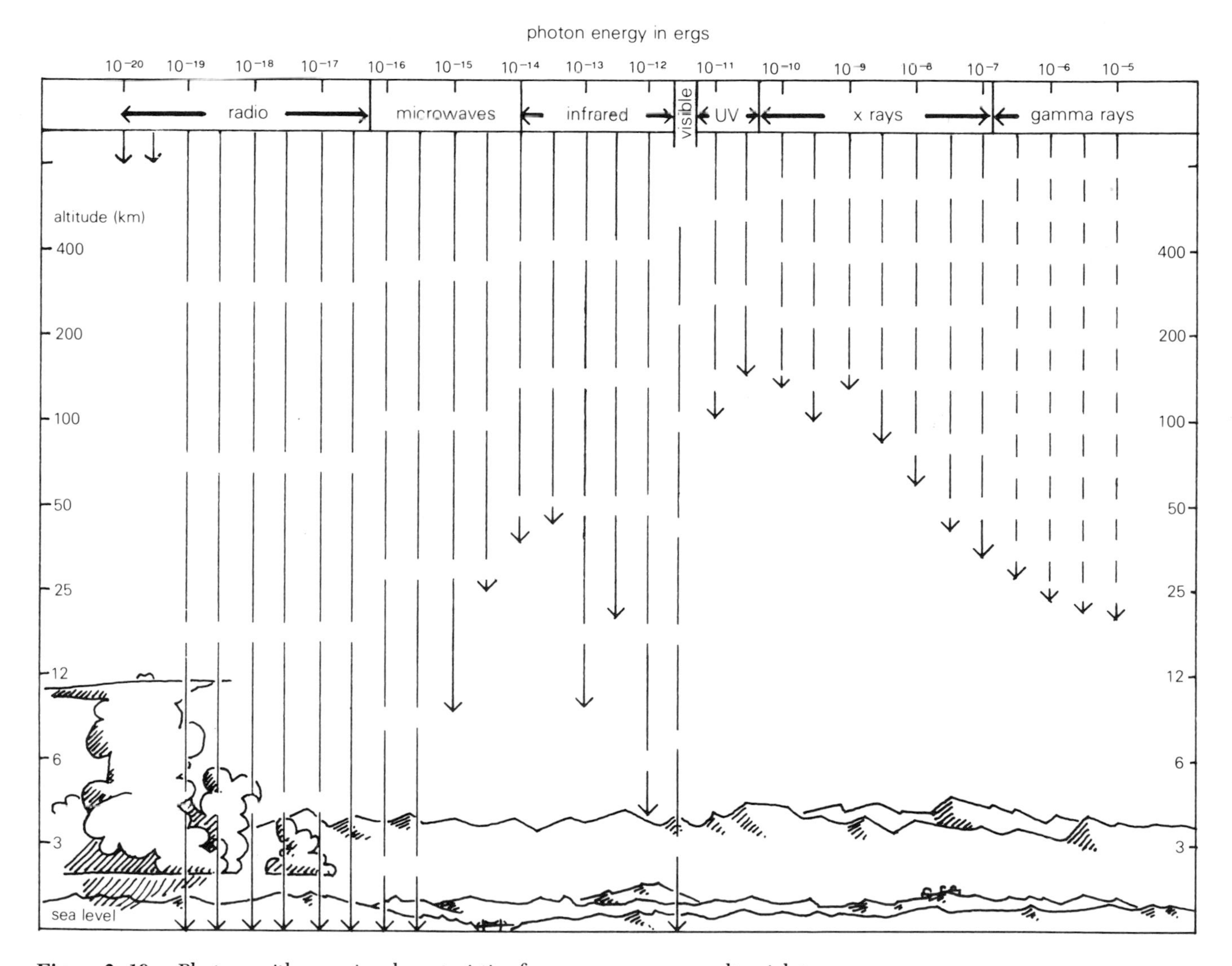

Figure 2–19. Photons with energies characteristic of gamma rays, x rays, ultraviolet light, infrared light, and microwaves usually cannot reach the surface of the Earth, because they are absorbed by ozone, oxygen, water vapor, and other molecules in the Earth's atmosphere. Only those photons with visible-light and radio-wave energies pass through the atmosphere freely.

in the absence of our atmosphere. A similar refractive effect is used by spectacle- and telescope-makers to focus light by passing it through properly shaped lenses. The atmospheric refraction is a mild effect, but it does prevent us from seeing objects as clearly as we could if we had no atmosphere. Not only does our atmosphere refract or bend the light rays that pass through it, but in addition the amount of this refraction is a tiny bit different along every path through the atmosphere, and even changes rapidly with time because the atmosphere is in constant motion (Figure 2–20). As a result, two light rays that start out at two slightly different moments will suffer a different series of bendings as they pass through our atmosphere. The variation in the amount of bending from one path to another makes stars "twinkle" or dance about in their apparent location from one moment to the next. The effect of the atmosphere's motions that causes this twinkling has been called "seeing" by astronomers. "How's the seeing?" one will ask the other in a professional manner. "Five to seven," the other will answer, not quoting betting odds but rating the "seeing" on a scale of one (worst) to ten (best). Thus astronomers reassure themselves that their education was worthwhile.

To eliminate the effects of seeing on visible-light photons, we once again can go above the atmosphere. Astronomers would like to see the U. S. Government launch the planned "Large Space Telescope" that would carry a gyrostabilized observing platform into orbit above the Earth's atmosphere. Such a telescope, by avoiding the undesired effects of seeing, could detect galaxies a hundred times fainter than the faintest we can see with Earth-bound telescopes.

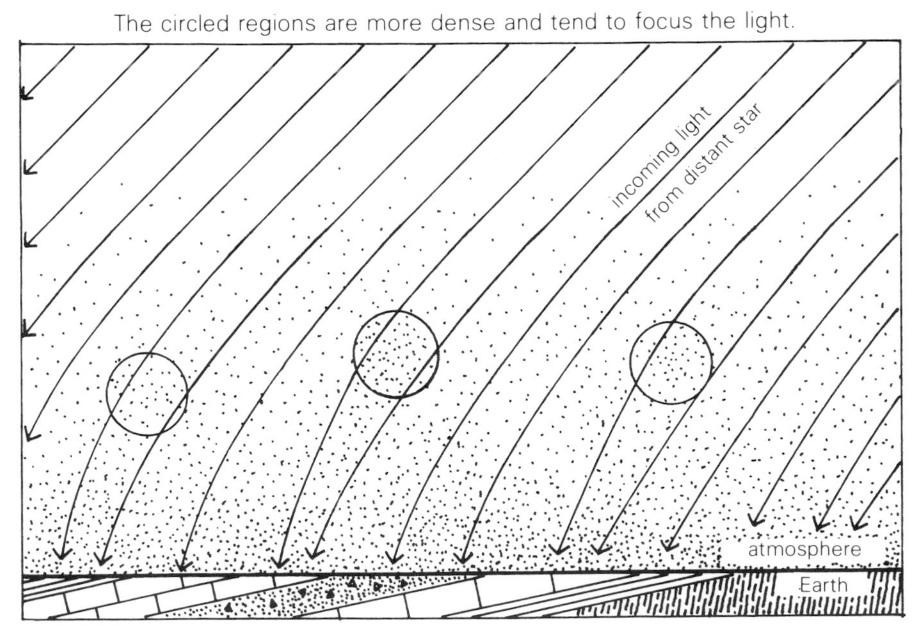

Figure 2–20. Our atmosphere acts as a lens and bends (or refracts) light rays passing through it. Because of small-scale turbulence in the atmosphere, the path taken by a light ray will vary from moment to moment. This causes the light from pointlike sources, such as stars, to "twinkle."

For otherwise unobservable wavelengths, astronomers now have satellites designed for x-ray detection, such as the Uhuru satellite, and for ultraviolet detection (the Copernicus satellite, Figure 2–21), as well as special high-flying airplanes to detect infrared photons that would be absorbed by water-vapor molecules in the lower atmosphere. The Large Space Telescope, which also would carry detectors to measure x-ray and ultraviolet photons, could help to resolve many burning cosmological questions, such as the problem of whether the universe ever will cease to expand and begin to collapse, which we shall discuss in the next chapter.

Figure 2–21. The Copernicus satellite, launched in 1972 (here shown in an artist's sketch), carries detectors to measure the numbers of ultraviolet and x-ray photons that reach it from different sources on the sky. Ground stations direct the pointing of the satellite, and a complicated tracking system keeps the detectors pointed in a given direction as accurately as any ground-based telescope.

The Four Kinds of Forces

Four kinds of forces rule the world of physics, four forces that govern the interaction of all the elementary particles that form the universe. Our present knowledge allows us this four-fold classification of forces, though someday we may see how all four types of forces are interrelated, if we can find the "unified field theory" that will explain where forces come from, not just how they express themselves. But as far as we can now tell, elementary particles can interact in four different ways: through gravitation, through electromagnetism, through "strong" interactions, and through "weak" interactions. The first pair, gravitational and electromagnetic forces, are easily observable at large distances, whereas the "strong" and "weak" forces dominate the submicroscopic world of elementary particles.

1. Gravitational forces are the weakest of forces, yet they govern all situations that involve great distances and large masses, because gravity always attracts and never repels. Every particle in the universe attracts every other particle by gravitation. The strength of this force, the amount of the attraction, depends only on the masses of the particles and the distances between them. For any pair of particles, the gravitational force between them is proportional to the mass of the first particle times the mass of the second particle, divided by the *square* of the distance between the particles' centers. Notice that the amount of the gravitational force that one particle exerts on another exactly equals the amount of gravitational force that the second particle exerts on the first. The two particles each attract the other with the same amount of gravitational force. However, a particle's ability to *react* to a force depends inversely on its mass. The same force that will accelerate a car to 100 kilometers per hour will not make a truck go that fast. For example, the sun's gravitational force attracts the Earth, and the Earth's (exactly equal) gravitational force attracts the sun. However, the sun's mass is 300,000 times the Earth's mass, so the sun resists this amount of force 300,000 times more than the Earth does. As a result, the Earth orbits around the sun, not the other way around (Figure 2–22). In truth, both the Earth and the sun orbit around the "center of mass" of the Earth-sun system (forgetting the other planets for a moment). This center of mass is 300,000 times closer to the sun's center than to the Earth's center. Since the Earth-to-sun distance, 150 million kilometers, is only 200 times the sun's radius, this center of mass lies well inside the sun, almost at its center.

   Because gravity always attracts and never repels, larger and larger masses inevitably exert more and more gravitational force. This ever-growing property of gravity allows it to win out over the other three forces, especially over electromagnetism, for large objects like

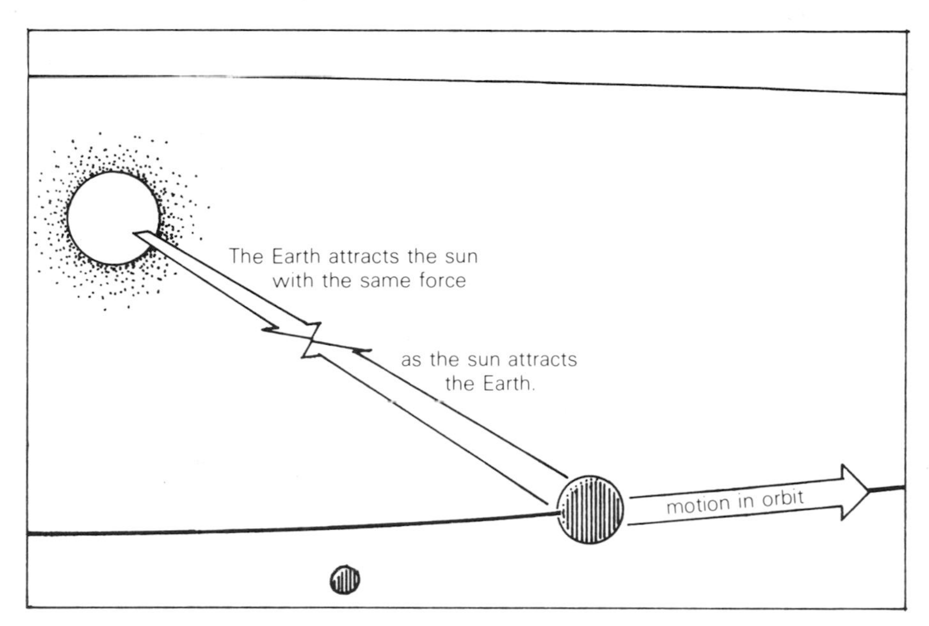

Figure 2–22. The gravitational force that the sun exerts on the Earth exactly equals in strength the gravitational force that the Earth exerts on the sun. Each of these gravitational forces attract one body toward the other. Because the sun resists the Earth's gravitational force some 300,000 times more than the Earth resists the sun's equal gravitational force, the Earth orbits around the sun and not the sun around the Earth.

the sun, the Earth, stars, and galaxies, each of which contain more than 10^{50} elementary particles (mostly protons, neutrons, and electrons). At the level of elementary particles, a different balance of forces applies, and gravity plays an insignificant role.

2. Electromagnetic forces rule the world of atoms. Every atom contains a nucleus with positive electric charge, around which orbit negatively charged electrons. The nucleus' positive charge comes from the protons, because the neutrons have no electric charge. Because electrons each have a mass 1836 times less than a proton's, they do the orbiting while the protons and the neutrons stay still. All this happens because of electromagnetic forces of attraction between the protons and the electrons, not because of gravitational forces. A proton and an electron *do* attract each other gravitationally, but the electromagnetic force between them exceeds the gravitational force by more than 10^{39} times! Therefore, electromagnetic forces determine what happens between the electrically charged protons and electrons in an atom. These forces lead to atoms with electron orbits of about 10^{-8} centimeters. Since the individual protons and electrons have sizes near 10^{-13} centimeters, the electron orbits are thousands of times larger than the sizes of protons, neutrons, or electrons.

Electromagnetic forces attract particles with opposite electric charge, but they repel particles with the same electric charge. The amount of the electromagnetic force between two particles depends on the charge on one times the charge on the other, divided by the square of the distance between them. This dependence is similar to the strength of gravity, except that we multiply the particles' electric charges and not their masses. Since electric charge can be positive or negative, while mass can be only positive, electromagnetic forces can attract or repel, while gravitational forces can only attract. Any large group of elementary particles will tend to have a balance of positive and negative charges. If it doesn't, the particles that provide an excess of one charge will attract particles with the opposite charge by electromagnetic forces until such a balance exists. Tiny atoms already have established equality between positive and negative charges, and in larger masses, such as our own bodies or the Earth, this balance is almost perfect. If two human beings each had one trillionth of a percent more positive than negative charges in their bodies, the electromagnetic repulsion between them would allow one person to propel the other against the Earth's gravity to the moon and beyond. In fact, the balance of charges in our bodies is far less than a trillionth of a percent's difference.

Human bodies contain about 10^{29} elementary particles; the sun and other stars have about 10^{58} particles. In a ratio of multiples, human beings stand halfway between atoms (a few elementary particles) and stars. Within our bodies, atoms form long chains of molecules because of electromagnetic forces. Most of our biological functions work through the same forces. In contrast, we all feel the gravitational force of the Earth. When we contract our muscles and leap in the air, the electromagnetic forces in our bodies, which make our muscles work, temporarily overcome the Earth's gravity. This offers proof that electromagnetic forces are stronger than gravity, since 10^{29} particles in our body can, for an instant, overcome the 10^{53} particles in the Earth.

3. The electromagnetic forces between protons and electrons govern the structure of atoms. However, if only electromagnetic and gravitational forces (the latter insignificant in nuclei) were present, then no nuclei could exist, because the protons' mutual electromagnetic repulsion would prevent them from forming any kind of nucleus. Another kind of force, called the *strong force,* operates among protons and neutrons (but not among most other, less massive elementary particles, such as electrons, photons, neutrinos, and antineutrinos). This strong force works only for extremely small distances, up to a few times the size of a proton or a neutron; that is, for distances only up to a few times 10^{-13} centimeter. Like

gravitational forces, *strong forces always attract*. Strong forces are far stronger than gravity at small distances such as those in atomic nuclei, but their strength falls off so rapidly with increasing distance that even for a distance of 10^{-8} centimeter (the size of an atom) they are totally insignificant. Because every "nucleon" (proton or neutron) attracts every other nucleon through strong forces, these forces provide the glue that holds together the nucleus in every atom more complicated than hydrogen (which has only one proton and no neutrons in the nucleus). The entire structure of matter from atoms on up depends on these strong forces, which are unknown in the larger world since their effects appear only over distances less than a few times 10^{-13} centimeter.

4. Weak forces are another kind of interaction among elementary particles, and like strong forces they have an effect only for distances less than a few times 10^{-13} centimeter. Also like strong forces, weak forces seem to operate only for some types of particles: They work only when at least one charged particle, and at least one neutrino or antineutrino, is involved.[6] As their name implies, weak forces are much weaker than strong forces, and the effects of weak forces tend to be masked by the operation of strong forces. We have discussed the other three kinds of forces in terms of stable situations, such as the Earth orbiting the sun (gravitation), electrons orbiting an atomic nucleus (electromagnetism), or the continued existence of the nuclei of atoms (strong forces). Weak forces, however, appear only when elementary particles change from one kind to another. The best example of such a change comes from neutrons. When a neutron does not form part of a nucleus—that is, when the neutron's distance from other neutrons and protons far exceeds its size—the neutron will change into a proton, an electron, and an antineutrino after about 15 minutes (Figure 2–23). This change occurs because of weak forces, and the reverse change of a proton, electron, and antineutrino combining to form a neutron also can occur as a result of the weak forces involved in this process. Neutrons within atomic nuclei do not change or "decay" into protons, electrons, and antineutrinos, and the reason that the neutrons do not decay comes from the strong forces among the various nucleons in the atomic nucleus. These strong forces in the nucleus dominate the situation and overcome the tendency of weak forces to make neutrons decay.

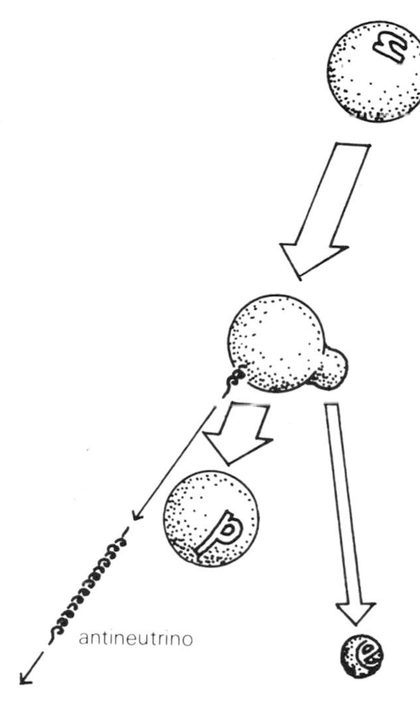

Figure 2–23. A neutron that is not part of an atomic nucleus will turn into three other particles—a proton, an electron, and an antineutrino—after about 15 minutes. This "decay" of the neutron into three other particles is the result of weak forces.

[6]Because weak forces always involve at least one charged particle, some physicists feel that electromagnetic forces and weak forces must somehow be intimately connected, part of the same superforce. This is part of the search for the "unified field theory" of all four forces that we mentioned previously.

Strong forces do not affect electrons, neutrinos, or antineutrinos, so when a proton, an electron, and an antineutrino meet, no strong forces are at work, and the effect of the weak forces can appear. The decay of free neutrons into protons, electrons, and antineutrinos is the reason why there are very few neutrons floating around by themselves, although there are plenty of protons and electrons that are not attached to any atoms. In addition to promoting this neutron decay, weak forces also appear in other kinds of interactions among nuclei. As we shall see in our discussion of how stars liberate energy (Chapter 7), weak forces play an intimate role in the basic energy-generation processes of the universe.

The four types of forces—no more, no less—seem to do it all in the universe. As far as we know, every interaction among particles can be described in terms of one or more of these four types, which themselves may be different manifestations of a single metaforce. Physicists take the search for a unified field theory seriously, although it is not an easy one (Einstein spent his last fifteen years at it). The unified field may stand as a metaphor for our search for a coherent way of life, but to scientists such a theory *would* show how the four forces in the universe are related.

Summary

Light waves, radio waves, x rays, and gamma rays all consist of *photons*, which are particles with no mass that always travel (in a vacuum) at the speed of light, 300,000 kilometers per second. Each photon has a characteristic energy of motion, which in turn specifies the photon's frequency and wavelength. The photon's frequency (number of oscillations per second) varies in direct proportion to the photon's energy, while the wavelength (distance between successive wave crests) varies as one over the photon's energy. The frequency times the wavelength always equals the same number, namely the constant speed of light.

When we receive photons from any source, we always find that the photons arrive at the speed of light, no matter how the source is moving relative to ourselves. However, the relative motion of a photon source does affect the photons' energies: Photons from a source in relative motion toward us have *larger* energies, whereas photons from a receding source have *smaller* energies, than we find from the same source if it is at rest with respect to ourselves. The amount by which this "Doppler effect" changes the photons' energies, frequencies, and wavelengths depends on the relative velocity of the photon source. We can use this dependence to measure the relative velocity of such a source along our line of sight (toward us or away from us).

Photons have no mass, but most other elementary particles do have some mass. Ordinary "matter" consists of electrons, protons, and neutrons, arranged into atoms that each have the less massive electrons in orbit around a nucleus of more massive protons and neutrons. *Electromagnetic* forces hold these atoms together: The protons and electrons, with opposite signs of electric charge, attract one another. *Strong* forces hold the protons and neutrons together in an atom's nucleus. The other two types of forces are *gravity*, which dominates the other forces only for large masses (such as the Earth), and *weak* forces, which govern certain reactions among elementary particles, such as the decay of a neutron into a proton, an electron, and an antineutrino.

Questions

1. What are photons? What distinguishes one kind of photon from another?
2. If two photons have wavelengths of 1 centimeter and 10 centimeters, respectively, which photon has more energy? By how much?
3. In order to interact with a particular photon, what size should an object be?
4. The two stars Polaris and Sadir (one of the stars that form Cygnus, the Swan) each emit about the same number of photons with the same energies each second. However, Polaris is 200 parsecs away from us and Sadir is 250 parsecs away. Which star appears brighter to us? By how much?
5. The two stars Aldebaran (in Taurus, the Bull) and Eltanir (in Draco, the Dragon) each emit photons at the same energies. However, Aldebaran has a relative motion along our line of sight at 55 kilometers per second away from us, whereas Eltanir's relative motion is toward us at 27½ kilometers per second. How will this relative motion affect the photons that we observe from Aldebaran and Eltanir? Is the size of the change in the photon energies the same for the two stars? Why?
6. What kinds of elementary particles are in the nucleus of an atom? Do electromagnetic forces tend to make these particles attract or repel one another? What holds such a nucleus together?
7. Consider a motionless neutron that turns into a proton, an electron, and an antineutrino. What kind of forces are involved in this change? How does the total electric charge after the change compare with the total electric charge of the neutron?
8. When a particle, say an electron, collides with its corresponding antiparticle, in this case a positron, what kinds of particles emerge from the col-

lision? Do these particles have mass? What happens to the energy of mass in the original colliding particle and antiparticle?

9. An atom of carbon-13 has six protons and seven neutrons in its nucleus. How many electrons will be in orbit around the nucleus?
10. Suppose a photon collided with a carbon-13 atom and knocked away one of its electrons, thereby "ionizing" the atom. Would this leave a net positive or a net negative electric charge on the ion that remains?
11. Suppose a photon hits a hydrogen atom that has its electron in the smallest allowed orbit and knocks the electron into the fourth-smallest allowed orbit. Will the photon gain or lose energy as a result?
12. What will happen to the hydrogen atom described in Question 11, once its electron has been knocked into the fourth-smallest orbit, if the atom is allowed to sit quietly by itself?
13. How do the atoms and molecules in our atmosphere affect the photons that try to penetrate it?

Further Reading

V. F. Weisskopf, *Knowledge and Wonder* (Doubleday Anchor, New York, revised edition, 1966).

H. Alfven, *Worlds-Antiworlds* (W. H. Freeman and Co., San Francisco, 1966).

D. Goldsmith and D. Levy, *From the Black Hole to the Infinite Universe* (Holden-Day, Inc., San Francisco, 1974).

T. Page and L. Page (Eds.), *Starlight* (Macmillan, New York, 1967).

L. Goldberg, "Ultraviolet Astronomy," *Frontiers in Astronomy* (W. H. Freeman and Co., San Francisco, 1970).

The Expanding Universe

Chapter **3**

"One thousand mahayugas—4,320,000,000 years of human reckoning—constitute a single day of Brahmah, a single kalpa · · · I have known the dreadful dissolution of the universe, I have seen all perish, again and again, at every cycle. At that terrible time, every single atom dissolves into the primal, pure waters of eternity, whence all originally arose."

Ancient Indian myth, quoted in *Myths and Symbols in Indian Art and Civilization*, by H. Zimmer

The universe has been expanding for the past fifteen or twenty billion years. We know this fact, but the uncertainty in the age of the expansion reflects our lack of detailed knowledge. Today the most crucial question for the future—whether the universe someday will stop expanding and start contracting—remains unanswered until we can make better observations to understand more about the evolution of the universe.

Astronomers found that the universe is expanding by measuring the distances and velocities of more and more distant objects. This series of observations, which revealed to us the general structure of the universe, lasted more than a century after the time when the first distance to another star was determined. We shall simply outline the methods of observation here, reserving a full explanation of distance measurements for Chapter 8, because we first want to understand the development of the universe before we follow the details of its present and future behavior.

Hubble's Law

Our belief that the universe is expanding rests on the central observational fact that galaxies are moving away from us, with speeds that increase in direct proportion to the galaxies' distances from us. Galaxies are huge gatherings of billions of stars, with some gas and dust mixed diffusely among the stars. The correlation between other galaxies' distances from us and their speeds of recession was discovered by Edwin Hubble, in 1929. We now call this relationship "Hubble's law": The velocity of a galaxy's recession from us equals a constant, H (the Hubble constant), times the galaxy's distance from us.

Why does Hubble's discovery, that recession velocity equals H times distance, mean that the entire universe must be expanding? Doesn't this "law" simply indicate that all galaxies are receding from us? The answer to this question is that if we inhabit an *average* galaxy in an *average* region of the universe, then every observer, anywhere, should see all the other galaxies moving away from that observer: *What we observe must represent an average sample of reality.*

The history of astronomical thought reflects an ever-stronger conclusion that the Earth does not occupy a special position in space. The Earth orbits around an average star, which in turn revolves around the center of the Milky Way galaxy, a typical "giant spiral galaxy" that forms part of a small cluster of galaxies, some quite similar to the Milky Way. Other galaxy clusters include galaxies such as our own, and all galaxies share a basic resemblance, as they are composed of millions or billions of stars that are much alike. When we measure the velocities of galaxies moving toward us or away from us, using the Doppler effect described in the previous chapter, we find that a very few galaxies are approaching us but almost all of them are receding from us.

Hubble succeeded in estimating the distances to some galaxies by finding stars in them that were like stars in our own galaxy whose distances and absolute luminosities were relatively well determined. From the comparison between the *apparent* luminosities of the stars in our own galaxy and the fainter stars in the other galaxy Hubble was able to determine the ratio of the stars' distances from us, because the apparent luminosity decreases with the square of the distance, as we discussed on

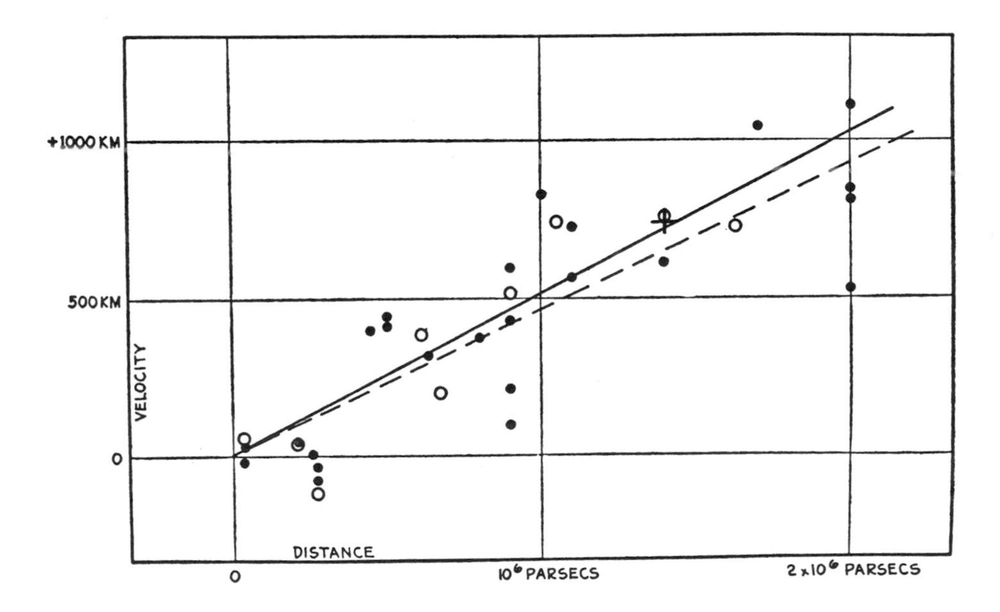

Figure 3–1. Hubble's first plot of the distances to galaxies against their velocities of recession was made in 1929. This graph showed a fair amount of scatter around the straight line that Hubble drew to represent the average behavior of galaxies [from the *Proceedings of the National Academy of Sciences*, **15**, 168 (1929)].

Page 9. Thus if two stars have the same absolute luminosity, but the star in our galaxy has an apparent luminosity a million times that of the star in another galaxy, we can conclude that the star in our Milky Way is a thousand times closer to us. Once Hubble found some distances to other galaxies in this manner and compared these distances with the galaxies' velocities toward us or away from us, he could plot the graph shown in Figure 3–1. Hubble realized that although the points on the graph of velocity versus distance do not form a perfect straight line, there *is* a general tendency for the recession velocity to increase with the galaxy's distance from us. The scatter in Hubble's graph arises from the fact that each galaxy has some random motion of its own, which adds to, or subtracts from, the general trend. This random velocity amounts to a few hundred kilometers per second, which for the galaxies nearest to us sometimes can overcome the general trend, so that some of the galaxies closest to us, such as the Andromeda galaxy, in fact are approaching us. These galaxies are represented in Figure 3–1 by velocities less than zero (meaning velocities of approach).

If Figure 3–1 doesn't seem very convincing, we should look at a graph that includes more distant galaxies. Within two years after Hubble's original discovery, in 1929, astronomers had measured the distances to galaxies fifteen times farther away than the farthest shown in Figure 3–1. These galaxies have recession velocities as large as 20,000 kilometers per second, and produce an unmistakable velocity-distance relationship (Figure 3–2). The random velocities of a few hundred kilometers per second produce only small deviations from the general relationship. The value of H can be found from the observations of galaxy distances and recession velocities, and the best value determined now is H equals 50 kilometers per second for every million parsecs of distance. Thus a galaxy ten million parsecs away shows a recession velocity of 500 kilometers per second, whereas a galaxy a billion parsecs away has a recession velocity of 50,000 kilometers per second. (When Hubble first discovered the universal expansion, he derived a value for H of 550 kilometers per second for every million parsecs of distance. Since then we have found new information about the stars and gas clouds used to measure distances, and have adjusted the distances accordingly. Astronomers are not too embarrassed by making errors of this size, because the universe is extremely large and confusing.)

Hubble's law, which summarizes observations of galaxies' distances and recession velocities, has been verified by more than forty years of observation, so astronomers have a great deal of confidence in it. As astronomers look to more and more distant galaxies, they become unable to see single galaxies, although they can observe galaxy clusters that contain hundreds or thousands of individual galaxies, spread out over "only" a few million parsecs. The most distant galaxy cluster for which the Doppler red shift has been measured shows a recession velocity of 140,000 kilometers per second, almost half the speed of light. *All* galaxies, except the very nearest, are moving away from us, never toward us. If we assume that our vantage point in space represents an average point of view, the conclusion that we draw from Hubble's law is this: On the largest distance scales that we observe, namely those between clusters of galaxies, *every point in the universe is moving away from every other point, with a speed that is proportional to the distance between the two points.* In swallowing this statement we have at least two objections to consider. First, how can every point in the universe move away from any other point? Second, doesn't this rule

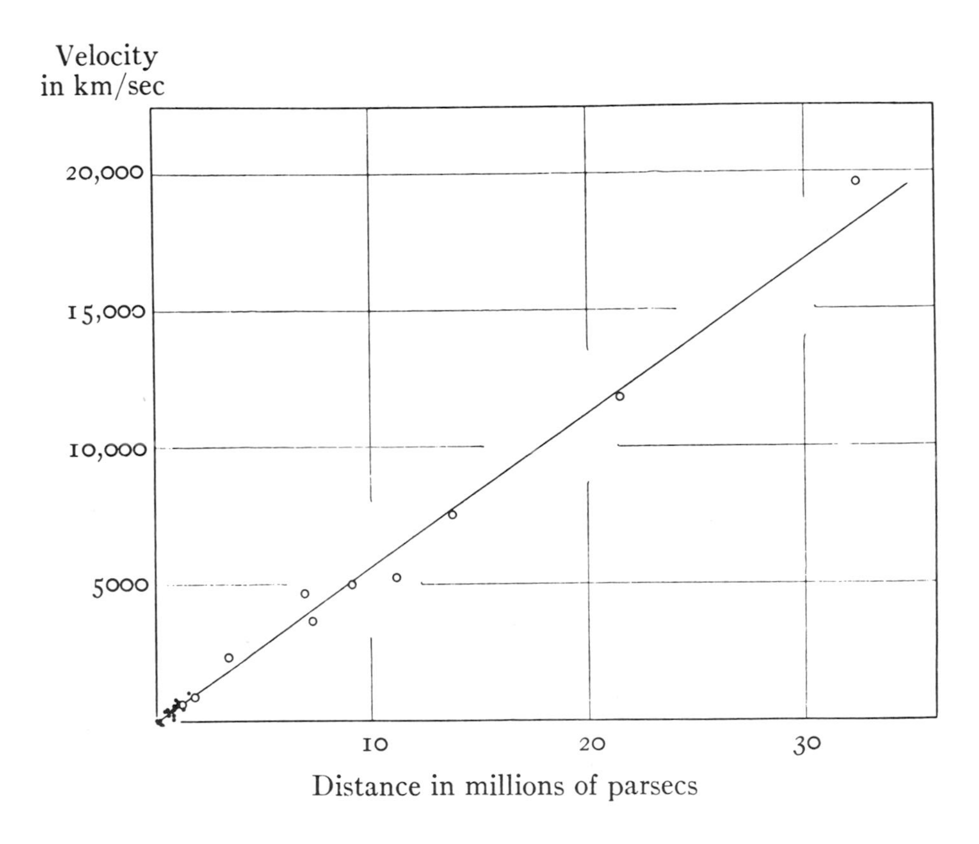

Figure 3–2. In 1931, Hubble (with the help of Milton Humason) had extended his graph of distances versus velocities to galaxies fifteen times farther away. The straight-line relationship that we now call "Hubble's law" had become much more evident. Since then, however, astronomers have reestimated the distances to galaxies so that, although the relationship is still a straight line, the factor by which we multiply a galaxy's distance to obtain its velocity of recession has been changed. [Graph from the *Astrophysical Journal,* **74,** 43 (1931); courtesy of the University of Chicago Press.]

mean that some points are moving away from each other at speeds greater than the speed of light? We shall deal first with the former objection, which lies at the center of our difficulties in understanding the expansion of the universe.

Models of the Universe

How can the universe expand everywhere? How can every point be moving away from every other point? We shall examine some models for the universe that try to show how this can happen, but none of these models will satisfy us completely, because we live in and are part of the universe. Our own minds cannot easily picture a *model* of the universe, which by our own definition means everything that exists.

Imagine a large balloon with gold coins glued to its surface (Figure 3–3). The coins represent clusters of galaxies and the *surface* of the balloon represents the universe. Now if we blow up the balloon, the coins

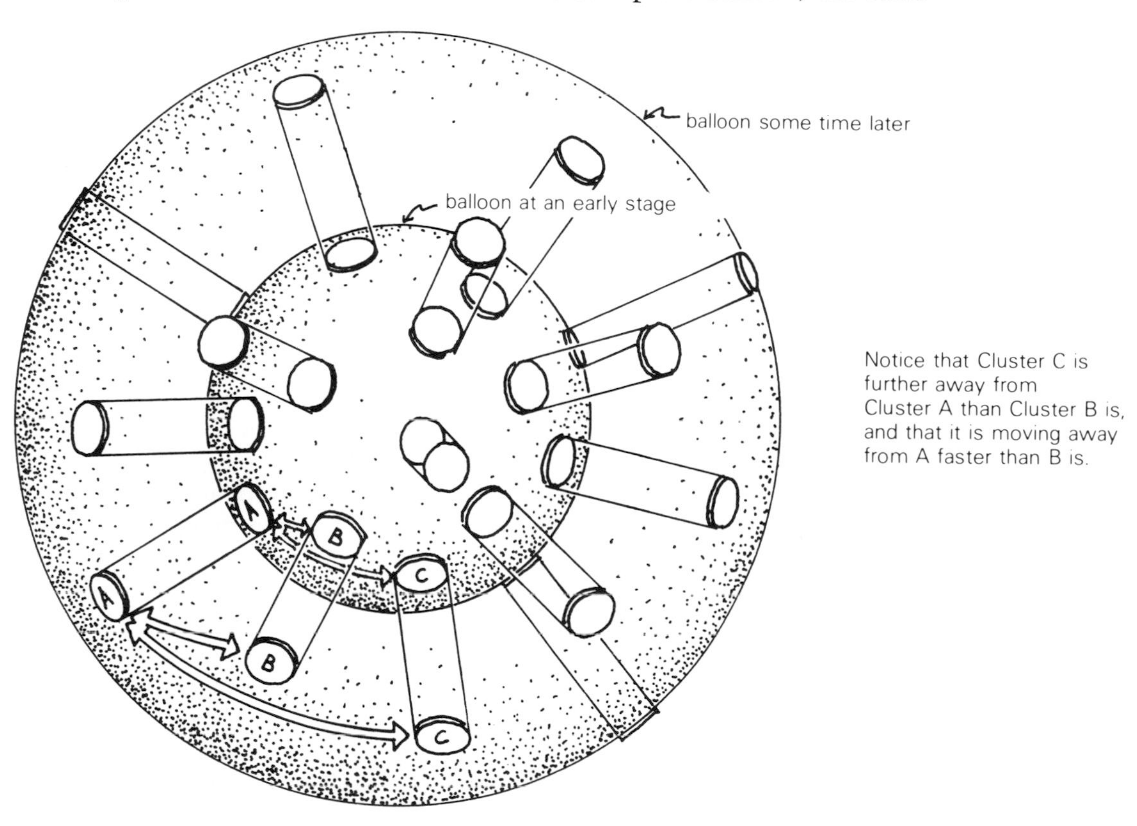

Notice that Cluster C is further away from Cluster A than Cluster B is, and that it is moving away from A faster than B is.

Figure 3–3. To picture the universe as best we can, imagine gold coins (galaxy clusters) glued to the surface of an expanding balloon. As the balloon expands, all of the coins move away from one another with speeds proportional to the distances between them.

all start to move farther away from each other. In fact, their velocities relative to one another, as measured along the surface of the balloon, follow their own Hubble's law: The relative recessional velocity of any two coins increases in proportion to the distance between them. Notice that although the balloon's surface, which represents all the space in the universe, does expand, the coins (clusters of galaxies) do not. As a result of the universe's expansion new space constantly appears, and it makes no more sense to ask where the space came from than it does to try to locate from where the extra space on the balloon's surface appeared.

The balloon model of the expanding universe reduces the three dimensions of real space to the two dimensions of the balloon's surface, so that we can grasp the concept of an expanding surface that has no center and no edges. All points on the balloon's surface move away from each other, everywhere. The real, expanding universe has no edges and no center. However, although we have an extra (third) dimension to use in examining the balloon, we lack this advantage in trying to picture the expanding universe. We are part of the universe, immutably inside, and could no more escape mentally than we could physically to "see" how the universe looks from outside. But by giving some thought to simpler, more comprehensible models of the expanding universe, we can begin to grasp its basic properties even without complete understanding.

Let us consider the balloon model a bit further. If the universe is expanding, it must have been smaller in the past. If we think backwards far enough in time, we reach a moment when the universe had zero size and occupied no space at all! Astronomers call this moment of beginning the "primeval fireball" or the "big bang," and their best estimates for the age of the universe put this big bang fifteen to twenty billion years ago. For comparison, the Earth and the sun are about 4½ billion years old, life appeared on Earth more than 3 billion years ago, and human beings less than 3 million years ago. Figuratively speaking, between fifteen and twenty billion years ago, someone took a giant breath and started the expanding universe. Or we may imagine that something gave the universe a mighty kick and sent its pieces flying in all directions. Since the first moment of the big bang, our particular part of the universe has been expanding away from every other part, or (if we prefer), the other parts have been expanding away from us.

The balloon model reminds us that the universe has no center. Of course, an expanding balloon *does* have a center, but this center lies in another (third) dimension, and not on the balloon's surface. When we model the universe by the outside of an expanding balloon, we use our three-dimensional intuition to examine a two-dimensional surface. If *all* of three-dimensional space resembles the *surface* of an expanding balloon, then any "center" of the expansion could not be on the three-dimensional "surface," but would instead lie in a fourth dimension of space, and the great thinkers of our day believe that space has only three dimensions. Surely a fourth dimension would be interesting to us, since (for example) it would allow us to walk through walls the way a child steps over a two-dimensional hopscotch box. In plain truth, however, we seem to have only three dimensions to think in. When we think about the original big bang, our minds try to get outside it, so that we feel that the big bang must have had some definite location in space. But if we remember that *all* of the universe, including all of space, was *inside* the big bang, we can begin to believe that the expanding universe did not have one central point. The entire universe expands all over.

What about expansion velocities greater than the speed of light? Everything that we know about the universe comes from particles and forces that never travel faster than light.[1] Hence we can draw around each point in the universe a sphere that represents the farthest "horizon" to which we can see. This horizon lies at a distance from us of the speed of light, c, times the age of the universe. Nothing beyond this horizon can affect us now, because no force from those nether regions has had time to reach us. It is reassuring to know that as the universe grows older, not only does the horizon of events that cannot influence us grow farther away, but also *the fraction of the universe within our horizon constantly increases* (Figure 3–4). That is, not only does the volume within a distance given by c times the age of the universe increase, but also the ratio of this volume to the total volume of the universe increases. Thus as the universe grows older, we (or another observer anywhere in the universe) can interact with a progressively larger and larger fraction of the universe, always broadening our horizons as we receive new information from regions that used to be just beyond the former horizon

Our knowledge of the universe encompasses a greater and greater part of the universe as more time elapses since the big bang. Now what about faster-than-light velocities? As we see increasingly more distant parts of the universe, aren't we going to encounter larger and larger velocities? Isn't there some part of the universe moving away from us at a speed greater than the velocity of light? The answer is no; there is no part of the universe moving away from us at a speed greater than light's. The direct proportionality between distance and recession velocity embodied in Hubble's law works only for velocities that are not close to the speed of light, such as 10% or 40% or 75% of the speed of light. Einstein's theory of relativity showed scientists that because the speed of light represents

[1]Physicists have postulated that particles called "tachyons," which *always* move faster than the speed of light, may in fact exist. Such particles have not yet been found, and furthermore, if they do exist, it may turn out that they cannot interact with ordinary particles (which always move more slowly than light).

the upper limit on any velocity, strange things ("strange" as far as our intuition is concerned) start to happen when an object approaches the speed of light. For one thing, velocities do not simply add; as we saw in our discussion of the Doppler effect (Page 21), if we send a light ray from a source approaching us at 1000 kilometers per second, we observe the light to arrive at speed c, not $c + 1000$ kilometers per second. As a result of the ways that velocities near the speed of light combine with one another, it turns out that no two points in the universe have a relative velocity greater than the speed of light.[2]

How Large Is the Universe?

Just how large is the universe? Is it infinite? Some people feel intuitively that the universe *must* be infinite, since any alternative would apparently involve some barrier or unpassable "end" to space. However, the universe *can* be finite and still have no end, just as the expanding balloon has no boundaries on its surface. Since the inside and outside of the balloon are the result of an extra (third) dimension, if we want to use the balloon model for the universe, we should picture ourselves as flat creatures made to slide around the balloon's *surface*. Then we can see that we might indeed live in a universe that always expands, but still remains finite in extent. *All* of three-dimensional space might be "curved," as the two-dimensional surface of a balloon is curved, so that space eventually bends back onto itself. Such curved space is called "closed" space, because you can go only a finite (though perhaps enormous) distance in it before you return to your starting point. In an "open," infinite universe you can go on and on literally forever.

The truth is that we don't know whether the universe is closed or open, finite or infinite, but we hope to find out soon. The question of whether the universe will ever stop expanding is closely linked to whether the universe is finite or infinite, so in the long run there is a "practical" reason for wanting to find the "curvature" of space in the universe. Toward the end of this chapter (Page 79) we shall describe how we can answer the unresolved question of whether the universe is finite or infinite by making more accurate observations.

---

[2]If we observe a galaxy in one direction with a recession velocity v_1 and a galaxy in the opposite direction with recession velocity v_2, then as seen from Galaxy 1, Galaxy 2 will have a recession velocity of

$$\frac{v_1 + v_2}{1 + \frac{v_1 \times v_2}{c^2}}$$

Hence if v_1 is 90% of the speed of light, c, and so is v_2, then Galaxy 1 will observe Galaxy 2 to have a recession velocity equal to 99.5% of the speed of light. Galaxy 2 will see the same recession velocity for Galaxy 1.

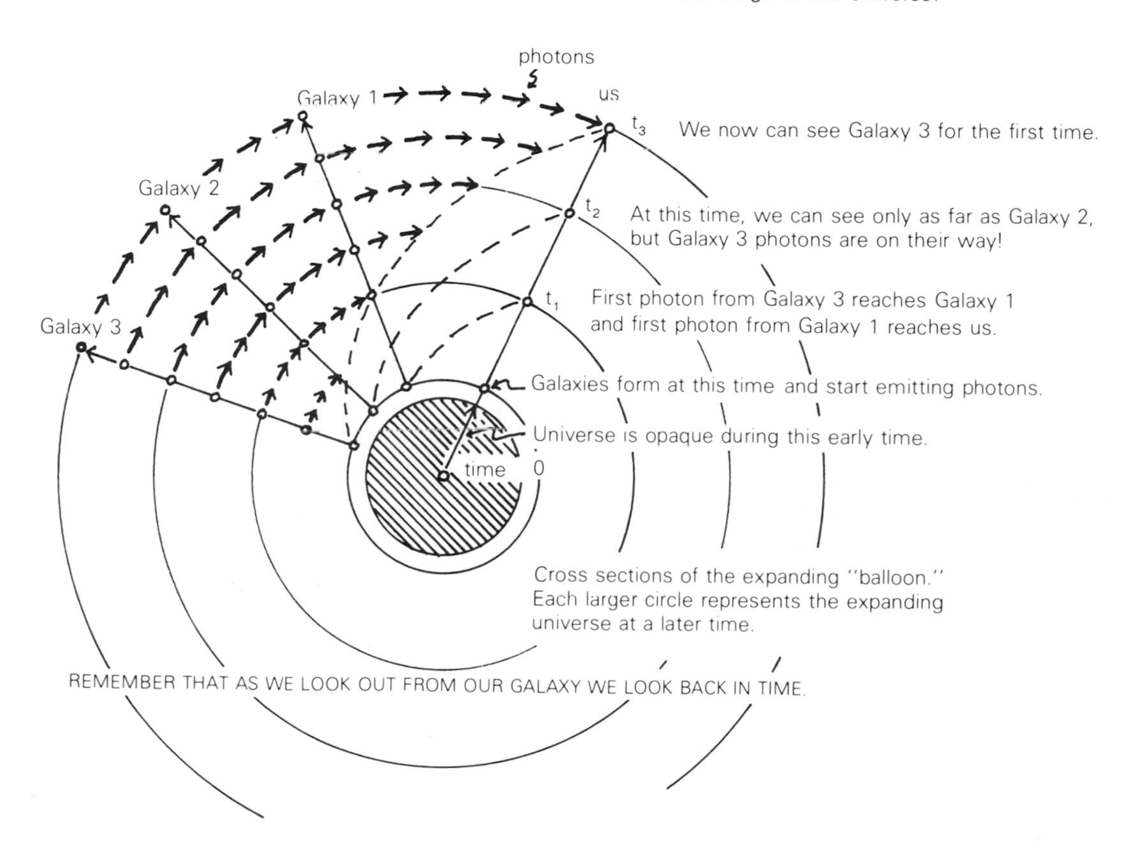

Figure 3–4. As the universe grows older, we can see increasingly more distant objects, because their light now has had time to reach us. Using the balloon model, we can see how the fraction of the universe that we can see actually increases as more time elapses since the big bang.

We can understand the past more easily than the future, though there may be less value in it. When the universe was younger, there was less space: In our model, the coins on the surface of the balloon were closer together. Because the universe had less volume, the average density of matter (mass per unit volume) was larger in the past than it is today. Figure 3–5 shows how the average density of matter has decreased steadily since the initial moment of the big bang, when the universe (for an infinitesimal moment) had an incredibly tiny volume and an enormously high density. From Figure 3–5 we can see that within one minute after the big bang, the average density of matter in the universe had decreased to 10^{-5} gram per cubic centimeter, which is less than the density of the Earth's atmosphere. A thousand years later, the average density in the universe was 10^{-15} gram per cubic centimeter, equal to the best vacuum on Earth, and after fifteen billion years, the density is only about 10^{-31} gram per cubic centimeter, less than the density of one proton per cubic meter. These figures are approximate because we don't know the present value (or the past value) of the average density of matter in the universe to within a factor of about ten. The average density of matter in the universe might be ten times larger, or five times smaller, than 10^{-31} gram per cubic centimeter.

Whatever the exact value may be for the present average density of matter, Figure 3–5 shows an unmistakable trend: The density has been decreasing as the universe expands. As the density of matter decreases, so does its temperature. The temperature of a group of particles (protons, hydrogen atoms, oxygen molecules, or whatever) measures the *average energy of motion per particle.* If we want to use the temperature to measure the energy of motion conveniently, we must use the "absolute" scale of temperatures. At zero on the absolute scale, there is *no* energy of motion per particle. Thus absolute zero in fact is absolute: Nothing can be colder than absolute zero, because no particles can have less energy of motion than zero energy. In terms of more familiar temperature scales, absolute zero corresponds to −273

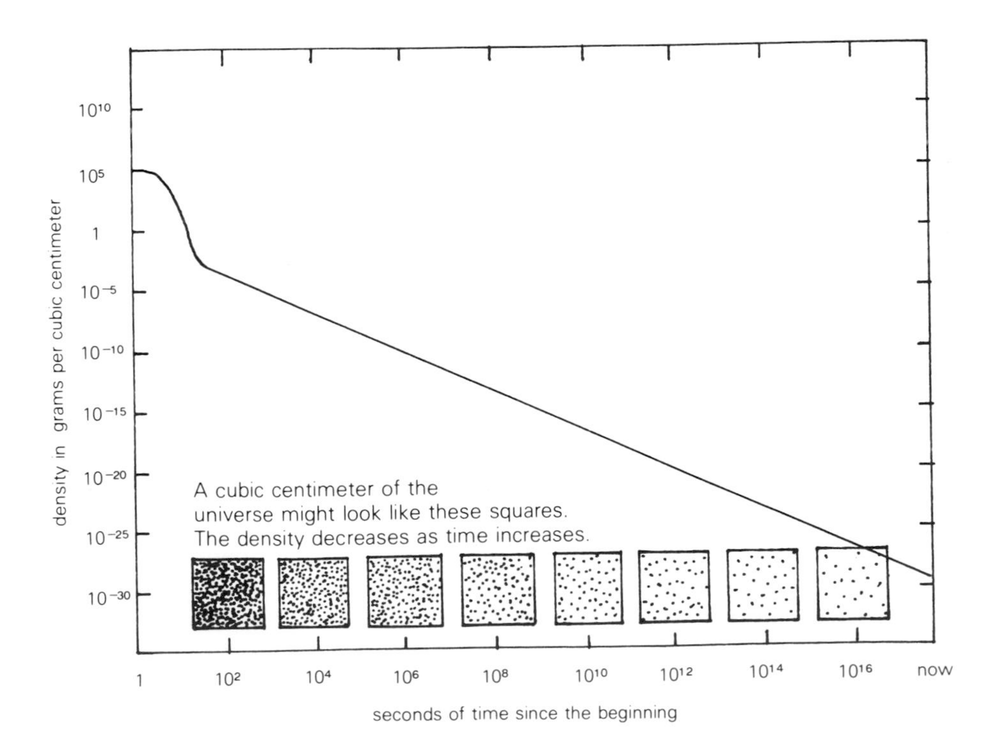

Figure 3–5. The average density of matter in the universe has been decreasing ever since the big bang occurred.

degrees Centigrade (Celsius) and to −459 degrees Fahrenheit. In the absolute temperature scale, water freezes at 273 degrees (zero degrees Centigrade, 32 degrees Fahrenheit) and boils at 373 degrees (100 degrees Centigrade, 212 degrees Fahrenheit). Because of the direct relationship between the temperature measured on the absolute scale and the average energy of motion per particle, astronomers and physicists use the absolute temperature scale. For example, the center of the sun has a temperature of about 13 million degrees absolute, whereas the sun's surface has a temperature of about six thousand degrees absolute. Because the temperature at the center is two thousand times the surface temperature, we know that the average energy of motion of a particle at the sun's center is about two thousand times greater than the average energy of motion of a particle at the sun's surface.

The average energy of motion per particle measures the level of activity around us. We live in what is by human standards a violent universe, because most of the matter in the visible universe has an absolute temperature of millions of degrees. By comparison, the Earth's surface, with temperatures of about 300 degrees absolute, represents a haven of languid motion, though particles on Earth do move more rapidly than particles in those frigid regions of space with temperatures of only a few degrees above absolute zero. The universe now has regions of vastly different densities and temperatures: stellar interiors at millions of degrees; stellar surfaces at many thousands of degrees; planetary surfaces and dense gas clouds at hundreds of degrees; diffuse gas clouds at tens of degrees or less. Typically (but not always), higher densities tend to produce higher temperatures. The sun's interior has a density of 100 grams per cubic centimeter, whereas its surface layers have a density of about 0.1 gram per cubic centimeter. Our atmosphere's density is about one thousandth of a gram per cubic centimeter, whereas gas clouds in space have densities of 10^{-10} to 10^{-23} gram per cubic centimeter. In the expanding universe, the average temperature of matter has decreased as the average density of matter decreased (Figure 3–6). This decrease in temperature was analogous to the way that steam will cool off (and condense into water) as it expands from the spout of a teakettle.

The Early Universe

During the early years of the expansion, particles of matter filled the universe almost uniformly, so that the average density was almost the same everywhere. Thus the distribution of matter was smooth, or "homogeneous," as opposed to the inhomogeneous, or clumpy, distribution of matter that we see today, when relatively dense stars and galaxies appear among the much less dense gas and dust. The question of how the smooth and homogeneous early universe became the inhomogeneous universe that we see today remains hard to answer. One thing does appear certain, however: Lumps of matter such as stars and galaxies must have formed only *after* the particles in the universe had combined into atoms of hydrogen and helium. This formation of atoms occurred some 300,000 years after the big bang, when the universe had an age less than one ten-thousandth of its present age, but nonetheless these first 300,000 years had a tremendous effect on all that came later.

When the big bang first began, there was light and there were particles. During the first seconds after the big bang, the entire universe was filled with matter at such an enormous density (greater than 10^5 grams per cubic centimeter) and such a high temperature (more than 10^9 degrees absolute) that no particle could maintain itself for more than a tiny fraction of a second. Instead, particles collided with each other almost instantaneously, and with such enormous energies of motion that they turned into other particles or became photons that had no mass. Colliding protons could become pi-mesons and neutrons, colliding neutrons might turn into electrons, positrons, and neutrinos, and so on. Furthermore, photons often became particle-antiparticle pairs, which in turn might collide with each other (turning back into photons and neutrinos and antineutrinos) or with other particles. Within this seething mass of matter and energy, every type of particle and antiparticle came into being briefly before some high-energy collision changed

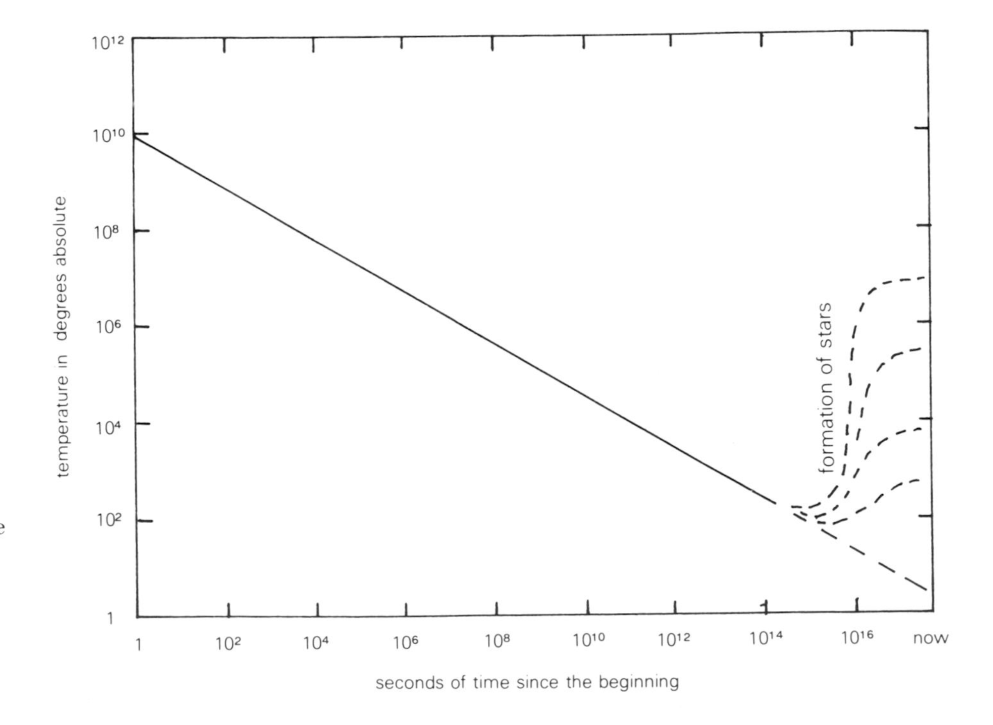

Figure 3–6. The average temperature of matter in the universe has decreased as the density decreased. About 300,000 years after the big bang, the average temperature of matter was about 3000 degrees above absolute zero. Since then, various lumps of matter at different temperatures have formed, so that the "average" temperature of matter covers a wide range of different temperatures.

the picture within a tiny fraction of a second. The first few minutes of the universe's expansion had more collisions that caused one kind of elementary particle to change into another kind than have occurred during the fifteen or twenty billion years since then. Whatever may have happened during these first minutes, whatever fantastic clash of primal particles becoming other forms of particles may have occurred, we know that these transformations among the different kinds of elementary particles ceased about a thousand seconds—say about fifteen minutes—after the big bang. By that time, the basic texture of the universe had been established, and everything that has happened since then has changed the mixture of various kinds of particles in the universe only slightly.

What was the basic mixture of particles that emerged from the universe's first fifteen minutes? Judging from what we can see in our sun, in other stars, and in cosmic debris that has been captured within the solar system, we conclude that almost all of the particles with mass in the universe are either protons, electrons, or helium nuclei (two protons plus two neutrons). In addition, a huge number of photons, neutrinos, and antineutrinos (none of which has any mass) travel among these particles at the speed of light. For every particle with mass in the universe there are a hundred million photons.

The neutrons in the universe mostly occur in helium nuclei, where two protons and two neutrons bind together by strong forces. These

Table 3–1
Formation of Different Elements During the First Few Minutes After the Big Bang

Element	Atomic number (number of protons)	Maximum number of nuclei (per 10^{12} protons) produced during the first few minutes after the big bang[a]	Present number of atoms (per 10^{12} hydrogen atoms) in the solar system and in stars like the sun
Hydrogen	1	10^{12}	10^{12}
Helium	2	8×10^{10}	8×10^{10}
Carbon	6	2×10^{6}	4×10^{8}
Nitrogen	7	10^{5}	1.2×10^{8}
Oxygen	8	6×10^{4}	7×10^{8}
Neon	10	300	1.1×10^{8}
Magnesium and all elements heavier than magnesium	12 or more	2500	1.2×10^{8}

[a] The number of nuclei of a given element that was produced during the first few minutes after the big bang depends on the density of the universe during those first few minutes. This density, like the present density of the universe, cannot be determined accurately, but we can calculate the *maximum* number of nuclei that could have been produced with the most favorable value of the density that was possible for producing nuclei of a particular element.

helium nuclei emerged during the first fifteen minutes after the big bang (Table 3–1). At that time there was one helium nucleus for every eight protons, approximately. Nuclei with two or three nucleons (rather than the four nucleons in ordinary helium) came out of the big bang in smaller amounts, while every kind of nucleus with more than four nucleons was extremely unlikely to have been formed through the particle collisions of the first fifteen minutes. Helium nuclei are especially stable and cannot be made easily into larger nuclei by collisions. Calculations show that at most two in a million helium nuclei became some larger nucleus during the initial violence after the big bang. This fraction is so small that we can conclude that just about every nucleus that is more massive, and thus more complicated than helium, such as the nuclei of the carbon, nitrogen, and oxygen atoms in our bodies, apparently did not form during the universe's first fifteen minutes. Instead, these nuclei were made billions of years later inside *stars* that exploded and spread their remains through space (see Chapter 8).

We have concluded that the early universe (aged from fifteen minutes to a few billion years) contained mostly protons, electrons, and helium nuclei, along with photons, neutrinos, and antineutrinos. Since protons and neutrons have a mass more than 1800 times an electron's mass, the electrons composed only a tiny fraction of the total mass in the universe. Almost all of the mass in the early universe was in protons (75%) and helium nuclei (25%), because there was one helium nucleus for every eight protons, and each helium nucleus has about four times a proton's mass. Some scientists find it strange that the number of particles (protons, neutrons, electrons) completely dominates the number of antiparticles (antiprotons, antineutrons, antielectrons), and they propose that the universe consists of matter and antimatter in equal amounts, but not in the same place. If a particle and its antiparticle meet, they will be annihilated and turn their combined energy of mass into the energy of motion of photons, neutrinos, and antineutrinos, as happened countless times during the universe's first fifteen minutes. Perhaps, though, we inhabit a region of space in which particles happened to outnumber antiparticles slightly, and the particles we see are all that remained after the particle-antiparticle pairs had been annihilated. Other regions of space originally might have had a slight excess of antiparticles over particles, so that these regions now contain only antiparticles. To us a galaxy of stars, all of which were made of antiparticles, would appear identical to a galaxy made of particles, such as our own Milky Way (Figure 3–7). The laws of gravity are the same for particles and antiparticles; strong forces would bind antiprotons and antineutrons into antinuclei; and since photons and antiphotons are identical, the light from such an antigalaxy would look just like the light from our own galaxy. Therefore, we may live in a universe with the same number of

Figure 3–7. The galaxy M 101 (number 101 in the catalogue compiled by Charles Messier) lies in the constellation Ursa Major. There is no sure way to tell whether this galaxy consists of matter or antimatter because the photons emitted by matter and antimatter are identical.

particles and antiparticles, or the universe may, in contrast, consist primarily of particles, as our own galaxy does. In either case, we believe that the first fifteen minutes of the universe processed all the matter into protons (or antiprotons), electrons (or antielectrons), and helium nuclei (or antihelium nuclei), together with the massless photons, neutrinos, and antineutrinos that also emerged from the enormous clash of particles.

The question of the possible matter-antimatter symmetry of the universe leads to the question of its positive-negative charge symmetry. As far as we can tell, the total number of positive electric charges (one per proton, two per helium nucleus, and so forth) equals the total number of negative electric charges (one per electron) in the Earth, in the sun, and in our entire galaxy. Presumably any antigalaxy also would have a total electric charge of zero, with the positive electric charges of the antielectrons balancing the negative electric charges of the antiprotons and antihelium nuclei. This symmetry between the total positive and negative charge seems to give the universe a total electric charge of zero. Because the total electric charge remains unchanged during particle collisions (as, for example, when a proton with positive charge and an antiproton with negative charge annihilate to produce two photons, each with no charge), the total zero charge in the universe has always been zero. Why the universe began with a zero total charge has not been explained, but it is a nice round number.

What has changed is the grouping of charges, first into nuclei (such as helium nuclei), and later into atoms. As the universe expanded and cooled, the positively charged protons and helium nuclei attracted and eventually captured electrons into orbits around them to form atoms. This key step in the universe's history, which occurred about 300,000 years after the big bang, played a crucial role in allowing the universe to achieve its present configuration, and the time of the formation of atoms also left behind a relic that today provides us with proof that the big bang really happened. This relic consists of the photons that filled the early universe and now carry the signature of the epoch when atoms formed.

The Microwave Background

Let us consider what happened to photons as the universe expanded. The first fifteen minutes produced a huge number of photons as particles collided with each other rapidly, and this is why there still are a hundred million photons for every particle with mass in the universe. In the early universe, the photons zipped around at the speed of light until they collided with a particle such as a proton or an electron. Recall that the universe is expanding, so that every particle is moving away from every other particle. The universal expansion causes photons from far-away sources to have large Doppler red shifts, which we discussed at the beginning of this chapter. The Doppler shift arises from the relative motion of the photon source away from us, which gives the photon as we observe it a smaller energy and a longer wavelength than it would have if the photon source were not receding from us. Now suppose that we were seated on a proton in the early universe, say a thousand years after the big bang, and that we observed a photon produced during the first fifteen violent minutes. Such a photon, as observed by us on the proton, would have a large Doppler shift arising from the fact that we are moving rapidly away from the point where the photon was produced, no matter where that was. Such photon red shifts provide us with a vivid reminder that the entire universe is expanding, everywhere. No matter in what direction we may look, any photon from the early universe will have a large Doppler shift to lower energies, and the longer the photon has been traveling (that is, the farther away the photon source is), the greater the photon's red shift will be. Photons observed now that come from the first fifteen minutes of the universe have been traveling for fifteen or twenty billion years. The Doppler effect reduces their energy by a factor of a hundred million, so that they have only 10^{-8} of the energy and 10^8 times the wavelength they would have if the universe were not expanding. The photons that arrive from the universe's first fifteen minutes come from a source—the entire universe at that time—which is moving relative to us at a speed of 99.999999% the speed of light. Of course, the source no longer exists in the same form, but the photons it produced do exist, and they have been traveling through space—all of it—since the first fifteen minutes after the big bang.

To return to our proton, notice that if we were observing the photons from the first fifteen minutes only a thousand years after the big bang, the Doppler effect from the universe's expansion would reduce the photons' energies 10^4 times, rather than the 10^8 times that we observe now. The size of the reduction in the photons' energy caused by the Doppler effect is important for the following reason. After the universe's first fifteen minutes, photons filled the universe, 10^8 of them for each proton, and the average energy of each photon was about 10^{-7} erg. This energy per photon, although tiny in human terms, is enormous for elementary particles. For comparison, the energy of mass of an electron, the electron's mass times c^2, is about 10^{-6} erg, which is only ten times larger than the energy of motion that every photon had at the time when the basic mixture of particles had emerged (fifteen minutes after the big bang). This rough equality of energy per photon and an electron's energy of mass arises from the fact that many of these photons had just been made by electron-antielectron annihilations.

To bind together a proton and an electron in a hydrogen atom requires far less energy than the energy per photon fifteen minutes A.B.B. (after the big bang). Even in the smallest and most tightly bound orbit of a hydrogen atom, the electron can be knocked completely loose from the proton by an energy input of only 2×10^{-11} erg, which is thousands of times less than the energy per photon at a time of fifteen minutes A.B.B. During the first few thousand years A.B.B., most of the photons speeding through space had enough energy to ionize any hydrogen or helium atoms they might meet, even though the Doppler shift reduced their energy by hundreds or thousands of times as seen from the atom's point of view. As a result, any atoms that formed during this period lasted only a short time (hundreds or thousands of years) before some photon completely freed the electrons from their orbits around atomic nuclei.

But after about 300,000 years, the Doppler effect finally left the photons unable to destroy atoms in this way. At this time, the photons produced by the universe's first fifteen minutes hit the atoms with less than 10^{-11} erg of energy as a result of the increased Doppler effect. The increasing time since the photons were produced inevitably led to a greater speed of the photon source (the first fifteen minutes, now vanished) with respect to the observer (a hydrogen or helium atom that had formed by electromagnetic attractive forces). Once most of the photons had energies less than 10^{-11} erg, they could not ionize either hydrogen or helium atoms. Nor could these photons "excite" the electrons in the atoms into larger orbits, since the energy needed to excite these atoms from the smallest electron orbit to the next smallest is about ¾ of the energy needed to ionize the atom completely. Therefore, once the average photon energy fell below 10^{-11} erg, the photons could no longer prevent hydrogen and helium atoms from forming, and in fact they could not even excite the electrons in these atoms into different orbits. Deprived of the chance to ionize or to excite the atoms, the photons thus did not interact with the atoms at all, and since that time the photons have passed among the atoms in a state of mutual ignorance. The atoms can't tell that the photons are there, and the photons can't communicate their presence to the hydrogen and helium atoms, because none of the photons has enough energy to excite or ionize an atom.

Still, the photons are there, and in fact they have been observed. The universal flood of photons (still 10^8 for every proton in the universe) bathes the Earth from all directions. In 1956, George Gamow predicted that these photons must exist, and in 1965, scientists at the Bell Telephone Laboratories in New Jersey observed some of them. To do this, they used a radio antenna (Figure 3–8) to collect the photons, which interacted with a crystal in the receiver (not, of course, made of hydrogen or helium atoms, but instead capable of interacting with photons of very low energy). These interactions generated a response that was amplified millions of times to allow verification of the detection of photons that had last interacted with matter when the universe was only 300,000 years old.

Why did these scientists use a *radio* antenna? The answer comes from the continuing expansion of the universe. Fifteen minutes A.B.B., when the photons were made, each photon had an energy of about 10^{-7} erg. When the Doppler effect had reduced their energy to 10^{-12} erg, so that the photons no longer could disrupt hydrogen or helium atoms, the universe was 300,000 years old. Since then, the Doppler effect has reduced the average energy of the photons a thousand times more, to 10^{-15} erg. Such photons belong to the radio band of energy, frequency, and wavelength (see Table 2–1).

Figure 3–8. This radio antenna at Bell Telephone Laboratories (Holmdel, New Jersey) was used for the first observation of microwave background photons.

The time when photons no longer had energies large enough to ionize hydrogen or helium atoms is called the "time of decoupling." Before then, the photons and the particles with mass were coupled together by their continued interaction. An atom would form, and almost immediately a photon would ionize it, giving up energy to do so. In this way, the photons and the matter (protons, electrons, and helium nuclei), had their destinies linked together, as their mutual interactions allowed them to influence each other. After the time of decoupling, the photons and the particles with mass have followed their destinies separately.

Matter in the universe now has become highly clumped into stars and galaxies. Such clumping was impossible before decoupling, when photons from all directions constantly bombarded atoms and broke them apart. This photon bombardment exerted a strong homogenizing effect on the matter: Any clump that started to form would be spread out once more as the photons destroyed its constituent atoms. Only after the sea of photons decoupled itself from the particles with mass could there be any coherent gathering of these particles into larger and larger clumps, which eventually grew into the galaxies of stars that we observe today, billions of years after the time of decoupling. Only the photons remain unchanged from that time, Doppler shifted to be sure, but still spread diffusely and evenly through the universe, still coursing through space amid the more recently made photons from stars such as our sun.

The photons that fill the universe as a relic from the time 300,000 years A.B.B. form a "microwave background" to all the other photons that appeared later. We call this a "microwave" background because most of the photons in it have energies typical of microwaves and the longest infrared photons (see Table 2–1). Microwave photons often are used for long-distance telephone communications, and this is why the Bell Telephone Laboratories had the right antenna to discover the microwave background photons. (As a matter of fact, the scientists discovered the background pretty much by accident, while they were testing their newly installed antenna.)

If we could observe the total spectrum of the photons in the microwave background, we would expect the number of photons at each wavelength to look like one of the curves drawn in Figure 3–9. Figure 3–9 also shows the measurements of the spectrum of photons in the microwave background. We cannot determine the number of photons with relatively short wavelengths (relatively large energies) from Earth, because our atmosphere absorbs these photons, and we have not yet sent the necessary detectors above the atmosphere.

The observed spectrum of the photons in the microwave background seems to fit a well-known spectral pattern, namely the spectrum of photons from an "ideal radiator." This is a fancy name for a group of particles that are in energy equilibrium with the photons among them. The particles (atoms, for instance) emit as much energy per second in the form of photons as they receive from the photons they absorb. The early universe provides an extremely good example of an ideal radiator, since the photons and the other particles (protons, electrons, and helium nuclei) constantly exchanged energy back and forth. The particles formed atoms that absorbed energy from photons, thereby breaking apart and producing new photons in the process of recombination. A particular "ideal radiator" will have a photon energy spectrum like that shown in Figure 3–9. The characteristic shape of the spectrum is the same for all ideal radiators, and the *temperature* of the ideal radiator determines where the peak (maximum number of photons) will occur. If we use energy terms, and consider the photon spectrum as the number of photons with various energies, then the peak energy (the energy with the greatest number of photons) will be directly related to the temperature. The peak energy (in ergs) always equals 4×10^{-16} times the temperature in degrees absolute. When the micro-

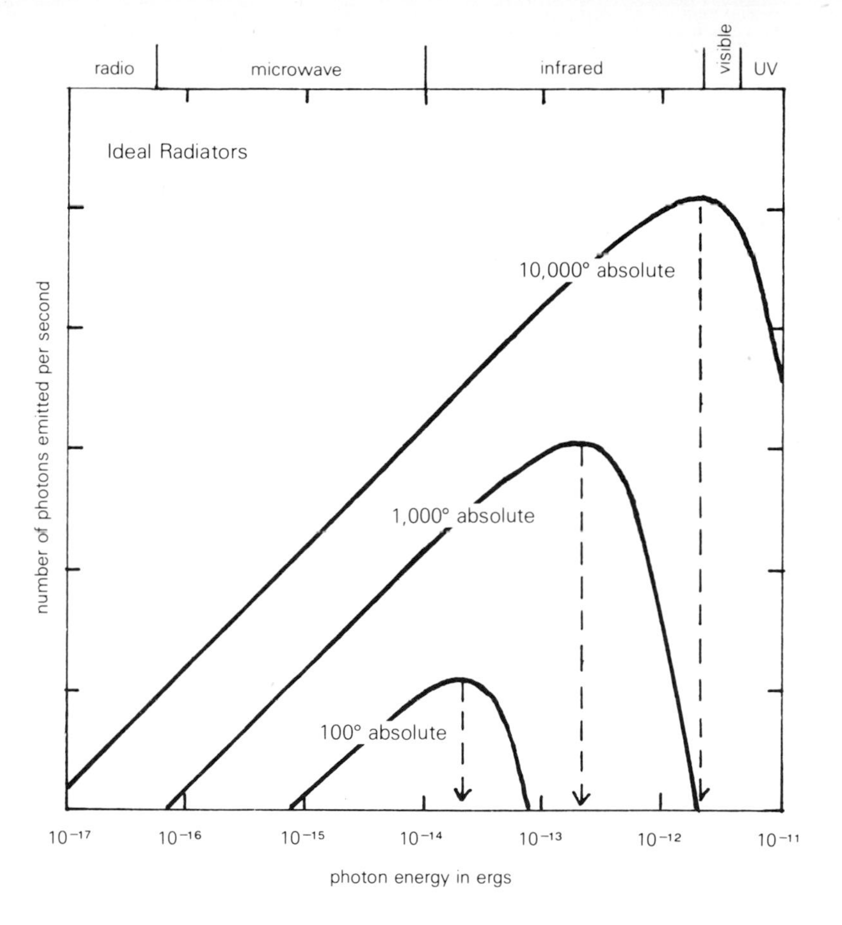

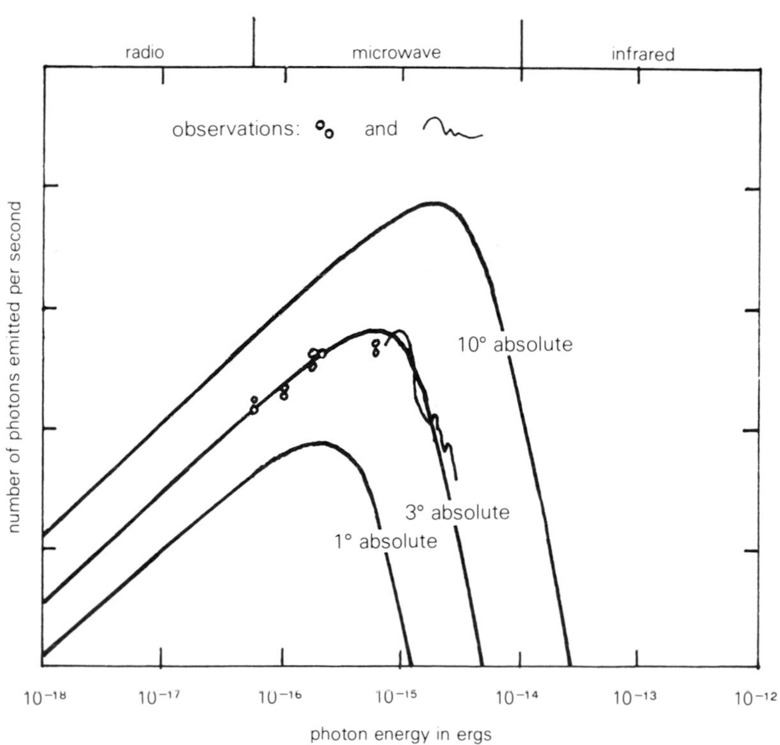

Figure 3–9. An "ideal radiator" will emit photons with an energy spectrum of a characteristic shape, as shown in the top drawing. The peak of the spectrum occurs at a photon energy that is proportional to the temperature of the ideal radiator. The greater the temperature, the greater will be the energy at which the peak of the photon energy spectrum occurs. The bottom drawing shows the measurements (represented by the dots and by the wiggly line) that have been made of the number of photons of various energies in the microwave background. We can see that at least for those energies at which we can observe the background photons, their energy spectrum corresponds fairly well to the energy spectrum of photons emitted by an ideal radiator with a temperature of three degrees above absolute zero.

wave background first decoupled from the matter, 300,000 years A.B.B., the photons in it had a peak energy of 1.2×10^{-12} erg. Some of the photons had energies greater than this, some had less energy, but the peak was at 1.2×10^{-12} erg. Almost none of the photons had energies greater than 2×10^{-11} erg, the energy needed to ionize a hydrogen atom. As the universe expanded, an observer would find that the energies of the photons in the microwave background all change in the *same* proportion, so that the shape of the spectrum remained the same—the shape characteristic of an ideal radiator (Figure 3–9). All the observed photon energies now have decreased a thousand times since the moment of decoupling, so the peak energy in the spectrum of background photons now occurs at 1.2×10^{-15} erg, instead of the 1.2×10^{-12} erg that characterized the spectrum at 300,000 years A.B.B. This change in the peak energy is equivalent to saying that the temperature of the ideal radiator that produced the photons (the entire universe) has decreased from 3,000 degrees at the time of decoupling to a characteristic temperature of three degrees now. That is, the photons that form the microwave background now have an energy spectrum the same as the spectrum of an ideal radiator at a temperature of three degrees above absolute zero.

The microwave background of photons that decoupled from the particles in the universe 300,000 years after the big bang seems to prove that the big bang really happened more or less as we have described. Before the discovery of these photons, some astronomers had suggested that although the universe indeed is expanding, perhaps new matter (created out of nothing, just as the entire universe somehow "started" out of nothing) would appear at exactly the rate necessary to compensate for the decrease in density. In this "steady-state" model of the universe, matter continues to appear in the amount needed to hold the average density in the universe constant despite the expansion, which would decrease the average density if no new matter were created. As a result of this appearance of new matter, the steady-state model of the universe would always present the same appearance (at least as far as the average density of matter in the universe is concerned), and there would have been no "big bang" fifteen or twenty billion years ago. The steady-state universe would have been the same forever.

The best argument against the validity of the steady-state model of the universe comes from the photons in the microwave background. Although some astronomers believe that these photons can be explained as arising from something other than the time of decoupling, 300,000 years A.B.B., most astronomers have decided that the microwave background provides a convincing refutation of the steady-state model of the universe in favor of the big bang.

Additional evidence in favor of the big-bang theory over the steady-state theory of the universe comes from the amount of helium, relative to hydrogen, that we find in the universe today. The fact that one quarter of the universe's mass consists of helium nuclei, while almost three quarters consists of hydrogen nuclei (protons), fits with astronomers' calculations of the fusion of protons into helium nuclei during the first seconds after the big bang. These calculations, as we mentioned previously, also imply that almost no nuclei heavier than helium should have been formed during the first seconds of the universe, when temperatures were high enough to fuse hydrogen into helium. Of course, the fact that the observations of helium-to-hydrogen abundance ratios in nearby stars and galaxies fits with the calculations made by astronomers does not provide *proof* that the big bang actually occurred; for that, the microwave background provides a firmer footing.

The Future of the Universal Expansion

Observations of distant galaxies tell us that the universe is expanding. Our curiosity naturally prompts us to ask whether this expansion will continue forever, thereby spreading the universe out farther and farther. Or will the expansion someday reverse itself, thereby causing the universe to contract and perhaps to rebound through another "big bang"? The answer to this question now lies just at the limits of what astronomers can hope to discover.

How could we ever find out whether the universe will someday stop expanding? The possibility of predicting the future of the universe arises from our ability to look into the past, which governs the future. When we look at faraway clusters of galaxies, we inevitably are looking backwards in time. Light waves and radio waves travel through space at a speed of 300,000 kilometers per second, or ten trillion kilometers per year. For astronomical distances the number of kilometers grows so large that we measure in parsecs (one parsec equals 3¼ light years or 35 trillion kilometers) in "kiloparsecs" (thousands of parsecs), and in "megaparsecs" (millions of parsecs); see Page 4. The Andromeda galaxy, the nearest large galaxy to us, is half a megaparsec (0.5 Mpc) away; thus its light takes 1.6 million years to reach us, and we see it as it was during the early phases of human evolution on Earth. The giant galaxy called M 51 (Figure 3–10) has a distance from us of ten megaparsecs; its light takes 33 million years to travel this distance to us; thus we see it as it looked during the early years of mammals. The most distant cluster of galaxies yet measured is 3000 megaparsecs (about ten billion light years) away, and the light we receive from this galaxy cluster left long before the sun or the Earth existed.

When we observe a cluster of galaxies more than a thousand megaparsecs away, we look back in time by more than three billion years, which is a significant fraction of the age of the universe. If we could observe objects at greater and greater distances, we could look back in time as far as we chose, until we reached the time of the big bang. We

Figure 3–10. The "Whirlpool Galaxy," M 51, has a distance of about ten million parsecs (ten megaparsecs) from us.

do this, in fact, when we observe the photons in the microwave background, which come from a time only 300,000 years A.B.B.

But why do we want to observe more and more distant objects? These observations could tell us how the expansion rate of the universe has changed over billions of years. Hubble's law for the universe's expansion gives an object's velocity of expansion away from us as a constant times the distance. Hubble derived this law from observations of relatively "nearby" galaxies, those within a thousand megaparsecs of our own galaxy. At larger and larger distances, we expect to find deviations from Hubble's law. These deviations arise partly from the way that velocities near the speed of light do not add to one another directly (Page 60), and partly from the fact that the universe has not always had the same overall rate of expansion.

The original big bang gave the universe an enormous shove that started everything flying away from everything else. Ever since then, each chunk of matter in the universe has continued to expand away from every other chunk. Sometimes chunks have managed to clump together in minor objects such as stars, galaxies, and galaxy clusters, but on the largest distant scales, those between galaxy clusters, everything still moves away from everything else. The only force that tends to slow the expansion is gravity. Each chunk of matter attracts every other chunk, and this mutual attraction tries to pull the various parts of the universe together again.

The future of the universe's expansion hinges on the struggle between the outward rush left over from the big bang and the attractive gravitational forces among the various components of the universe. This competition between the two tendencies, outward expansion and mutual attraction, does change the speed of the expansion, and someday the gravitational forces could stop the universe's expansion and begin a universal contraction.

To determine whether this will really happen, we must measure two universal characteristics: first, the average density of matter in the universe; second, the "deceleration parameter" that characterizes how much the expansion is slowing down from its former values. To measure the average density of matter in the universe, we can try to count all the galaxies in some region around us that we hope typifies the universe as a whole. We then can estimate the mass per galaxy and the size of the region that contains the galaxies. Dividing the total mass (number of galaxies times the mass per galaxy) by the volume that contains these galaxies will then give us the average density of matter in this volume. From this method of estimating, we calculate that the average density of matter in the universe is about 10^{-31} gram per cubic centimeter—the same density as we would have if a building ten kilometers on a side contained a single ping-pong ball. On the average, space is mighty empty.

We hope to measure the second universal characteristic, the "deceleration parameter," by observing faraway galaxies and by using this information from ages past to determine the changes in the universe's expansion. Figure 3–11 shows how we could plot the velocities of galaxies away from us against their distances, as Hubble did when he discovered Hubble's law. For the nearer galaxies, the clusters' velocities and distances follow the direct proportionality of Hubble's law. But as we observe more and more distant galaxies, we find that this straight-line relationship does not continue indefinitely, because we are looking back into an age when the expansion rate was different. Our present ability to estimate galaxies' distances and velocities has taken us about to the point where the various possibilities (eventual contraction versus infinite expansion) branch away significantly from the line of Hubble's law.

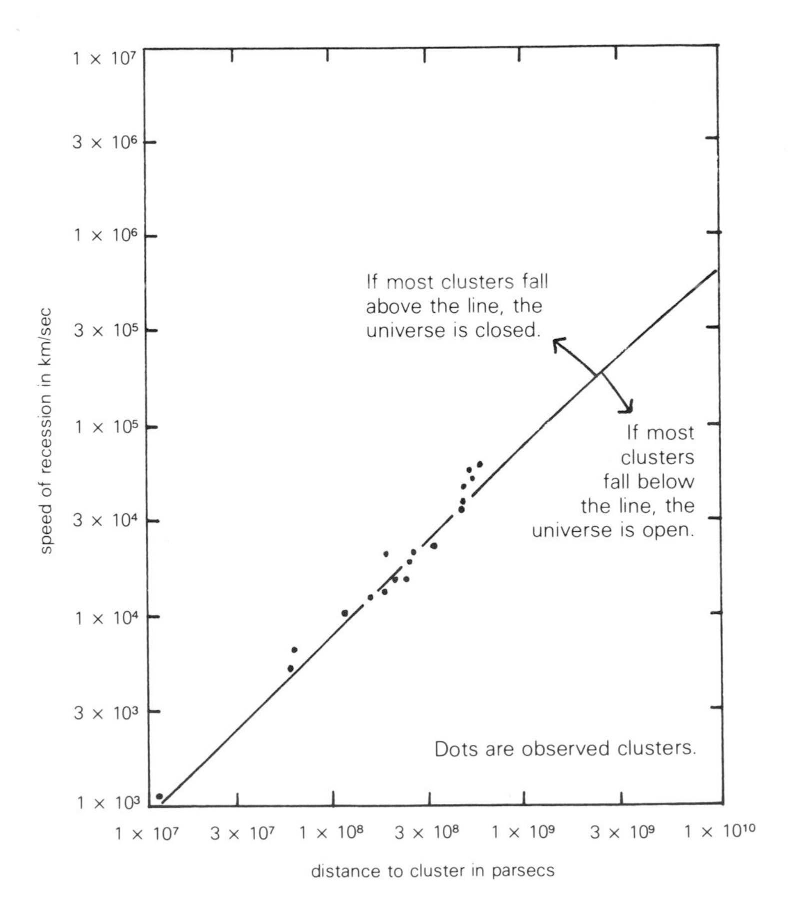

Figure 3–11. If we graph the Doppler shifts of galaxies, which give their speeds of recession away from us, against the galaxies' distances, which are (roughly) estimated from their apparent brightnesses, we can hope to see whether the trend corresponds to that expected for a closed (finite) or for an open (infinite) universe. If the observed relationship between galaxies' recession velocities and their distances falls above the solid line, the universe must be finite. If the relationship turns out to lie below the solid line, the universe must be infinite. So far, the trend indicates that the universe is finite, and that the universe eventually will begin to contract, but our data are not accurate enough for us to be sure of this.

The reason why we must know *both* the average density of matter and the "deceleration parameter" to determine the future of the universe is as follows. In addition to gravitational forces, which dominate the interactions of large chunks of matter separated by large distances, astrophysicists from Einstein on have speculated that an additional force that does not depend on the amount of mass may influence the behavior of the universe over the largest scales of distance. This additional force has the name "cosmological constant" because it appears as a constant term in a famous equation that Einstein derived.[3] If the "cosmological constant" turns out to be zero, then there is no additional force. But if the constant is some positive or negative number other than zero, this means that the way in which two huge regions in the universe interact with each other depends on more than the amount of mass in them and the distance between them. The new kind of force that would exist if the cosmological constant is not zero could not have any significant effect on objects separated by distances smaller than those between clusters of galaxies, but as we go to larger and larger distance scales, the effect of a nonzero cosmological constant would become more and more evident, since this constant tells us how the different parts of the universe affect one another. When we look at faraway galaxies to determine the deceleration parameter, we also are comparing parts of the universe separated by billions of light years, so our observations bring us information on whether the cosmological constant differs from zero. Unfortunately, these observations cannot yet answer this question, but if they were a little bit more accurate, we could determine the whole shebang—the deceleration parameter, the cosmological constant, and the much greater mystery that they control.

If we accurately knew the present average density of matter in the universe and the present value of the deceleration parameter, we could answer two important questions: Will the universe expand forever? And is the universe infinite? In general, the larger the average density of matter in the universe, the easier it will be for the mutual gravitational forces to overcome the continuing effects of the big bang. Also, the more "deceleration" of the universe that has taken place since the big bang, the more likely it is that the universal expansion will "decelerate" to zero and begin to contract. A set of equations, derived by theorists who devote their lives to these matters, shows how the two basic numbers, the present average density and the present deceleration parameter, give us the answer not only to the contraction question but also to the finite versus infinite universe question.

---

[3]This famous equation was not $E = mc^2$ but another equation entirely. Einstein left behind many great equations.

Whether the universe is finite or infinite, and whether it will start to contract sometime, indeed can be determined once we know the two key numbers. The trouble is that we can only estimate them, because our observations do not extend as far as we would like, and we have difficulties in determining the distances to faraway galaxy clusters accurately. What is worse, we might be surrounded by a region of the universe that does not fairly represent the entire universe, so that our determination of the density and deceleration parameter from what we can see might not answer the questions for all of the universe. However, since we have no reason to think that we are being fooled at a cosmic level, we may assume that we occupy an average region of space. To support this assumption, we find that the number of photons in the microwave background at a particular energy is the same for any direction. This gives us some reassurance that the universe is homogeneous and that our position within it is an average one. If this is the case, then our measurement of the present average density of matter and deceleration parameter can reveal the future of the entire universe.

If we look at what we have found out so far and compare the numbers with the theoretical possibilities, we cannot resolve the future course of the universe. As of now, we can't tell whether the universe is finite or infinite, or whether it eventually will cease expanding. Because of the difficulty of estimating distances and masses, our current knowledge of the average density of matter in the universe has an uncertainty by a large factor. The present density of matter might be five times less than our best estimate, 10^{-31} gram per cubic centimeter, and it could be considerably greater. How much greater depends on how much matter exists that does not radiate photons, as the stars in galaxies do. We shall discuss this problem further in the next chapter, but we can say now that our density estimate could be at least ten times too small.

Likewise, the deceleration parameter cannot yet be determined accurately enough to resolve the two key questions. In this case we are limited mainly by the difficulty of seeing extremely distant clusters of galaxies with sufficient clarity to determine their general nature and thus to estimate their distances. If we can recognize the general type of galaxy, we can compare its apparent brightness with the apparent brightness of a similar galaxy whose distance we already know (Page 20). To measure galaxies' velocities of recession by the Doppler effect is somewhat easier, though still difficult. Astronomers hope that a large space telescope, almost as large as the biggest ones on Earth, can be placed into orbit above the Earth's atmosphere. Such a telescope, directed automatically from ground stations, could avoid the atmospheric blurring and absorption that limit ground-based telescopes, and should be capable of detecting galaxies as much as a hundred times fainter than the faintest we can see now. This greater sensitivity of detection would give us the capacity to resolve, perhaps once and for all, the questions of the size and the future of the universe.

If the density and deceleration numbers we could find by using a large space telescope should indicate that the universe eventually will contract, we would have to consider the possibility that the universe has had many "oscillations": first a big bang, then expansion, slowing of the expansion, eventual reversal into a contracting phase, a rushing together toward enormous densities, a bounce, another big bang, and so on (Figure 3–12). If this is the correct cosmological model of the universe, then we simply happen to live during one of the expanding phases along a cycle of oscillation, which someday will turn into a con-

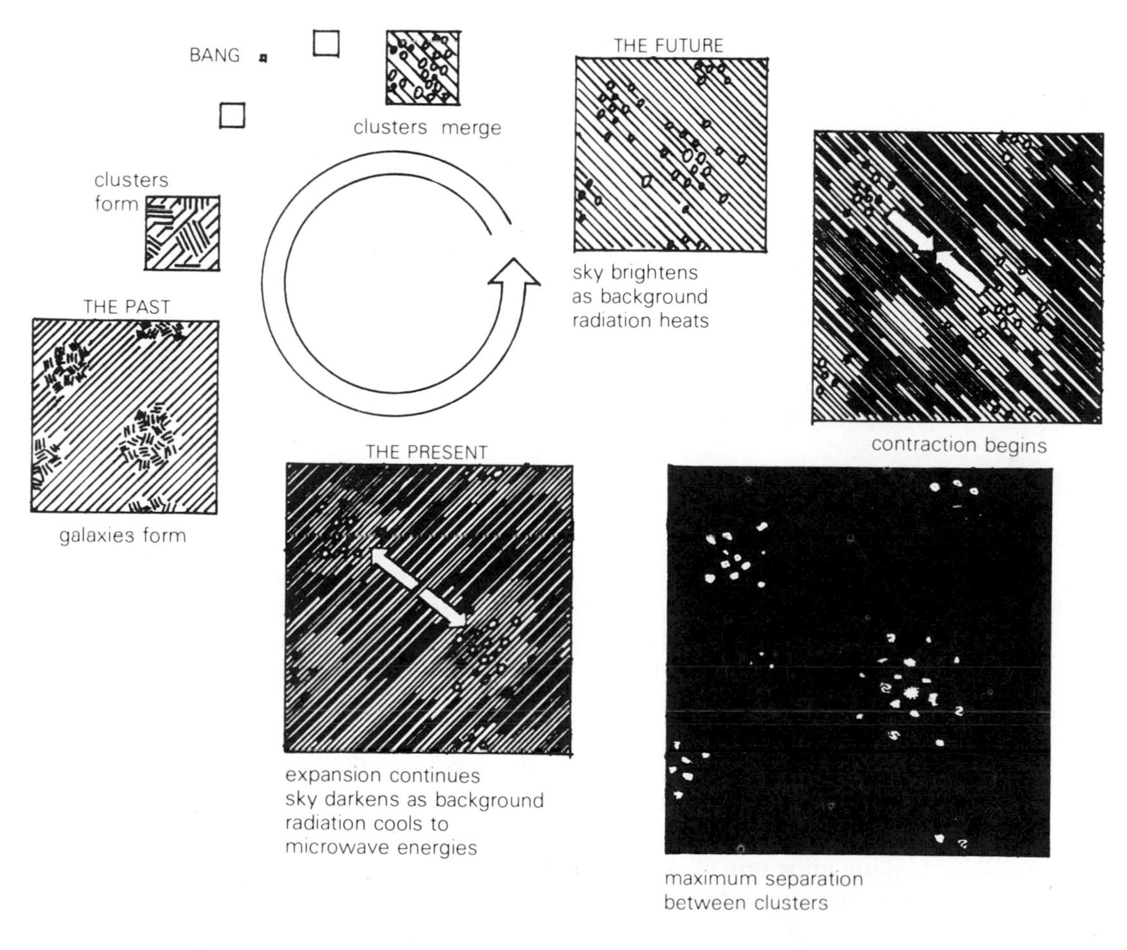

Figure 3–12. If the universe eventually will contract, it is reasonable to suppose that it has had many oscillations, first growing larger, then smaller, then (after a big bang) larger again. When the universe is expanding, the characteristic energies of the photons observed in the microwave background keep decreasing. After the universe begins to contract, the background photon energies increase. As the contraction rushes toward the immense densities of the next big bang, the increasing photon energies make the entire sky brighter and brighter.

tracting phase that will recycle our triumphs and failures into another big bang. However, if the universe is destined to expand forever, no such universal recycling will occur, and what we do now may reverberate forever.

The question of whether the universe is finite (closed) or infinite (open) also has some philosophical importance. If the universe is infinite, there are an infinite number of stars and galaxies, within which an infinite set of possibilities become reality. An infinite universe would contain every event realizable under the laws of nature. Thus most likely an infinite universe would include planets with every kind of life nature might allow, speaking (or whatever) an infinite variety of languages, with their own kings and queens. But sometimes Mary would succeed Elizabeth, and sometimes Elizabeth would succeed Mary, no matter how many names and aspects they might have. This book and every other kind of communication could exist in every conceivable altered form in every "language." The point about infinity is that it's not just a lot, it is all the way up there, and an infinite universe makes every possibility really happen somewhere. This mind-boggling aspect of an infinite universe may serve to balance our reluctance to conceive of a finite universe. This also might be reality—three-dimensional space curved around something like the surface of an expanding balloon. In our efforts to understand cosmology, we seem destined to much hard work at obtaining additional information before we can discover how much we don't know.

Summary

We can conclude from astronomical observations and our own logic that *the universe has been expanding for the past fifteen to twenty billion years.* We observe that almost all galaxies are moving away from ourselves, with speeds that increase in proportion to their distances from us. If we inhabit an average region of space, then all galaxies should see the same proportion (Hubble's law) between other galaxies' recession velocities and distances. Thus the entire universe must be expanding, and from the proportionality constant that relates the velocities to the distances, we can determine the age of the expansion.

The enormously high temperatures in the first seconds of the universe produced all varieties of elementary particles from tremendous numbers of high-energy collisions. After the first few minutes, the basic mixture of particles that we see today had been established. Photons emerged in huge quantities from the first few minutes, because of the many collisions that occurred. Until about 300,000 years after the "big bang" that began the expansion, these photons had enough energy to break apart any atom that formed temporarily. However, the increasing Doppler effect that arises from the universal expansion kept reducing the photon energies. After 300,000 years of expansion, the photons hitting an atom would not arrive with enough energy to break the atom apart; thus, since that time the photons have passed among the atoms with almost no interaction. These photons, produced in the first few minutes after the big bang, have continued to lose energy (as we observe them), and today they form a background of microwaves that has been detected and measured with radio antennas.

To determine whether the universe ever will stop expanding, we must look backwards in time by observing more and more distant galaxies, whose light takes longer and longer to reach us. Such observations can tell us how the relationship between galaxies' distances and their velocities of recession has changed with time, and thus how the universal expansion will change in the future. In addition, measurement of the average density of matter in the universe will tell us how well the universe can "pull itself together." At the present time, both our observations of the most distant galaxies and our determinations of the average density of matter are not quite accurate enough to decide the question of whether the universe ever will stop expanding and start contracting. Similarly, these observations fall short of being able to tell us whether the universe is closed (finite) or open (infinite).

Questions

1. Why do we conclude that the entire universe is expanding, merely because we observe that most galaxies are moving away from us?
2. Why are some galaxies approaching us rather than receding from us? How far away are these galaxies?
3. The average recession velocity of the galaxies in the Coma cluster is 7000 kilometers per second, whereas the average recession velocity of the galaxies in the Corona Borealis cluster is 21,000 kilometers per second. Which galaxy cluster is farther from us, and by how much?
4. Why does the average density of matter in the universe affect whether the universe ever will stop expanding and start contracting?
5. What was the temperature in the universe three minutes after the big bang? Three months after the big bang? Three hundred thousand years? (See Figure 3–6.)
6. What temperatures do we find in the universe now? Why is there such a wide range of temperatures?
7. Do any parts of the universe consist mainly of antimatter? How can we tell?

8. What does the universal microwave background consist of? Where did the particles in this background come from?
9. How long after the big bang did the universe as a whole produce great numbers of photons? How long did these photons keep destroying any atoms that formed?
10. Why did the photons become unable to destroy atoms?

Further Reading

G. Gamow, *The Creation of the Universe* (Bantam Books, New York, 1961); *One, Two, Three . . . Infinity* (Bantam Books, New York, 1961); "The Evolutionary Universe," *New Frontiers in Astronomy* (W. H. Freeman and Co., San Francisco, 1975).

A. Sandage, "The Red Shift," *New Frontiers in Astronomy* (W. H. Freeman and Co., San Francisco, 1975).

P. J. E. Peebles and D. T. Wilkinson, "The Universal Fireball," *Frontiers in Astronomy* (W. H. Freeman and Co., San Francisco, 1970).

A. Webster, "The Microwave Background Radiation," *New Frontiers in Astronomy* (W. H. Freeman and Co., San Francisco, 1975).

Plate 1. The great spiral galaxy in Andromeda, which is ▶ half a million parsecs away, is quite similar to our own Milky Way galaxy. Two smaller satellites of the giant spiral appear above and below it in the photograph.

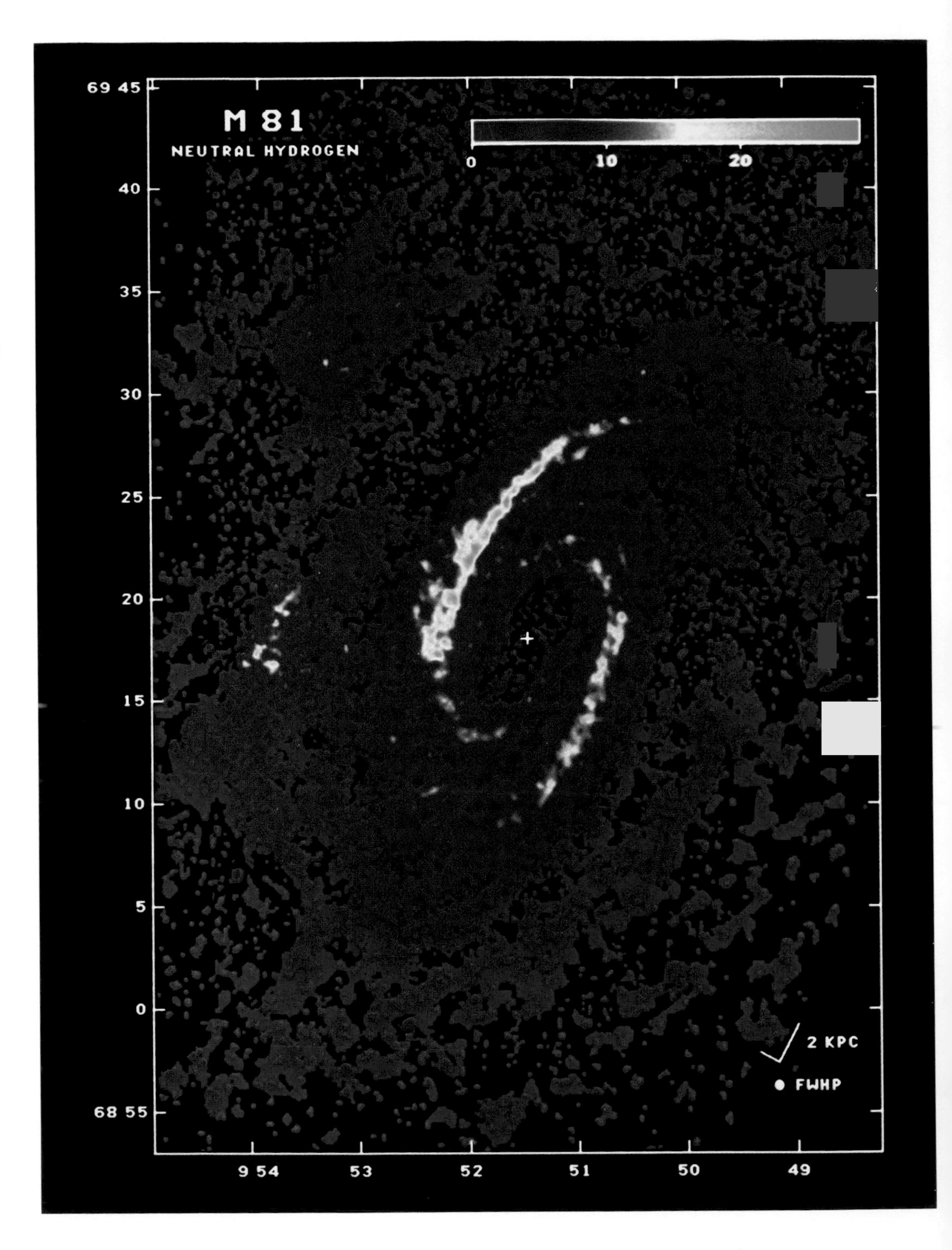

Plate 2. A map of the 21-centimeter radio-wave photons emitted by the spiral galaxy M 81. The colors indicate the number of photons emitted by various parts of the galaxy, with yellow showing the most intense emission and deep purple the least intense emission. Most of the 21-centimeter photons come from the galaxy's spiral arms, where the density of hydrogen atoms is highest. Notice that relatively little emission comes from the central regions of the galaxy. This shows that these regions contain few hydrogen atoms in comparison with the number of hydrogen atoms in the spiral arms.

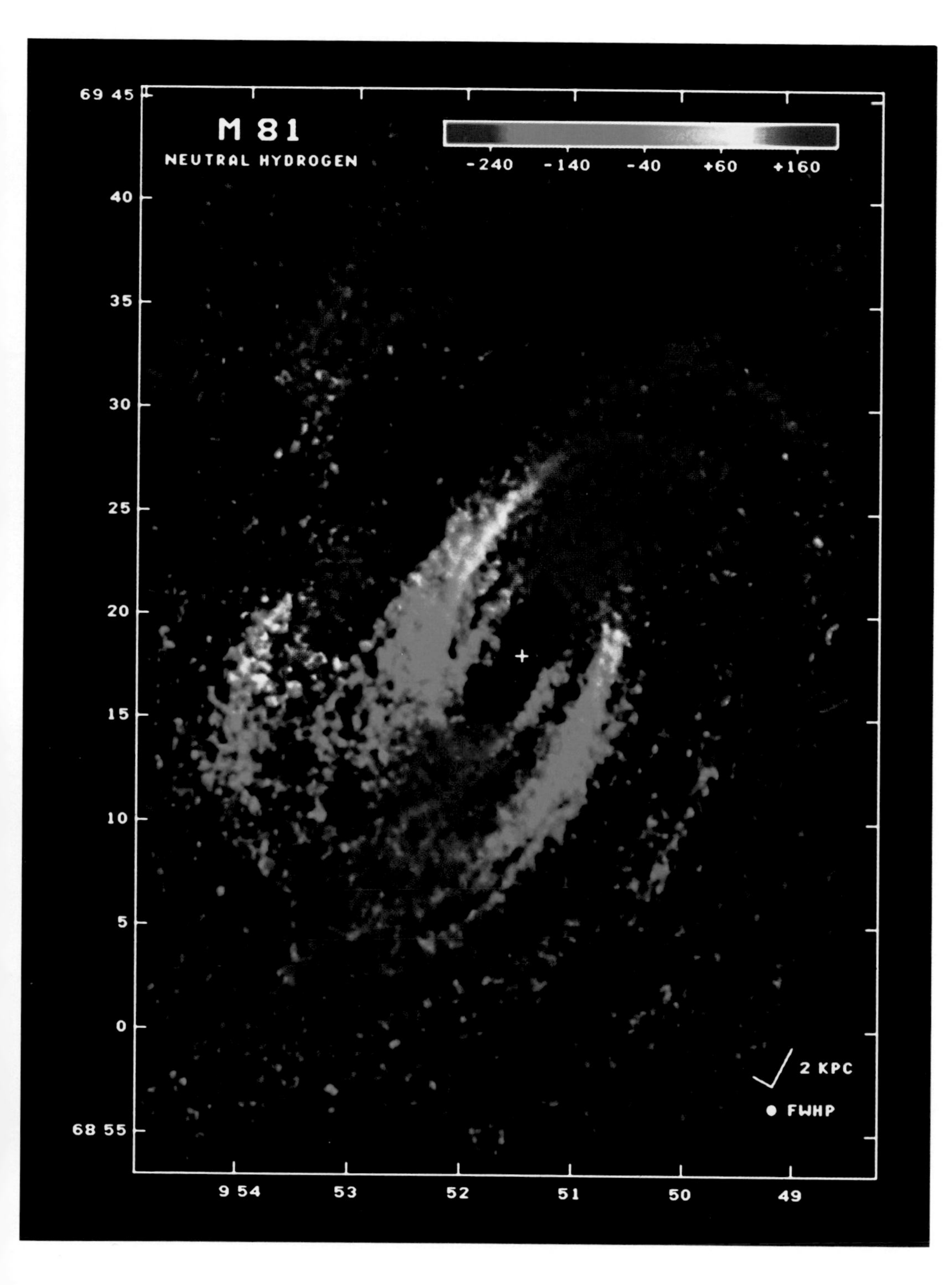

Plate 3. Another map of the 21-centimeter radio-wave photons from the galaxy M 81 uses color coding to show the velocity of the hydrogen atoms that produce the photons. The brightness of the various points in the map represents the number of 21-centimeter photons, whereas the color represents the Doppler shift of the photons: Blue regions are approaching us, and red regions are receding from us. The scale on the map gives the color coding for the measured velocities along the line of sight in kilometers per second. This map shows that the galaxy is rotating; the lower portions are approaching us and the upper portions are receding.

Plate 4. The spiral galaxy NGC 7331 in Pegasus (above) appears more nearly edge-on to us than does the Andromeda galaxy. Notice the bluer light from the arms, which shows that the young, hot stars are concentrated there.

Plate 5. The Trifid Nebula in Sagittarius (right) is an H II region that shows dark dust lanes of absorption across its face. The small dark spots or "globules" in the nebula are believed to be regions where stars are now forming.

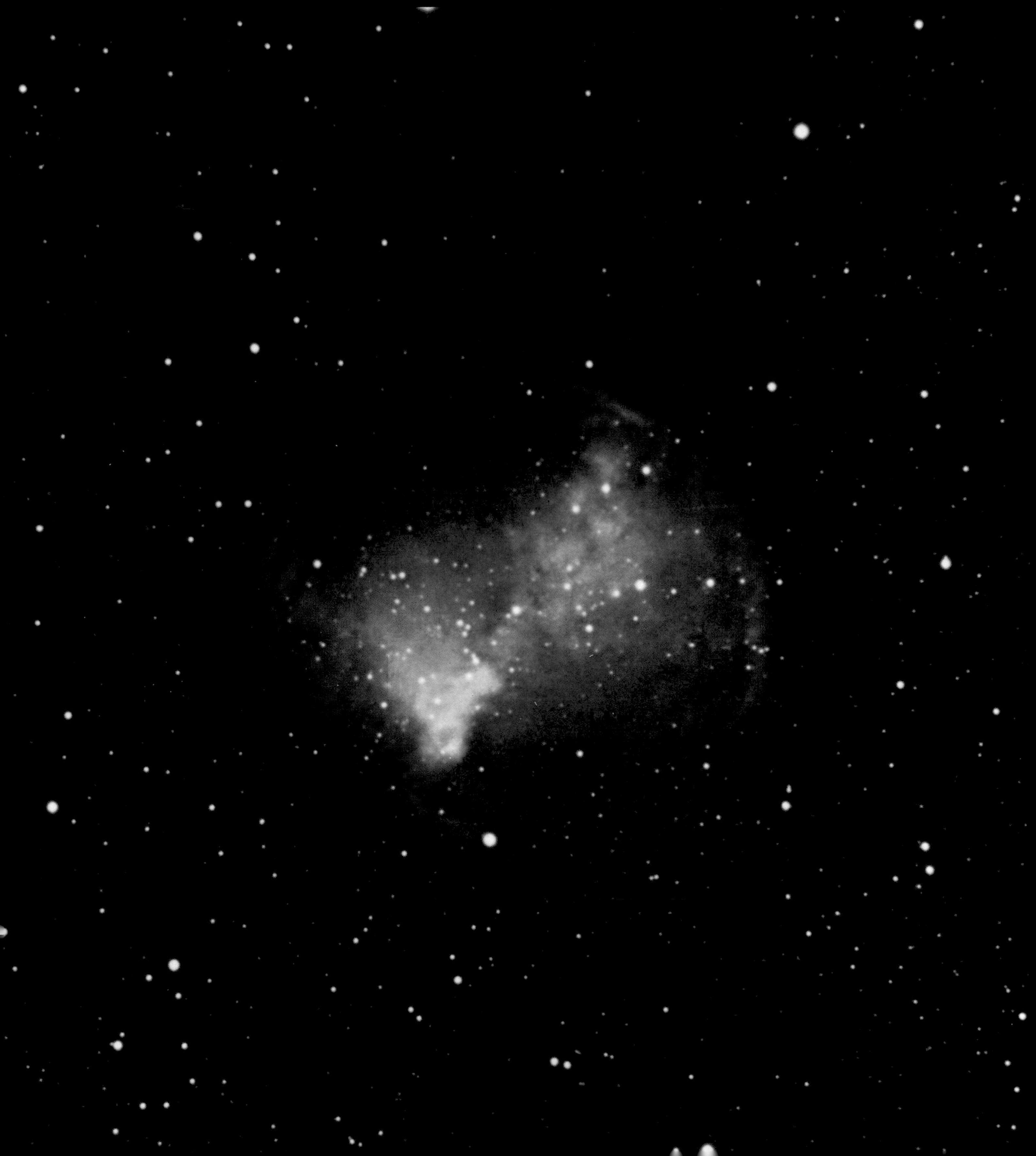

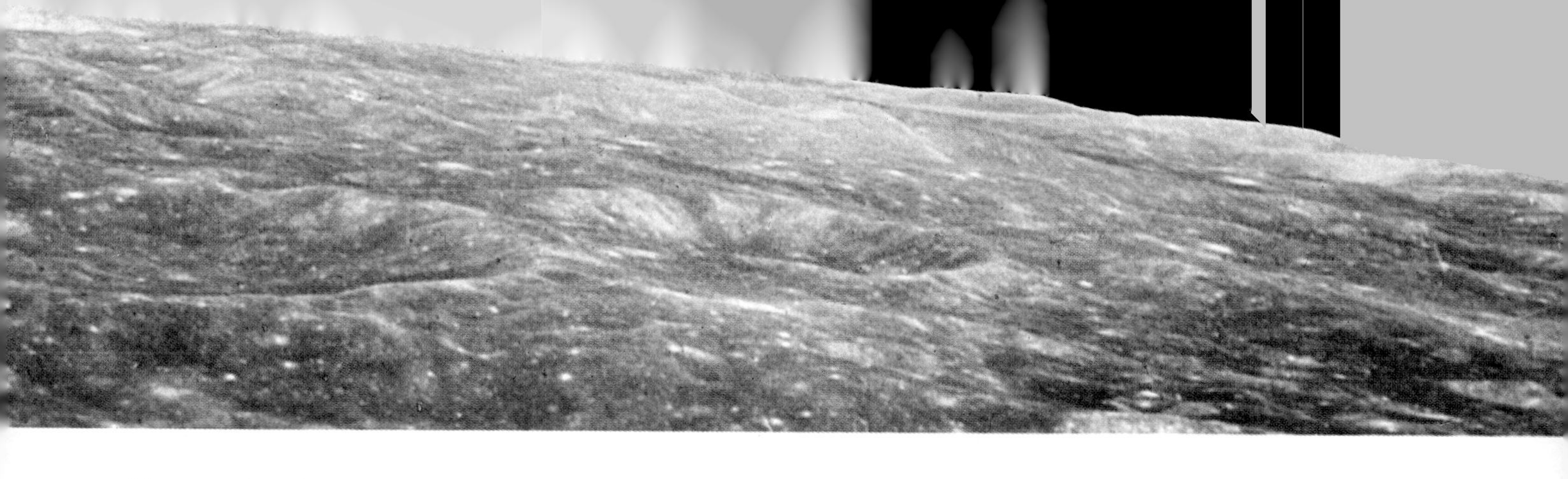

Plate 6. This planetary nebula, called the "Dumbbell Nebula," is located in the constellation Vulpecula (left). The hot blue star at the very center of the nebula supplies the photon energy (in ultraviolet photons) that makes the entire nebula shine by ionizing the hydrogen atoms around the star.

Plate 7. The Earth rising over the lunar surface, as photographed by the Apollo-8 astronauts. The sun was setting over western Africa as this photograph was made. Because the Earth's diameter is four times the moon's, the full Earth (as seen from the moon) appears four times larger than the full moon.

Plate 8. The planet Jupiter as photographed by Pioneer-10. The multicolored bands around the planet are parallel to its equator, and arise from circulation patterns that reflect Jupiter's rapid rotation. The Great Red Spot can be seen at the lower right edge of the photograph.

The Origin of Galaxies

Chapter **4**

"... vortices of dust and fire, swirling out of the ultimate spaces and heavy with perfumes from beyond the worlds."

H. P. Lovecraft

Within the expanding universe, matter appears in clumps called stars. Averaging over all space, the density of matter inside a star exceeds the density of matter between the stars by about 10^{30} times. Thus, for example, most of the mass in our Milky Way galaxy resides in the galaxy's hundred billion stars, but these stars occupy such a tiny fraction of the galaxy's volume that they all could fit into the space between the sun and the five stars nearest the sun without bumping into one another.

The clumping of mass into stars also involves the clumping of stars into galaxies, and of galaxies into galaxy clusters. Galaxies vary in sizes and shapes, but each has a fairly compact arrangement of from one million (10^6) to almost a trillion (10^{12}) individual stars. Figures 4–1 and 4–2

Figure 4–1. This photograph shows a typical spiral galaxy, NGC 2903 (number 2903 in the New General Catalogue of galaxies), in the constellation Leo.

Figure 4–2. This elliptical galaxy, NGC 185, is in the constellation Andromeda. Elliptical galaxies contain almost no diffuse gas or dust.

show examples of spiral and elliptical galaxies, which are the most common types. Spread among the stars in spiral galaxies such as our Milky Way we often can detect diffuse gas and dust (Figure 4–3), but the stars have most of the galaxy's mass. If we spread all of the mass in stars evenly throughout a typical galaxy, the average density of this spread-out matter would be about 10^{-24} gram per cubic centimeter, which is equal to about one elementary particle in each cubic centimeter. Low as this density may appear, it still is thousands of times greater than the density of matter between galaxies.

Instead of being isolated in space, galaxies tend to appear in clusters. A typical galaxy cluster such as the Coma cluster (Figure 4–4) contains several hundred or even several thousand galaxies. If we now imagine the mass in the galaxies spread diffusely through the entire cluster, the

Figure 4–3. The spiral galaxy M 104 (in Virgo), seen edge-on, shows the dust band in the central plane that absorbs some of the light from the stars in the galaxy.

density of matter within a galaxy cluster would be less than the average density of matter inside a single galaxy, but still hundreds of times greater than the density of matter between the clusters of galaxies. The gas between galaxy clusters has so little density, in fact, that astronomers are not sure that they have detected it, but they already are confident that the density of any such gas must be far less than the average density of matter within a cluster of galaxies.

Condensation of Protogalaxies

In describing the clumping of matter into stars, of stars into galaxies, and of galaxies into galaxy clusters, we have reversed the order in which this clumping occurred. Matter in the universe originally spread uniformly through space, and clumps began to form a billion years or so after the big bang. When matter began to condense, the largest clumps

Figure 4–4. The Coma cluster of galaxies, located in the constellation Coma Berenices, contains more than a thousand galaxies. This photograph shows only the center of the cluster.

formed first, and subclumps formed within them later. Thus "protoclusters," the regions that became galaxy clusters, formed first; "protogalaxies," or galaxies-to-be, condensed later; and "protostars," clumped within the protogalaxies, condensed after that (Figure 4–5). Finally, as protostars condensed to the size of present-day stars, small clumps called protoplanets formed around them. This sequence of clumping from the largest first to the smallest last makes sense once we consider the forces that govern the clumping process.

The expansion of the universe works against the formation of any clump of matter, because the expansion tends to move all particles farther from all other particles. The forces that might overcome this expansionist tendency in some local region of space are gravitation, electromagnetism, and the strong force. Of these three, electromagnetism and the strong force are unlikely to help. Electromagnetic forces between two particles can be attractive or repulsive, depending on whether the particles have opposite electric charge or the same electric charge. Any large region of space will contain almost equal numbers of positive and negative charges; if it did not, the charge that dominates soon would attract a compensating amount of charge of the opposite sign. Any volume that contains more than, say, 10^{24} elementary particles almost surely will have a total electric charge of zero and thus will exert no electromagnetic force on other particles. But since 10^{24} elementary particles have a mass of about one gram, we cannot hope to make such regions into a clump with a large mass using electromagnetic forces.

Strong forces have no hope of making large clumps because they act only over distances of a few times 10^{-13} centimeter. Thus strong forces can hold atomic nuclei together (and indeed, strong forces are responsible for clumping on the size level of elementary particles), but they cannot cause clumps even the size of an atom (10^{-8} centimeter).

In short, gravity must be the force that overcame the universal expansion in the struggle to form clumps of matter. Although gravitational forces between two elementary particles are small, they always increase as the number of particles increases. This additive power of always-attractive gravitational forces allows them to dominate the structure of the universe. Stars hold themselves together by their own gravity: Each particle in the star feels the combined gravitational forces from all the other particles that keep the star's parts from flying off into space. Within a galaxy, the individual stars each move at speeds of hundreds of kilometers per second, but the stars stay in the galaxy because the hundred billion or so other stars combine to hold them by gravitational forces. Similarly, the galaxies in a galaxy cluster, each sailing through space in complicated orbits around the center of the cluster, hold each other as cluster members by their mutual gravitational attraction.

THE HIERARCHY OF CLUMPS

1. clusters of galaxies
10^7 parsecs across

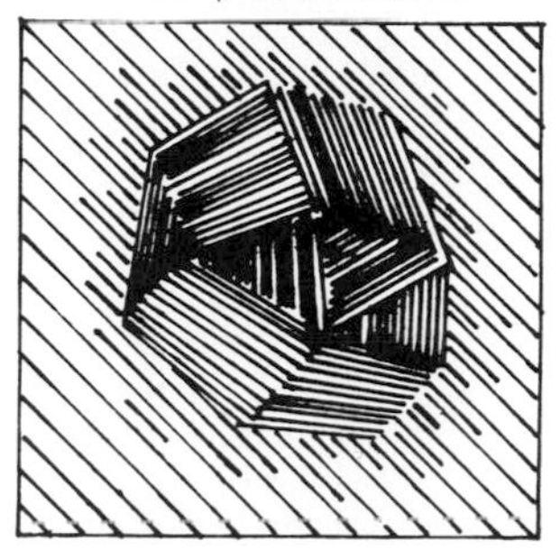

2. individual galaxies
10^5 parsecs across

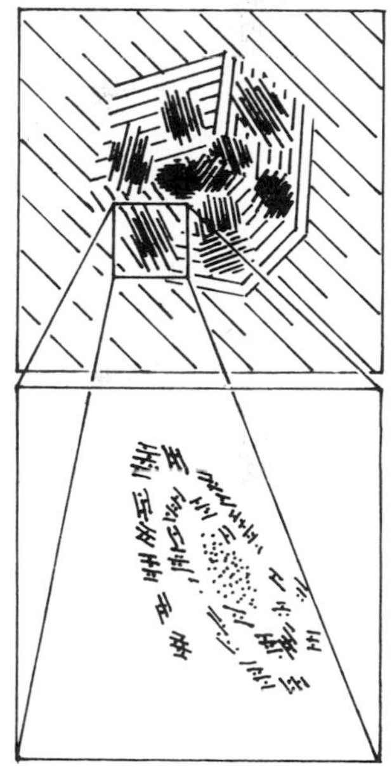

3. star clusters born
out of clouds
1 to 100 parsecs across

Figure 4–5. When matter in the universe began to form clumps, the largest clumps (protoclusters of galaxies) formed first, the next largest (protogalaxies) formed next, and smaller clumps (protostars or protoclusters of stars) formed after that.

The original protoclusters somehow must have become gravitationally bound together well enough to overcome the disruptive tendency of the universal expansion. In addition to the expansion's attempts to spread particles apart, the random motion of individual particles tends to disrupt any clump of matter. For example, molecules in the Earth's atmosphere constantly dart around at speeds of a kilometer per second, and a few of them manage to escape from the Earth's gravity every moment. Inside the sun, protons and electrons zip past each other at speeds of 30,000 kilometers per second, but the sun has enough mass to keep the particles together and not let them all fly apart into space. We use temperature to measure the average energy of motion of particles (Page 62); thus higher temperatures indicate that each particle is moving with a larger random velocity. This definition applies easily to a "gas," which we define as a mixture of particles that are free to dart about randomly among each other, rather than being linked together in larger structures such as the particles in a liquid or a solid.

The primordial gas of the early universe had a constantly decreasing temperature. At the time when atoms could form and remain undestroyed, the temperature was nearly 3000 degrees absolute. Until this time, the temperature was far too high to allow condensations to form. The constant push from photons that interacted with the temporarily formed atoms before the time of decoupling exerted a further homogenizing influence that broke up any clumps in the matter. The fact that we observe no change in the strength of the microwave background from different directions in space shows that no large changes in the density of matter from place to place existed at a time 300,000 years after the big bang.

Once hydrogen and helium atoms formed, the temperature of the particles continued to decrease, and as time went on, each atom had a smaller and smaller random velocity. We can compare this decrease of random velocities to what happens inside an automobile engine when the spark plug ignites the mixture of gasoline and air in the cylinders. The release of energy from this ignition gives each molecule a large random velocity, and some of the molecules strike the piston to force it downwards. As the volume inside the combustion chamber increases, the random velocity of each molecule decreases. Similarly, as the volume of the universe increases, the random velocity of each particle decreases, so the temperature of the particles at any given time also decreases.

Suppose that within the expanding universe, a small perturbation in the local density occurs, so that a particular region has a density of matter slightly greater than the average density at that time. Will this density perturbation "grow"; that is, will the density's deviation from the average steadily increase? Or will the perturbation be smoothed away by the homogenizing effects of the expansion and of the random velocity of each particle? In this situation, gravity tends to increase the density within the perturbation, because the region with a density greater than average has a better chance of attracting more particles than do the regions of average density. Gravitational forces therefore struggle to collect matter into clumps, fighting against the smoothing effects of expansion and temperature. At any moment in the universe's history, two basic facts determine what happens to a small perturbation in the local density. First, perturbations are more likely to become important clumps at lower temperatures, because the particles have smaller random velocities. Second, a perturbation of a given amount, for example, with a density one percent larger than average, will grow more readily for larger sizes of the region involved. This happens because larger regions contain more mass to work as a unit in trying to hold on to more and more particles.

The result of the struggle of gravitation against temperature and universal expansion has been calculated to be this: At any time A.B.B., which can be characterized by a certain temperature of matter and a certain rate of expansion, there exists a specific critical size for a region where the density has a small perturbation. Density perturbations in regions *larger* than this critical size will grow, whereas perturbations over smaller regions will disappear as the expansion and random-velocity effects smooth them away.

This description of density perturbations has ignored the additional homogenizing effect from photons. Before the energy per photon decreased below the amount needed to ionize atoms, the interaction of photons with matter would have prevented any clumping. However, after atoms formed (300,000 years A.B.B.), the interactions of photons with atoms ceased almost entirely, so our description makes sense for times past the moment of decoupling. At this time of decoupling, the critical size for a density perturbation to grow was about ten parsecs. Any perturbation that increased the density by, say, one percent in a region larger than ten parsecs would be likely to grow into an ever-increasing contrast in density, whereas density perturbations in smaller regions would have been smoothed away. The total mass inside a region ten parsecs across at the time of decoupling amounted to about one millionth of the mass in our galaxy, or a hundred thousand times the mass of the sun. We should remember that the universe used to be much denser than it is now. Today an average region of the universe ten parsecs across contains less than one thousandth of the mass in the *sun*.

We have concluded that at the time of decoupling, when the photons stopped interacting with the matter, density perturbations of a small amount in any region larger than ten parsecs would grow into steadily more important perturbations. Since the temperature of the matter

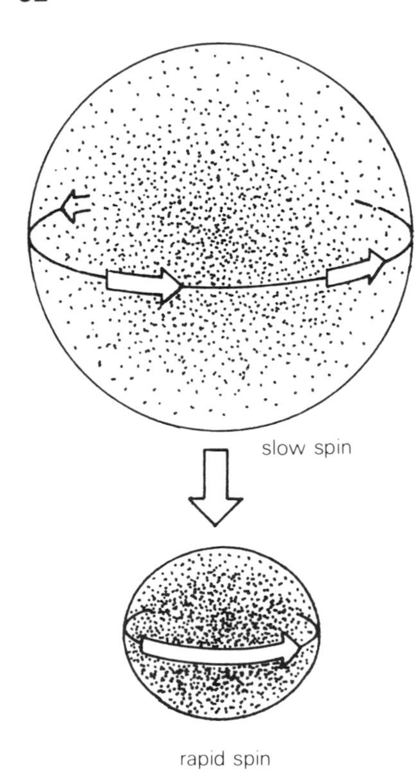

Figure 4–6. A contracting gas cloud will spin faster as it contracts. The cloud's "angular momentum" varies in proportion to its rate of spin times the square of its size perpendicular to the axis of spin. The angular momentum stays constant for an isolated object. Thus if the cloud contracts, its rate of spin must increase.

continued to decrease, perturbations in still smaller regions (say one parsec across) could grow later. The lower temperature then would not have such a strong smoothing-out effect. However, the perturbations that began growing as soon as possible (from the time of decoupling) had a head start on the others. Such perturbations with a head start were the larger ones that each contained a mass much greater than a star's mass.

But how did such perturbations occur in the first place? Astronomers have no good answer to this question. In all honesty, the calculations that we have described refer to a simplified model of the universe. The calculations do show that any regions larger than ten parsecs at the time of decoupling would have their density perturbations grow. Thus a perturbation of, say, one percent in the density of a region a million parsecs across might increase, so that the density inside the perturbed region would be, say, twice as great as the average density of the universe after a billion years had passed. What we care about is the ratio of the density inside the perturbation to the average density in the universe; the universal expansion tends to reduce all densities, but a perturbation can oppose this tendency once it has become more significant than about a one percent increase. A typical galaxy such as our Milky Way (30,000 parsecs in diameter) has a density of matter more than a million times greater than the average density of matter in the universe. The fact is that although gravitational forces indeed can make density perturbations become more pronounced, they do so only slowly. Even the billions of years that passed after the time of decoupling may not have been enough for gravitational forces to make galaxy clusters and individual galaxies. Since galaxies and galaxy clusters *have* formed, and are already billions of years old, their very existence is one of the many problems that our present understanding of the universe has not solved. Where the original perturbations came from, and what processes besides gravity helped make their density become millions of times greater than the average density, remain unanswered questions. We can at least take comfort in the fact that the theory does suggest that the condensations with the best chance to form at the time of decoupling were those with masses approximately equal to the masses of galaxies. At least these masses of growing density perturbations had to be hundreds of thousands of times greater than the masses of stars, which we believe formed later, after the galaxies had begun to condense.

The Importance of Angular Momentum

We have seen that gravitation is the most important factor for any clump of matter to condense to greater densities. Unless a density perturbation occurs over a certain minimum size, and includes a certain minimum mass, it will not continue to grow under the influence of gravitational forces from the matter within the region of increased density. Let us suppose that a region of the minimum required size does begin to contract under the influence of its self-gravitational forces. The next governing factor to consider is the condensation's amount of spin or "angular momentum."

The angular momentum of a spinning object measures the object's rate of spin times the *square* of the object's size perpendicular to the axis of spin (Figure 4–6). If a contracting clump of matter that, for example, eventually forms a galaxy does not spin when it begins to contract, most likely it never will spin, even as a more compact galaxy. But if the contracting clump does have some spin, the clump will spin more rapidly as it contracts. The importance of angular momentum arises from the fact that the amount of angular momentum in a spinning clump stays the same as the clump contracts, provided that the clump remains relatively isolated and does not, for instance, collide with other clumps.

Because the amount of angular momentum remains constant in an isolated object, if the object becomes ten times smaller, it will spin a hundred times more rapidly. If the Earth, for example, were to shrink suddenly to one half its present size, its rate of rotation would quadruple, so that days would be only six hours instead of twenty-four. Contracting galaxies obey the same angular-momentum rules, which indeed are familiar to those of us on Earth who enjoy figure skating or acrobatic diving. A high diver who wants to spin more rapidly will contract her body to a smaller size (Figure 4–7), but to slow her motion and enter the water smoothly she will extend her body to a maximum size.

A clump of matter contracting into a protogalaxy may have started rotating from encounters with neighboring clumps of matter. These

Figure 4–7. To spin more rapidly a high diver will contract her body and thus take on a smaller size. Because her angular momentum stays constant, the diver will spin faster than when her body is fully extended.

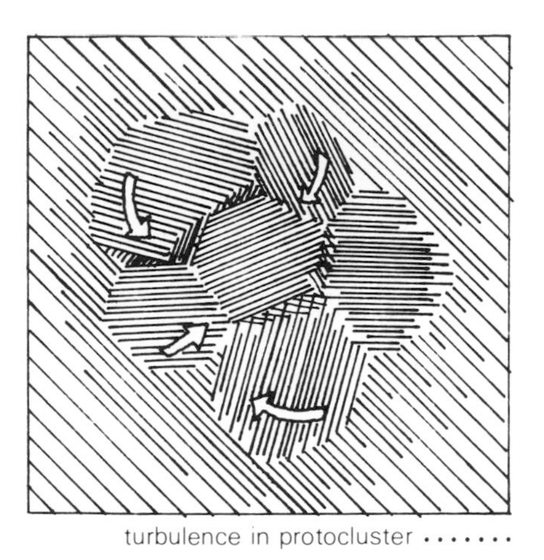

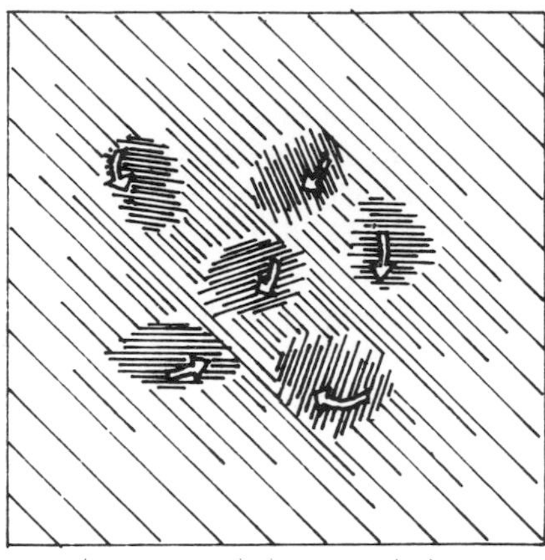

Figure 4–8. When protogalaxies began to form, each clump of gas that produced a galaxy might have begun to spin. The total spin of all the clumps combined could remain zero if the directions and amounts of the spins balanced one another.

other clumps each pulled the contracting cloud in various directions, and the result may have been that each clump began to spin (Figure 4–8). The spin of one clump in one direction can be compensated by the spin of another clump in the opposite direction. Then the total spins add to zero, and for the universe as a whole no spin has been created; instead, the spin has been distributed among the various clumps, which spin in opposite directions.

If a contracting protogalaxy began to spin as it became an identifiable clump, then its amount of angular momentum will stay constant as long as it remains relatively far from other clumps. Then the protogalaxy will spin more and more rapidly as it contracts to a smaller and smaller size. Consider a protogalaxy with a diameter of a hundred thousand parsecs that rotates once every two billion years. The protogalaxy will contract to become a galaxy with a diameter of thirty thousand parsecs like the Milky Way. As the protogalaxy contracts, its rate of spin will increase so that its amount of angular momentum, given by the rate of spin times the square of the size, stays the same. The final galaxy then will spin about ten times more rapidly than the protogalaxy, so it will rotate once every two hundred million years (instead of every two billion years).

When a protogalaxy contracts it tends to spin more rapidly. This means that each part of the protogalaxy revolves more quickly around the center of the clump. The increasing rate of spin has an important effect on the protogalaxy, because the final rate of spin determines the general form that the galaxy will have. Since the final spin rate depends on the initial spin rate of the protogalaxy (because the angular momentum stays constant, so the spin rate must increase to balance the decrease in size), the galaxy's final form depends on the initial spin rate of the clump that contracted to form the galaxy.

Figure 4–9. The giant elliptical galaxy M 87 is the largest galaxy in the Virgo cluster of galaxies.

Types of Galaxies

There are two general types of galaxies: elliptical and spiral. Elliptical galaxies, such as M 87[1] (Figure 4–9), have a spherical or slightly flattened, "elliptical" shape (Figure 4–10). In contrast, spiral galaxies are much flatter than elliptical galaxies. Seen from the side (Figure 4–11), spirals appear extremely thin, with a central bulge like a pie plate, or (as

[1]Galaxies designated by M and a number, such as M 87 or M 51, appear in the list of "nebulae" made by Charles Messier in the 1780's. Messier was a French amateur astronomer who liked to search for comets, and made his famous list of nebulae to catalogue those objects that definitely were *not* comets. Today Messier's cometary discoveries are forgotten, but many of the nebulae on his list turned out to be spiral and elliptical galaxies that we still designate by Messier's numbers.

Figure 4–10. The elliptical galaxy NGC 205 is a satellite of the giant spiral galaxy in Andromeda (see Plate 1).

Figure 4–11. The spiral galaxy NGC 4565 in Coma Berenices shows the central bulge of stars within the flattened disk that typifies spiral galaxies.

Figure 4–12. Seen at an angle, spiral galaxies such as M 81 in Ursa Major show a swirling, spiral pattern with bilateral symmetry. Plates 2 and 3 show maps of the 21-centimeter radio emission from this galaxy.

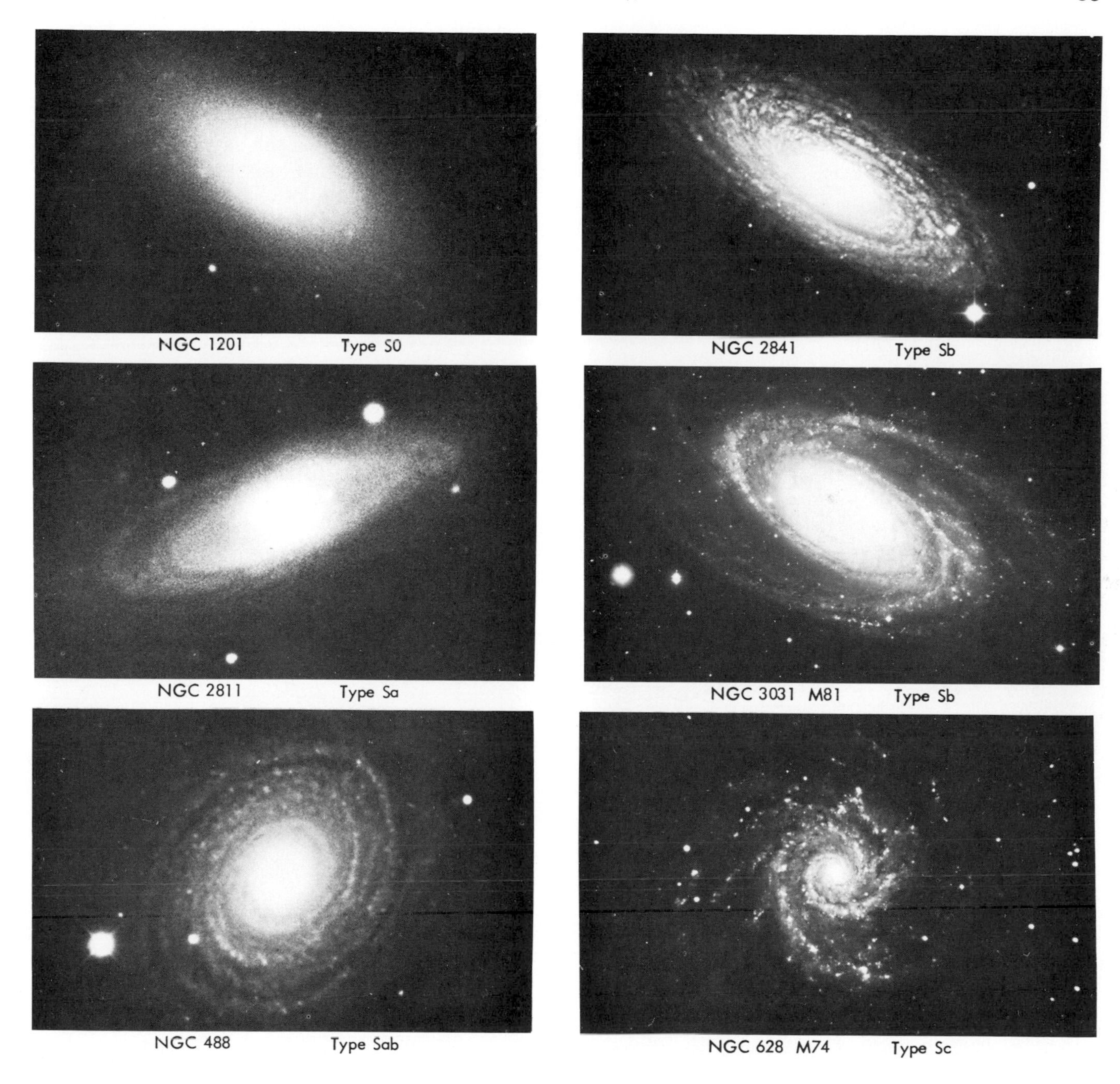

Figure 4–13. Six different spiral galaxies show the differences in tightness of the winding of the spiral arms. The Sab spiral NGC 488 in Pisces has much more tightly wound spiral arms than the Sc spiral NGC 628, which also is in Pisces.

the poet Diane Ackerman expressed it) like a python that recently has had a full meal. Seen from "above" or "below," rather than edge-on, a spiral galaxy (Figure 4–12) resembles a vast whirlpool with swirls called "spiral arms" that give the galaxy type its name. The tightness of the winding of the spiral arms allows us to make a subclassification of spiral galaxies. Type Sa has the most tightly wound spiral arms, type Sb are more loosely wound, and type Sc have the loosest spiral arms (Figure 4–13). The Milky Way's galaxy type probably is Sbc, which means intermediate between types Sb and Sc. Of course, we cannot see the Milky Way galaxy from outside, because we are located within it, near the galaxy's median plane of symmetry but nearer to the galaxy's edge than to its center (Figure 4–14).

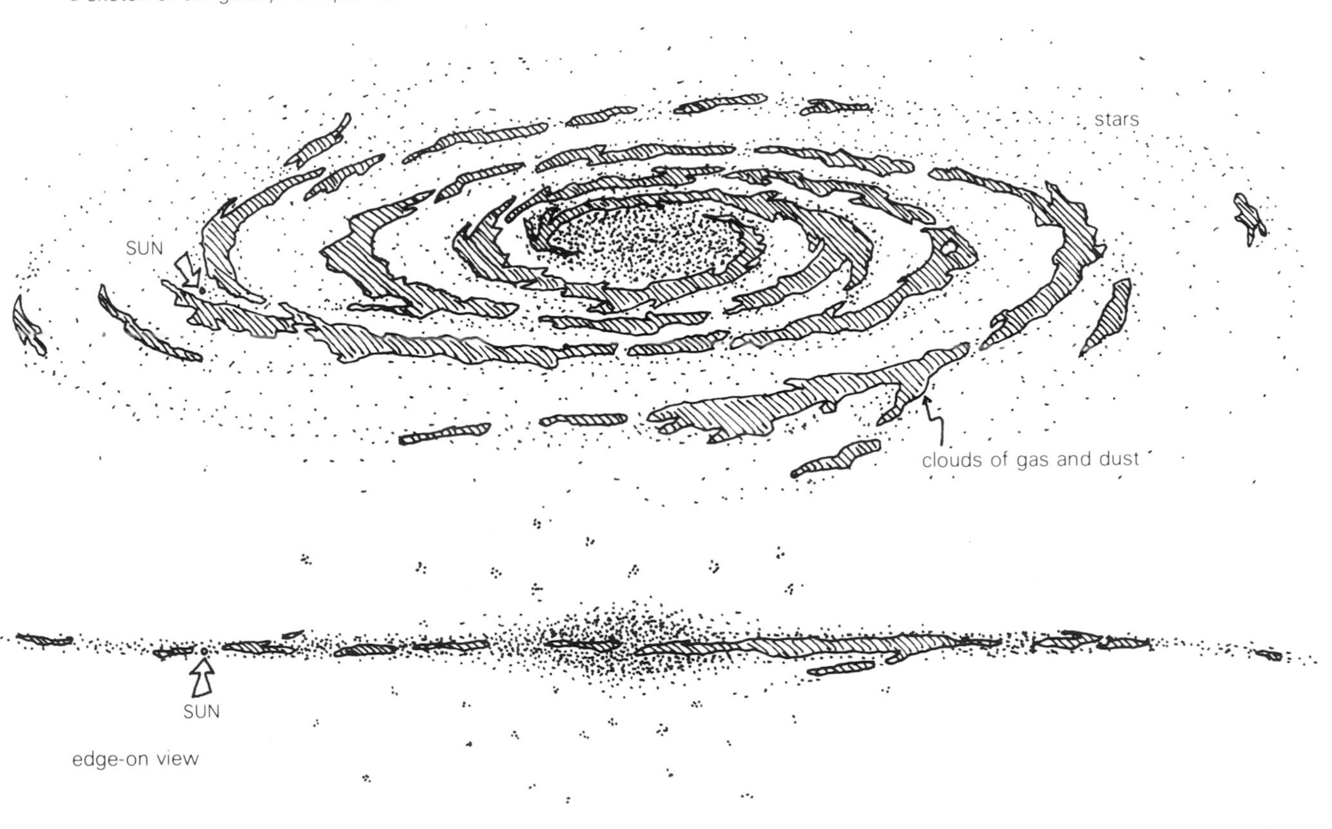

Figure 4–14. In the Milky Way galaxy, our sun lies near the plane of symmetry called the median plane. The sun is closer to the galaxy's edge than to its center. As we can see from photographs of other galaxies, there is no sharply defined outer edge to a spiral galaxy; the stars thin out progressively away from the center.

An additional subclassification of spiral galaxies are the "barred spirals." These galaxies have a flat shape like ordinary spirals, but they have an extended "bar" or spindle-shaped concentration of stars in their central regions (Figure 4–15). Barred spirals are designated by the letters SB, and they are classified as SBa, SBb, and SBc in the same way as spiral galaxies, by the tightness of their spiral-arm patterns (Figure 4–16).

A class of galaxies called S0 forms a transition from spirals to ellipticals. S0 galaxies are almost identical in general shape with spiral galaxies, but they show little or no spiral-arm pattern (Figure 4–13).

Elliptical galaxies are classified by their degree of flattening, from E0 (no flattening) to E7 (highly squashed). Even an E7 elliptical galaxy, however, is not nearly as flat as a spiral galaxy (Figure 4–17). Elliptical galaxies have no spiral arms or other special structural features. The density of stars in elliptical galaxies increases toward the galaxies' centers (as indeed the density does in spiral galaxies as well), but elliptical galaxies appear to be basically simple conglomerations of stars with

Figure 4–15. The barred spiral galaxy NGC 1300 lies in the constellation Eridanus.

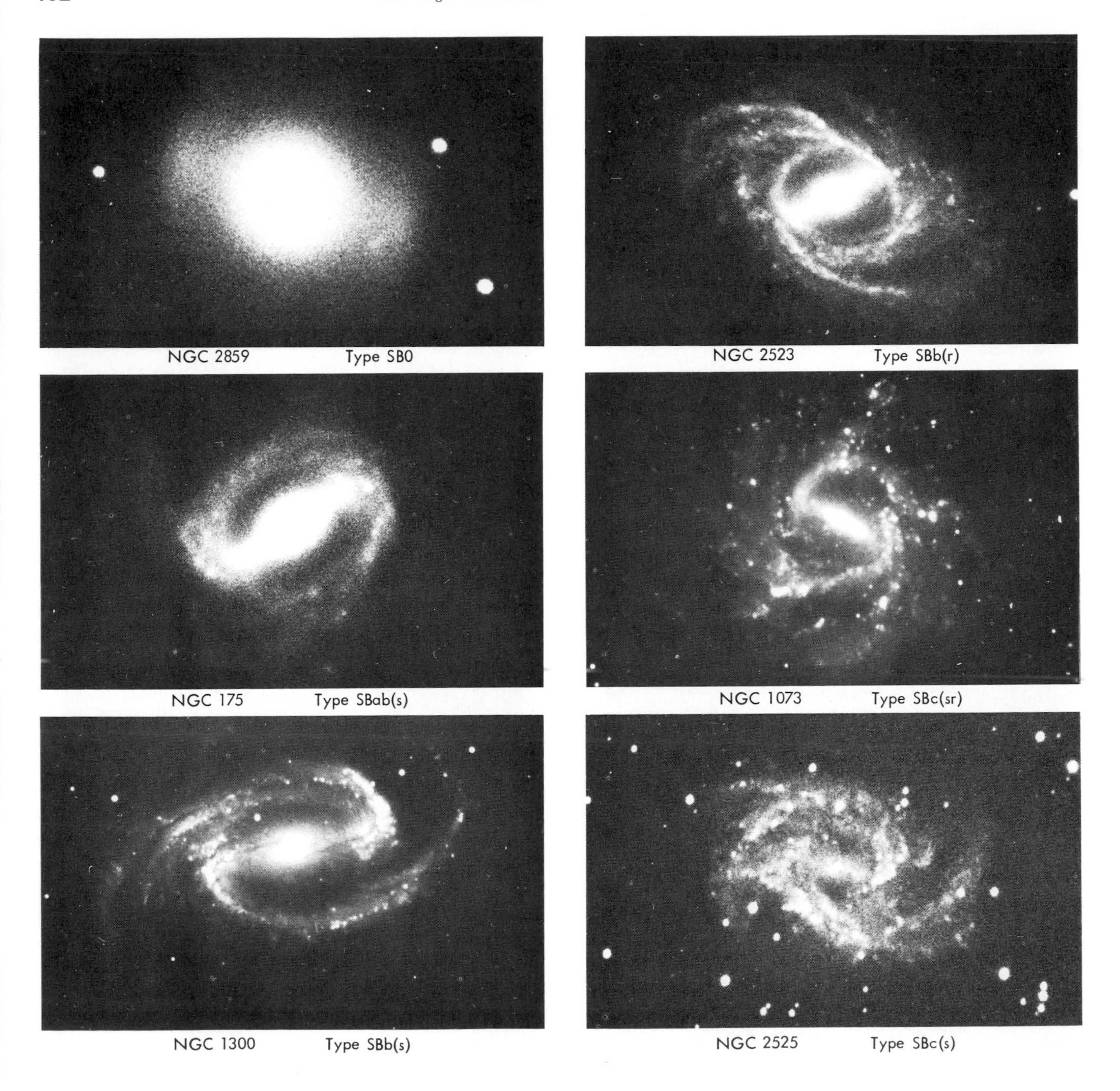

Figure 4–16. Six different barred spiral galaxies show a wide variation in appearance and in the tightness of the winding of the spiral arms.

no special features such as spiral arms. Also, elliptical galaxies contain almost no gas and dust between the stars in the galaxies. On the contrary, in spiral galaxies gas and dust within the spiral arms appear distinctly. We can observe the gas in spiral galaxies in two ways. First, we can detect the photons of 21-centimeter wavelength emitted by the hydrogen atoms that form most of this gas. Hydrogen atoms each occasionally reverse the sense of their electron's spin (Page 37), thereby producing a low-energy photon; the sum of these photon emissions makes a detectable quantity for radio astronomers. Second, we can see the dust in the spiral arms, because it absorbs some of the light from the stars within galaxies. If we look at the "Milky Way" on the sky as it runs through the constellation Cygnus, we see that the Milky Way appears to fork into two branches. In fact, the stars in our galaxy do not split into two branches, but instead a dust lane in front of some of the stars absorbs their light and gives the appearance of a bifurcation in the Milky Way. When we look at other galaxies edge-on, we have the best chance of seeing the absorption that the dust in spiral galaxies produces, because most of this dust lies in the median plane of the galaxy (Figure 4–11). Figure 4–18 shows the central regions of the Milky Way galaxy.

When we try to determine what causes some protogalaxies to become spirals while others become ellipticals, we find that the most important factor seems to be the amount of spin or angular momentum in the contracting protogalaxy. Protogalaxies with little angular momentum became elliptical galaxies, whereas those with more angular momentum flattened into spiral galaxies.

In both spiral and elliptical galaxies, individual gas clumps later contracted into protostars and still later into stars. Each galaxy's stars now orbit around the center of the galaxy because of the gravitational force from all the other stars. We can think of this combined force as pulling each star toward the center of the galaxy. Instead of falling into the galactic center, each star follows an orbit that may pass near the center but avoids going all the way in; many stars have nearly circular orbits. A star in its orbit balances the pull of gravity by its own momentum. The star's "momentum" simply represents any object's natural tendency to follow a straight-line path at a constant speed unless outside forces act on the object. For example, a rocket launched straight up may continue to go straight up until it falls back to Earth. But if the rocket travels in space far from any other object, once it accelerates to a certain speed and stops firing it will continue through space in a straight line, with a constant velocity, until some outside force grows strong enough to change its direction or its velocity.

Stars in galaxies move in orbits around the galactic center that are like the planets' orbits around the sun (Chapter 9), only much larger. In these orbits, each star's momentum constantly allows it to avoid being

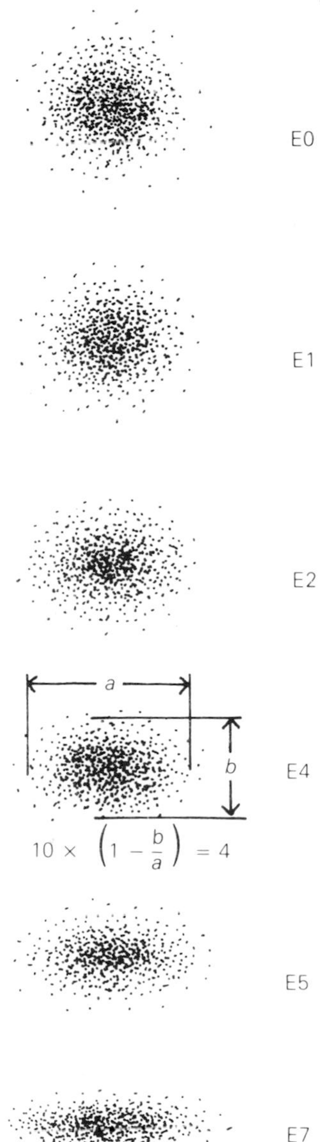

Figure 4–17. These representative shapes of elliptical galaxies show the elongation of an E7 type compared with, say, type E0 or E1. If we let a represent the galaxy's long dimension and b its short dimension, then the galaxy's elongation number is $10 \times (1 - b/a)$. For example, if $b/a = 0.6$, the galaxy is type E4.

Figure 4–18. The Milky Way in the direction of the constellation Sagittarius. The dark patches in the photograph are caused by the absorption of light by interstellar dust.

pulled into the center of the galaxy. In spiral galaxies, most stars move in nearly circular orbits around the center. Our sun, for example, follows an almost circular path around the galactic center at a speed of 250 kilometers per second and a distance of 10,000 parsecs from the center. Each orbit takes about 250 million years, so the sun has made perhaps twenty orbits since it condensed. Although most of the sun's motion in the Milky Way consists of this circular orbit around the center, the sun also bobs up and down through the galaxy's median plane as it circles (Figure 4–19). The circular part of the sun's motion occurs at a speed of 250 kilometers per second, while the superimposed up-and-down oscillation gives the sun an added speed of about 10 kilometers per second in directions perpendicular to its main orbit. Each time the sun circles once around the galactic center it bobs up and down four or five times, reaching a maximum distance of only about 200 parsecs from the galaxy's median plane.

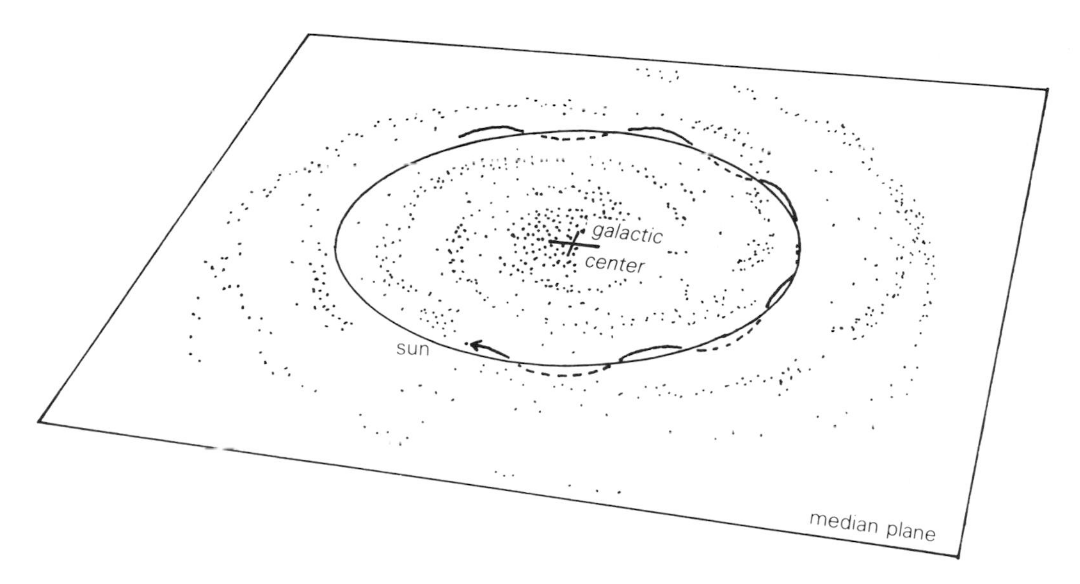

Figure 4–19. As the sun circles around the galactic center, it also oscillates up and down through the median plane of our galaxy several times during each orbit.

The protogalaxies that became spiral galaxies had a great deal of angular momentum. The individual clumps that became protostars each acquired some of the galaxy's spin, so that the stars in spiral galaxies now circle the galactic center (Figure 4–20). Stars closer to the center move more rapidly in their orbits, while stars farther away move more slowly. Thus the galaxy does not rotate as a rigid object like a wheel would, because its outer regions have a slower rotation velocity than the inner regions do. The general rotation pattern in spiral galaxies has given each star a roughly circular orbit with enough momentum to keep the star orbiting around the center and thus to keep the galaxy from contracting further. The stars' random up-and-down oscillations oppose the tendency for the stars to occupy precisely the median plane of the galaxy.

The total gravitational force on a star near the median plane of a spiral galaxy pulls the star both toward the center and toward the plane (Figure 4–21). The star opposes the pull toward the center with its momentum in its circular orbit, which keeps the star permanently away from the center. In contrast, the star responds to the gravitational pull towards the median plane by moving through the plane. As the star approaches the median plane, it picks up enough speed and enough momentum to carry it through to the other side, where it slows down,

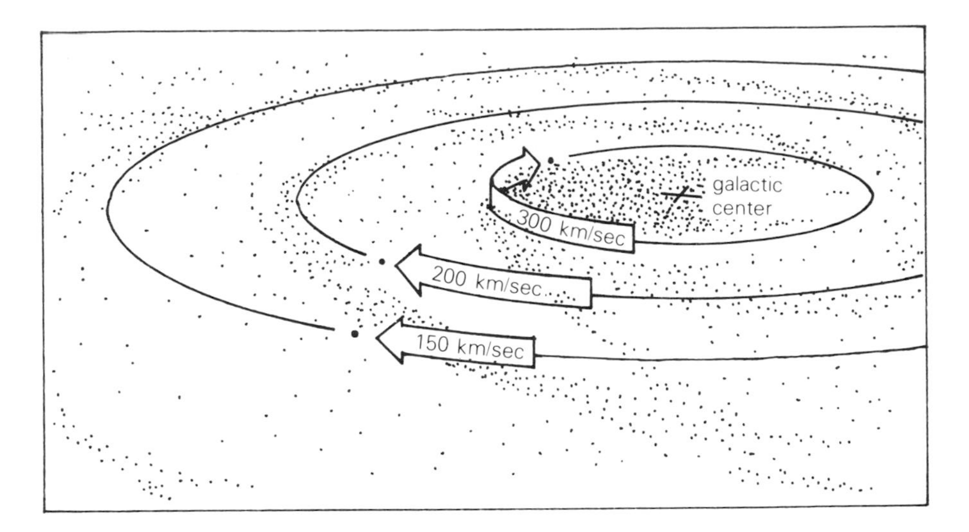

Figure 4–20. Most of the stars in a spiral galaxy move in almost circular orbits around the galactic center. The stars closer to the center move along their orbits more rapidly than the stars further from the center do.

eventually reverses direction, and begins the oscillation in the opposite direction. Even in a giant galaxy, stars are spaced so far apart that they almost never collide with one another. The stars' oscillations back and forth through the median plane give the galaxy an overall thickness in the direction perpendicular to the plane. Because the stars' circular velocities around the galactic center far exceed the additional oscillation velocity perpendicular to this circular motion, spiral galaxies have diameters far greater than the galaxies' thicknesses. The reason that spiral galaxies are fifty or a hundred times wider than they are thick is that an average star's circular orbital velocity far exceeds its up-and-down oscillation velocity.

Elliptical galaxies formed from clumps with less angular momentum than spiral galaxies. Ellipticals do rotate, but their rotation doesn't dominate their structure like the rotation of spiral galaxies. In elliptical galaxies, as in spirals, an individual star has a motion compounded of an orbit around the galactic center with some bobbing up and down along the orbit (Figure 4–19). But in elliptical galaxies, the orbits do not all lie near the same plane, as they do in spirals. In addition, these orbits often are highly elongated ellipses rather than the near-circular orbits of stars in spiral galaxies. As a result, the overall rotation of elliptical galaxies becomes somewhat hidden by the fact that each star's orbit takes it in a different direction. The total of all the stars' orbits adds up to a rotation of the total galaxy, but does not produce as coherent a rotation as we find in spiral galaxies. The more highly flattened ellipticals (types E4 through E7) have a greater angular momentum than the less flattened ones (types E0 through E3), and this larger angular momentum has

Figure 4–21. A star slightly above or below the median plane of a spiral galaxy feels the effect of gravitational forces pulling it both toward the center of the galaxy and toward the median plane.

been responsible for the (slightly) greater resemblance of the orbital motions of the stars in the more flattened ellipticals to the motions of stars in spiral galaxies.

The larger spiral and elliptical galaxies have similar masses and sizes. Each contains 10^{11} or 10^{12} stars, and has a diameter of about 30,000 parsecs. The amount of angular momentum in the original protogalaxy apparently determined whether the galaxy became a spiral or an elliptical.

In addition to spirals and ellipticals, there are "irregular" galaxies. About ten percent of galaxies show neither the disklike structure of spirals nor the smooth structure of ellipticals. Two good examples of "irregular" galaxies are the two galaxies called the "Magellanic Clouds" that orbit around our own giant Milky Way spiral (Figure 4–22). Each of the Magellanic Clouds[2] consists of a loose grouping of about 10^{10} stars, while other irregular galaxies may have from 10^6 to 10^{10} stars. Irregular galaxies thus typically have total masses that are many times less than the mass of the Milky Way, or of other giant spiral and elliptical galaxies.

Astronomers also have found some "dwarf" galaxies with elliptical or spiral forms. Among these dwarf galaxies, ellipticals are far more common than spirals, but among the giant, "standard" galaxies with 10^{11} to 10^{12} stars, ellipticals and spirals are almost equal in number.

We have seen how protogalaxies must have contracted from original density perturbations. This contraction took billions of years to form the galaxies we see today, which have masses of one million (10^6) to one

[2]The Magellanic Clouds owe their name to Ferdinand Magellan, whose sailors noted them during Magellan's expedition around the world in 1520. These galaxies cannot be seen from northern latitudes such as those in Spain or Portugal because they are too close to the south celestial pole. Magellan's crew at first thought the galaxies were clouds, but then noticed that these "clouds" always appeared in the same place on the sky, relative to the stars. The Magellanic Clouds are in fact the closest galaxies to our own Milky Way.

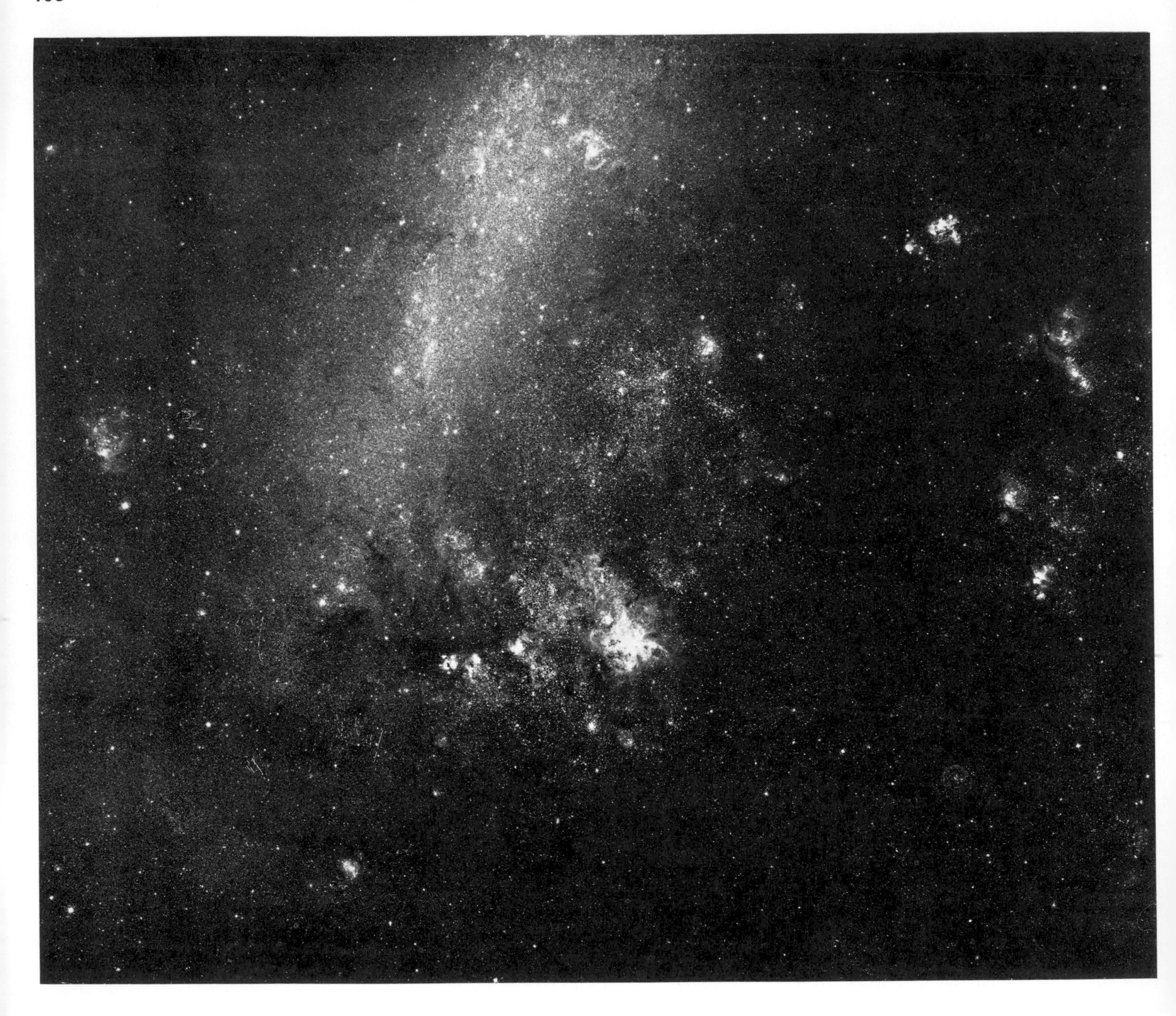

Figure 4–22. The Large Magellanic Cloud, a satellite of our own Milky Way galaxy, is about sixty thousand parsecs from the Milky Way.

trillion (10^{12}) times the sun's mass. The amount of spin or angular momentum in the protogalaxies seems to have determined whether they became spirals or ellipticals. Irregular galaxies apparently are the youngest galaxies, and may change their structure as they age. However, spirals will remain spiral and ellipticals will remain elliptical as they grow older. When we consider the Milky Way galaxy in detail (Chapter 6), we shall discuss how the characteristic spiral-arm pattern might arise in a flattened galaxy like our own.

Motions of Galaxies in Clusters

Let us now look at a typical large cluster of galaxies, the "Coma cluster" in the constellation Coma Berenices (Figure 4–4). A photograph of the cluster shows more than a thousand spiral and elliptical galaxies. The total mass of the Coma cluster equals about 10^{15} times the sun's mass. Such large clusters of galaxies form the basic units of the universe on the greatest distance scales. These clusters appear to hold themselves together by the mutual gravitational attraction among the individual galaxies, which in turn owe their structure to the mutual attraction among their individual stars. Just as the stars' motions in each galaxy balance the combined pull from other stars, the galaxies' individual motions give them the momentum to balance the total gravitational force from the other galaxies in a cluster. The individual members of a galaxy cluster orbit around the center of mass of the cluster, because the total gravitational force from the rest of the cluster averages out to be the same as the pull from the cluster's mass if it were all concentrated at the position of the center of mass (Figure 4–23). To be precise, the total gravitational force on a given galaxy in a cluster is the same as the effect from a mass at the cluster's center equal to the total mass of all the galaxies *closer* to the center than the particular galaxy is. The galaxies

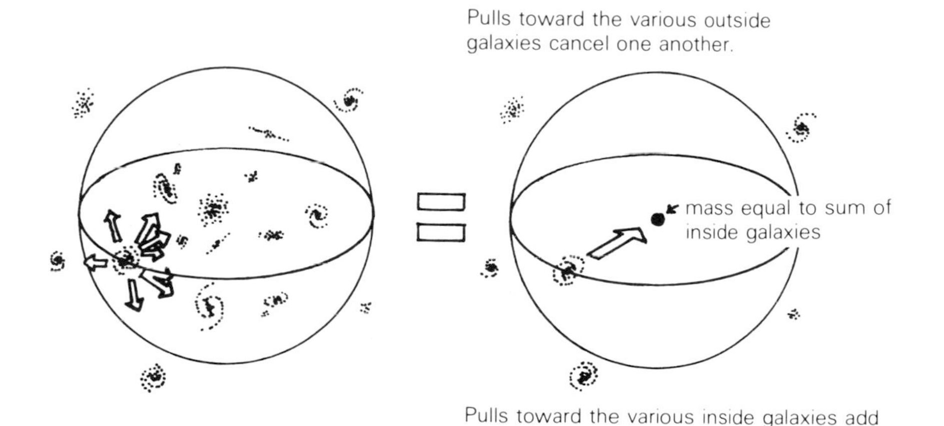

Figure 4–23. The total gravitational force on a galaxy that is a member of a galaxy cluster represents the sum of the forces from the various other members of the cluster. (Galaxies outside the cluster are too far away to exert a significant gravitational force.) This sum of forces is the same as the gravitational force from a mass at the cluster's center that is equal to the amount of mass closer to the center than the galaxy in question.

farther from the center pull in different directions, and their total gravitational effect on the particular galaxy comes out to be zero.

We now can understand the general distribution of matter in the universe. Matter comes in clumps called stars, which themselves clump into galaxies. Each galaxy contains many millions or billions of stars. In spiral and irregular galaxies, gas and dust spread among the stars contribute a small fraction of the galaxy's total mass. Elliptical galaxies, however, contain almost no gas or dust. Galaxies themselves clump into galaxy clusters with hundreds or thousands of members. As the universe evolved, the clumping process went from large to small: first the protoclusters formed, then the protogalaxies, and last the protostars. Because elliptical galaxies contain almost no gas or dust, we may conclude that they have finished forming stars. In spiral galaxies most of the stars formed billions of years ago, but some protostars still are condensing out of the remaining gas and dust. These young, bright stars form only in the spiral arms and thus make the arms much brighter than the rest of the galaxy. Irregular galaxies consist largely of young stars, because many protostars still are condensing from gas and dust clouds in these galaxies. In contrast to the formation of stars in galaxies, the formation of galaxies seems to have ceased long ago. Whatever matter may lie between galaxy clusters (and little or none has yet been detected), we have not yet seen any evidence that galaxies have formed from protogalaxies anywhere during the past few billion years. However, any honest astronomer will tell you that we don't really know what a protogalaxy in the early stages of formation would look like, so that this absence of evidence does not provide compelling evidence that protogalaxies have not formed during the past few billion years.

Measuring the Velocities of Galaxies

Hubble's law summarizes our observations of galaxy clusters: Each cluster is moving away from us with a speed proportional to the cluster's distance from us. To understand how Hubble's law can be verified, we must consider how we measure the velocities and distances of galaxy clusters.

The velocity toward us or away from us can be found using the Doppler effect (Page 22). Motion of a light source toward us will increase the energy of each photon we detect, whereas motion away from us will decrease the energy of each photon. Galaxies are made of billions of stars, and although each star has its own peculiarities, in general, one star is much like another. Most stars contain about the same mixture of various kinds of atoms in their outer layers, the most abundant kinds being hydrogen, helium, carbon, nitrogen, oxygen, and neon. As photons produced deep in the stars' interior pass outwards, the atoms in the

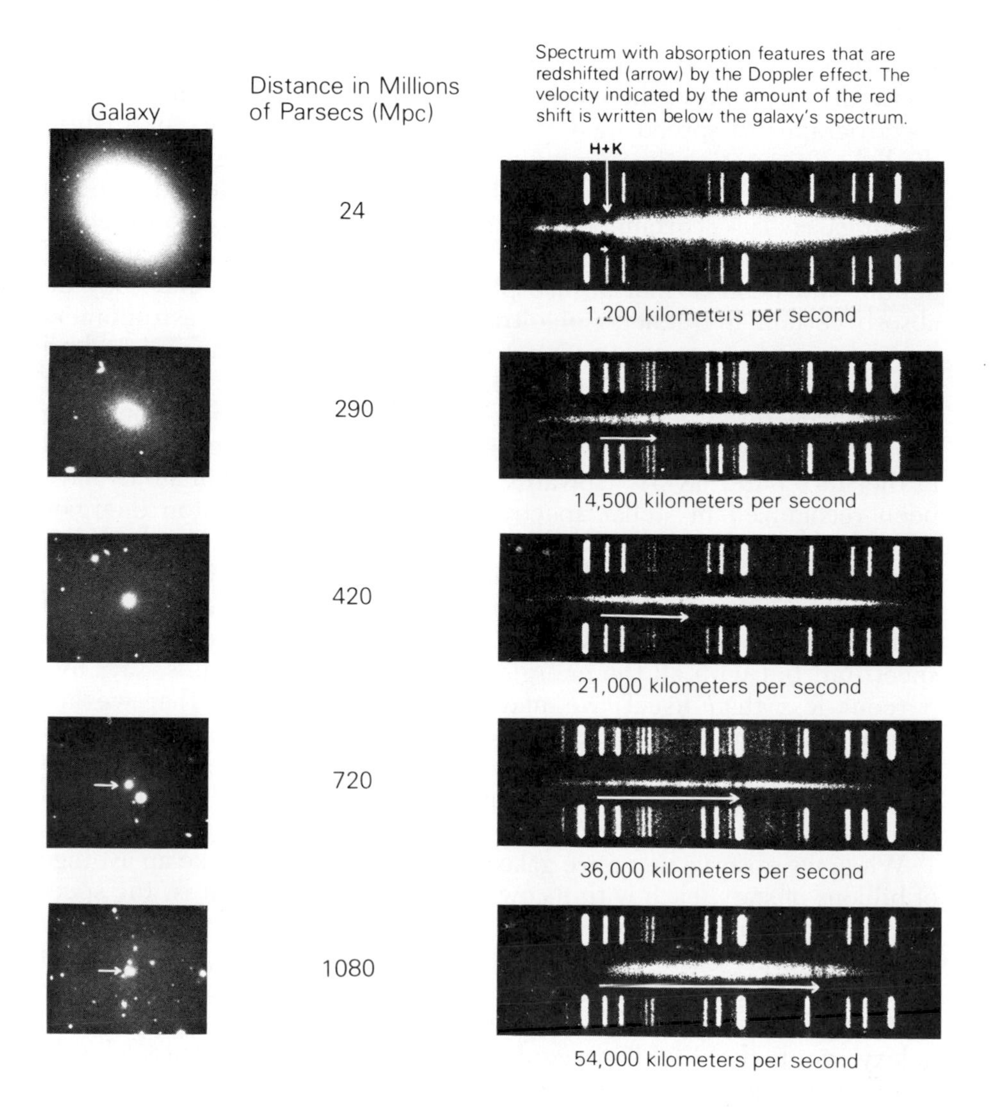

The distances to the galaxies shown here have been estimated by using a Hubble constant equal to $H = 50$ kilometers per second per Mpc.

Figure 4–24. The spectra of light from various galaxies shows the familiar absorption lines caused by calcium ions, but shifted by different amounts toward longer wavelengths (lower energies) as the result of the Doppler effect.

stars' outer layers absorb some photons with particular energies. The energies of the photons absorbed are precisely those needed to excite the atoms' electrons into larger orbits. Photons with other energies get by the atoms unhindered and radiate outward into space. The light from stars thus tends to lack the photons with exactly the right energies to excite the electrons of the most common atoms. When astronomers use spectrographs to spread light into its various colors, they can analyze the light by counting the number of photons of each energy (or each frequency or wavelength). They usually find that the visible light from a star or galaxy lacks certain photons, namely those that were absorbed by atoms or ions of hydrogen, carbon, nitrogen, and oxygen. The atoms of helium and neon do not absorb photons of visible light, but they do absorb certain frequencies of ultraviolet light. By now, astronomers have observed the energy spectra of the light from perhaps a hundred thousand individual stars. Most of these stars absorb photons in the way that we have described, so their spectra show that certain photons with energies by now completely familiar to astronomers have been absorbed in the stars' outer layers. Upwards of fifty different kinds of atoms have been recognized in stellar spectra by the particular photon energies they absorb, so that not just hydrogen, carbon, nitrogen, and oxygen have been detected, but also sodium, calcium, iron, nickel, magnesium, sulfur, phosphorus, aluminum, titanium, potassium, and a host of other elements. If we find the usual pattern of absorption dips in a star's spectrum, but with all the energies of the absorbed photons, say, five percent less than usual, we may reasonably conclude that we are observing a normal star that is moving away from us at five percent of the speed of light. For speeds much less than the speed of light, the fractional change in energy equals the relative speed of the source given as a fraction of the speed of light.

When we observe an entire galaxy, our observations make an average of billions of stars, each with its own motion in space. That is, the spectrum of light from a galaxy consists of billions of individual contributions from stars. Since each star emits its own photon spectrum, which shows the effects of the number of atoms of each element and the temperature in the star's outer layers, we might expect that the total spectrum of the galaxy would be highly irregular. However, because stars closely resemble one another in their composition and temperature (we discuss different kinds of stars in Chapters 7 and 8), the spectrum of light from an entire galaxy still shows the familiar pattern of photon absorption at particular energies (Figure 4–24). We also should note that each star has its own motion in a galaxy, so that the light from each star has its own particular Doppler shift. Because of these individual Doppler shifts, the galaxy's spectrum may become hopelessly blurred as the photon absorption from one star's light spectrum falls at an energy different from the dip in another star's spectrum (Figure 4–25). Luckily, though, in most galaxies the velocity of any individual star relative to the center of the galaxy will not exceed a few hundred kilometers per second, which is only a few tenths of a percent of the speed of light. Therefore the resultant blurring of the absorption dips in the total spectrum that we observe will spread a given dip over only a few tenths of a

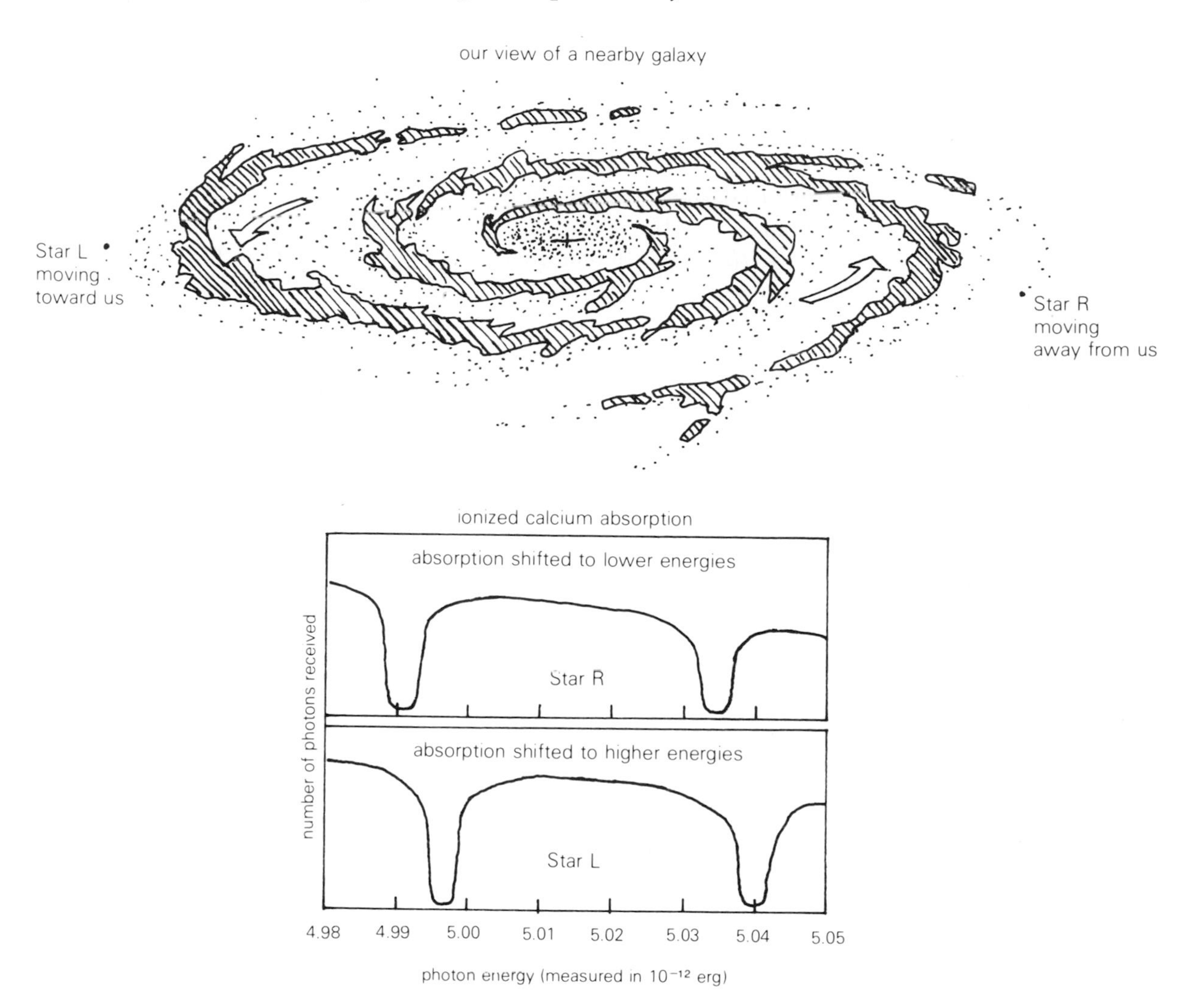

Figure 4–25. Because stars within a galaxy are moving with respect to each other, the Doppler shift can move the absorption dip from one star's spectrum to an energy different from that for another star's absorption dip. In most galaxies, however, the absorption dips all occur at about the same energy, because the velocity of one star with respect to another is small compared to the speed of light.

percent of the entire spectrum (Figure 4–26). The amount of this blurring of absorption features in fact provides a good way to determine the average random velocity of a group of stars, but the blurring does not stop us from recognizing the familiar pattern of absorption features in the spectrum of light from most galaxies.

We can conclude that to measure the speed with which a galaxy is moving toward us or away from us, we can spread the galaxy's light into colors and see which photon energies have been absorbed. We almost always see a familiar pattern, which reflects the ordinary amount of ordinary stars, but the entire pattern has been shifted to lower energies. If the pattern has been shifted so that all the energies are three or eight or twelve percent less than usual for stars in our own galaxy, we conclude that the galaxy is moving away from us at three or eight or twelve percent of the speed of light. The difference in velocities among the stars *within* a particular galaxy is always much less, perhaps a tenth or two tenths of a percent of the speed of light.

our view of a distant galaxy

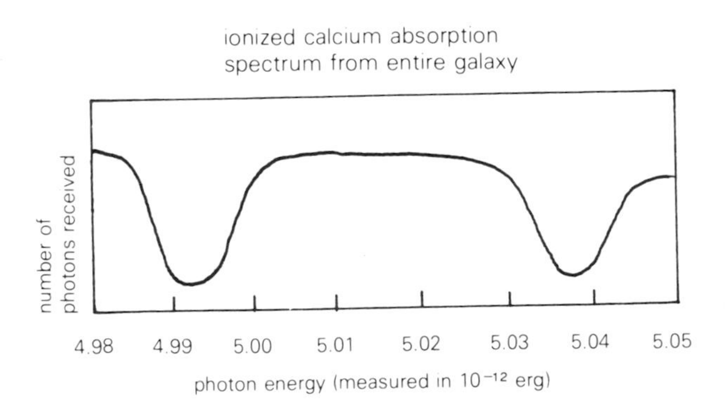

Figure 4–26. When we observe the spectrum of light from a galaxy, we observe the photons from billions of stars. The spectrum of the light from most of the brighter stars may have a dip at a certain photon energy caused by the absorption of photons with that particular energy in the stars' outer layers. Then the sum of all the spectra of the stars in a galaxy will show an absorption dip that averages the contributions from the spectra of the individual stars.

In this way, astronomers can measure the speed of a galaxy relative to ourselves in the direction along our line of sight to the galaxy. They can measure this speed even when they cannot observe individual stars within a galaxy, or even individual galaxies within a galaxy cluster.

Measuring the Distances to Galaxies

Now that we know how to measure a galaxy's velocity, we cannot stop ourselves from asking: How do we find the galaxy's distance from us? To do this, we use the method of comparing the apparent brightness of two objects that we think have the same true brightness, but have different distances from us. As we discussed on Page 20, an object's apparent brightness decreases as the *square* of its distance from us. Thus if two galaxies have the same true brightness, as M 51 (the Whirlpool galaxy, Figure 3–10) and M 31 (the Andromeda galaxy) do, but M 51 is ten times farther away than M 31, then the farther galaxy will appear a hundred times fainter than the nearer one. In our attempts to determine the distances to galaxies and galaxy clusters, we always try to compare the apparent brightness of two sources that we think have the same true brightness. Furthermore, we must know the distance to the nearer source for this brightness comparison to help us. If we do know this distance, then we can use the fact that the ratio of the distance of the farther source to the distance of the nearer source will be the square root of the ratio of apparent brightnesses, with the fainter source always farther away. For example, the sun appears about a trillion (10^{12}) times brighter to us than the star called 40 Eridani does. Once we know that the sun has about the same true brightness as 40 Eridani, we can conclude that the sun must be one million (10^6; the square root of 10^{12}) times closer to us than 40 Eridani is. Hence the sun is far more important to us than 40 Eridani is, although this star (five parsecs away) has become about the best-studied triple-star system.

To find the distances to other galaxies, astronomers rely heavily on two types of "standard" sources that should have about the same true brightness wherever they appear. The first kind of such sources are variable stars, which are stars whose light fluctuates in brightness periodically. By timing the period of the light variations, astronomers can determine what type of variable star they are observing. In some special types of variable stars, called Cepheid variables (after the first one to be discovered, the star Delta Cephei), the period over which the star's light fluctuates from bright to faint and back again is related to that star's true brightness: The longer the period of light fluctuation, the greater is the true brightness of the Cepheid variable star. A related class of stars, called RR Lyrae variables (named, oddly enough, after the first one of these to be discovered, the star RR Lyrae), can be recognized by characteristic features in the stars' photon spectra. By finding the distance to Cepheid variables and RR Lyrae variables in our own galaxy,[3] astronomers have established the true brightness of these stars. If we assume that the same type of star will have the same true brightness in other galaxies, then we are in business: If we observe a variable star in another galaxy that, for example, fluctuates in light output in the same way as Cepheid variables in our own galaxy, then we assume that the other galaxy's star has a true brightness equal to that of a Cepheid variable star in the Milky Way with the same period of brightness fluctuation. The ratio of the stars' apparent brightnesses then gives the square of the inverse ratio of their distances as we described above. This method has been used to find the distances to the nearest galaxies, such as the Andromeda galaxy, which was the first galaxy to have its distance accurately measured when Edwin Hubble found Cepheid variables in it, in 1923.

Unfortunately, the Cepheid variables and RR Lyrae variables do not have enough true brightness to be seen at enormous distances. Only for those galaxies closer than a few million parsecs can we use them as standard sources for distance measurements. For more distant galaxies, astronomers needed a brighter kind of standard source, and they found it in the form of "H II regions." H II regions are areas in which stars are still forming. Inside the condensing clouds of gas and dust, some young stars have begun to shine with great intensity. The stars will not last long at their great brightnesses, but for a few million years they can light up the cloud with an immense true brightness. The nearest H II region in our own galaxy is the Orion Nebula (Figure 4–27), which appears as the middle "star" in Orion's sword. Inside the Orion Nebula, four young, bright stars radiate large quantities of photons that have enough energy per photon to ionize the atoms around them. Since the gas consists mostly of hydrogen atoms, this ionization turns the gas from hydrogen atoms, called "H I," into ionized hydrogen, called "H II." The hydrogen ions, which are protons, try to recombine with the free electrons, but soon after they do so another photon ionizes them once again. However, each recombination does produce at least one photon (Page 35), and we see some of these photons when we observe the entire gas cloud shining with energy emitted by the stars inside it.

H II regions are especially helpful to us because they have large true brightnesses. The Orion Nebula's absolute brightness exceeds the absolute brightness of our sun by about a hundred thousand times, so although it is five hundred parsecs away, it appears as bright as the stars in the Big Dipper, which are only twenty-five parsecs away. Other H II regions have even greater true brightnesses, so they can be seen even

---

[3]Unfortunately, no Cepheid variable star on RR Lyrae variable is close enough to us for us to measure its distance by the parallax shift. More roundabout methods must be used to establish the distances of these stars even inside our own galaxy.

in galaxies that are tens of millions of parsecs away. Of course, the fact that different H II regions have different true brightnesses makes it difficult to use them as "standard" sources. However, astronomers have noticed that the *largest* H II regions in a galaxy always seem to have about the same *size*, namely a few hundred parsecs across. (Compared with these giant H II regions, the Orion Nebula, which is only a few parsecs across, seems tiny indeed.) The apparent angular size of an H II region in another galaxy will decrease in proportion to the galaxy's distance from us. In a galaxy ten million parsecs away, the angular diameter of the largest H II regions should be about six seconds of arc, whereas in a galaxy thirty million parsecs away, the angular diameter will be two seconds of arc. For comparison, the apparent angular diameter of the sun or the moon is eighteen hundred seconds of arc, whereas the apparent angular diameter of the planet Jupiter is about thirty seconds of arc. The method of measuring the apparent angular sizes of H II regions has been used to estimate the distances of galaxies as far as a hundred million parsecs away. Since the different H II regions in various galaxies probably have some variation in their true sizes, even

Figure 4–27. The Orion Nebula contains several thousand solar masses of gas illuminated by young, bright stars within the gas. In addition, a great amount of absorbing dust hides the left-central parts of the nebula from our view.

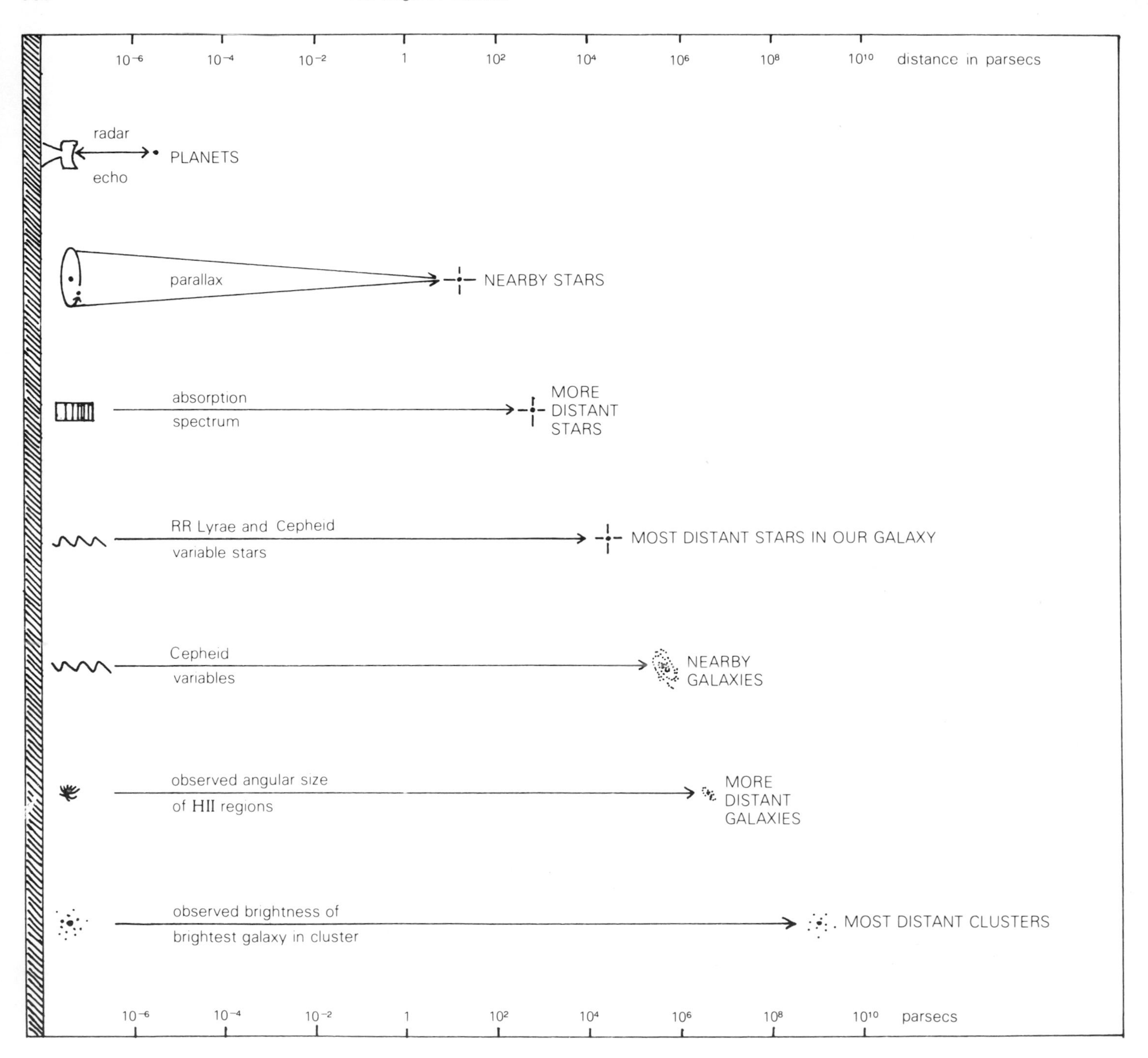

10⁻⁶
10⁻⁴
10⁻²
1
10²
10⁴
10⁶
10⁸
10¹⁰
distance in parsecs
radar
echo
PLANETS
parallax
NEARBY STARS
absorption
spectrum
MORE
DISTANT
STARS
RR Lyrae and Cepheid
variable stars
MOST DISTANT STARS IN OUR GALAXY
Cepheid
variables
NEARBY
GALAXIES
observed angular size
of HII regions
MORE
DISTANT
GALAXIES
observed brightness of
brightest galaxy in cluster
MOST DISTANT CLUSTERS
parsecs

among the very largest such regions, we must be ready to accept some inaccuracy in the distance measurements made by the method we have described above. But it's good to have a method that works, even without great accuracy, for galactic distances up to a hundred Mpc.

For galaxy clusters more than a hundred Mpc away, we use a new "standard" source: galaxies themselves. We have determined the distances to galaxies in several dozen clusters closer than a hundred Mpc by the methods described in the preceding paragraphs. We then can find the absolute brightness of these galaxies, because we know their apparent brightness and their distances from us. It turns out that the few brightest galaxies in a big cluster, especially the brightest elliptical galaxies, always have almost the same true brightness, which is equal to forty billion times the true brightness of our sun. Suppose now that we photograph a galaxy cluster farther from us than a hundred million parsecs. We can compare the apparent brightness of this cluster's brightest elliptical galaxy with the apparent brightness of the brightest elliptical in a cluster whose distance we determined by using the method of variable stars and H II regions. If the brightest elliptical in the faraway cluster appears, say, nine times fainter than the brightest elliptical in the Coma cluster, then the faraway cluster must be about three times farther from us than the Coma cluster is, if the two elliptical galaxies really do have the same true brightness. Since the distance to the Coma cluster is about a hundred Mpc, the more distant cluster would have a distance of three hundred Mpc.

Astronomers thus have worked out a way to find the distances to farther and farther objects, as summarized in Figure 4–28. The size of the Earth's orbit can be determined by measuring radar echoes from the sun and planets, because the echoes return with some delay after making a round trip to the object and back at the speed of light. Stars are much too far away for us to use the radar-echo technique, but we

◀ **Figure 4–28.** A schematic representation of the various ways to measure distances to different types of astronomical objects. For the nearest objects (planets and satellites), we can use radar-echo techniques to measure distances. The closest stars, too far away for radar echoes, show parallax shifts that allow their distances to be measured by trigonometry. These stars can be compared with more distant stars that show the same type of spectrum and presumably have the same true brightness; the comparison of apparent brightnesses then gives the ratio of distances. RR Lyrae and Cepheid variable stars can be observed throughout our galaxy and in the nearest other galaxies. Their light fluctuation's characteristics point out the (already known) true brightness of the star, and again we can use the comparison of true brightness and apparent brightness to find the stars' distances, and thus the distances to nearby galaxies. For more distant galaxies, we use the apparent size of the largest visible H II regions, which decreases as the distance increases. Finally, once we know the true brightness of a "typical" giant galaxy, we can compare the apparent brightness of a faraway galaxy with the true brightness that we believe the galaxy must have to find the distance to the farthest galaxy clusters.

can measure the distance to the nearest stars in our galaxy by the parallax shift, as described on Page 3. We then can compare the apparent brightness of distant stars in our galaxy with the apparent brightness of nearby ones, which, as seen from the details in their photon spectra, all ought to have about the same true brightness. The ratio of apparent brightnesses then yields the square of the ratio of distances. Some of the stars whose distances can be found in this way lie close to Cepheid variables and RR Lyrae variables in our own galaxy, so we can find the distance to these "standard" variable stars in the Milky Way. We also can use the method of comparing apparent brightnesses to find the ratio of distances to similar variable stars in other galaxies. For more distant galaxies we can compare the apparent sizes of the largest H II regions, which appear smaller in angular extent if the galaxies are farther away. Finally, we have found that the brightest elliptical galaxies in a large cluster of galaxies all have about the same true brightness, so we can use these elliptical galaxies as a new set of "standard" sources. By comparing the apparent brightness of the brightest ellipticals in a distant cluster with the apparent brightness of the most prominent elliptical in a "nearby" cluster, we once again can find the ratio of the two clusters' distances from us.

This brief description of the many rungs on the ladder of distance measurements shows that at each step errors may creep into our estimates of the distances to faraway galaxies. (Thus, for example, the *estimated* distances to galaxies such as those in the Coma cluster have increased by a factor of ten since 1929.) Astronomers have developed ways to check the methods outlined here, but these methods are at least as uncertain as the distance estimates themselves. As a result of the cumulative uncertainties, the distance to a cluster of galaxies that is found to be, say, 500 Mpc away has an accuracy of at best forty percent. Thus the galaxy cluster could be 300 Mpc or 700 Mpc distant from us without surprising many astronomers. Objects progressively nearer to us have progressively more accurately known distances, until we can proudly announce that we know the distance to our moon to better than one centimeter! This contrasts sharply with an error of 200 Mpc for a galaxy cluster 500 Mpc away from us, but it gives astronomers something to work on.

The greatest single uncertainty in measuring the distance to extremely distant galaxies comes from the fact that we are looking at a galaxy that is significantly *younger* than the galaxies most familiar to us, such as the Milky Way and the Andromeda galaxy. If we observe a galaxy one or two billion parsecs away, we are looking back three or six billion years in time. Although astronomers think that they know a fair amount about how the appearance of stars changes with time (Chapter 8), when it comes to adding everything together and talking about an entire galaxy of a hundred billion stars, most astronomers admit the likelihood of sizable errors. If we observe a galaxy in the Coma cluster, we see the galaxy as it was about four hundred million years ago—not very long, in comparison with the ten billion years or so that most galaxies can call their lifetimes so far. But if we compare one of the brightest galaxies in a distant cluster—say one that is a hundred times fainter than the brightest galaxy in the Coma cluster—with these not-so-old snapshots, we run into trouble. The faraway galaxy might be about ten times farther from us than the Coma cluster, so its light might be four billion years old. Then the changes in the galaxy's light output that arise from the evolution of the stars within the galaxy can spoil our simple assumption that the brightest galaxy in a large cluster always should have about the same true brightness. This assumption could be true *now* but not for all time (not for four billion years ago compared with "now," for example), and then the "evolutionary effects" on the light output from galaxies would add a large error to our attempts to determine just how far away these faint clusters of galaxies are. This problem enters the determination of these great distances in a fundamental way, since we never can alter the fact that an object seen at a great distance reveals only how it was, not how it is.

So we must live with uncertainty in our estimates of galaxies' distances. By comparing our observations of galaxies' velocities with our determinations of their distances, we can verify Hubble's law. Within

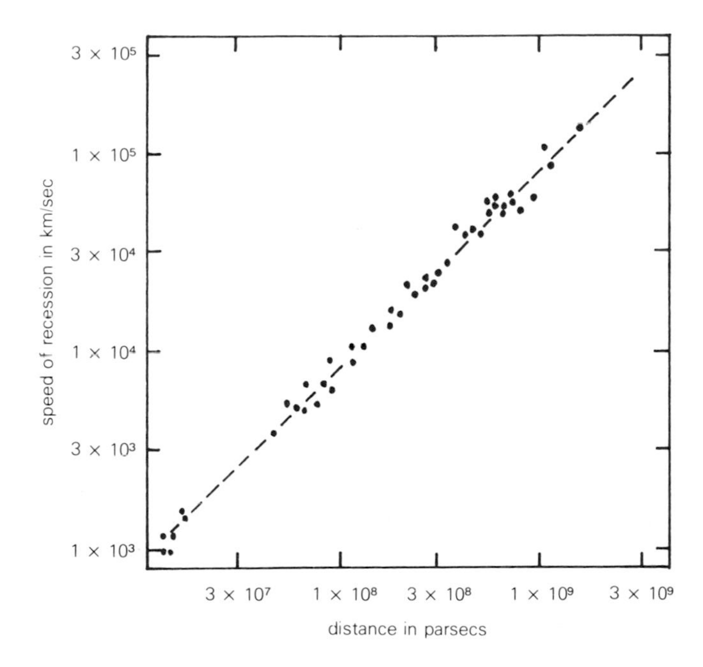

Figure 4–29. The relationship between galaxies' distances and speeds of recession have been determined by years of patient observation by Dr. A. Sandage and his colleagues.

our limits of accuracy, an object's velocity away from us varies in proportion to its distance from us, for any object as far away as the nearest cluster of galaxies. If our faith in this relationship remains firm, we can turn the relationship around and then *use* Hubble's law to find the distances to faraway objects. Suppose that we observe a strange object with a photon spectrum that reminds us of the spectra of stars in galaxies. If the object appears so strange that we cannot tell anything about its distance by looking at it, we can use Hubble's law: The distance must be proportional to the amount of the Doppler shift. Figure 4–29 shows Hubble's law as determined by many observations of galaxies and galaxy clusters. A strange object whose spectrum appears to indicate a recession velocity of, say, 50,000 kilometers per second should have a distance of one billion parsecs, on the basis of Hubble's law alone, and without reference to any other facts about the object. We shall look at one such kind of strange object, the quasars, in the next chapter, and shall see whether the determination of distances from velocity measurements, using Hubble's law to relate the velocity to the distance, will make sense to us or not.

Summary

We can see that matter in the universe is no longer spread evenly through space, but instead appears in clumps called stars, which themselves clump into galaxies and galaxy clusters. The process of condensation and subcondensation that led to what we see today must have started after the photons no longer exerted a homogenizing effect on the matter. That is, clumping must have begun in earnest only after the time of "decoupling," some 300,000 years after the big bang.

Gravitational forces seem to be the only ones capable of forming large clumps of matter, so we may conclude that these clumps somehow became bound together as separate entities by their own self-gravitation. The same self-gravity then started to pull the clump toward a smaller size and a larger density of matter. As this happened, the clump's original rate of spin (if it were not zero) would increase, because the clump's total angular momentum (mass times the rate of spin times the size squared) would stay constant. The constancy of angular momentum implies that a smaller size must be linked to a larger rate of spin.

Spiral galaxies appear to have formed from protogalaxies that acquired more angular momentum than elliptical galaxies did, and as a result, spirals are much flatter than ellipticals, though both types of galaxies do rotate. In a typical giant spiral galaxy such as the Milky Way, which spins around once every few hundred million years, most of the stars and the diffuse gas move in almost circular orbits around the galactic center. As these parts of the galaxy circle the center, they combine their orbital motion with a smaller, up-and-down oscillation through the median plane of the galaxy.

Questions

1. Why do galaxies tend to appear in clusters rather than as single, isolated units?
2. Why do stars tend to group into galaxies rather than being spread out evenly through space?
3. What kinds of forces produced galaxy clusters, galaxies, and stars? Why didn't the other kinds of forces play the dominant role in this process?
4. The average random velocity per particle decreased as the temperature in the universe decreased, during the first few million years after the big bang. What effect did this decrease in random velocity have on the ability of particles to form clumps?
5. What are the major differences between spiral and elliptical galaxies? What are the chief similarities between these two galaxy types?
6. Why should the amount of angular momentum within a protogalaxy affect the kind of galaxy that condenses from this protogalaxy?
7. How can we observe the diffuse gas and dust that lies between the stars in a spiral galaxy, such as our own Milky Way?
8. A large galaxy may contain a hundred billion stars, each with its own particular velocity. How can we hope to measure the motion of an entire galaxy along our line of sight, using the Doppler effect, if we cannot observe each of these hundred billion stars individually to find the motion of the individual stars?
9. The galaxies M 81 and M 83 have almost the same true brightness, but M 81 appears to be four times as bright as M 83. Which galaxy is closer to us? By how much?
10. The galaxies M 87 and NGC 7793 appear to have almost the same brightness, but M 87 is four times farther away from us than NGC 7793. Which galaxy has the greater true brightness? By how much?
11. Why do astronomers find it useful to observe the brightest individual stars in another galaxy?

12. What is an H II region? Why are they useful for measuring the distances to other galaxies?

Further Reading

M. Rees and J. Silk, "The Origin of Galaxies," *New Frontiers in Astronomy* (W. H. Freeman and Co., San Francisco, 1975).

H. C. Arp, "The Evolution of Galaxies," *New Frontiers in Astronomy* (W. H. Freeman and Co., San Francisco, 1975).

A. Sandage, *The Hubble Atlas of Galaxies* (Carnegie Institution of Washington, D.C., Publication No. 618, 1960).

H. Shapley, *Galaxies* (Harvard University Press, Cambridge, Mass., 3rd ed. revised by Paul Hodge, 1972).

Radio Galaxies, Exploding Galaxies, and Quasars

Chapter **5**

"Amid whose swift half-intermitted burst
Huge fragments vaulted like rebounding hail,
Or chaffy grain beneath the thresher's flail."

Samuel T. Coleridge, *Kubla Khan*

Most of the matter in the universe that we can see is located in galaxies, and most of these galaxies are spirals or ellipticals. Of all the galaxies we have observed, a small fraction show peculiar characteristics that have demanded further study. Such galaxies include "radio" galaxies, which emit large amounts of radio waves; "Seyfert" galaxies, which are named after the astronomer who discovered that their central regions are highly active; and "exploding" galaxies, which show evidence of tremendous outbursts in their relatively recent past. In addition to these strange galaxies, a new class of objects called "quasi-stellar radio sources," or "quasars," was uncovered in 1963. Quasars are so strange that astronomers cannot yet decide whether they are a special kind of galaxy or an entirely new kind of object for us to understand. Quasars may be the most distant objects ever observed, or they may be the mysterious products of exploding galaxies. We shall look at the various classes of peculiar galaxies and then see how quasars might be related to them.

Radio Galaxies and Exploding Galaxies

The science of radio astronomy started as the hobby of an American radar engineer named Grote Reber, who built a backyard telescope to focus radio waves and a detector to measure them. Most of the radio waves that Reber detected came from the sun (which emits quite faint radio waves but happens to be relatively close to us) or from the central regions of our own galaxy, which are two billion times farther from us than the sun is. In the central regions of the Milky Way, electrons moving at nearly the speed of light through magnetic fields produce photons of synchrotron radiation (Page 39) that have the frequencies of radio waves. Karl Jansky of the Bell Telephone Laboratories was the first person to realize that the photons Reber had detected came from the galactic center. Like Reber, Jansky engaged in radio astronomy as a hobby, and it took twenty years after Jansky's first observations for the Bell Laboratories to begin serious radio astronomy.

Although the Milky Way galaxy does produce plenty of radio-wave (radio) photons, astronomers do not consider our galaxy to be a "radio galaxy." They reserve this name for galaxies that emit a *large fraction* of their total energy as photons of radio wavelength. The Milky Way hardly qualifies as a radio galaxy because less than one millionth of the energy that it emits appears in the form of radio photons. Most of the galaxy's energy output comes in photons of visible, infrared, and ultraviolet light frequencies. A radio galaxy such as Cygnus A, one of the first sources that Reber detected (Figure 5–1), emits about as much energy (or more, in the case of Cygnus A) in radio photons as it does in visible-light photons, in contrast to a "normal" galaxy such as the Milky Way, which emits one millionth as much energy in radio waves as it does in visible light.

Radio galaxies apparently owe their tremendous outflow of radio photons to violent events occurring within them. Ordinary galaxies con-

Figure 5–1. The radio galaxy Cygnus A is the most intense radio source in the constellation Cygnus.

sist mostly of stars, which produce huge quantities of higher-energy photons in visible, infrared, and ultraviolet light waves. Only a small fraction of a star's energy output emerges as low-energy radio photons. If diffuse hydrogen gas exists in a galaxy, it will emit the photons of 21-centimeter wavelength described on Page 36, but these radio photons will never carry a sizable fraction (as much as a thousandth) of a galaxy's total energy production. If we observe that a galaxy emits as much, or even more, energy in radio photons than in light-wave photons, we can be sure that something besides stars must be responsible for most of the galaxy's energy output. The radio photons from most radio galaxies arise from synchrotron radiation. To produce photons by the synchrotron process, the source must contain many charged parti-

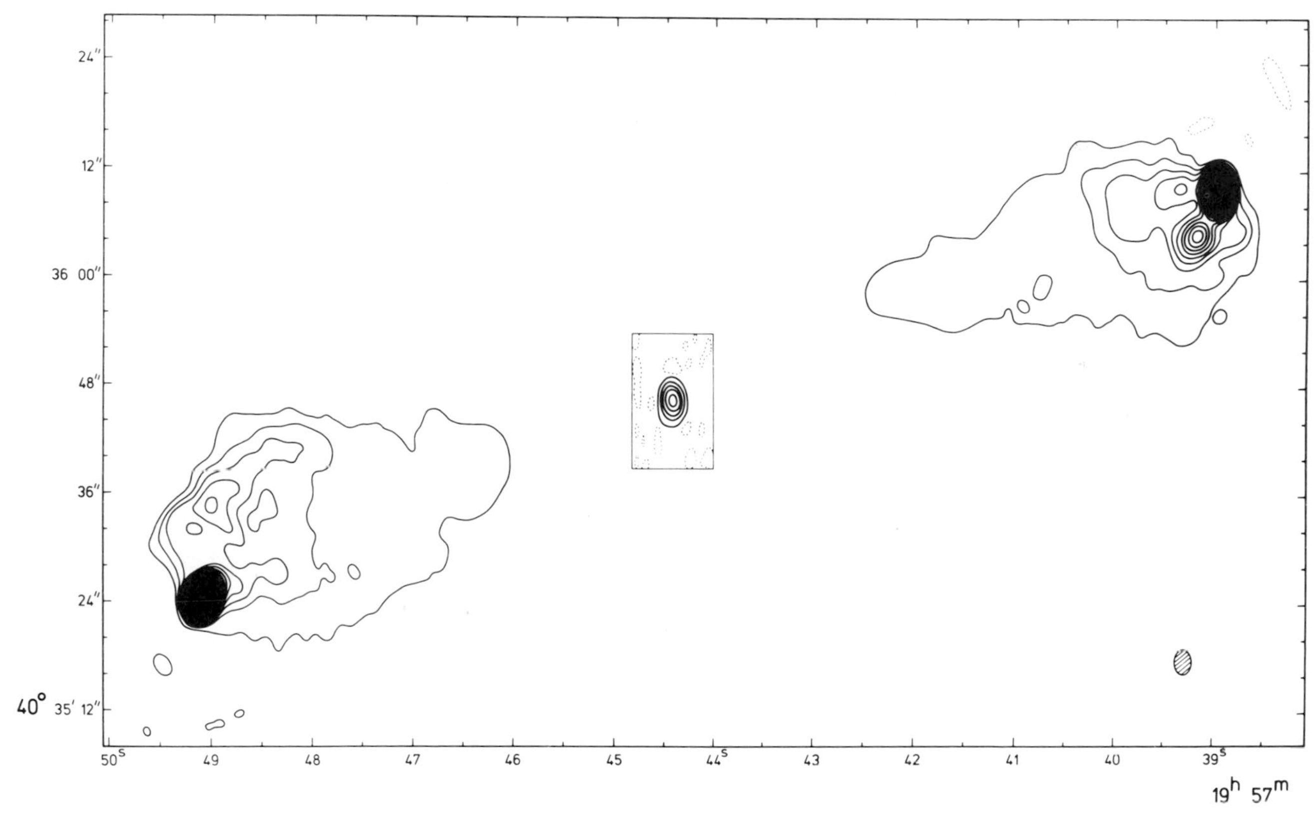

Figure 5–2. A map of the radio emission from Cygnus A and its surroundings. The radio emission is most intense where the contour lines are closest together: The two dark ellipses represent regions where the emission is so intense that contour lines could not be drawn accurately. The visible part of Cygnus A, shown in Figure 5–1, would fit snugly into the contour lines within the box at the center of the figure.

cles (probably electrons) that are moving *at almost the speed of light* in magnetic fields. Hence radio galaxies almost surely contain large numbers of fast-moving particles that somehow have been accelerated to almost the speed of light.

When we study radio galaxies in detail, we usually find that the radio photons do not come from the central regions of the galaxy. Instead, the radio photons are produced in large regions far from the center of the galaxy and indeed often outside the galaxy's stars (Figure 5–2). Radio "maps" of the photons from most radio galaxies show that two "clouds" of radio emission, symmetrically located on either side of the visible galaxy, produce most of the radio emission. In a typical radio galaxy, the two clouds emitting radio photons each are about a hundred thousand parsecs across, and are separated by one hundred thousand to one million parsecs. The galaxy itself spans a mere 30,000, or so, parsecs.

If the synchrotron process produces the radio photons, then we must conclude that the charged particles with velocities near the speed of light occupy the regions mapped by our observations of radio emission. We may ask ourselves: Why should these highly accelerated particles lie so far from the centers of the radio galaxies? How have the particles been accelerated to nearly the speed of light?

Astronomers do not have reliable answers to these questions, but they do have a number of theories. The first theory came from the suggestion that Cygnus A is actually a pair of colliding galaxies. Space is so large that galaxies collide only rarely, but if they did, almost all the stars in the galaxies would miss each other. In an ordinary galaxy, stars are like a swarm of bees with a kilometer distance between each pair of bees, so that two swarms can pass through each other without any bees (or stars) colliding. But if the galaxies contain clouds of diffuse gas, the clouds are likely to collide with other clouds when the galaxies pass through each other, because these clouds often are larger than the average distance between stars and are fairly numerous. Thus when two galaxies collide, their collision might leave behind two regions where gas clouds have slammed into one another (Figure 5–3). Even though the gas spreads out diffusely in the clouds, the atoms in the gas can affect each other, and the result of the collision could be the violent acceleration of particles in the diffuse gas. Such an acceleration could strip electrons from atoms and could make the electrons move fast enough to produce photons of synchrotron radiation for millions of years. The energy carried away by these photons eventually would decelerate the electrons to the point where they no longer emit synchrotron radiation.

This theory of colliding galaxies sounds fine, but in fact galaxies simply do not collide often enough for the theory to explain most radio galaxies. We can calculate the fraction of galaxies that have collided

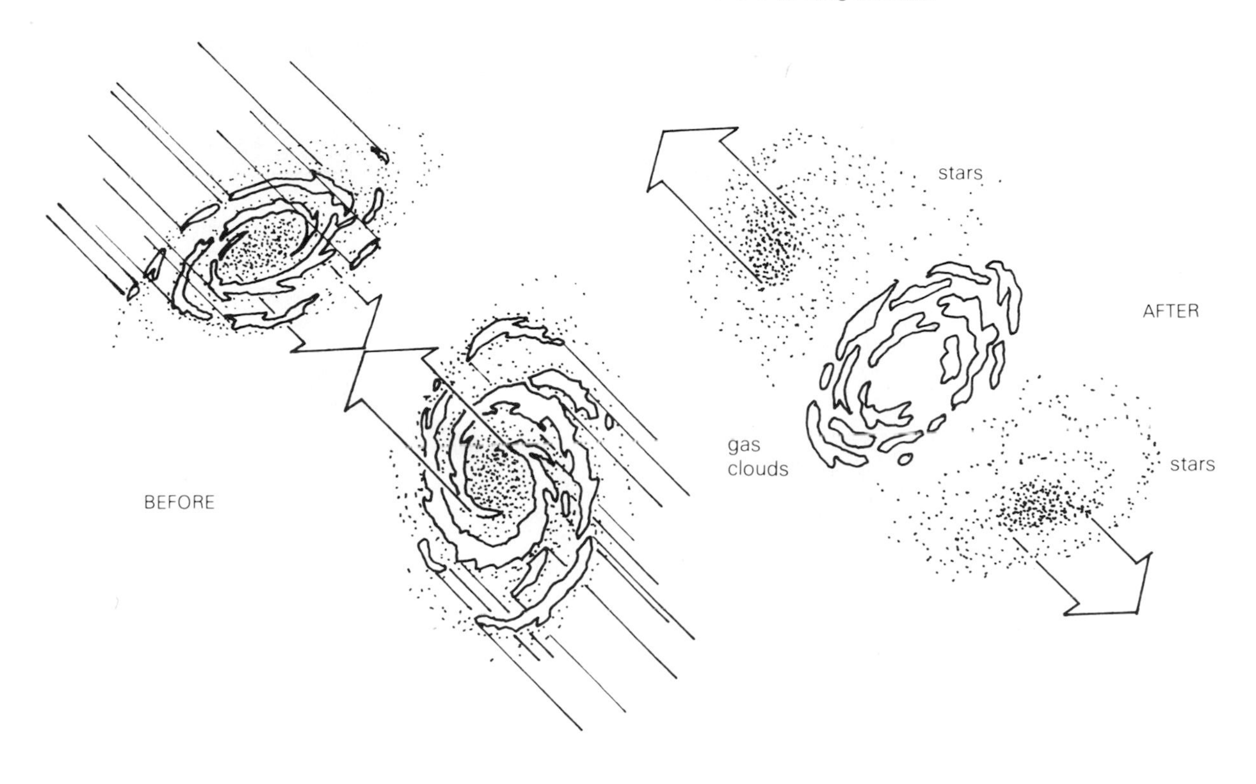

Figure 5–3. If two galaxies were to collide, their stars all would miss one another, but the gas between the stars would be swept out of the galaxies and left behind as a large clump.

"recently," say during the past ten million years. This fraction falls far short of the fraction of galaxies, about one in ten thousand, that are radio galaxies. Also, the theory predicts radio emission from one region, not two.

Another interesting theory to explain radio galaxies links them to explosions near the centers of unusual, "peculiar" galaxies. Somehow these explosions might shoot out clouds of particles moving at nearly the speed of light. This theory gains support from the fact that many radio galaxies appear to be quite unusual in visible-light photographs. For example, Centaurus A (Figure 5–4), a powerful radio galaxy, has a strange dark mass of gas and dust as a sort of girdle around an otherwise ordinary elliptical galaxy. If clouds are shot out from galaxies, they will expand as they move farther from the center and should continuously emit synchrotron photons if they contain many electrons moving near the speed of light. Perhaps the reason that the number of radio photons

Figure 5–4. The radio source Centaurus A is the peculiar galaxy NGC 5128, which looks something like an elliptical galaxy with a huge inner tube of gas and dust wrapped around it. The radio emission from Centaurus A is most intense in regions well outside the visible galaxy.

often peaks near the outer boundary of the emitting clouds (Figure 5–2) is that particles are encountering some intergalactic gas (if such gas exists) that slows the cloud and thus tends to pile up particles near the outer edge of the cloud of fast-moving particles.

These two suggested explanations for how the photons from radio galaxies arise should serve to show that astronomers' difficulties in understanding nature are matched by their fertile imaginations. We should bear in mind that most galaxies are not racked by violent explosions that produce great clouds of radio emission. Photographs of such ordinary galaxies look much like one another; having seen one, you've

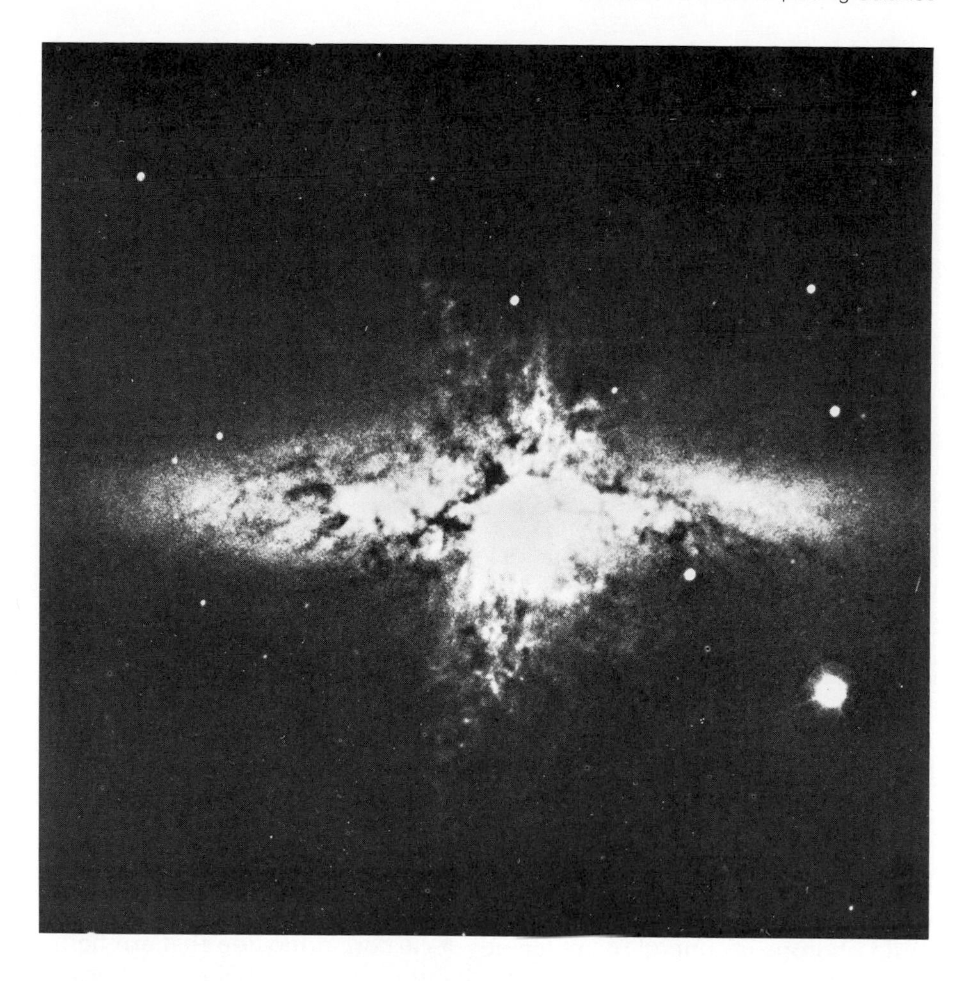

Figure 5–5. Within the irregular galaxy M 82 an explosion near the center apparently has occurred within the past few million years.

seen them all, so this book shows only some typical representatives. But because of their violent nature, radio galaxies each look different from the rest, as if each explosion had altered the galaxy in a different way. The central parts of the irregular galaxy M 82 (Figure 5–5) appear to have exploded a few million years before the picture was taken. We can judge this time since the explosion from the speed at which the gas filaments still are streaming outward from the center of M 82 (measured by the Doppler effect) and the distance they have traveled. Remember that the galaxy's light takes 13 million years to reach us, so when we talk about how the galaxy looks "now," we really mean how it looked 13 mil-

Figure 5–6. A photograph of the giant elliptical galaxy M 87, taken with a short exposure to show only the brightest, innermost regions of the galaxy. We can see the "jet" of material that apparently has been ejected from the center of the galaxy.

lion years ago, and when we speak of "a few million years ago," we mean a few million years before 13 million years ago. M 82 also emits radio photons produced by molecules of carbon monoxide that are being expelled outward from the galactic center, like the filaments (mostly made of hydrogen gas) that we see in Figure 5–5. The general speed outward of the filaments is only a few hundred kilometers per second, but within them huge numbers of electrons have been accelerated to almost the speed of light.

Another example of an explosion within a galaxy appears in the giant elliptical M 87 (Figure 5–6), which is the largest galaxy in the Virgo cluster of galaxies, about ten million parsecs away. If we take a short-exposure photograph of the galaxy, so that we see only the brightest, most central regions, we can detect a jet of material with several "knots," or condensations, along it that is being shot outward from the central regions. This jet is also a source of radio photons. Astronomers haven't yet figured out why such explosions might occur in radio galaxies, but they are fairly well convinced that such explosions are responsible for the huge regions of radio emission that we observe in radio galaxies.

Figure 5–7. The Seyfert galaxy NGC 4151 has an especially small and bright central nucleus, here somewhat blurred by the long exposure of the photograph. This nucleus emits great amounts of radio, infrared, visible-light, and x-ray photons.

Most galaxies that emit large amounts of energy in radio photons also emit a large number of infrared and microwave photons. These photons each have energies greater than the energy of radio photons, but less than the energy of visible-light photons. For example, the exploding galaxy M 82 emits ten times more energy in infrared photons than our galaxy does in photons of every energy (yet M 82 has only one tenth the mass of our galaxy). In fact, most of M 82's energy output appears as infrared photons, although we still call M 82 a "radio galaxy" because it emits so many radio photons in comparison to a "normal" galaxy. Somehow the explosion in M 82 has produced this vast outpouring of infrared and radio photons, which clearly does not come from stars.

Seyfert Galaxies

A class of peculiar galaxies clearly related to exploding galaxies are called "Seyfert" galaxies (Figure 5–7). These galaxies, first discovered by Carl Seyfert, resemble ordinary spiral galaxies, but their central, "nuclear" regions pour forth tremendous amounts of energy. These nuclear parts of Seyfert galaxies may emit more energy per second in infrared and radio photons than the rest of the galaxy does in all types of photons. Since Seyfert galaxies are no more massive than our own galaxy, once again we have a huge, unexpected flow of energy to explain. Even more puzzling, most of the energy from a Seyfert galaxy emerges from a nucleus that is at most a few tens of parsecs across, compared with the thousands of parsecs of an entire galaxy. The several dozen Seyfert galaxies that astronomers had discovered by the 1960's had them puzzled, until quasars came along and left them baffled.

Quasars

Astronomers knew about Seyfert galaxies in the 1940's, about radio galaxies in the 1950's, and about the explosion in M 82 by 1962. In the year 1963, astronomers met with objects far stranger than the previously known "peculiar" galaxies. Quasi-stellar radio sources, called quasars for short, briefly stood the world of astronomy on its ear, and astronomers have not yet entirely recovered their equilibrium. Quasars are either the most distant objects in the universe that we can see, by far the most powerful producers of photon energy though much smaller than galaxies, or else quasars are the herald of new laws in physics waiting to be understood by continued observations that will suggest and test new astrophysical theories.

Quasars owe their full name, quasi-stellar radio sources, and their discovery to the radio photons that they emit. As radio astronomers began to measure the photons from every part of the sky that reached their telescopes, they tried to see which sources of radio photons could be matched with sources of visible-light photons. By comparing the positions on the sky of the radio sources with visible-light atlases of the sky, they found that most sources of radio photons are either the peculiar radio galaxies we have discussed, or are recognizable radio sources within our own galaxy, many of them the remnants of exploding stars called supernovae (Chapter 8). But in addition to the radio galaxies and the radio sources within our own galaxy, some sources of radio photons seemed to be ordinary stars. At least they looked like stars when photographed (Figure 5–8). However, when the light from these "stars" was spread into its spectrum of energies, the results did not look like the spectrum of an ordinary star.

First, the spectra showed *emission* of photons at certain energies, rather than the *absorption* of certain photons characteristic of starlight. Second, astronomers could not recognize the emission frequencies among the usual patterns known from stars. Third, there indeed were some frequencies at which photons had been absorbed, but these too showed an unfamiliar pattern of frequencies. Fourth, the objects were emitting a greater ratio of blue light to yellow light than typical stars do. Astronomers use this ratio (the "color index") as a convenient way to classify a star without examining its entire spectrum, and on this basis the quasars are clearly "bluer" than ordinary stars.

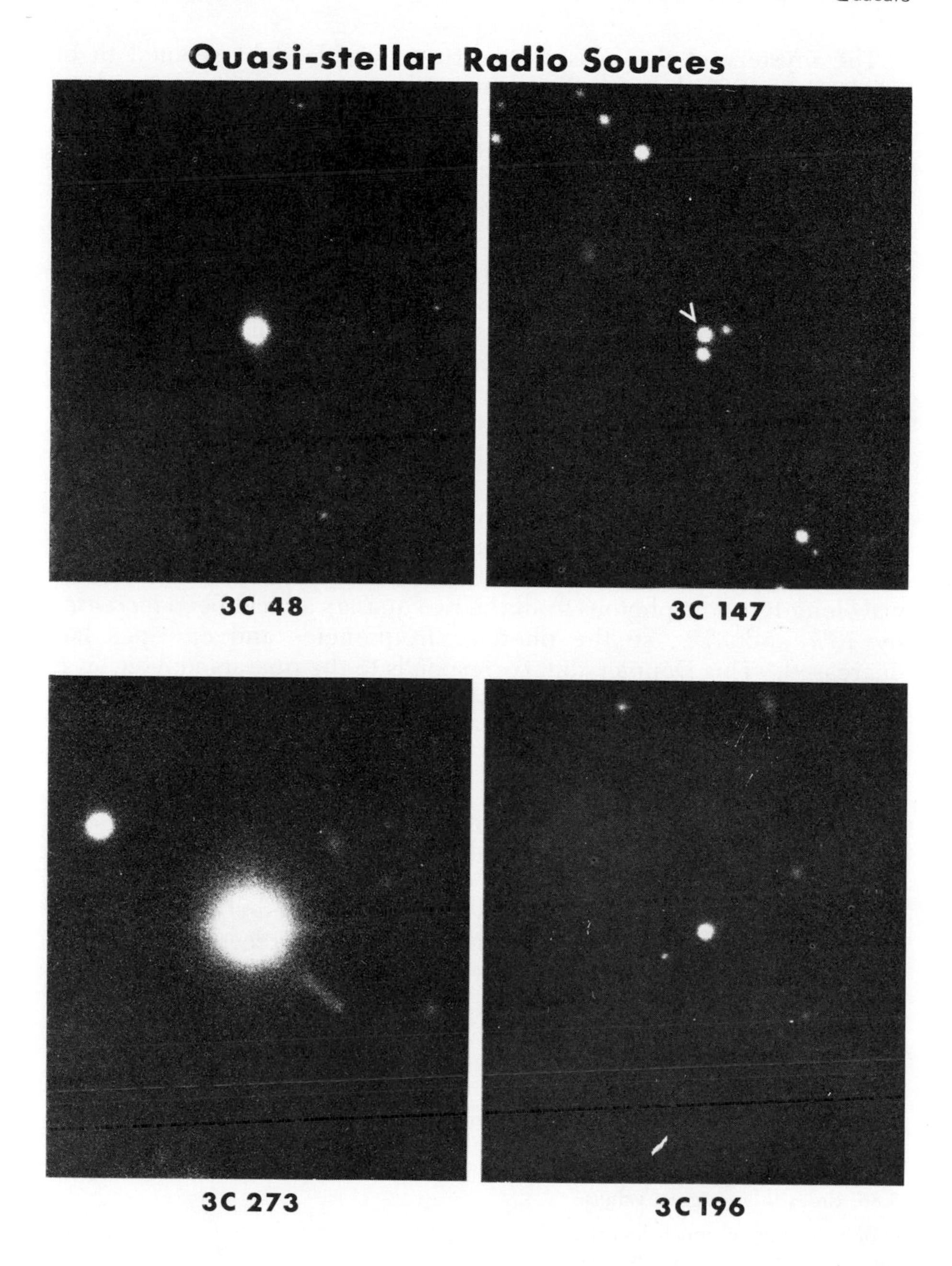

Figure 5–8. Four well-known quasi-stellar radio sources, which look much like stars on visible-light photographs. One of the nearest quasars, 3C 273, does have an accompanying jet somewhat reminiscent of the jet at the center of M 87 shown in Figure 5–6.

The mystery of the quasars' spectra of visible light seemed to be resolved in 1963, but this apparent resolution brought an astonishing revelation. Astronomers realized that the quasars' spectra showed the emission of photons at frequencies familiar from their studies of the emission of photons by hydrogen atoms. The emitted photons in quasar spectra had the frequencies characteristic of a hydrogen atom's electron jumping from the third-smallest, the fourth-smallest, or the fifth-smallest orbit into the second-smallest orbit (Page 32), but the frequencies all had been reduced substantially by the Doppler effect. Thus the frequencies of the photons emitted from quasars appeared unfamiliar, but the ratios of these frequencies to one another were recognizable as the familiar ratios for hydrogen atoms that emit photons when their electrons jump from a larger orbit into the second-smallest orbit. The frequencies of the photons emitted when the electron jumps to the smallest orbit of a hydrogen atom are ultraviolet photons and thus have energies too large to appear in the visible-light region.

The first quasars whose spectra could be understood in this way showed Doppler "red shifts" of 16% and 37%. That is, the familiar wavelengths of the photons from the two quasars all had been *increased* by 16% and 37%, so the photons' frequencies and energies had decreased.[1] This Doppler shift corresponds to the quasars moving away from us at enormous velocities, 14½% and 30% of the speed of light, respectively.[2] If we interpret these Doppler shifts as arising from the general expansion of the universe, then these quasars must lie at fantastic distances from us, about 800 million parsecs and 1.6 billion parsecs, respectively. The most distant galaxy ever measured has a Doppler red shift that increases the photon wavelengths 46% over their original wavelengths, meaning that the galaxy has a distance of two billion parsecs and a recession velocity of 36% the speed of light. The first two quasars to be recognized turned out to have recession velocities almost as great as the farthest galaxy's, yet the quasars' apparent brightnesses far exceeded the brightness of the most distant known galaxy. These two quasars, called 3C 48 and 3C 273 (numbers in the Third

---

[1]The amount of the red shift usually is expressed as a percentage increase in the wavelengths. If all the wavelengths increase by, for example, 16%, they become 1.16 times the original values. Then the photons' energies and frequencies all become 1/1.16, or 0.86, times their original values.

[2]For velocities, v, much less than the speed of light, c, the algebraic relationship for the Doppler shift to a new frequency f' compared to an original frequency f is f'/f equals $1-(v/c)$. Since frequency (f) times wavelength (w) always equals a constant (c), f'/f always equals w/w'. The complete algebraic relationship for all velocities is

$$\frac{f'}{f} = \sqrt{\frac{1-(v/c)}{1+(v/c)}}$$

Cambridge Catalogue of radio sources), appear thousands of times brighter than the most distant known galaxies, yet they lie almost as far away. Furthermore, quasars are clearly smaller than galaxies, because even the most distant galaxies are large enough to look fuzzy (Figure 5–1). In contrast, quasars look like points of light, as stars do. Any fuzziness in the pointlike appearance of quasars can be shown to be photographic blurring (Figure 5–8), so astronomers called them quasi-stellar objects (or, as the young daughter of one woman who observes quasars called them, crazy stellar objects).

The news that quasars are intrinsically much brighter than galaxies, but smaller than galaxies and incredibly far away, spread rapidly through the network of astronomers, who hurried to photograph more radio sources to see if they too were "quasi-stellar." Indeed many were—today about two hundred quasars have been detected—and many of them showed still more fantastic properties than the first two quasars. Astronomers now have found quasars whose spectral features (those wavelengths at which an especially large number of photons have been emitted or absorbed) show Doppler red shifts of up to 353%! That is, the photon wavelengths all have been increased by 353%, or 3.53 times, so that the wavelengths we observe all are 453%, or 4.53 times the original wavelengths. For quasars such as these, spectral features originally in the ultraviolet, like the photons produced in hydrogen atoms when the electron jumps into the smallest orbit, now appear in the red-light part of the visible spectrum. The quasar with a Doppler red shift of 353% is moving away from us at 92% the speed of light, and should be something like five billion parsecs away!

As for the quasars' small sizes, as the 1960's went by, astronomers found more and more puzzling results, because they were able to set smaller and smaller upper limits to the possible sizes of quasars. Without knowing it astronomers had photographed quasars many times before 1963, but the quasars had passed for ordinary stars on their photographic plates. Astronomers now searched through their libraries of photographs and discovered that the light output from quasars had varied. They studied quasars more closely and found that some quasars can double their light output in less than a day. This observational fact tells us something about the size of quasars. *Whatever region produces their light cannot be more than one light day across*, at least in the case of those quasars whose light varies this rapidly. Figure 5–9 represents the region that produces most of a quasar's light—perhaps a group of exploding stars, perhaps one super-star, perhaps something entirely new. If this region were so large that light must take more than a day to cross it, we could not see sharp changes in the light on a time scale as rapid as one day. Because, if the light output at the source did change in less than a day, we would see the change from the nearer part of the

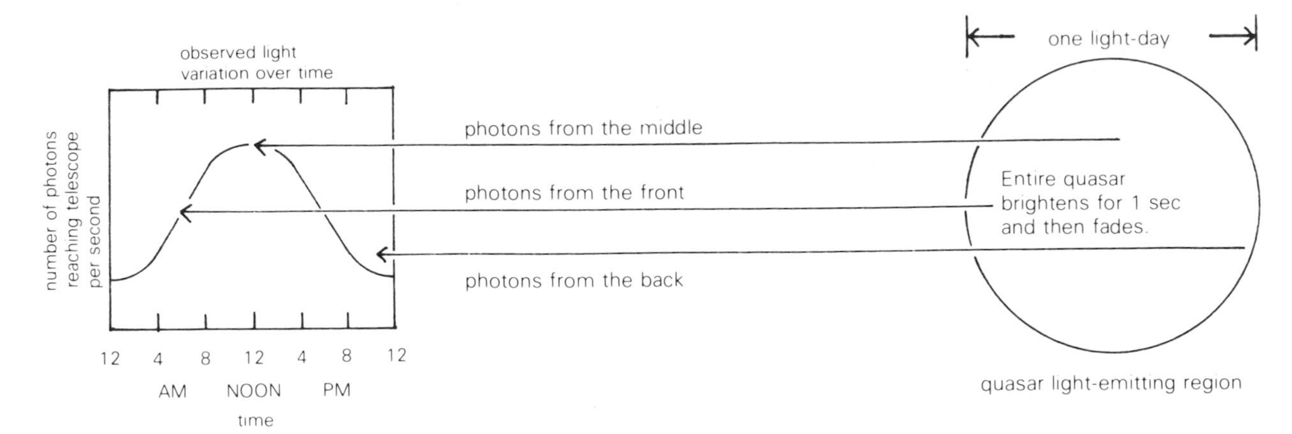

Figure 5–9. Whatever region produces most of a quasar's light cannot be larger—as measured in light-travel time—than the period over which the quasar varies its output. Suppose that a quasar suddenly increased all over in brightness in a very short time, say one second. If the quasar were more than one light-day across, we then would see the change from the front of the region more than a day sooner than the change from the back of the region, and the sudden change in the quasar's output would be smoothed into a slow, small change that took longer than a day from our viewpoint.

region more than a day earlier than the change from the farther part. The finite length of time that light takes to cross the region would have to be added, in different proportions, to the time that light from the entire quasar takes to reach us. The result would be that a rapid variation in output at the quasar would be smoothed out into a slow variation as we observe it. Thus we can conclude that at least some quasars have sizes no larger than the distance light travels in a day. This distance of one "light day," 1/365 of a light year, equals 25 billion kilometers, or about twice the size of our solar system, and far less than the average distance between two neighboring stars in our galaxy. Thus quasars apparently produce more energy than a galaxy in a region about the size of our solar system!

When astronomers measured the infrared radiation from quasars, they found that quasars, like Seyfert galaxies, are producing huge amounts of energy in infrared and microwave photons.[3] The quasar 3C 273, for example, emits ten times more energy as infrared photons than it does in visible-light or radio photons. Thus still *more* energy from quasars had to be explained by astrophysical theories, and theorists went to work.

---

[3]Because of the difficulties in building good infrared detectors that astronomers had during the early 1960's, infrared astronomy did not really get started until the late 1960's and early 1970's, making it one of the youngest branches of astronomy.

At first, astronomers agreed that quasars look like an extreme case of Seyfert galaxies, with their outer parts missing. The nuclei of Seyfert galaxies produce huge amounts of energy in infrared, visible-light, microwave, and radio photons from a region less than ten parsecs across. Quasars produce something like a hundred times more energy than a Seyfert galaxy in the same kinds of photons, from regions not ten parsecs but one one-hundredth or one one-thousandth of a parsec in diameter. Even though we really don't know how Seyfert-galaxy nuclei produce their energy, they clearly belong to recognizable galaxies. Perhaps quasars, which are much brighter, much more distant, and much more compact than the Seyfert galaxies we can see, are themselves the tremendously active centers of galaxies or of protogalaxies. Thus quasars would be some sort of galaxies that had explosive outbursts years ago before settling down into more normal sorts of galactic life, and only now do we see the photons from these outbursts.

Until about 1968, most astronomers believed this hypothesis, that quasars are mysterious sources of tremendous amounts of energy from within a small region of space, that the photons from quasars have traveled the farthest, and are thus the oldest, of all the photons we see from various individual energy sources in the universe (save only the microwave background radiation). In this conception of quasars, the shifts in the frequencies of intense photon emission or absorption in quasars' spectra arise from the Doppler effect. This Doppler effect in turn shows that the quasars must be moving away from us at enormous velocities, which apparently could arise only from the overall expansion of the universe. Hence quasars must be billions of parsecs away, and must have much greater true brightnesses than galaxies do to be so easily observable on Earth.

Are Quasars the Farthest Objects?

The interpretation of our observations of quasars as proof that they are extremely distant objects may not be correct. Astronomers may have overestimated the distances to quasars by a factor of a hundred, which would mean that they judged quasars' true brightnesses ten thousand times too high. The possibilities of making errors of this size give astronomers a feeling of power unmatched by other sciences. But if quasars do turn out to be a hundred times closer to us than we thought, they provoke new questions of physics that cannot be answered with our present knowledge. (Of course, our present knowledge cannot resolve all our questions about quasars even if they are as distant as their Doppler shifts would indicate, but the idea of placing quasars a hundred times closer to us raises, as we shall see, fundamental problems in physics as opposed to *ordinary* problems.)

An American astronomer named Halton Arp has provided most of the observational evidence that quasars are not billions of parsecs away. In

1961, Arp started to make a catalogue of peculiar-looking galaxies, the sort that are likely to be sources of large numbers of radio and infrared photons. Such peculiar galaxies included Seyfert galaxies, exploding galaxies, and other unusual galaxies. Arp found that these peculiar-looking galaxies seemed to have more quasars located close to them on the sky than would be expected by chance. For example, the galaxy NGC 7413 (Figure 5–10) has a peculiar bulge on one side. Near this bulge, we can see the quasar 3C 455 (marked by an arrow). The quasar's position on the sky is only 23 seconds of arc from the position of the center of the galaxy. This angular distance is one eightieth of the angu-

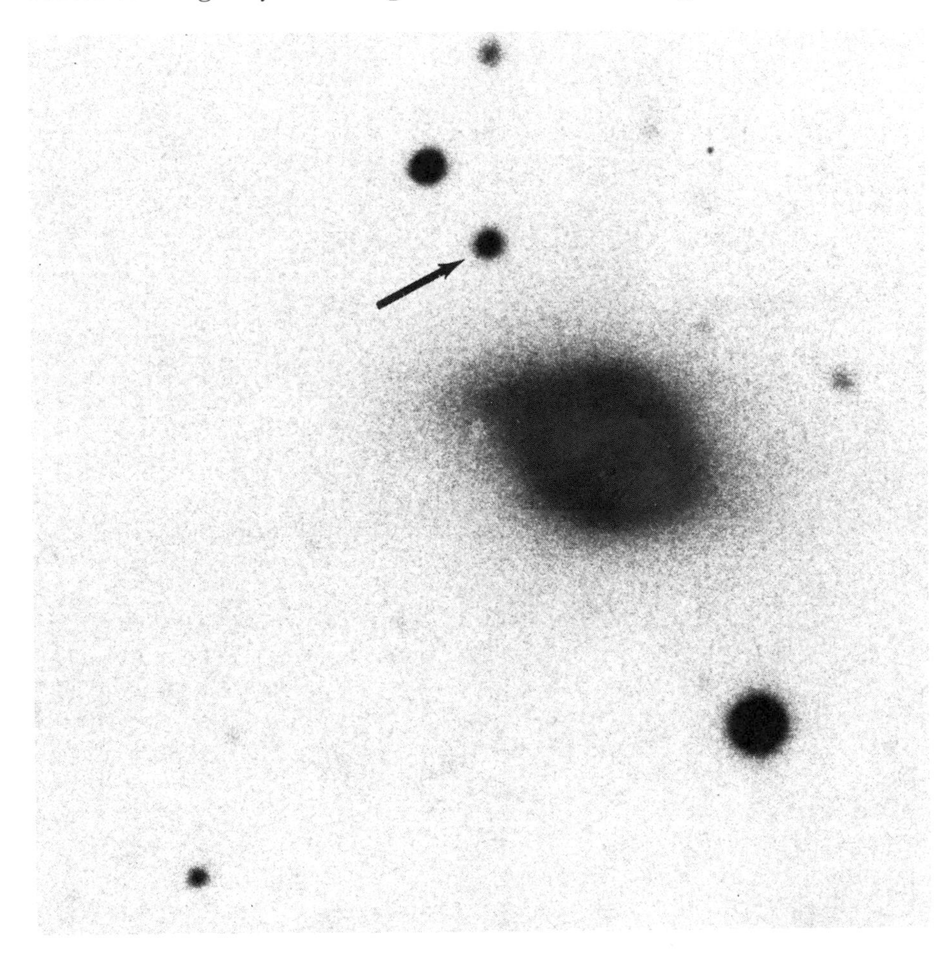

Figure 5–10. The galaxy NGC 7413 shows the quasar 3C 455 close to it. This is a negative print, rather than a positive one, so the galaxy and quasar appear as dark areas. Astronomers prefer such negative prints because they usually show a bit more detail than positive prints do.

What we see in a telescope:

may actually be:

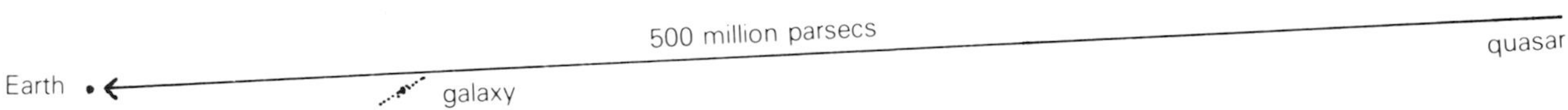

Figure 5–11. We might see the quasar close to the galaxy on the sky, while in reality the quasar lies much farther away from us than the galaxy does. In this case, the apparent closeness of the quasar to the galaxy would arise purely by chance. But there seem to be too many quasars close to radio galaxies and peculiar galaxies on the sky for us to explain all of these associations as the result of chance.

lar diameter of the full moon as seen from Earth. If the relative nearness of the galaxy and the quasar as we see them is not the accidental result of a chance alignment (Figure 5–11), then the quasar and the galaxy must be relatively close to one another in space. Otherwise we would be seeing just a coincidental lining up of a galaxy a few tens of millions of parsecs away with a quasar a few billion parsecs away, a hundred times farther from us than the galaxy. Arp's arguments hinge on the statistics of whether the closeness of the position of some quasars and some peculiar galaxies on the sky can be accidental or not. Statistics rarely say the same thing to everyone, and this problem cannot be resolved easily one way or the other. Besides, if statistics should say, for example, that there is only one chance in a hundred that the closeness in position is accidental, does this mean we must conclude that peculiar galaxies and quasars are truly close to one another in space? Astronomers have conducted lively debates on the question of where quasars really are. This debate continues today, and may not be settled for a long time, so we should look at both possibilities.

If quasars have the distances that their Doppler shifts indicate, they are the most powerful sources of photon energy known, yet this energy arises from a region the size of the solar system. But if quasars actually are close in space to peculiar galaxies, what causes the large shifts of the frequencies and wavelengths in the photons they emit? If these shifts arise from the Doppler effect, this means that the quasars all are moving away from us at a large fraction of the speed of light. Since the quasars' velocities are unrelated to Hubble's law, why aren't any of them moving toward us? And how does the quasar hold together while something accelerates it to these enormous velocities? Nothing we know about could remain so compact while being accelerated to almost the speed of light; this would be like trying to accelerate a jet plane to a speed of one million kilometers per hour with a hydrogen bomb as the motor.

If quasars are about a hundred times closer than we first assumed, then their true brightnesses are ten thousand (one hundred squared) times less than we thought. This conclusion follows from the fact that the apparent brightness of an object decreases as the square of its dis-

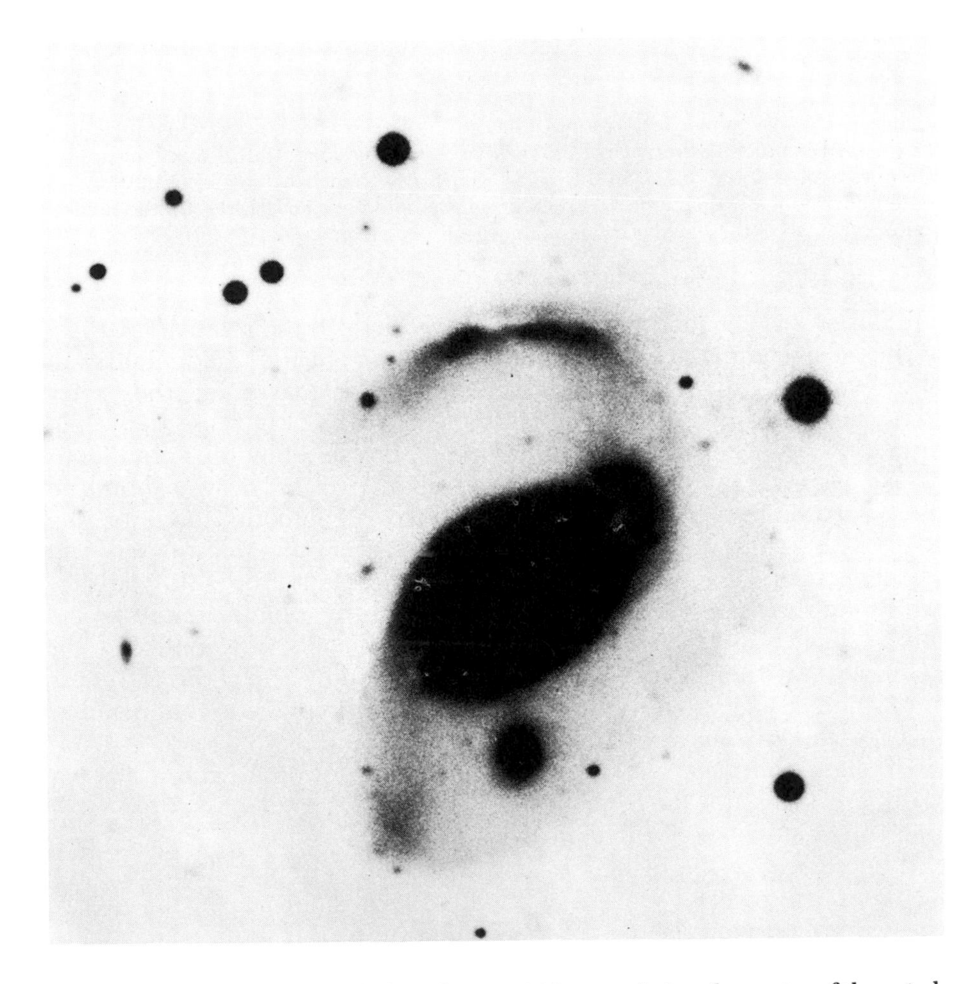

Figure 5–12. The quasar called Markarian 205 lies just below the center of the spiral galaxy NGC 4319, as shown on this negative print. If both the quasar and the galaxy have the same distance from us, we can see that the quasar's brightness falls somewhat below the brightness of an entire galaxy. This holds true for the ratio of absolute brightnesses of the quasar and the galaxy, as well as for their ratio of apparent brightnesses, so long as the galaxy and the quasar are the same distance from us.

tance from us. If quasars are "only" tens of millions of parsecs away, their true brightness would "only" equal the true brightness of a moderately large galaxy (Figure 5–12). We still would have no good explanation of how such an object could produce the energy corresponding to this much true brightness. However, the problem of explaining the quasars' energy output would not loom as large as it does if the quasars are the objects with the greatest true brightness. The big problem that arises if quasars are close to galaxies is to explain their enormous red shifts. As far as we know, only the universal expansion can produce the tremendous velocities, close to the speed of light, of one object relative to another object that lies billions of parsecs away. If the quasars' red shifts come from the Doppler effect, and quasars are less than a million parsecs from galaxies with far smaller red shifts, we can't understand how quasars could have such enormous velocities, and always directed away from us. But if the red shifts do not arise from the Doppler effect, what could produce them? Two suggestions are: first, a gravitational red shift, and second, an unknown cause of the red shift.

To take the first idea first: Any photon that has to fight against gravity will lose some of its energy in doing so. Gravity affects not only particles with mass, such as the sun and the Earth, but also particles with no mass at all, such as photons. Newton's law of gravity, that the gravitational force varies as the product of the masses divided by the square of the distance between them, is valid only for particles that have mass. Photons, which have no mass, still feel the effects of gravity. Photons that pass close by a massive object such as the sun are bent from their usual straight-line paths into curved paths (Figure 5–13). Photons that try to leave the surface of the sun feel the sun's force of gravity too. They must fight their way from a region of stronger gravitational pull (the sun's surface) to regions of lesser gravitational pull (far from the sun). As the photons from the sun move outward into space, they always travel at the same *speed*, but they lose a tiny fraction of their energy. Every photon escaping from the sun's surface loses the same fraction of its original energy. The photon's loss of energy corresponds to a decrease in its frequency and an increase in its wavelength, by the same fractional amount for each photon. This *gravitational* red shift arises from the loss of energy as photons escape from a gravitational field, and it mimics the effect of a Doppler red shift that arises from the motion of a photon source away from the observer. Both the Doppler red shift and the gravitational red shift change all the photons' energies by the same fractional amount.

For photons escaping from the sun, the gravitational red shift amounts to only one half of a hundredth of a percent. This miniscule decrease in the photons' energies can barely be measured, and the tiny amount of the fractional change shows that the sun's gravitational field is

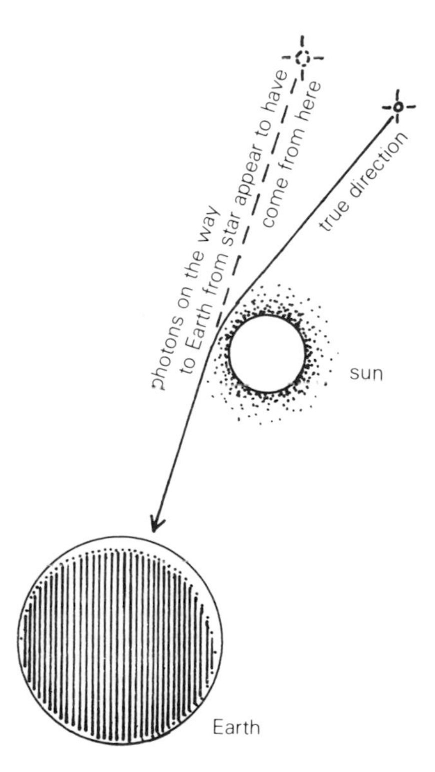

Figure 5–13. Photons that pass close by the sun are bent toward the sun by the sun's force of gravity.

extremely weak in universal terms. The universal measure of the effect of the gravitational field of force at a star's surface is the star's mass divided by its radius. The larger this ratio is, the larger is the fraction of its energy that a photon loses in escaping from the gravitational field of force. On this basis, the gravitational force at the sun's surface doesn't amount to much, and the gravitational field of the Earth is weaker still, but these gravitational forces do produce a reduction in the energies of photons that escape outwards. Some stars have more intense gravitational forces at their surfaces than the sun does, because their ratio of mass to radius far exceeds the ratio for the sun, but even these stars reduce the energy of escaping photons by only about one percent. If we were to imagine that quasars produce gravitational red shifts that reduce the energy per photon by ten percent or more, then the quasars must have ratios of mass to radius much greater than that for any star. In fact, this ratio would have to be a hundred thousand times greater than the sun's ratio of mass to radius for gravitational red shifts to explain quasars' spectra.

Suppose that a quasar had a radius equal to the sun's. Then to produce a gravitational red shift of 50% (that is, to increase all the photon wavelengths to 1½ times their original values, thus to decrease all the photon energies to ⅔ of their original values), the quasar would have to have a mass 100,000 times the sun's mass. Remember that even if quasars are near galaxies and thus only a few tens of millions of parsecs away, quasars still produce a billion times more energy each second than the sun does. We hardly can believe that a single object with a mass of "only" a hundred thousand solar masses would produce a *billion* times the energy output of the sun. But if a quasar had a mass a billion times the sun's mass, it would produce a gravitational red shift of 50% in the photon wavelengths if its radius were ten thousand times the sun's radius. This radius is slightly less than the maximum radius allowable on the basis of the quasars' light variations. Could such objects with a billion solar masses somehow form close to peculiar galaxies, which themselves have masses perhaps one hundred billion times the sun's mass? Present theories show that any such object could not maintain itself, but instead would collapse under its own gravitational force in a time of about one day to form one of the "black holes" we will describe in Chapter 12. We cannot completely rule out the possibility of gravitational red shifts in quasars, but we can note that despite the temptation to have *some* theory to believe in, astrophysicists do not consider this a likely explanation. Gravitational red shifts themselves deserve closer attention, which we shall postpone until the final chapter of this book.

Now to consider the category of "other theories." Some astrophysicists have concluded that quasars' red shifts can be explained neither as Doppler shifts nor as gravitational red shifts. The Doppler-shift explanation either puts the quasars so far away that their energy outputs are beyond understanding (especially in view of their relatively small sizes), or else it demands an unavailable method by which quasars could be accelerated away from us at nearly the speed of light in some sort of galactic explosions. The gravitational-red-shift explanation calls for an extremely massive object compressed into a relatively small volume. Calculations show that such a massive object almost certainly would collapse to almost no size at all and stop emitting photons (see the black holes of Chapter 12).

As a result of these theoretical difficulties, some of the bolder (more foolhardy?) theoreticians have decided that quasars must operate through laws of physics that somehow differ from the laws that we have found to be an accurate summary of our observations on and near the Earth. Until now the laws of physics found by experiments within the solar system seem to have a universal character. Other stars, and indeed other galaxies and even galaxy clusters, produce photons with energy spectra that can be understood easily as arising from familiar atoms and ions, but affected by velocity-caused Doppler shifts. Galaxies orbit around each other in ways that the usual laws of gravity seem to explain. The rules of physics deduced on Earth appear to work throughout the universe, so far. Perhaps, though, we might have overlooked something because of our limited experience. Perhaps quasars provide an exception to the "laws" of nature that we now use to summarize what we have observed, and perhaps quasars eventually will reveal how the laws of physics should be modified. But before we hurry to invent a new "law" of physics to explain quasars, we should recall that the old laws of physics have served us well. They owe their usefulness to their ability to summarize thousands of observations in a single statement; thus before we add on more statements, we should ask ourselves how many observations convince us that the new law is necessary. Eventually we may conclude that to understand quasars, we must indeed determine new ways of looking at the nature of physical reality. Einstein, for example, convinced his contemporaries that instead of the total mass and the total energy each remaining a constant in the universe, the total of mass plus energy (since the mass carries a certain energy of mass, given by $E = mc^2$) represents the constant quantity. Until we have to make such a large adjustment in our thinking, we can say that observations of quasars cannot be explained easily either by assigning the quasars distances of billions of parsecs, or by concluding that quasars are close to peculiar galaxies and therefore are only tens of millions of parsecs away. This dilemma excites astronomers, because they recognize that the most important questions demand a new context of physical "reality" in which they can consider the question. The answer often follows naturally once we know the right way to ask the question.

Summary

Some galaxies emit far greater amounts of radio photons than ordinary, "normal" galaxies do. These radio photons do not come from stars in the galaxies, rather they come from violent events in the vicinity of the radio galaxies. In many cases, the radio photons are produced in two regions far outside the visible galaxy, located nearly symmetrically on either side of the galaxy's center. Other radio galaxies show more centralized outbursts, such as the jet in the middle of the elliptical galaxy M 87, or the apparent explosion inside the irregular galaxy M 82. Radio galaxies also produce huge numbers of infrared photons, a characteristic that they share with Seyfert galaxies. Seyfert galaxies show intense emission of visible-light and infrared photons from a small central region, which astronomers believe must be the site of some kind of violent motions.

Quasars appear as points of light, rather than being fuzzy like galaxies, but they also produce vast quantities of radio and infrared photon energy. The visible-light spectra of quasars show the largest Doppler shifts yet observed. If these Doppler red shifts arise from the overall expansion of the universe, then quasars are the most distant objects known and the most powerful sources of photon energy. However, many quasars seem to lie close to peculiar-looking galaxies. If this is not a projection effect, and if the quasars are truly close to the galaxies in space, then quasars are about a hundred times nearer to us than we thought, and their enormous Doppler shifts remain a mystery yet unresolved by astrophysics.

Questions

1. What is a radio galaxy? Why don't most galaxies produce large amounts of radio photons?
2. From where does most of the radio-photon emission from radio galaxies arise? How might radio photons be produced in these regions?
3. What is a "Seyfert galaxy"? How are they similar to radio galaxies? How are they different?
4. What is a quasar? How can we be sure that a quasar is not a star?
5. What is unusual about the quasars' spectra of visible-light photons?
6. If a quasar varies noticeably in brightness during a time of one month, what upper limit can we place on the size of the light-emitting region of the quasar, as measured in light years? How large is this upper limit, measured in parsecs?

7. Why do some astronomers believe that quasars are not the most distant objects yet observed? Why do other astronomers believe that quasars *are* the most distant objects yet observed? Which interpretation of quasars' distances best explains our observations of quasars' visible-light spectra?

8. If quasars are not the farthest objects yet seen, how far away are they? How do the absolute brightnesses of quasars then compare with the absolute brightnesses of galaxies?

9. The quasars 3C 147 and 3C 196 show red shifts in their visible-light spectra, presumably arising from the Doppler effect, of 0.55 and 0.87, respectively. That is, all the wavelengths in these quasars' spectra have been increased by 55% and by 87%, respectively, from the wavelengths we would see if the two quasars were at rest with respect to ourselves. How have the frequencies of the photons from the quasars been changed by the Doppler effect? How have the photon energies been changed? If quasars obey Hubble's law as galaxies do, which quasar is farther away? By about how much?

10. How fast are the two quasars described in Question 9 moving with respect to ourselves, if we use the approximate formula $f'/f = 1 - v/c$? In what direction are they moving? What velocities do the two quasars have, if we compute v using the accurate formula $\frac{f'}{f} = \sqrt{\frac{1 - (v/c)}{1 + (v/c)}}$? (See footnote on Page 22.)

11. The largest radio galaxy known is 3C 236, which shows two main components of radio emission separated by five million parsecs. If this galaxy's radio emission arises from particles that were shot out from the galaxy between them at almost the speed of light, what is the minimum time that must have elapsed since the particles were ejected?

Further Reading

G. B. Field, H. Arp, and J. Bahcall, *The Redshift Controversy* (W. A. Benjamin, Inc., Reading, Mass., 1973).

F. D. Kahn and H. P. Palmer, *Quasars* (Harvard University Press, Cambridge, Mass., 1967).

R. J. Weymann, "Seyfert Galaxies," *New Frontiers in Astronomy* (W. H. Freeman and Co., San Francisco, 1975).

F. Hoyle and G. Burbidge, "The Problem of the Quasi-Stellar Objects," *Frontiers in Astronomy* (W. H. Freeman and Co., San Francisco, 1970).

G. Burbidge and M. Burbidge, *Quasi-Stellar Objects* (W. H. Freeman and Co., San Francisco, 1967).

R. Strom, G. Miley, and J. Oort, "Giant Radio Galaxies," *Scientific American* (August, 1975).

Chapter 6

The Milky Way and the Local Group of Galaxies

"I never was attached to that great sect,
Whose doctrine is, that each one should select
Out of the crowd a mistress or a friend,
And all the rest, though fair and wise, commend
To cold oblivion, though it is in the code
Of modern morals, and the beaten road
Which those poor slaves with weary footsteps tread,
Who travel to their home among the dead
By the broad highway of the world, and so
With one chained friend, perhaps a jealous foe,
The dreariest and the longest journey go."

P. B. Shelley

The galaxies we know best are the ones closest to us. In particular, we spend our lives inside the Milky Way galaxy, which contains all the stars visible to our unaided eyes. What we know about stars comes mostly from studying the relatively nearby stars in the Milky Way. However, our location inside the Milky Way galaxy hinders us from getting a clear look at the structure of our galaxy. For this reason, nearby galaxies can reveal a great deal about how the stars in galaxies orbit around each other, as well as how the different types of galaxies vary in size and composition.

The Local Group

The Milky Way belongs to a small cluster of galaxies called, for obvious reasons, the Local Group. The total known membership of the Local Group includes seventeen galaxies, which together form a miniature galaxy cluster whose numbers fall far below the hundreds or thousands of galaxies in a large cluster. Some astronomers believe that the Local Group may be a subgrouping, part of the Virgo cluster that contains thousands of galaxies. The center of the Virgo cluster lies in the direction of the constellation Virgo, at a distance from us of twenty million parsecs. Since the cluster seems to be about twenty million parsecs in radius, the Local Group indeed may form an outlying clump in the Virgo cluster. Astronomers still are working overtime to determine whether galaxy clusters are divided into subclusters or grouped into superclusters, but so far they have not resolved these important questions. Figure 6–1 shows the unusually rich variety of galaxy types in the Hercules cluster of galaxies, about 200 Mpc away.

If we imagine the entire universe to be a giant bowl of chicken soup, the Virgo cluster of galaxies would be a grain of rice in the soup, and the

Figure 6–1. The central regions of the Hercules cluster of galaxies show a large number of different types of galaxies. The cluster has more than a thousand member galaxies.

Local Group would be a small speck near the edge of the rice grain. Part of this speck would be the Milky Way galaxy, and one of the atoms (10^{-8}-centimeter radius) could represent our solar system. This model does not fit perfectly, but it serves to remind us of the scale of sizes in the universe.

The seventeen galaxies in the Local Group fall into the three basic categories of spirals, ellipticals, and irregulars. Ten are ellipticals, all of which are dwarf galaxies far smaller than the Milky Way. Four are irregular galaxies, of which two are satellites of the Milky Way galaxy. Three are spirals, namely the Milky Way, the Andromeda galaxy (Page 7), and the smaller spiral M 33 (Figure 6–2). Compared to the average distribution of galaxy types, the Local Group is deficient in large ellipticals, since it has none with even ten percent of the mass of the Milky Way. In the Local Group, we find more dwarf ellipticals and irregulars than we usually see in large clusters, but this can be understood from the

Figure 6–2. The Sc galaxy M 33, in the constellation Triangulum, is one of the three spiral galaxies in the Local Group.

fact that dwarf galaxies are harder to see, compared to large spirals and ellipticals, so at larger distances many of them would be invisible to us.

In terms of mass, the Local Group consists mainly of the Milky Way and the Andromeda galaxy, plus their satellites, the smaller spiral M 33 and six other galaxies of much smaller mass. The dwarf elliptical galaxies have an undistinguished look, as shown by the galaxy NGC 147 (Figure 6–3). Such galaxies each have a mass ten times to a hundred thousand times less than the Milky Way or the Andromeda galaxy. They contain from a million to ten billion stars in a spheroidal distribution, which is densest at the center and thins out toward the edges. As in all elliptical

Figure 6–3. The dwarf elliptical galaxy NGC 147 in Cassiopea is representative of this class of galaxies.

galaxies, almost no gas or dust can be seen among the stars. Irregular galaxies such as the Magellanic Clouds (Figure 4–22) or NGC 6822 show huge clouds of gas and dust where stars are still condensing. In irregular galaxies, a far larger proportion of stars are young stars than in spiral and elliptical galaxies.

The two irregular satellites of our own galaxy, called the Magellanic Clouds, have contributed a great deal to our understanding of stars' intrinsic behavior. These two galaxies are the Milky Way's nearest neighbors, located only sixty thousand parsecs away, just about ten times closer than the Andromeda galaxy. Although the Magellanic Clouds are farther away from us than any of the stars in our galaxy, their large sizes give them a larger apparent diameter on the sky than that of the full moon, so from southern latitudes where the Clouds can be seen, they appear as vague "fluffy" objects that we might take for atmospheric clouds at first glance. What makes the Magellanic Clouds helpful to astronomers is the fact that they contain many kinds of stars all about the same distance away from us. Astronomers always have had difficulties finding the distances to stars and galaxies (Chapter 4). The Magellanic Clouds offer them a chance to observe thousands of individual stars close enough to be seen separately (with telescopes) yet far enough away that the stars form a coherent, localized grouping. Even if we cannot determine the precise distance to the Clouds, we can be sure that if one star in them *appears*, say, twice as bright as another, the first star really *is* twice as bright. That is, the ratio of the apparent brightnesses of the stars in one of the Magellanic Clouds accurately reflects the ratio of the stars' true brightnesses.

The most important discovery to come from the Magellanic Clouds has been the excitement over Cepheid variable stars. These stars vary regularly in brightness over periods from several hours to several days. In the early 1900's, Henrietta Leavitt found that the Cepheid variable stars in the Small Magellanic Cloud showed a correlation between their apparent brightnesses and the amount of time they took to go through one cycle of brightness change. Figure 6–4 shows this correlation: The brighter stars take longer to go through an up-and-down cycle in apparent brightness. Thus Cepheid variable stars with the same period of brightness variation must have the same *true* brightness. Leavitt realized that this correlation could be used to determine the distances to Cepheid variables in other galaxies. Suppose we observe a Cepheid variable star in the Andromeda galaxy with the same period of brightness variation as a Cepheid variable in the Small Magellanic Cloud, but the apparent brightness of the Andromeda star is eighty times less than the brightness of the star in the Small Magellanic Cloud. Since the apparent brightness of a star decreases as the square of the distance from us, we can conclude that the Andromeda galaxy is about nine

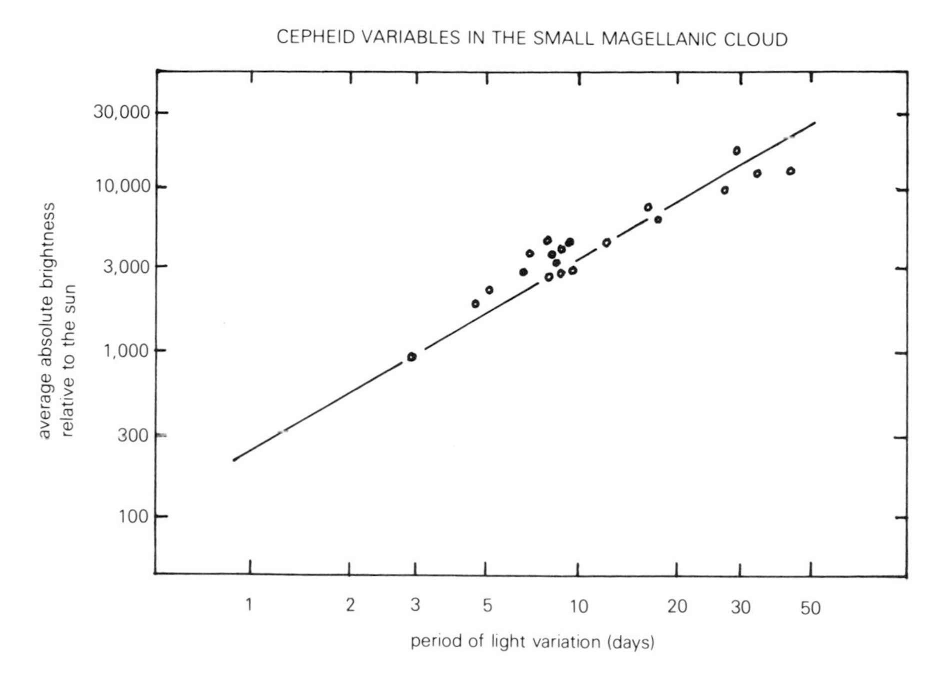

Figure 6–4. This graph shows the correlation between the absolute brightness of Cepheid variable stars, and the length of their cyclical periods of change in brightness, for the Cepheid variables observed in the Small Magellanic Cloud.

times farther away than the Small Magellanic Cloud. Leavitt's discovery, which enables us to relate any Cepheid variable star to a similar star in the Magellanic Clouds, has given us the means to determine the distances to the other galaxies in the Local Group (Table 6–1). In particular, we know the distance to the Andromeda galaxy, our typical giant spiral, from observations of the Cepheid variables that it contains. A great regret among astronomers is that no giant elliptical galaxy lies close enough to us to show a visible Cepheid variable star. The nearest giant elliptical, M 87 (Figure 5–6), lies twenty million parsecs away. At this distance none of the brightest Cepheid variables, large though their true brightness is, can be seen as individual stars, so astronomers have to rely on less certain methods (Page 116) to find the distance to M 87. Notice that these other methods, such as measuring the diameter of the largest H II regions in a galaxy, often start by using the Andromeda galaxy as a key standard galaxy. Thus the Cepheid-variable method for finding distances that Leavitt discovered forms the keystone in estimating the distances to all other galaxies.

The Milky Way galaxy itself contains a hundred billion stars, all of which remain in the galaxy because of the gravitational forces they exert on each other. As a result of these mutually attractive forces, stars orbit in circular or elliptical paths around the center of our galaxy (Chapter 4). Most of these orbits follow the same direction and lie in nearly the

Table 6–1
Members of the Local Group of Galaxies

Name	Galaxy type	Distance from Milky Way (in parsecs)	Diameter (in parsecs)	Mass (in solar masses)
Milky Way	Sbc		30,000	2×10^{11}
Large Magellanic Cloud	Irr	60,000	8,000	1×10^{10}
Small Magellanic Cloud	Irr	60,000	2,500	1×10^{9}
Andromeda Galaxy (M 31)	Sb	600,000	30,000	2×10^{11}
NGC 221	E2	600,000	2,500	1×10^{9}
NGC 205	E5	600,000	5,000	3×10^{9}
M 33	Sc	600,000	15,000	1×10^{10}
Ursa Minor System	E4	70,000	1,000	1×10^{5}
Sculptor System	E3	100,000	2,000	3×10^{6}
Draco System	E2	100,000	1,300	1×10^{5}
Leo II System	E0	220,000	1,500	1×10^{6}
Fornax System	E3	250,000	5,000	2×10^{7}
Leo I System	E4	300,000	1,500	3×10^{6}
NGC 6822	Irr	450,000	3,000	4×10^{8}
NGC 147	E6	550,000	3,000	1×10^{9}
NGC 185	E2	550,000	2,500	1×10^{9}
IC 1613	Irr	600,000	4,000	1×10^{8}

same plane, so our entire spiral galaxy has a coherent sense of rotation: The Milky Way turns in a definite direction around its rotation axis. However, the galaxy does not rotate as one solid body would. Instead, the stars close to the center of the galaxy complete an orbit in less time than the stars farther out. Our sun completes a nearly circular orbit around the center of the galaxy every 250 million years. The radius of the sun's orbit, ten thousand parsecs, places it farther out from the galactic center than most of the stars in the galaxy. A star closer to the center, say with an orbit five thousand parsecs in radius, takes only 125 million years to circle the galactic center. Stars with orbits fifteen thousand parsecs in radius take as long as 400 million years to complete an orbital cycle. In each of these orbits, the star moves at the speed necessary to balance the gravitational pull from the regions of the galaxy closer to the center than the star. The gravitational force from regions farther from the center than the star will average out to zero. These forces pull the stars in various directions and thus counteract each other (Figure 6–5), whereas the gravitational force from the inner regions always attracts the star toward the galactic center.

We have seen that most of the stars in the Milky Way lie in the same plane and circle the center in the same direction. Aside from this dominant structural pattern, we should notice two important facts in the

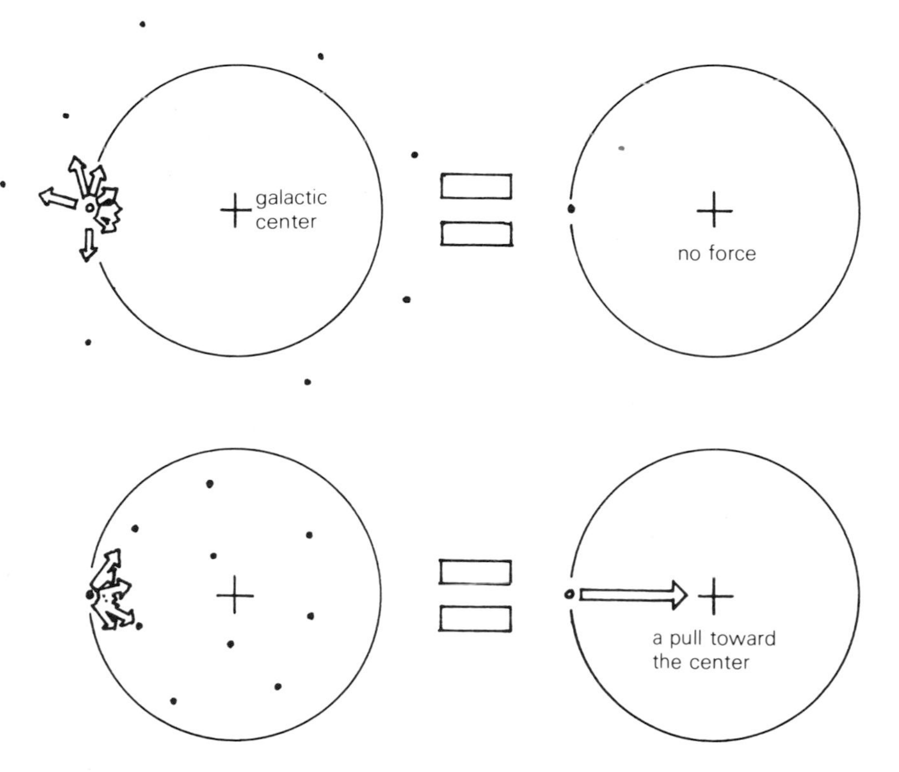

Figure 6–5. The gravitational forces on a star from the regions farther out from the galactic center than the star is will average out to zero, because these regions pull in opposite directions. The gravitational attraction from stars closer to the center will tend to add, thereby producing a net gravitational pull toward the center of the galaxy.

distribution of stars in our galaxy. First, the young stars and the interstellar gas appear predominantly in the spiral arms of the galaxy. Second, many stars appear in localized star clusters.

Spiral Structure in Galaxies

Photographs of spiral galaxies such as Figure 6–6 show the striking spiral arms that give such galaxies their names. Today astronomers are just beginning to understand how spiral-arm patterns could arise within a rotating disk of stars. Their hypotheses are discussed later in this chapter, but for now we must emphasize the key point that *spiral arms look more spectacular than they are*. From photographs of a galaxy such as the Sc spiral M 33 (Figure 6–2), we might conclude that the spiral arms contain most of the stars and most of the mass in the galaxy. In fact, in the galaxy M 33 and in other spiral galaxies the density of stars inside spiral arms does not greatly exceed the density of stars between the arms, but the spiral arms do contain most of the young, bright stars, and most of the interstellar gas. The same is true for the Milky Way galaxy, where we find that the distribution of young, hot, and extremely bright stars outlines a spiral-arm pattern. We also can map out the

Figure 6–6. The Sc galaxy NGC 5364 in Virgo shows prominent spiral arms.

distribution of neutral hydrogen atoms from the 21-centimeter photons they emit (Page 36). When we look along a particular line of sight in the plane of the galaxy (Figure 6–7), we receive large numbers of these photons from hydrogen atoms, but with wavelengths slightly Doppler shifted away from 21 centimeters. We can conclude that the Doppler shifts arise from the motions of the emitting atoms toward us or away from us; that is, from the difference in velocity around the galactic center between ourselves and each particular group of emitting atoms along the line of sight.

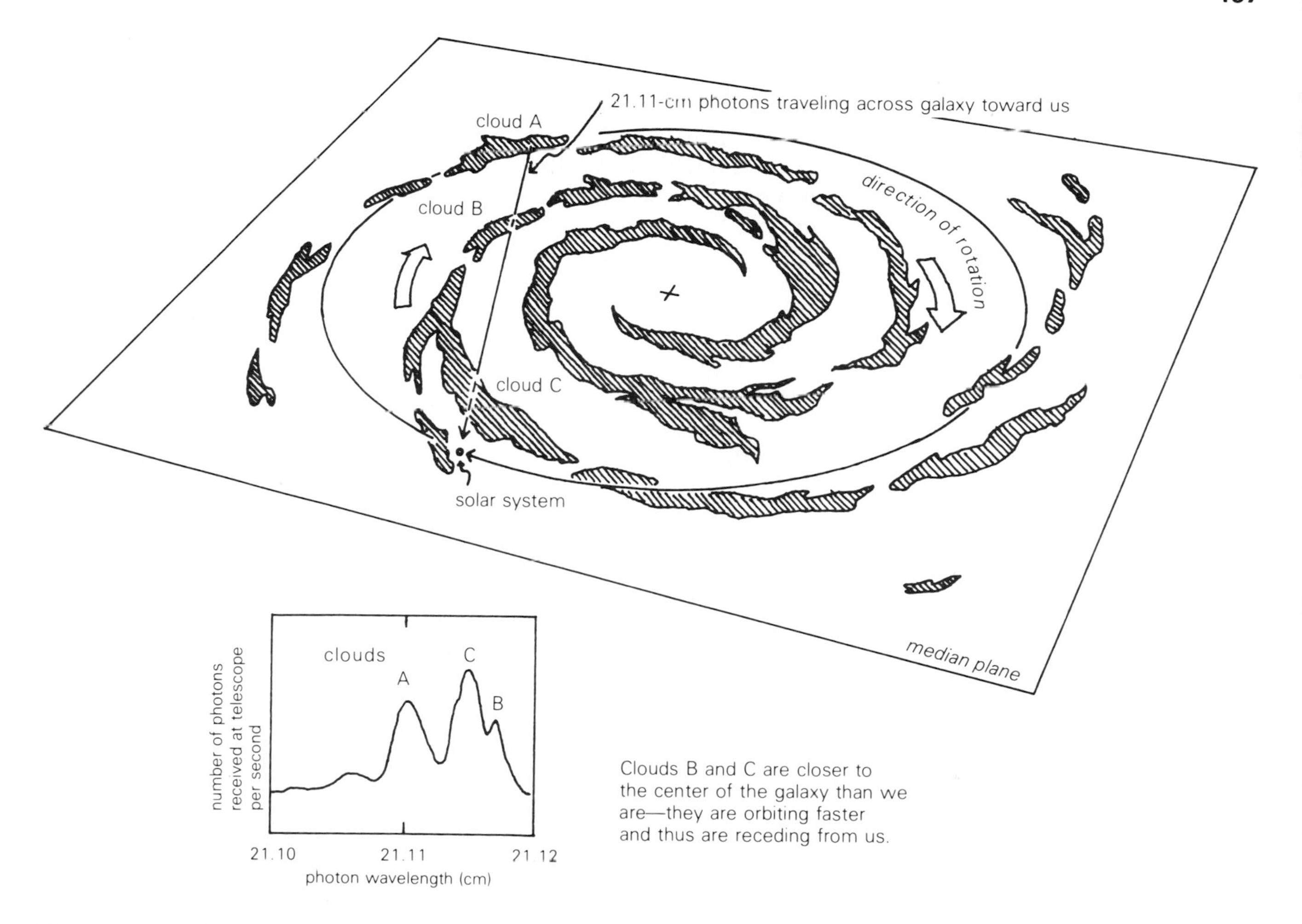

Figure 6–7. Hydrogen atoms emit photons of 21-centimeter wavelength along a line of sight in the plane of our galaxy. These photons will arrive with a Doppler shift in energy if the regions that emit them have a velocity of approach or of recession relative to ourselves.

From the fact that the inner parts of the galaxy rotate faster than the outer parts, astronomers can estimate the location of each group of atoms along a line of sight, if they assume that the atoms all move in circular orbits around the galactic center. The Doppler shift then allows the astronomers to determine how far away the atoms that emit at each frequency are located (that is, on what circular orbit the atoms are moving).

Maps of the density of hydrogen gas at various points in the galaxy made in this way (Figure 6–8) show concentrations of gas in streams that resemble the spiral arms seen in photographs. However, these streams appear to form circles rather than spirals, and they do not correspond well with the locations of young, bright stars that also delineate spiral arms. The distances to these young stars can be estimated by comparing their apparent brightness with the apparent brightness of a similar star that lies closer to us (Page 115), and the young stars are considered a better way to locate a galaxy's spiral arms than the maps of hydrogen gas made with the assumption that the atoms are moving in perfectly circular orbits.[1]

In the Milky Way galaxy, and in other spirals as well, young bright stars and interstellar hydrogen gas delineate the spiral arms, with the young stars being better indicators than the gas. But when we count all the stars that we can see within a few thousand parsecs of the sun and try to determine their distances, we find that many stars appear between the spiral arms. Our sun does seem to lie just inside a spiral arm, but many stars do not. Like the sun, these stars between the arms have ages of billions of years, rather than tens of millions or hundreds of millions of years that characterize "young" bright stars. Astronomers thought about this, made thousands of spectroscopic observations of individual stars, and came up with some fascinating conclusions.

Stellar Populations

On the basis of the fraction of their mass that consists of elements heavier than hydrogen and helium, stars can be grouped into two categories or "populations," which astronomers named Population I and Population II. Population I includes the young, hot stars, and in fact all the stars that formed during the past ten billion years or so (as far as we can tell). Population I stars have a higher fraction (perhaps one or two percent) of the heavy elements (heavier than helium) in their mass than most Population II stars do. Population II includes the most mature, older stars that formed about the same time that the older galaxies did. *Most stars in our own galaxy, and in almost every other large galaxy, are*

---

[1]Unfortunately, we cannot see the young, bright stars in our own galaxy if they are more than one or two thousand parsecs away, on account of the obscuration by interstellar dust (Page 177). This dust lies close to the plane of the Milky Way, just as the young stars do. When we look out of the Milky Way and not along the plane of our galaxy, we can observe galaxies outside our own without the problem of being located inside what we are trying to observe. Thus it is easier to map the spiral arms of another galaxy by observing the 21-centimeter photons emitted by hydrogen atoms than it is for our own galaxy. Similarly, we actually can get a more accurate picture of the distribution of the young, bright stars in another galaxy than we can for our own Milky Way. Maps of the hydrogen gas in nearby spirals such as M 33, M 31, and M 51 show good agreement between the location of the hydrogen gas and of the young, bright stars, both of which trace out the galaxies' spiral arms. See Plates 2 and 3.

Population I stars. The sun is a Population I star, but it is an old and dim member of this younger generation of stars. (Because the Population I stars are the younger stars and the Population II stars are the older stars, astronomers have been accused of naming these categories backwards.) Population I stars, which on the average are younger and brighter than Population II stars, show up much better on photographs like Figure 6–2, especially the spectacularly young and bright stars that formed within the past hundred million years.

Within the local regions of our galaxy, the bright stars Rigel (in Orion's foot), Bellatrix (Orion's dimmer shoulder), the three stars in Orion's belt, and the stars in the Orion Nebula, all are young, hot stars that are between one million and one hundred million years old. These

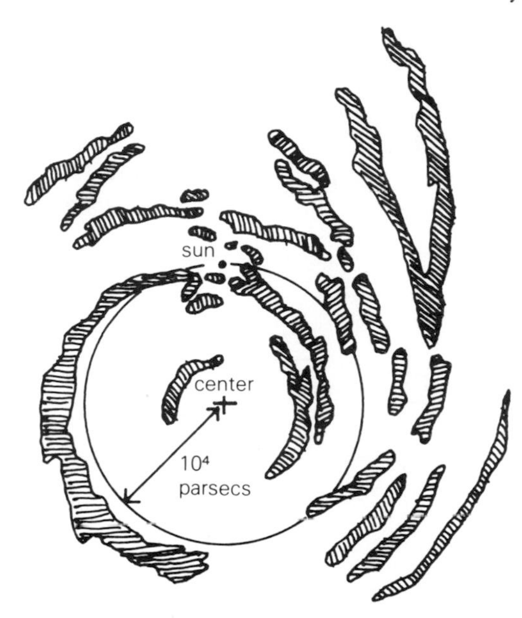

Figure 6–8. If we imagine that all the hydrogen atoms in our galaxy are moving in circular orbits, we can use the amount of the Doppler shift of the 21-centimeter photons that we observe to determine the location of the emitting atoms. Then we can make a map of the density of hydrogen atoms at each location, as shown here. These maps do not show the kind of spiral structure that we would expect (although they do reveal certain concentrations of hydrogen atoms). Thus we may conclude that the model of our galaxy in which all atoms move in perfectly circular orbits is not as accurate as we would like. This method of mapping the distribution of hydrogen atoms by the 21-centimeter photons that they emit does not work for the regions of the Milky Way that are directly toward the galactic center, or directly away from the galactic center, as seen from the solar system. The method fails because these regions do not show a correlation between the atoms' distances from us and their relative velocities toward us or away from us. In addition to this, not all of the Milky Way can be seen from the observatory where this map was made, so the left side of the map is incomplete.

stars have absolute brightnesses thousands of times greater than the sun's; thus even though their distances from us (150 to 500 parsecs) exceed the distances to most of the other stars we can see at night, they rank among the fifty stars with the greatest apparent brightness. Many other such young, bright stars lie farther away from us, so they are not visible among the brightest stars. Such stars outline the spiral arms of other galaxies as well as the Milky Way, making the arms stand out from the background of older, fainter stars. In fact, the average density of stars *inside* a spiral arm is only about double the density of stars *outside* a spiral arm. The young, hot stars, which can produce bright H II regions (such as the Orion Nebula), emphasize the spiral arms at the expense of the rest of the galaxy. Such a spiral arm, as defined by the young, bright stars within it, may be five hundred parsecs or more in diameter, and the spacing between spiral arms is about two thousand parsecs.

The spiral arm in which our sun now lies is called the "Orion arm" after the bright stars in the constellation Orion. The next arm out is the "Perseus arm," named after the star cluster *h* and χ Persei, which lies about two thousand parsecs farther out from the galactic center than the sun does (Figure 6–9).

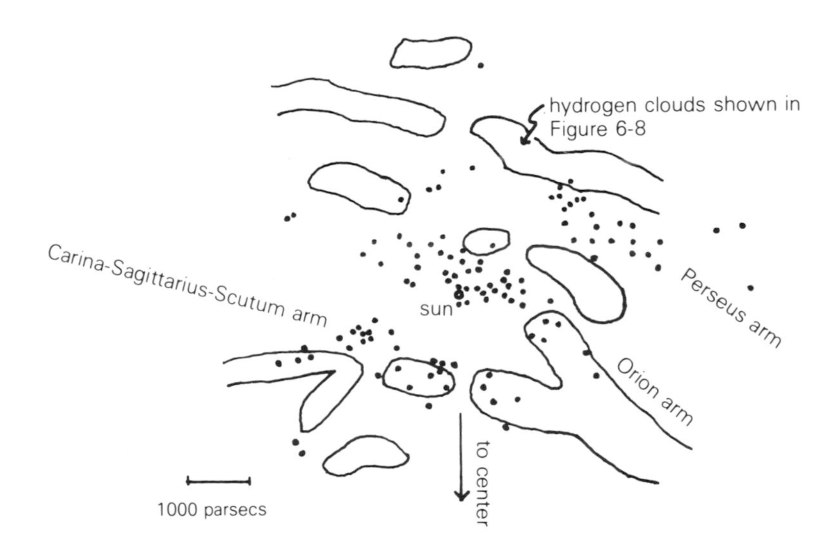

Figure 6–9. We can use our estimates of the distances to young, bright stars to map the spiral arms in the neighborhood of the sun. We find that the sun lies near the inner edge of a spiral arm, called the "Orion arm" because it includes the young stars that form most of the constellation of Orion. The next arm out from the sun in our general vicinity lies about two thousand parsecs away, and because it contains the open cluster *h* and χ Persei, it is called the "Perseus arm."

To sum up our present understanding of the two stellar populations in our galaxy: The older stars are Population II, and the younger are Population I. Astronomers make this classification on the basis of the fraction of the masses of stars made of elements heavier than helium. Population I stars, such as the sun, contain one or two percent of their mass as these heavier elements, whereas in Population II stars the heavy-element fraction is ten or a hundred times *less*. The Population I stars are concentrated heavily toward the galactic disk, where the youngest ones outline the spiral arms. Population II stars also concentrate toward the galactic disk, but not with the extreme degree of concentration of Population I stars. The bulk of the mass in stars in our galaxy resides in the older Population I stars such as our sun.

The distribution of stars of different ages can be determined for the nearest of the galaxies, where the brightest stars can be examined

Figure 6–10. The open cluster *h* and χ Persei consists of Population I stars.

individually. Elliptical galaxies apparently turned almost all of their matter into stars long ago, so that although ellipticals have both Population I and Population II stars, they have *no young Population I stars.* In spiral galaxies, Population II stars seem to have formed as the protogalaxy contracted toward a disk, whereas Population I stars formed (and continue to form) after the galaxy had assumed a disklike structure. Even irregular galaxies, which have the greatest fraction of young, hot Population I stars, contain some Population II stars. At the other extreme, some dwarf elliptical galaxies seem to contain no Population I stars (as far as we can tell) so they may represent cases where star formation

Figure 6–11. The globular cluster M 13 in the constellation Hercules contains about a million stars, all of which formed at about the same time, more than ten billion years ago. These stars all belong to Population II.

began early (more than ten billion years ago), turned all the protostar material into stars, and thus brought star formation to an everlasting halt.

Star Clusters

Surveys of the sky show that stars in our galaxy, and indeed in other galaxies, often group together in clusters. There are two distinct types of clusters: open clusters (also called galactic clusters) and globular clusters (Figures 6–10 and 6–11). Open clusters contain several hundred stars, whereas globular clusters contain hundreds of thousands or even millions of stars. Open clusters lie in the plane of the Milky Way galaxy, while globular clusters form a spherical distribution around the galactic center (Figure 6–12). Open clusters contain primarily Population I stars, whereas globular clusters consist exclusively of Population II stars. Finally, open clusters are loose aggregations of stars that may not persist as definite clusters, but the stars in globular clusters will remain held together by their mutual gravitational attraction.

From these distinctions between the two types of star clusters, astronomers conclude that globular clusters formed as our protogalaxy contracted into its present form many billions of years ago. This is confirmed by other measurements of cluster ages (Page 222). Open clusters, how-

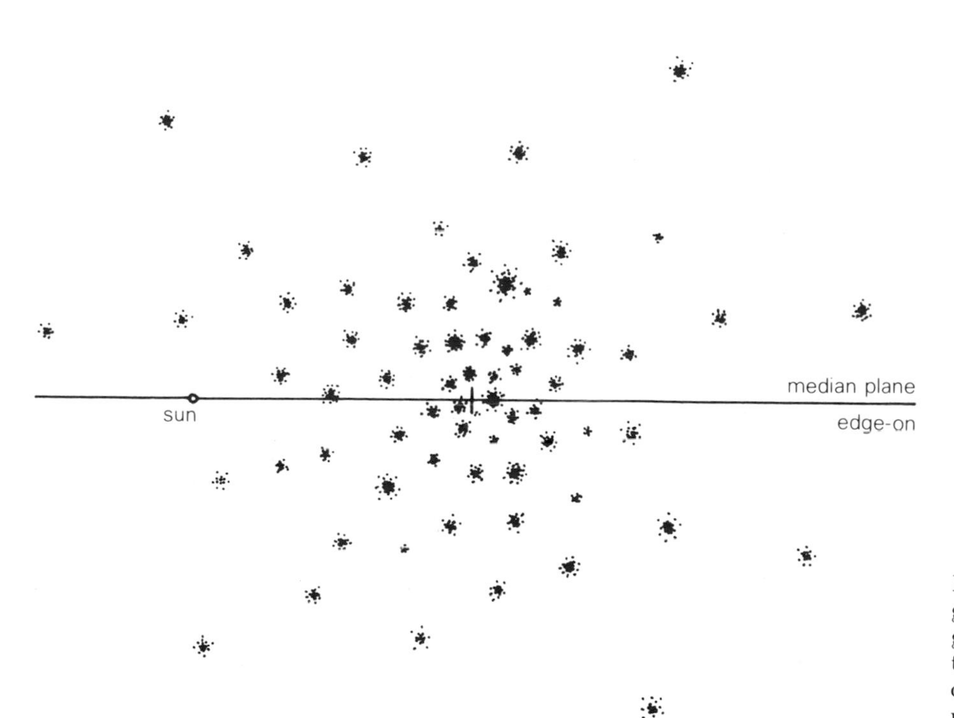

Figure 6–12. The distribution of globular clusters in the Milky Way galaxy shows a strong concentration toward the galactic center and a lack of concentration toward the median plane of the galaxy.

ever, must be groups of younger stars that formed less than a few billion years ago. The reason for this conclusion lies in the locations and the motions of these clusters, as well as in the age of the stars in them. Globular clusters do not appear concentrated close to the galaxy's central plane. As Figure 6–12 shows, more globular clusters appear close to the galactic center than far out toward the galactic boundaries, but these clusters may lie in any direction above or below the plane of our galaxy. This distribution of globular clusters in space provided the first clue to the location of the center of our galaxy, because a photograph taken in the direction of the center reveals a large number of globular clusters (Figure 6–13). Only the obscuration of our view by interstellar dust kept astronomers from recognizing the true location of the galactic center until about 1920.

Figure 6–13. A photograph taken in the direction of the galactic center (in the constellation Sagittarius) includes more than thirty globular clusters. Most of the brighter "starlike" images in the photograph in fact are globular clusters.

The motions of globular clusters can be estimated by measuring the Doppler shifts of the light from their stars. From this effect we can determine the clusters' radial velocities (toward us or away from us) relative to ourselves. We can use this information, together with our knowledge of the sun's orbit around the galaxy, to estimate the clusters' distances from us and to determine an approximate orbit for each globular cluster. We find that most of these clusters are not following circular orbits around the galactic center. Instead, their orbits are highly elongated (Figure 6–14), and at the orbits' farthest points the clusters may be twenty thousand parsecs or more from the galactic center. We can hypothesize that globular clusters formed as the Milky Way contracted, and before the protogalaxy had taken on a disklike form. The globular clusters' orbits thus reflect the distance from the center at which they formed (ten thousand to thirty thousand parsecs). These orbits are like the orbits of stars in elliptical galaxies: highly elongated, and without much unified sense of direction. And it appears that globular clusters formed before our galaxy "knew" that it would be a spiral, so the clusters move in paths like the orbits in an elliptical galaxy.

Globular clusters in other spiral galaxies such as M 33 and M 31 (the Andromeda galaxy) show the same distribution of orbits as do the globular clusters in the Milky Way. We can conclude that density perturbations with a mass of one hundred thousand to ten million times the sun's mass apparently had a tendency to become definite clumps as protogalaxies contracted, and within these clumps further perturbations later produced protostars and eventually made the Population II stars that we see today. By our best estimates, the Milky Way contains about five hundred globular clusters, which together include a fraction of a percent of the stars in our galaxy.

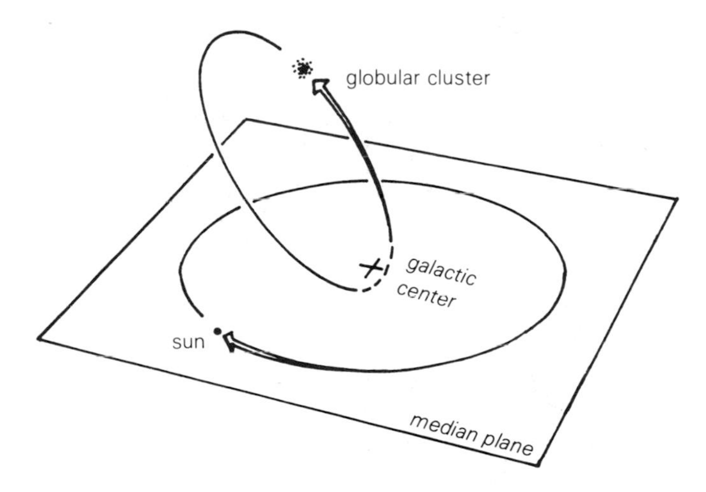

Figure 6–14. Globular clusters do not orbit around the center of the Milky Way in the same plane as most of the stars do. In addition, the orbits of globular clusters around the galactic center are highly elongated, unlike the orbits of most individual stars, which are nearly circular.

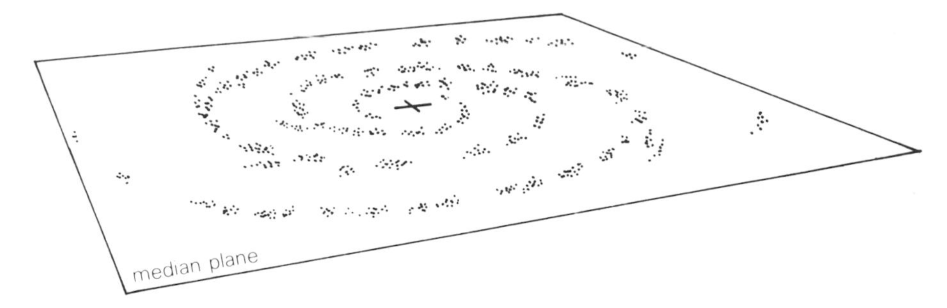

Figure 6–15. The open clusters (here represented by dots) are concentrated close to the median plane of the galaxy.

Globular clusters are easy to see because of the concentration of stars within them. A typical globular cluster such as M 13 (Figure 6–11) has a diameter of 10 parsecs and contains a million stars. The region near the sun that is ten parsecs in diameter contains only about a hundred stars, so the star density in a globular cluster exceeds the local star density by about ten thousand times. Even so, the stars have plenty of room to move around the cluster without colliding.

As opposed to globular clusters, open clusters clearly belong to the galactic disk (Figure 6–15). The most important open clusters (impor-

Figure 6–16. The Pleiades, an open cluster about three times as far from us as the Hyades. Notice the nebulosity in which these young stars are still immersed, which presumably is the remnant of the gas from which the stars formed.

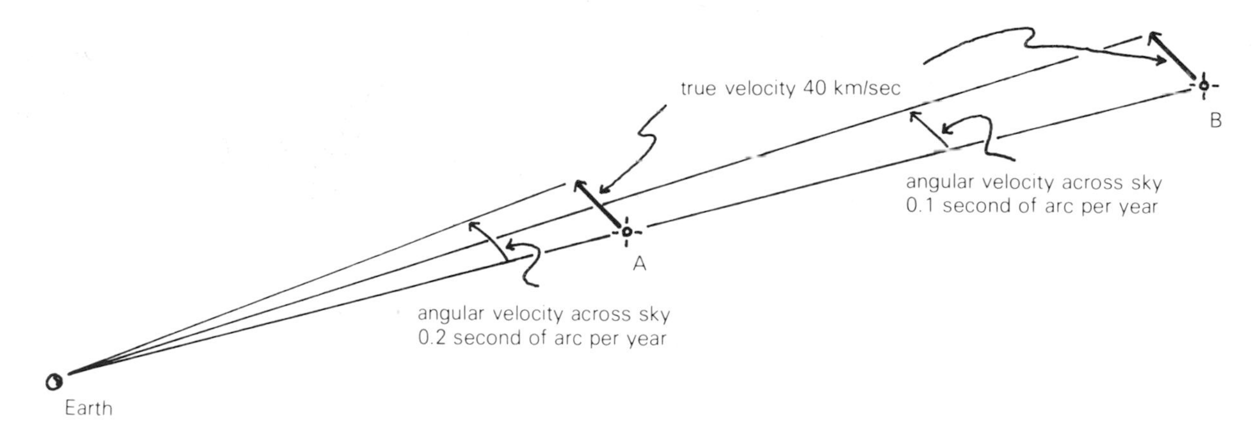

Figure 6–17. The amount of angular displacement, or "proper motion," that we measure for a given star depends both on the star's velocity through space and on the star's distance from us. If a star were twice as far away as it actually is, then the star's velocity would produce only half the proper motion that we now measure per year.

tant to *us*) are the Hyades, the Pleiades (Figure 6–16), and the double cluster called *h* and χ Persei (Figure 6–10).[2] These clusters consist of Population I stars, whereas globular clusters contain Population II stars. Open clusters have diameters that are about equal to those of globular clusters (five to twenty parsecs), but since they contain only a few hundred stars, their star density is a thousand times less than the density of stars in globular clusters. A "loose" open cluster in fact may be just slightly denser than the average density of stars. Nevertheless, astronomers are sure that the stars in an open cluster do not just happen to appear together; instead they conclude from the similarity of the star's ages that the stars were born together and move together.

The nearest open cluster, the Hyades, lies close enough to us (forty parsecs away) that we can measure the motion of its stars across the sky, relative to the background of much more distant stars. Photographs taken of the Hyades year after year allow us to determine the small amount that each individual star has moved, called the star's "proper motion," with respect to the background of more distant stars that do not appear to move. This proper motion can be measured only because the stars are near enough to us for their motions through space to be seen in only a few years' time. If the stars in the Hyades were ten times farther away, we would have to wait ten times as long for the stars' velocities in space to produce the same angular displacement on the sky (Figure 6–17). We also can use the perspective effects of our view of the stars' motions to find out something about their distance.

[2]How astronomers managed to name this cluster with one lower-case Roman letter (*h*) and one Greek letter (χ) is a story that would curl the hair of anyone who believes that scientists are masters of logical and orderly procedures.

We can combine our measurements of a star's proper motion *across* our line of sight with measurements from the Doppler effect of its velocity *along* the line of sight. The amount of proper motion depends both on the star's velocity and on its distance, but the Doppler effect depends only on the star's velocity. When we measure the proper motion and the Doppler shift for the stars in the Hyades, we find that they are all moving in about the same direction (Figure 6–18). This hardly could arise by accident, thus we believe that the stars were born together. However, the mutual gravitational force among the stars in the Hyades or in other open clusters cannot hold these clusters together for longer than a few hundred millions of years. While the clusters exist, though, we can use the measurements of both the proper motions and the radial

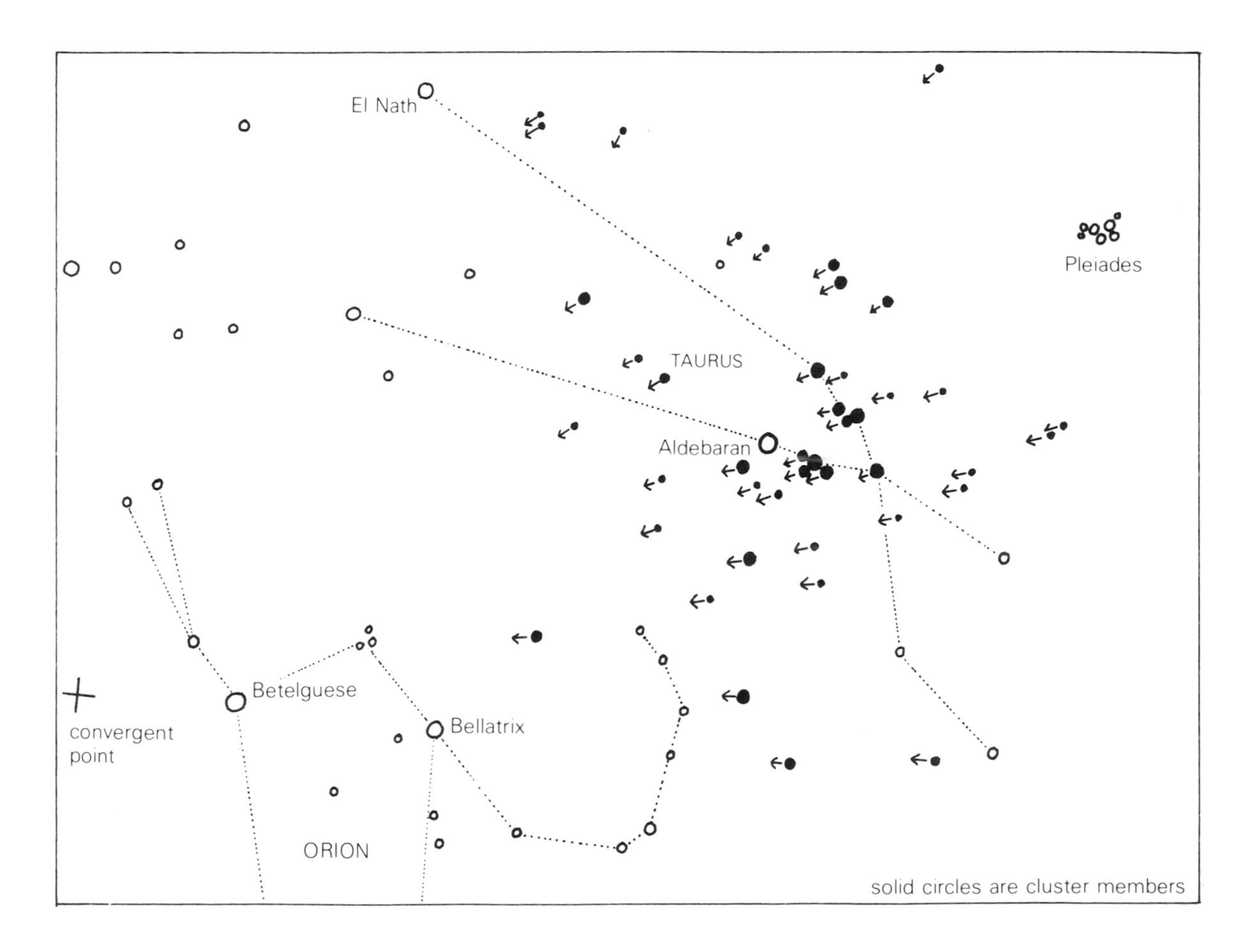

Figure 6–18. The proper motions that we measure for the stars in the Hyades all point to about the same place on the sky, called the "convergent point."

velocities of the stars in it to determine the clusters' distances. Figure 6–19 shows that if stars in an open cluster are moving through space in parallel paths, we shall see these paths projected on the sky not as parallel tracks, but as all pointing toward the same place, called the convergent point. From Figure 6–18, which shows the directions of proper motion of the eighty brightest stars in the Hyades, we can locate the convergent point for the Hyades cluster. This apparent convergence effect arises simply from the fact that we see the stars' motions projected on the sky. However, we can use our trigonometric abilities to find from the combination of a star's proper motion with its radial velocity and its distance on the sky from the convergent point, all three of which we can measure, the star's distance from us. This method basically relies on the fact that the Doppler shift measures velocity, whereas the proper motion measures velocity divided by a distance (so that larger distances produce smaller proper motions). Hence the combination of the two will yield the distance, if we know what we are doing.

The stars in the Hyades are about forty parsecs away, but the diameter of the cluster is about ten parsecs. In comparison, the Pleiades are 125 parsecs away, so they appear more compact (the seven visible stars of the total of more than a hundred span only the angular size of the sun on the sky), although they have about the same real diameter as the Hyades cluster does. Open clusters such as the Hyades, the Pleiades, and *h* and χ Persei will disintegrate as a cluster: They do not have enough mass for their component stars to hold one another in a cluster

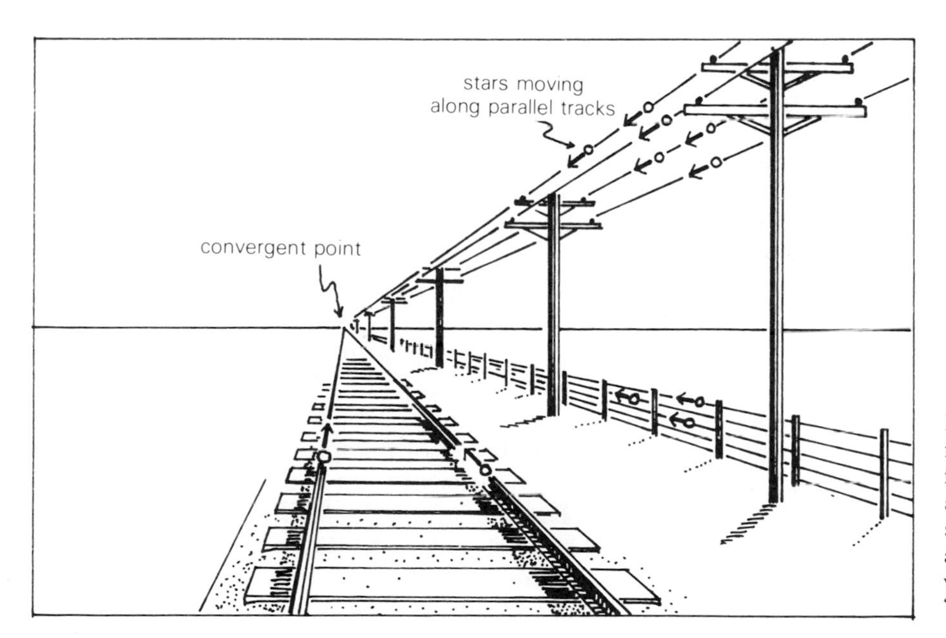

Figure 6–19. The converging effect in the Hyades' proper motions arises from the way that we see stars as apparently projected on a "celestial sphere" around us. Stars moving along parallel tracks through space will appear to us as heading for a "convergent point."

permanently, so the stars slowly will drift apart because of their individual (small) velocities relative to one another. This fact contrasts open clusters with globular clusters, which do have enough mass to stay together for billions of years, and provides additional evidence that the stars in open clusters, all of Population I, cannot be as old (with some rare exceptions) as the Population II stars in globular clusters.

An extreme example of open clusters are stellar *associations*, which consist entirely of quite young stars. The stars in associations are moving in generally the same direction, but they will separate after only a few million years, and the density of stars in an association is no greater than the average density of the stars around them. We can recognize these associations by the common ages and common motions of the stars within them. Five of the seven stars in the Big Dipper belong to such an association, which also includes some of the stars in the Big Dog (Canis Major) and the Little Dog (Canis Minor), which are far away from the Big Dipper on the sky. That is, the sun is almost in the middle of this particular association of a few dozen stars, without being part of it or affected by it. In a few tens of millions of years, as the motions of the various stars distort constellations such as the Big Dipper, these associations no longer will be recognizable, because the stars in them will tend to drift apart from each other along somewhat divergent paths through the galaxy.

Our Milky Way contains about twenty thousand open clusters, each with a hundred to a thousand stars. Thus about ten million stars in our galaxy belong to open clusters. Another hundred million or so stars belong to globular clusters, so we can conclude that both types of clusters together include less than one percent of the stars in the galaxy. Most stars, our own sun included, are not especially young nor extremely old, and are not parts of any particularly well defined clump. These stars simply orbit around the galaxy along with other stars, independent, yet part of an entire galactic system. They may have been born with other stars, but since then have drifted apart.

Motions in the Galaxy

The basic orbit of most of the stars and of the gas clouds that form our galaxy, except for the globular clusters, is an almost circular swing around the galactic center, together with an up-and-down bobbing motion that we described on Page 104. The orbital speed and orbital period depend on the distance from the galactic center. We can confirm this pattern by studying the rotation of other spiral galaxies such as the Andromeda galaxy (M 31). The density of stars is greatest in the innermost regions of the galaxy. These centrally located stars form a spheroidal bulge at the centers of spiral galaxies about five hundred parsecs in radius (Figure 6–20). Stars in this central bulge orbit around the center once every forty million years or so. The stars at distances of one to ten

Figure 6–20. The spiral galaxy NGC 891 shows the familiar central bulge because we see the galaxy edge-on.

thousand parsecs from the center all tend to move around the center with about the same speed (two hundred to three hundred kilometers per second), but since the more distant stars have farther to go they take longer to complete an orbit. A star in the Milky Way that is four thousand parsecs from the central bulge orbits the center once every hundred million years, while our sun, which is some ten thousand parsecs from the center, takes two hundred and fifty million years to complete an orbit. Stars still farther from the center than our sun not only have still farther to travel in their orbits, but in addition their velocities in orbit are less than the sun's. Therefore stars that are fifteen thousand parsecs from the center may take four or five hundred million years to finish each orbit.

Let us idealize the orbits of stars in our galaxy by assuming them to be perfect circles (Figure 6–21). For stars in orbits with a radius between nine thousand and fifteen thousand parsecs, the stars' velocities in orbit are related directly to the size of the orbits. The larger the orbit, the less is the velocity of a star in a circular orbit. We can use this direct relationship between the size of the orbit and the speed in orbit to determine whether a star that we observe really *is* moving in a circular orbit. To do this, we first must determine the star's distance from us and then use trigonometry to find the star's distance from the galactic

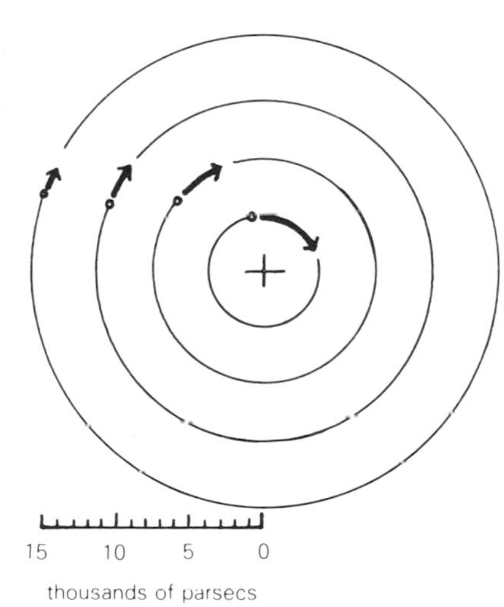

Figure 6–21. We can model the motions of stars in our galaxy fairly well by assuming that each star moves in a circular orbit. Stars farther from the galactic center tend to move more slowly along their orbits.

center (Figure 6–22). Next we can use the Doppler effect to measure the star's velocity along the line of sight. Again we can use trigonometry to estimate what the star's velocity in orbit is, if we assume that the orbit is a circle (Figure 6–23). Then we compare this orbital velocity with the velocity that a star must have if its orbit is a circle with a radius equal to the star's distance from the galactic center. (This second velocity can be calculated from the law of gravitational force.) If the two velocities are equal, then the star's orbit is indeed circular. But if the velocity derived from assuming the orbit is circular differs greatly from the expected velocity for a circular orbit with a radius that equals the star's distance from the center, then we must conclude that the star's orbit does not closely approximate a circle.

Recall that the sun's orbit around the galactic center is almost circular. A star near the sun moving in an almost circular orbit at the same distance from the center as the sun would have the same orbital velocity around the center as the sun does (Figure 6–24). Thus the velocity of such a star *relative* to the sun should be zero, so there should be no Doppler shift of the photon energies in the light from the star. Indeed,

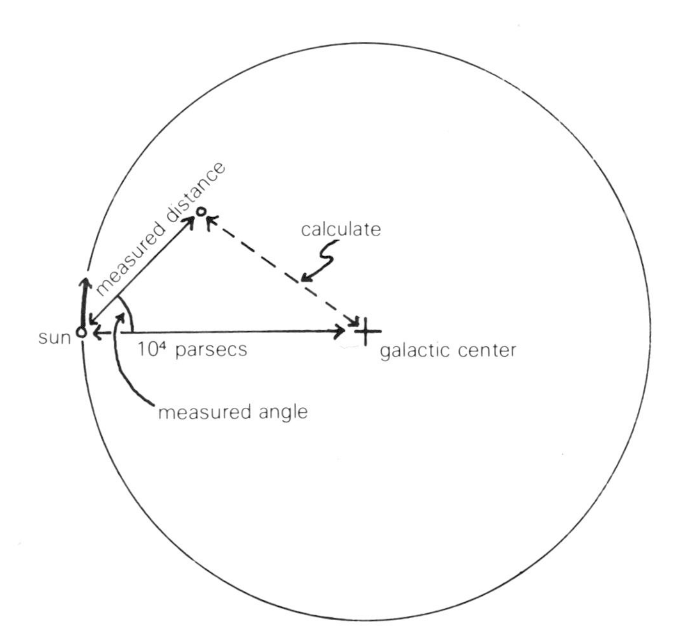

Figure 6–22. If we can measure the distance to a star in our galaxy, we then can use trigonometry plus our knowledge of the distance to the galactic center (10^4 parsecs) to find that star's distance from the galactic center. We can calculate the third side of a triangle from two sides and the angle between.

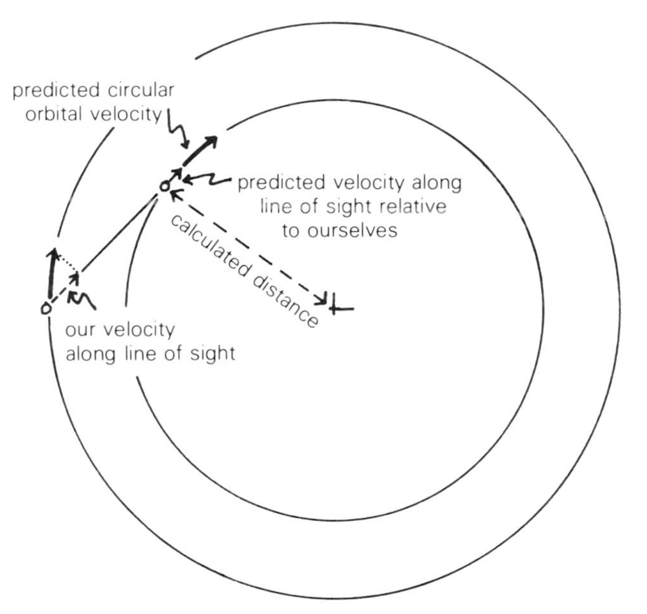

Figure 6–23. If we know a star's distance from the galactic center, then we know how fast the star should move in orbit *if* its orbit around the center is a circle. This allows us to calculate what the velocity of the star relative to ourselves, along our line of sight to the star, should be for a circular orbit.

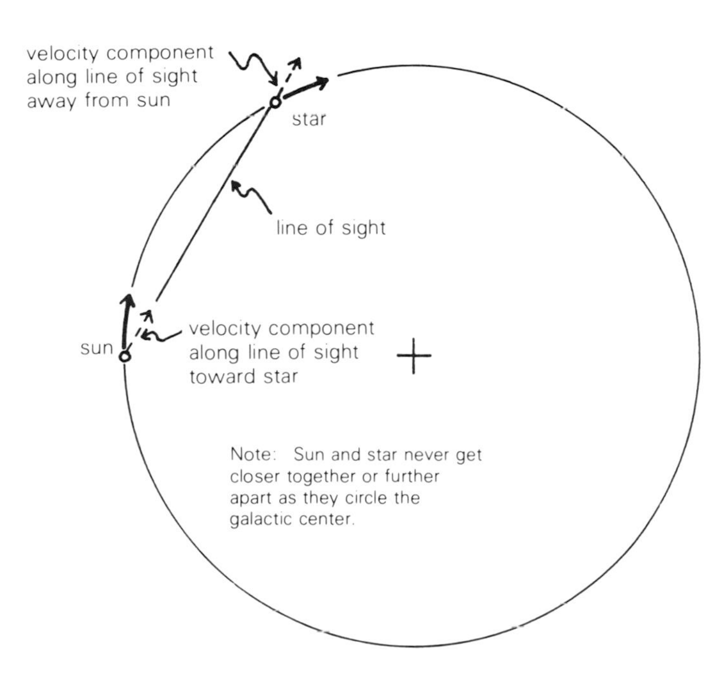

Figure 6–24. If a star has the same distance from the galactic center as the sun, and if the star is moving in a circular orbit around the center as the sun is, then the star should have zero velocity relative to ourselves if the star is fairly close to the sun.

most of the stars within a few hundred parsecs of the sun show radial velocities relative to the sun that rarely exceed twenty or thirty kilometers per second, and are often less. Since the velocity in a circular orbit of a star at our distance from the galactic center is about 250 kilometers per second, we can conclude that the variations by twenty or thirty kilometers per second are minor, and that all the stars are moving in approximately circular orbits around the center. Some relatively nearby stars, however, show a Doppler shift that indicates a radial velocity of fifty to one hundred kilometers per second, or even more, relative to the sun. These "high-velocity" stars (high velocities relative to ourselves) must be moving in elongated orbits around the galactic center (Figure 6–25). These elongated orbits give the high-velocity stars a different velocity in orbit than the orbital velocities of stars that move in almost circular trajectories, such as the sun, even if both the high-velocity star and the sun are the same distance from the galactic center. In fact, the orbital velocities of these stars moving in elongated trajectories are no greater, and may be less, than the sun's orbital velocity, but the stars do have a high velocity *relative to us*. High-velocity stars are Population II stars, which are much older than the sun and other Population I stars. The younger, Population I stars all have nearly circular orbits. This result confirms what we know about the elongated orbits of globular clusters, which consist of old, Population II stars and

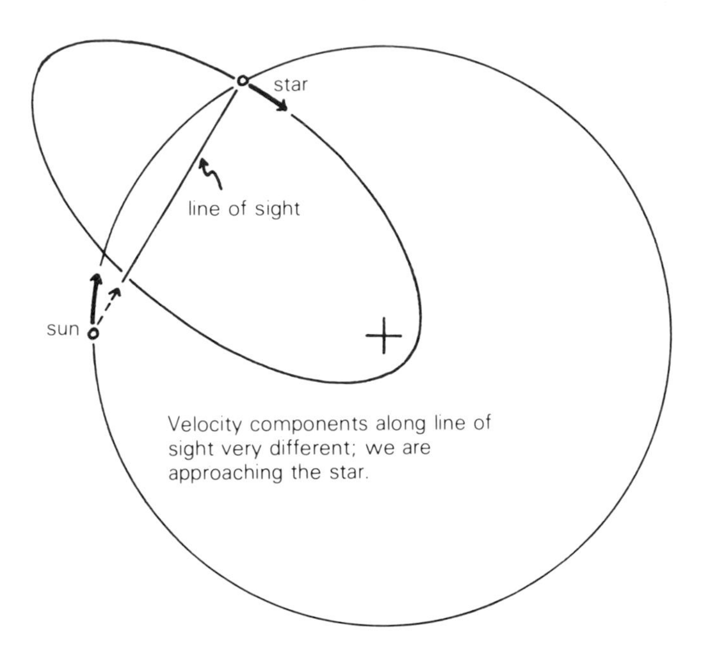

Figure 6–25. If a star is moving in a highly elongated orbit around the galactic center, then even when it has the same distance from the center as the sun does, it will not have the same velocity as the sun. Hence we will observe that the star's velocity with respect to ourselves is not zero.

would be typical "high-velocity" objects if they came near the sun. In fact, the globular cluster nearest to us is four thousand parsecs away, a hundred times farther than the Hyades.

Interstellar Matter

Among its hundred billion stars, our galaxy contains interstellar gas and dust spread out diffusely. Astronomers use the word "gas" to refer to collections of individual atoms or of small molecules, whereas "dust" particles contain a million or more atoms each, bound together as a solid body. These dust grains in interstellar space apparently consist mainly of carbon atoms linked together in long chains to form graphite (pencil lead). An average interstellar dust grain may be about one ten-thousandth of a centimeter in diameter, and the outermost layer of the grain consists of a coat of ice only a few atoms thick.

The interstellar gas consists mainly of hydrogen atoms, along with some atoms of helium, carbon, nitrogen, oxygen, neon, iron, sulfur, silicon, and so on. The elements heavier than helium appear in the interstellar gas in about the same proportions as they do in Population I stars. This fact lends support to the idea that stars condensed out of the interstellar gas. The younger generation of stars then accurately reflects the chemical composition of the gas from which they formed.

About half of the hydrogen atoms have paired to form hydrogen molecules (H_2), especially in those regions where the gas is densest. The other half of the interstellar hydrogen exists in the form of hydrogen atoms. These atoms emit photons of 21-centimeter wavelength, and as Figure 6–8 shows, we can use our observations of these photons to map out the distribution of hydrogen atoms throughout the galaxy. As we discussed on Page 158, these maps do not give a totally accurate picture of the distribution of hydrogen atoms in our galaxy, because to make the maps we must assume that each atom moves in a circular orbit around the galactic center. The fact that hydrogen atoms have random motions of their own, amounting to a few kilometers per second, is enough to introduce a significant distortion into the maps that we make with the simplifying assumption of circular orbits. Still, by long study of these maps of 21-centimeter photons from hydrogen atoms within our own galaxy and within other spiral galaxies, astronomers have found that the gas and dust in a spiral does follow the same spiral arms traced out by the young, bright stars.

These observations show that the density of hydrogen atoms is largest within the spiral arms, although there is some hydrogen gas spread between the arms as well. (See Plate 2.) Since hydrogen atoms contain most of the mass of the interstellar material (and hydrogen molecules are most of the remainder), we can be fairly sure that all of the gas and dust in our galaxy shows a spiral arm pattern like that of the young stars. It is worth remembering that there is a large amount of gas between the arms as well: The basic fact about interstellar matter in spiral galaxies is that it is concentrated toward the median plane of the galaxy, like the Population I stars in open clusters. This similarity in location confirms the idea that Population I stars formed relatively recently from interstellar gas, and that the leftover gas still fills the spaces between the younger stars.

When we observe the finer structure of the interstellar gas, we find that within a spiral arm the gas comes in clumps called "interstellar clouds." A typical interstellar cloud has a radius of one to one hundred parsecs and a density ten to a thousand times greater than the average density of the interstellar gas (about one atom per cubic centimeter). Interstellar clouds, which have densities of ten or a thousand atoms per cubic centimeter, thus are far more rarefied than the best vacuum we have made on Earth. Occasionally we find interstellar clouds whose density is more than a million atoms per cubic centimeter, but this density is still 10^{18} (one billion billion) times less than the average density inside a star. These dense clouds often contain not only individual atoms of hydrogen, carbon, and oxygen, but also molecules such as carbon monoxide, formaldehyde, and even methyl alcohol. Each of these molecules emits or absorbs photons with particular wavelengths, as atoms do, and we can detect them as a result of their particular emission or absorption patterns in photon spectra. These relatively complicated molecules, some of which contain a dozen or more atoms, can form from the atoms themselves by collisions within the gas, once the density has reached a million or so atoms per cubic centimeter. Such elementary molecules perhaps could link together to form the basic constituents of living organisms called amino acids (Chapter 11). The temperature in one of these dense interstellar gas clouds, which is about one hundred degrees absolute at the most, is far below the average temperatures on Earth (290 degrees absolute). Hence such clouds do not appear to be likely places for life to start, but we do not yet know enough to resolve this question.

Aside from the possibility of life, these dense clouds are interesting as likely sites for the birth of stars. Because the average density of matter in a star (about one gram or 10^{24} atoms per cubic centimeter) so far exceeds the density inside an interstellar cloud, any contracting cloud has a long way to go before it condenses into stars. However, some clouds seem to be forming stars right now. The Orion Nebula (Figure 4–27) has a cluster of young, bright stars called the "Trapezium" inside it that makes the cloud glow. Close by the Trapezium, a bright source of infrared radiation appears to be the result of a group of stars just beginning to shine (see Figure 11–4). The "cocoon" of dense gas around the stars prevents us from seeing them directly, but astronomers think that in only a few thousand years the stars' photons will push their cocoon aside and emerge into the larger universe around them.

We have ignored the interstellar dust grains until now, but these grains have an importance far greater than their low density in the galaxy would indicate. Dust grains contain less than one percent of the interstellar matter in the Milky Way, while the interstellar matter itself has perhaps two to five percent of the total mass of the galaxy. However, the dust particles are highly efficient absorbers of photons with visible-light and ultraviolet energies. Atoms and molecules can absorb only photons with definite, particular energies, but dust grains are likely to absorb photons of energy over a wide range of energies, because the grains' individual atoms can recoil in many subtle ways through their linkage to all the other atoms in the grain. As dust grains absorb photons, they preferentially absorb more blue and ultraviolet photons than they do the redder, lower-energy photons among the starlight that passes by them. Thus the light that passes among dust grains emerges *redder* than the light that entered, because more of the blue photons than the red ones have been absorbed. This interstellar reddening effect grows as light travels farther and farther through the galaxy. Thus the light from nearby stars undergoes almost no reddening, whereas for stars ten thousand parsecs away, the preferential absorption of the bluer photons completely changes the star's apparent color. Because the dust grains' average reddening effect per kiloparsec of path length can be estimated, and because the dust grains are fairly evenly distributed within the central plane of our galaxy, we can use the amount of reddening as a last resort to estimate the distances to stars whose distances we cannot find in other ways. If we think we know how red a star ought to be (from our study of the features in its photon spectrum), and if we can see how red the star actually appears, then from the difference between what we expect and what we observe, we can calculate how much interstellar reddening has affected the starlight in its passage to us. This gives us a rough estimate of the star's distance.

Because there are a great number of dust grains along the 10,000 parsecs that separate us from the center of the galaxy, the light from the innermost regions of our galaxy does not reach us at all. That is, the dust grains not only absorb more blue light than red light, they simply absorb so much of both that they prevent us from receiving the photons of any energy. The general absorbing effect of interstellar dust easily can be seen on photographs of other galaxies (Figure 6–20), as well as in the photograph of the globular clusters near the center of the Milky Way (Figure 6–13). Notice that the interstellar dust grains, like the interstellar gas and the young Population I stars, concentrate near the plane of the galaxy. This concentration of dust helps us to see out through our own galaxy to observe other galaxies. If we try to look through the plane of our galaxy, we encounter so many dust grains that we cannot see any other galaxies in these directions (Figure 6–26). But if we look out in directions perpendicular to the median plane, we can see faraway galaxies and galaxy clusters. The absorption of photons by interstellar dust grains creates a misnamed "zone of avoidance" on the

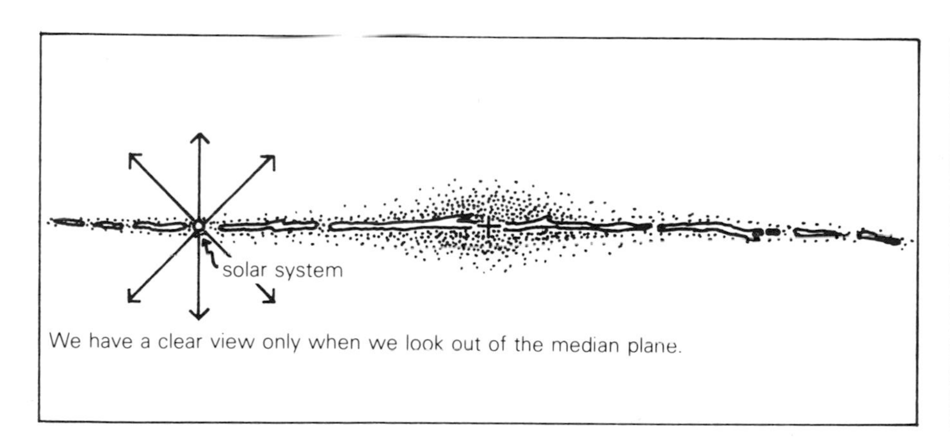

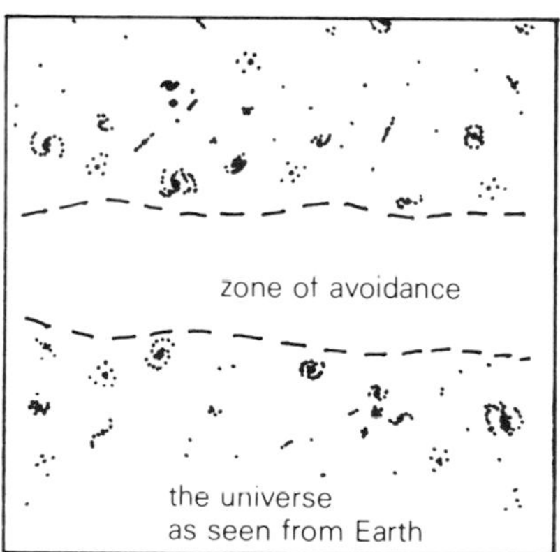

Figure 6–26. We cannot see out of our own galaxy when we look directly along the median plane, because the absorption of starlight by the dust in that plane prevents the photons from reaching us. This absorption produces a "zone of avoidance" on the sky in which we can see almost no other galaxies.

sky where almost no galaxies can be seen. It is not that the galaxies avoid these areas, rather that we cannot see them behind the screen of dust particles in our own galaxy.

In 1970, astronomers managed to detect two good-sized galaxies that are only a few million parsecs away, just outside the Local Group. Interstellar absorption by dust grains almost totally obscures our view of these "Maffei galaxies," named after their discoverer (Figure 6–27). This absorption of photons does not affect the galaxies, of course, only our view of them. Since the "zone of avoidance" covers about ten percent of the area on the sky, perhaps one galaxy in ten (even if bright enough to be seen clearly if there were no absorption) hides behind a shroud of interstellar dust grains within our galaxy.

Figure 6–27. One of the two "Maffei galaxies" that lie relatively close to the Milky Way but happen to be located in a direction that puts them in the zone of avoidance. Dust within our galaxy absorbs many of the photons from the Maffei galaxies, and therefore obscures our view of these galaxies. This photograph was taken in red light, which has a better chance of getting through the interstellar dust than blue light does.

Theories of Spiral Structure

The spiral-arm pattern of our galaxy and of many other galaxies poses a problem: Why does the spiral pattern persist? If all of the parts of the galaxy rotated *with the same rotational period*, then we might not have to explain the persistence of the pattern, although we still would have to explain its origin. But the outer parts of the galaxy take longer than the inner parts to circle around the center. A star five thousand parsecs out from the galactic center will orbit once in 125 million years, whereas our sun, ten thousand parsecs out, will take 250 million years to complete an orbit. The result of this difference in orbital period is that the galaxy's spiral arms should be wound up tight after a few orbits (Figure 6–28). That is, if the stars formed a spiral pattern a few billion years ago, by now the pattern would be wound up so tightly that it should be completely smeared out by the different times for stars at various distances out to complete an orbit. This situation is like a marching band that tries to perform a pinwheel maneuver. If the outermost musicians fail to make a circle in the same time as the innermost musicians, they will fall so far behind that after a couple of circles the pattern will be completely obliterated (Figure 6–29).

Yet we see spiral galaxies by the thousands, since almost half of all galaxies, including our own, are spirals. Therefore the spiral pattern cannot be a brief episode in a galaxy's existence, or else only a small fraction of galaxies would have spiral arms.

Theoreticians have suggested two possible explanations of spiral patterns in galaxies. Both of these theories emphasize that the position of the spiral *pattern* must be explained. Recall that spiral arms are defined

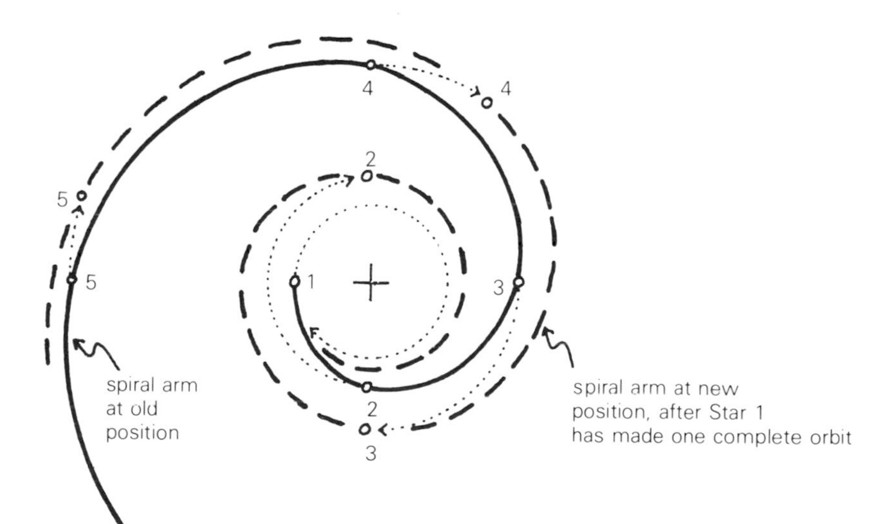

Figure 6–28. Because the outer parts of our galaxy move in their orbits around the center more slowly than the inner parts do, we would expect that the galaxy's spiral arms would be wound up tight after a few galactic rotations.

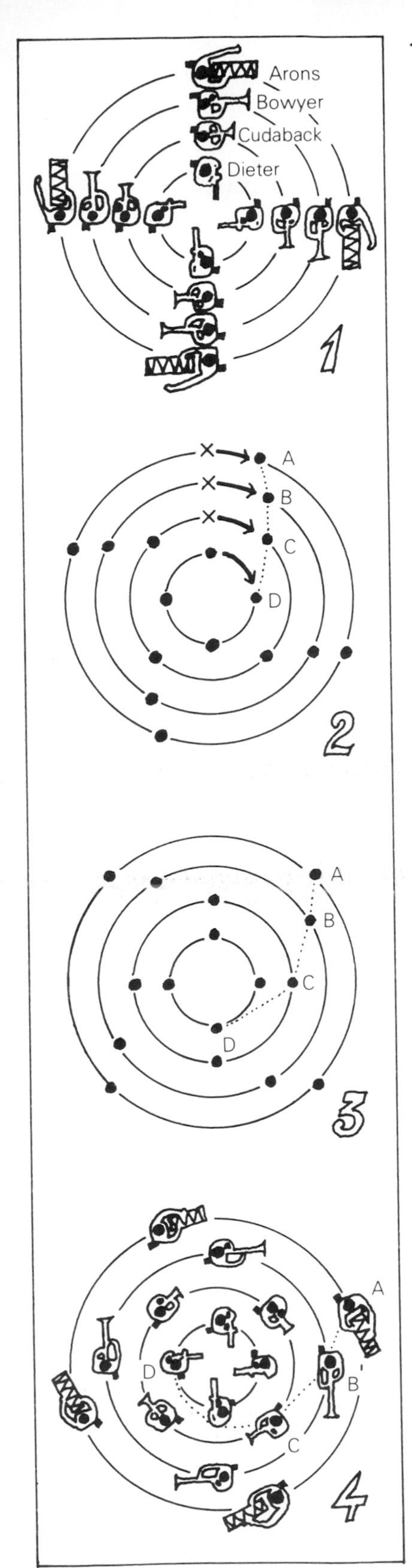

◀ **Figure 6–29.** If a marching band does not have its outer members move more rapidly than its inner members as it rotates, the pinwheel formation they begin soon will degenerate into confusion.

by the gas and the young stars they contain. Other stars spread throughout the entire galactic disk may be inside or outside of spiral arms. We may conclude that whatever causes spiral arms consists of a mechanism that collects gas and forms young stars. Hence astrophysicists of bold imagination have suggested that a wave pattern rotates around the center of a spiral galaxy, something like a water-wave pattern in a pond. The water-wave pattern moves outward from a disturbance, but the individual drops of water mostly bob up and down as the wave goes by. Similarly, in a spiral galaxy, the wave pattern circles the galaxy at a speed different from the individual stars' speeds. This wave pattern consists of alternating larger and smaller gas densities, just as a water-wave pattern consists of alternating wave troughs and wave crests. The density-wave pattern in spiral galaxies itself has a spiral form and rotates around the galactic center more slowly than the stars and gas orbit the center. Hence the stars and gas tend to overtake the pattern from behind. Once they are inside the high-density regions of the pattern, the stars are slightly more crowded together than they were outside it, and the gas too becomes slightly more dense. The word "slightly" here means by a factor of two or three. The actual amounts of density increase might be two or three times for the density of stars and about ten times for the gas. These increases are not large in comparison with the enormous range of densities found in different astronomical situations, but they could have an important effect. The increase in the density of gas and stars seems to trigger the formation of large gas clouds that condense into protostars and produce young, bright stars. This process takes only a few million years, a mere moment by the standard of the galactic rotation, which takes hundreds of millions of years. In these few million years, many young, bright stars will form, start to shine, and even begin to fade out because of their prodigious rate of energy liberation (Page 222). By the time that the pattern has moved a significant distance around the galaxy, relative to the stars, the stars that once outlined a spiral arm will be fading into obscurity, while a new group of stars is forming from the gas that recently entered the regions of higher density (Figure 6–30). The density-wave *pattern* persists, but the bright stars perish.

Calculations show that if a galaxy can start such a density-wave pattern with two spiral arms, the pattern will be stable and will last for billions of years. But how does such a spiral density-wave pattern get started? One theory assigns the origin of the pattern to close encounters between galaxies, whereas another theory suggests that the rotating central bulge of the spiral galaxy generates the density-wave pattern.

As for the first theory, we know that galaxies in clusters often pass close to one another even though actual collisions are quite rare. ("Often" here means every few billion years or less.) Such close encounters could generate galactic "tides" that become spiral density-wave patterns, as has been shown by computer simulations of these near misses (Figure 6–31). In this theory, the spiral pattern in the Milky Way was started most recently by a close encounter with the Large Magellanic Cloud about five hundred million years ago.

The second theory suggests that spiral galaxies generate their own density-wave patterns. Spiral galaxies have a central rotating spheroid of stars, or a bar-shaped spindle of stars in the case of barred galaxies. The stars in the central parts of these galaxies, five hundred or a thousand parsecs across in their total aggregate, all circle the center with the

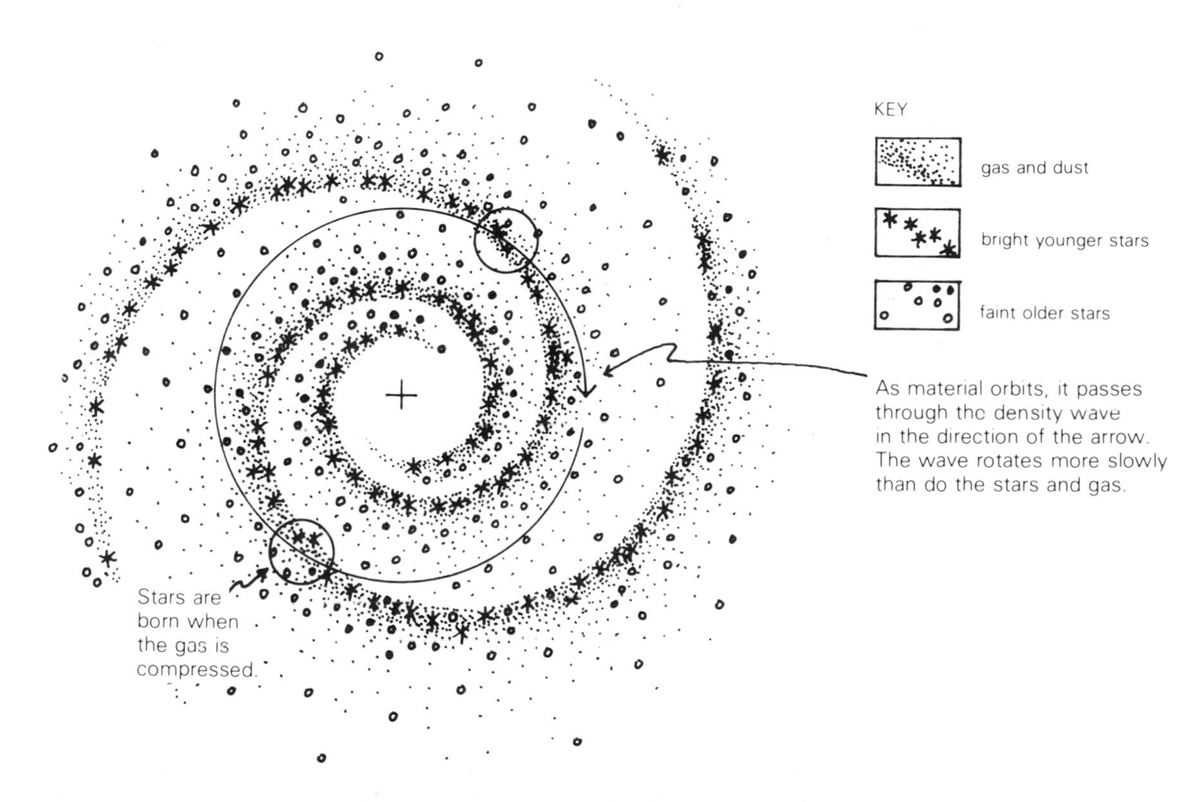

Figure 6–30. A schematic representation of the spiral structure in a galaxy like our own. The rotating density-wave pattern keeps compressing new clouds of gas to form new generations of young stars. Thus the stars' spiral structure echoes the structure of the rotating density wave.

same orbital period. This coherent motion may be able to generate spiral density waves that propagate outward from the central bulge. This theory could be more relevant for barred spirals than for ordinary spirals, because the rotating bar provides a natural source for a symmetric, two-armed density-wave pattern. Both of the theories of how the density-wave pattern began in spiral galaxies have their supporters among astronomers. As often is the case, one theory may be more relevant to the real universe than the other, or they both may be correct (one for spiral galaxies and the other for barred spirals), or both may be wrong.

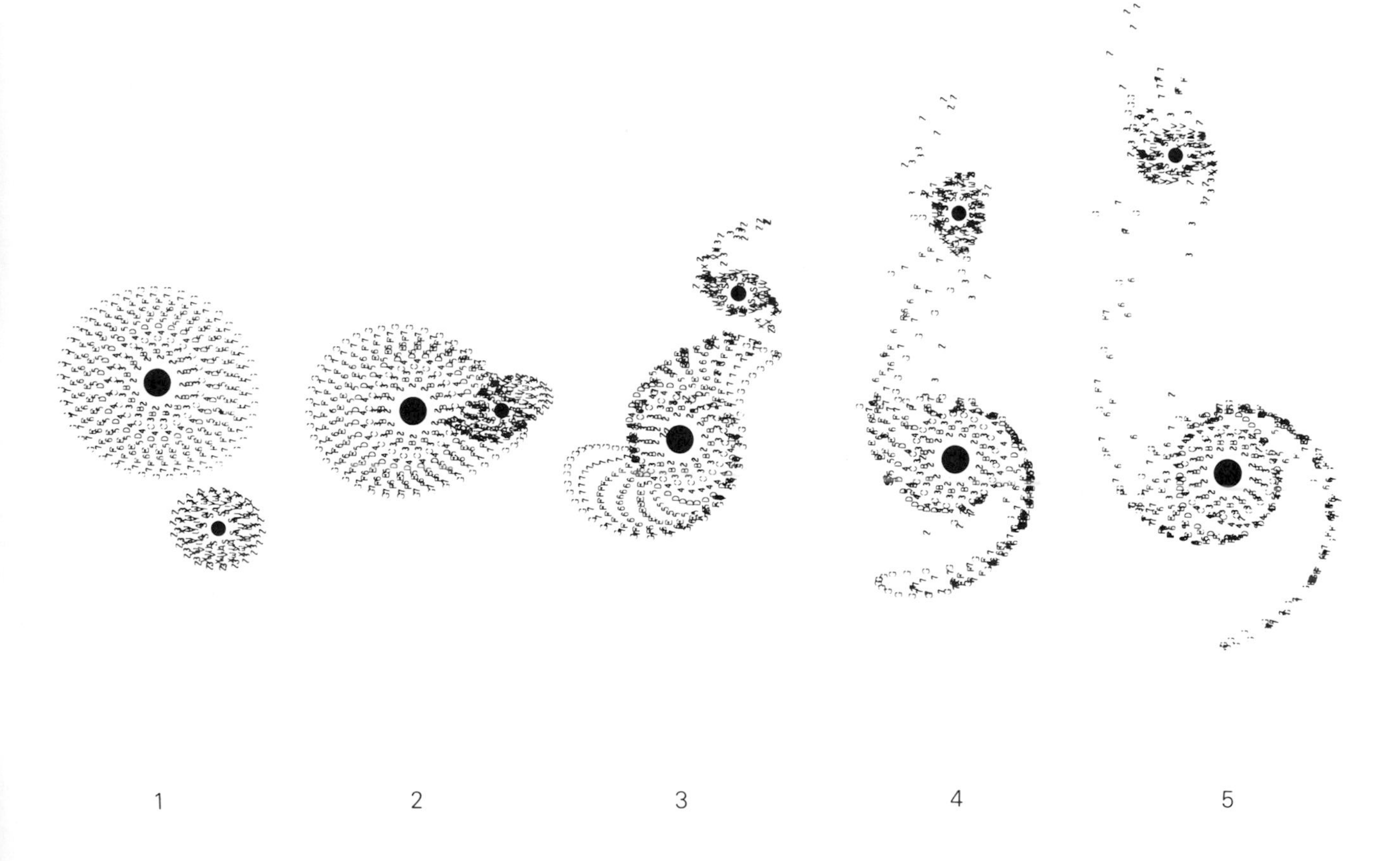

Figure 6–31. A computer simulation of a close encounter between two galaxies shows how the galaxies' gravitational forces act on each other to pull out bunches of stars into spiral-arm patterns. The figure shows five moments in time during and after encounter, and suggests that such near misses may generate spiral arms, at least temporarily. In addition, these encounters might start the density-wave pattern that would continue long after the galaxies had passed close to each other.

Summary

Our Milky Way galaxy belongs to a small galaxy cluster called the Local Group, which contains other giant spiral galaxies, some smaller elliptical galaxies, and several irregular galaxies. Observations of the stars in these galaxies have led astronomers to divide most stars into two general types or "populations." The stars of Population I are *younger*, and contain a larger fraction of elements heavier than helium than we find in the older, Population II stars. Population II stars appear in globular clusters, which apparently formed even before the galaxies had finished condensing, and also are found throughout the rest of a galaxy. Population I stars, which include the sun as a typical older member, contain all of the young, bright stars that outline the arms of a spiral galaxy. Such young Population I stars last only a few tens or hundreds of millions of years. Elliptical galaxies contain *no* such young Population I stars. However, most elliptical galaxies, which can be as massive as the largest spiral galaxies, do contain a large number of the older Population I stars along with their Population II stars.

In spiral galaxies, most stars orbit the galactic center in nearly circular paths and in almost the same plane, bobbing up and down through the median plane as they orbit. In contrast to elliptical galaxies, spiral galaxies contain clouds of gas and dust, together with a more diffuse background of gas spread among the stars and concentrated in the median plane. This gas, which is mostly hydrogen atoms and hydrogen molecules, also orbits around the galactic center, as do the dust clouds. The dust absorbs starlight so efficiently that we cannot see out of our own galaxy in directions along the median plane, because the sun lies almost in the middle of this plane.

Questions

1. What is the Local Group? Is it part of a Bigger Group?
2. Astronomers first discovered the relationship between the true brightnesses of Cepheid variable stars and the stars' periods of light variation *before* they knew the distances to any of these stars. How was this discovery possible, since we must know the distance to a star before we can calculate the star's true brightness from our observations of its apparent brightness?
3. Why does the large true brightness of Cepheid variable stars make them especially useful?
4. What are spiral arms? Why do young stars appear preferentially inside spiral arms?

5. Which are the older star clusters (on the average), globular clusters or open clusters? What are the key differences between these two types of clusters?
6. How does a stellar "association" differ from a star cluster?
7. Which stars in a spiral galaxy are moving the fastest along their orbits around the galactic center? Which are moving the slowest?
8. The sun's speed in its almost circular orbit around the galactic center is measured in hundreds of kilometers per second, whereas its speed in directions perpendicular to its basic orbit is measured in tens of kilometers per second. In this respect, the sun's motion is typical of most stars in the Milky Way. How does the appearance of a spiral galaxy such as our own reflect the ratio of the two velocities mentioned above?
9. Why do some stars in our galaxy that are close to the sun have a large velocity relative to the sun? Why do most stars that are near the sun have small velocities relative to the sun?
10. Which "population" of stars includes the "high-velocity" stars mentioned in Question 9? Why?
11. What does interstellar matter in our galaxy consist of? Is the interstellar gas spread evenly among the stars? Is the interstellar dust spread evenly? Why does interstellar dust "redden" the starlight that passes through it?
12. About how long does our galaxy take to rotate once? Why would we expect the spiral arms of our galaxy to be completely wrapped up after several rotations?

Further Reading

B. Bok and P. Bok, *The Milky Way* (Harvard University Press, Cambridge, Mass., 4th ed., 1974).

B. Bok, "The Arms of the Galaxy," *New Frontiers in Astronomy* (W. H. Freeman and Co., San Francisco, 1975).

M. Roberts, "Hydrogen in Galaxies," *Frontiers in Astronomy* (W. H. Freeman and Co., San Francisco, 1970).

Stars—Formation and Energy Liberation

Chapter 7

"Before you study Zen, mountains are mountains and rivers are rivers; while you are studying Zen, mountains are no longer mountains and rivers are no longer rivers; but once you have had enlightenment, mountains are once again mountains and rivers again rivers."

Old Zen proverb

Most of astronomy deals with stars, which are the basic units of the visible universe. Stars are self-contained sources of energy, held together by their own self-gravity. In space, we find single stars such as our sun, double stars orbiting around each other, triple, quadruple, and higher-multiple star systems. Star clusters may contain from several dozen to several million stars, and almost all stars, including these clusters, clump together in galaxies. These galaxies contain varying numbers of stars of different types, and part of astronomers' efforts are devoted to classifying the types of stars and of galaxies in a useful way. We have seen, for example, how the general classification of stars in our galaxy into Population I and Population II reflects the stars' ages, and the same classification proves useful for the stars in other galaxies as well. In this chapter, we shall examine first how stars may have formed, then how they resemble one another in their energy liberation, and finally how they differ from one another.

The Formation of Stars

If clusters of galaxies formed from density perturbations in the early universe, and galaxies formed from condensations within galaxy clusters, then surely we can believe that stars formed from subcondensations within the contracting galaxies. About ninety-five percent of the matter in our galaxy, and in other galaxies as well, appears in clumps. These clumps, which we call stars, have a narrow range of masses, from one tenth of the sun's mass to fifty times the sun's mass. The stars all emit some photons of visible light, although the number of photons emitted each second varies by a factor of a billion or more from star to star. Furthermore, stars emit these photons because they turn energy of mass into energy of motion by "thermonuclear" reactions among the elementary particles deep inside them. This basic characteristic of stars cannot be observed directly. We can determine some stars' masses and brightnesses from their distances, their motions, and their apparent brightnesses. But the question of how stars can liberate energy of motion, by a process entirely strange to our intuition, remained unanswered until 1939, and even then the answer had to come from long chains of deduction rather than by experimental proof.

We cannot now answer the question of how stars *formed* in exact detail, although astronomers have made fair progress in calculating and in observing the beginnings of stellar existence. In the chapter on the origin of galaxies (Page 91) we pointed to the fact that a clump of gas with a mass greater than a hundred thousand solar masses could hold together and condense under its own self-gravitation. Globular star clusters have total masses that equal or exceed a hundred thousand solar masses, and we believe (because of their elongated orbits around the

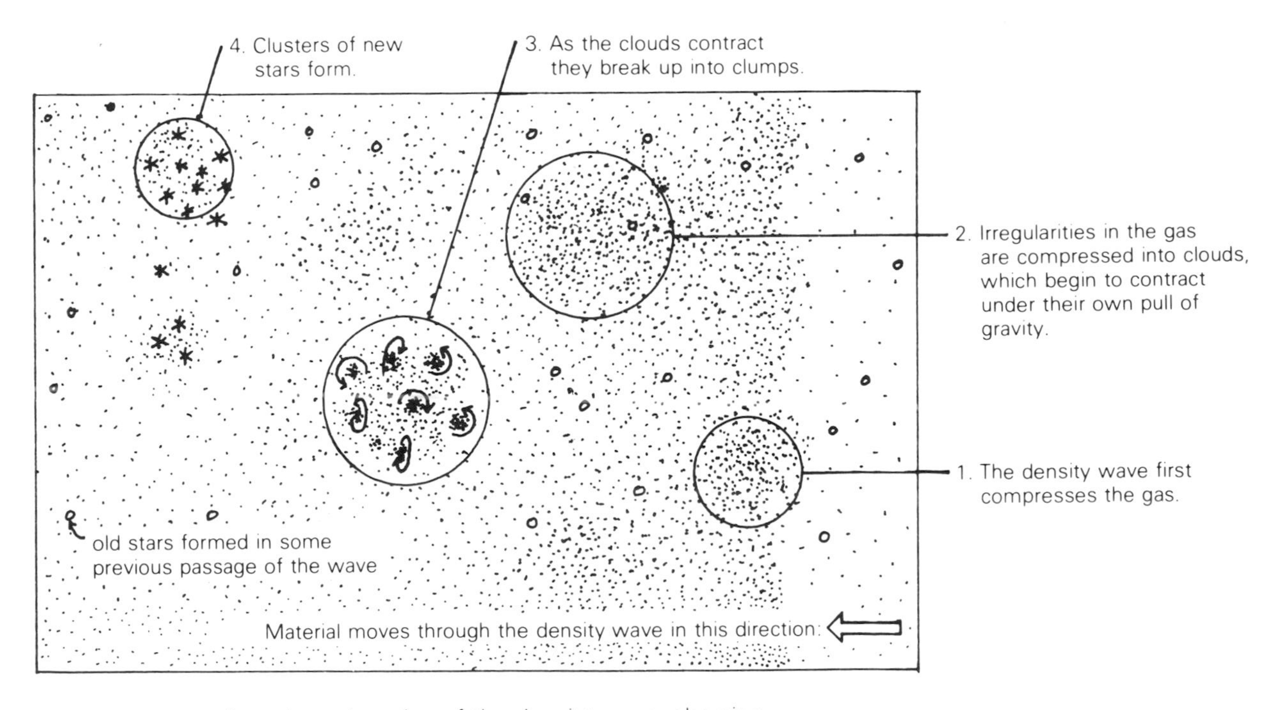

An enlarged section of the density wave, showing steps in the sequence of star formation.

Figure 7–1. As a gas cloud contracts, it can fragment into clumps that become protostars. Each of these clumps might have some spin, although the total spin could add to zero, as in the case of the larger clumps that became protogalaxies (see Figure 4–8).

centers of the galaxies to which they belong) that these clusters condensed during the early history of their galaxies, when the protogalaxies had not yet shrunk to their present sizes. We therefore conclude that globular clusters condensed from initial perturbations in much the same way as galaxies did, and in this sense the globular clusters are subgalaxies, with a millionth of a big galaxy's mass in each of them. Some dwarf galaxies, however, have masses no greater than the total mass of a large globular cluster, although the galaxies are much larger than the clusters and hence are more diffuse than any globular cluster.

How did perturbations with masses *less* than about a hundred thousand times the sun's mass form stars? Clumps with small masses have trouble condensing because their self-gravitational forces cannot pull the clumps together against the random motions of the individual atoms. These random motions are measured by the gas temperature: Twice as large an absolute temperature means that the particles each have twice as much energy of motion. The lower limit of a hundred thousand solar masses prevented less massive clumps from condensing during the protogalaxy phase, when the temperature of the atoms that might clump together still exceeded a thousand degrees. If the temperature were lower than that of the protogalaxy phase, then the random velocities of the individual atoms would be smaller, and clumps of smaller mass could form.

The key to forming the individual clumps that became stars therefore turns out to be the low temperatures within interstellar gas clouds. If an interstellar cloud ten parsecs across has an average density of a thousand atoms per cubic centimeter, it will have a total mass of ten thousand solar masses. When the temperature of the atoms in this cloud decreases to about twenty degrees above absolute zero, the cloud will be on the verge of instability. That is, if something ruffles the cloud a bit, perhaps the passage of a spiral density wave (Page 180), then the cloud will tend to collapse. The gravitational forces among the various parts of the cloud will attract each part of the cloud closer to the others. This collapse will take about a million years, and as the cloud collapses it can fragment into subcondensations that become protostars (Figure 7–1). The reason why such fragmentation can occur is that the higher-density regions now have a proportionately larger self-gravity that will enable individual clumps with star-sized masses to hold themselves together, against the tendency of random motions that would return the gas to a diffuse distribution.

The best theory of star formation for spiral galaxies appears to be that the passage of spiral density waves through dense interstellar clouds triggers the collapse of these clouds and their subsequent fragmentation into protostars. As we look around our galaxy, we can observe dark "globules" of gas that may well be protostars (Figure 7–2). The best evidence for the next stage of star formation, when the protostars turn into real stars, comes from the Orion Nebula, which we discussed in a preceding chapter (Page 116). The Orion Nebula, which is about five hundred parsecs away, has a mass a thousand times the sun's. Four extremely young, bright stars (the "Trapezium") within the nebula are ionizing the hydrogen gas around them, and as the ionized atoms recombine—only to be ionized once again—they each emit the photons that we can see and photograph. The nebula also contains a larger number of less massive young stars, which play no role in ionizing the gas around them. In addition to the stars, a cloud of gas with about two hundred solar masses is emitting large numbers of infrared photons. It

Figure 7–2. This gas cloud in the constellation Cygnus shows the dark concentrations, called "globules," that may be gas and dust clumps in the process of forming protostars.

is easy to speculate that this gas forms a cocoon around young stars that still are hidden from us. As the young stars produce photons, each photon that is absorbed by the gas around the star will tend to push this gas outwards. Eventually such a transfer of energy from the photons to the gas will thin out the gas around the star to the point where some of the photons can reach us directly, thereby allowing us a clear view of the star. Calculations show that the "cocoon" phase should last only about a million years after a star begins to emit photons inside a dense gas cloud, so the infrared source in the Orion Nebula should be no more than a million years old, which is extremely young in galactic terms. The collapse of the cloud that formed the stars also took only about a million years. Thus the stars are producing photons less than two million years after they started to form from a cloud with a density of merely a thousand atoms per cubic centimeter.

We also can observe slightly older stars, called "T Tauri stars" (after their best representative, the star T Tauri), which clearly are pushing material away from themselves, as shown by measuring the Doppler shift of the photons that this circumstellar material absorbs (Figure 7–3). Therefore T Tauri stars seem to be the last phases of "cocoon" stars that have not yet freed themselves from the remnants of the gas that formed them.

Most of the stars that form from collapsing gas clouds appear as separate individuals, far from their nearest neighbors. For example, the distance from the sun to its nearest stellar sister, Alpha Centauri, is almost a billion times larger than the sun's radius. About ten percent of the

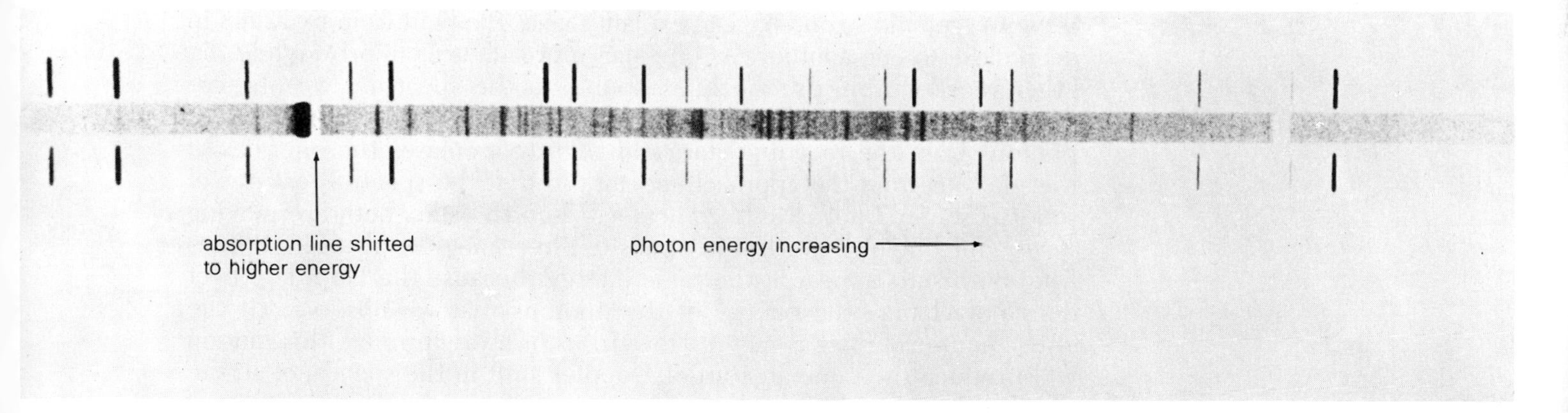

Figure 7–3. The spectra of photons emitted by T Tauri stars show absorption lines with a Doppler shift toward *increased* energy, indicating motion of the absorbing material toward ourselves. This photograph of the spectrum of a T Tauri star is printed as a negative, so the absorption lines are the white lines (arrow).

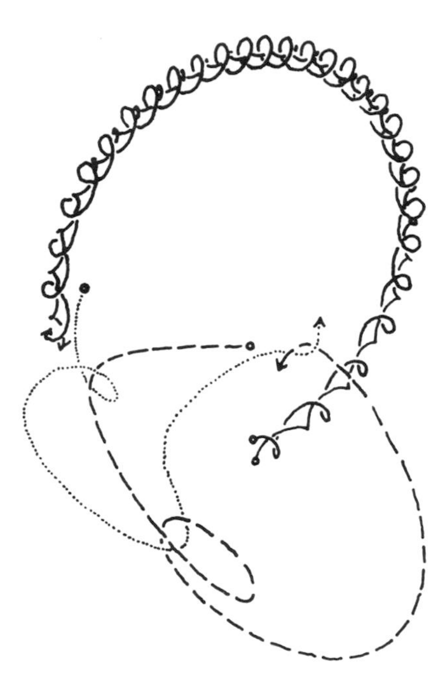

Figure 7–4. The various stars that form a multiple-star system all attract each other by gravitational forces. The result of these attractive forces appears in the complex orbits of the stars around one another, or more precisely around their common center of mass.

collapsing gas clouds, however, condense into double, triple, or higher-multiple star systems, in which the individual stars are separated by ten to a million times their own diameters. In comparison with the huge distances between separate stars, such stellar twins, triplets, quadruplets, and so forth crowd together into a tiny volume. These multiple-star systems apparently result from the fragmentation of a gas cloud late in the process of condensation, at a time when the contracting cloud had shrunk to only ten to a million times the present size of stars.

Within a multiple-star system, the individual members each orbit around their common center of mass. Each of the stars attracts all the others by gravitational forces, and from the complex pull of multiple forces the stars move along circular or elliptical orbits (Figure 7–4). In a double-star system in which the two stars have the same mass, each star will orbit around the common center of mass, which lies halfway between them. If one star has three times the mass of the other, the center of mass lies three times nearer to the more massive star than to the less massive star. A famous example of a double star is the middle star in the handle of the Big Dipper, called Mizar. On a clear night you can see a fainter star close to Mizar called Alcor, which is about one sixth as bright as Mizar. These two stars orbit around each other (actually around their common center of mass) as double stars do, but this is not the whole story. Mizar has a companion star so close to it that we cannot observe it directly; therefore we can only deduce that the companion exists from what we observe in the spectrum of the light from Mizar. This spectrum shows the absorption lines at frequencies characteristic of the photon energies that hydrogen atoms absorb (Page 34). But the lines periodically double, merge again into one, double again, then merge, and so on. We can explain this as the light from two stars in orbit close to one another. As one star approaches us in orbit while the other recedes from us, the lines double in the spectrum we observe (Figure 7–5). The Doppler effect slightly decreases the energies of the photons from the receding star, and slightly increases the energies of the photons from the approaching star, so that the spectral features of the two stars do not exactly coincide. When the stars both are moving *across* our line of sight, but in opposite directions, the absorption lines from both stars appear at the same energy, because the Doppler effect does not change the energy of the photons that we observe. (If the entire center of mass is moving toward us or away from us, this motion will produce the same amount of Doppler shift in the spectra of all the stars in the system. The *change* in the amount of the Doppler shift of one star's photons relative to another's is what makes the lines periodically appear double.) The amount of the largest Doppler shift, about forty kilometers per second, tells us how fast the stars move in orbits around their common center of mass, and the length of time between

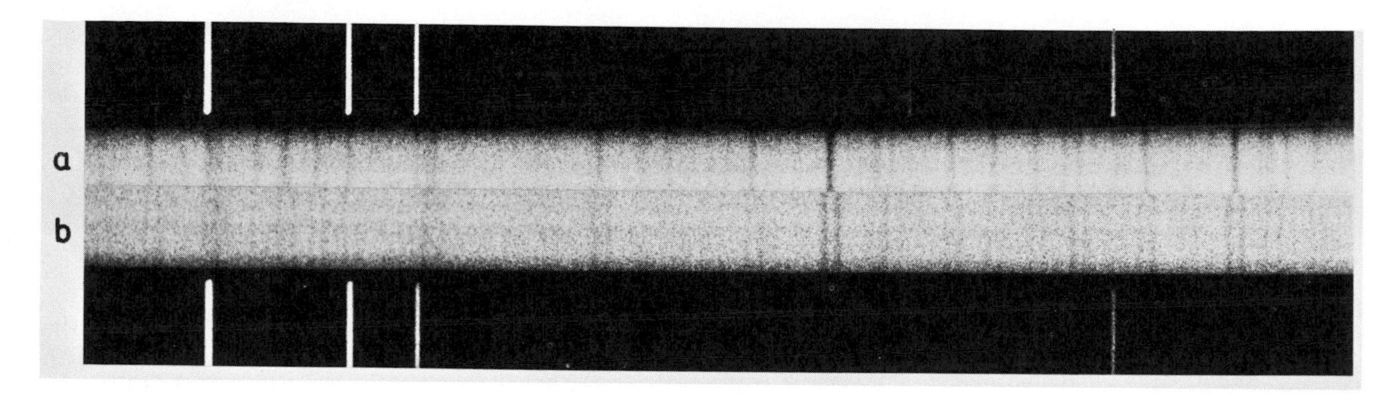

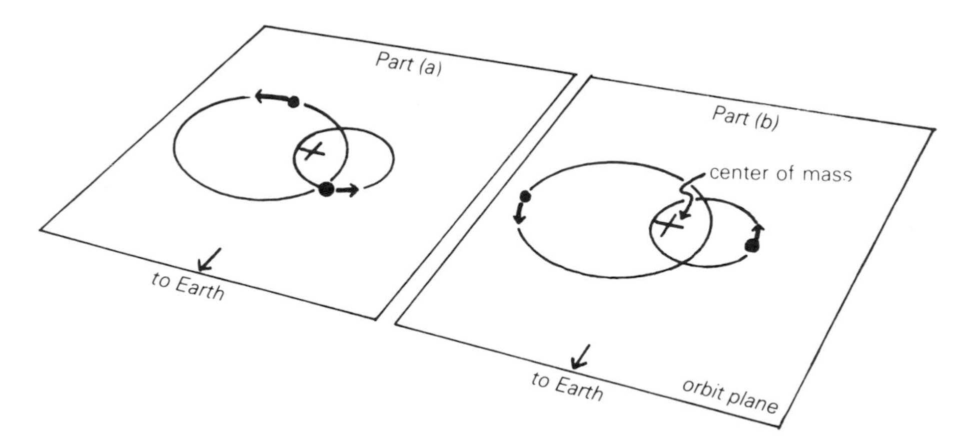

Figure 7–5. The spectrum of the light from a "spectroscopic double star" shows a periodic doubling of the absorption features as one star moves toward us while the other moves away from us, relative to the motion of the center of mass of the double-star system. This composite spectrum taken of the star Mizar at two different times shows (Part a) the single absorption lines that arise when both stars are moving across our line of sight, and also (Part b) the doubling of the lines that occurs when one star is moving toward us and the other away from us.

the doubling of spectral features, about ten days, tells us how long the stars take to move half way around their orbit (see Figure 7–5).

Many familiar stars are "spectroscopic" double stars, which cannot be seen as individual stars in a telescope because the stars are too close to each other. The closer together the stars are, the faster they must orbit around their common center of mass to keep from falling into each other. In such a case we can more easily detect their existence as two stars, because the amount of the Doppler shift in their spectra increases. The star Castor, the second brightest star in Gemini, shows three individual stars in a good telescope. *Each* of these three is a spectroscopic double star, so in fact there are six stars, and the three close pairs perform a complicated dance around each other as each of the pairs itself spins on its own cycle (Figure 7–6).

For some double stars, we on Earth happen to be looking just exactly along the plane of the stars' orbits around their center of mass. In this case, when one star passes in front of the other we observe an eclipse that happens again when the stars' positions in orbit are reversed (Figure 7–7). These "eclipsing" double stars were the first to attract attention. In particular, the famous star Algol ("The ghoul" in Arabic), in the constellation Perseus, shows a sudden decrease in apparent brightness every three days when the brighter, hotter star passes behind the fainter star of the pair. Eclipsing double stars have their eclipses only because we happen to be looking along the plane of their orbits, so the chances of our observing a particular double star as an eclipsing pair are small. The closer together the two stars are, the greater is the chance that we will see eclipses as they orbit around each other.

Energy Liberation in Stars

Whether stars come singly or in multiples, their basic property remains their liberation of energy of motion. How stars release energy remained an unsolved mystery until late in the 1930's, when scientists realized

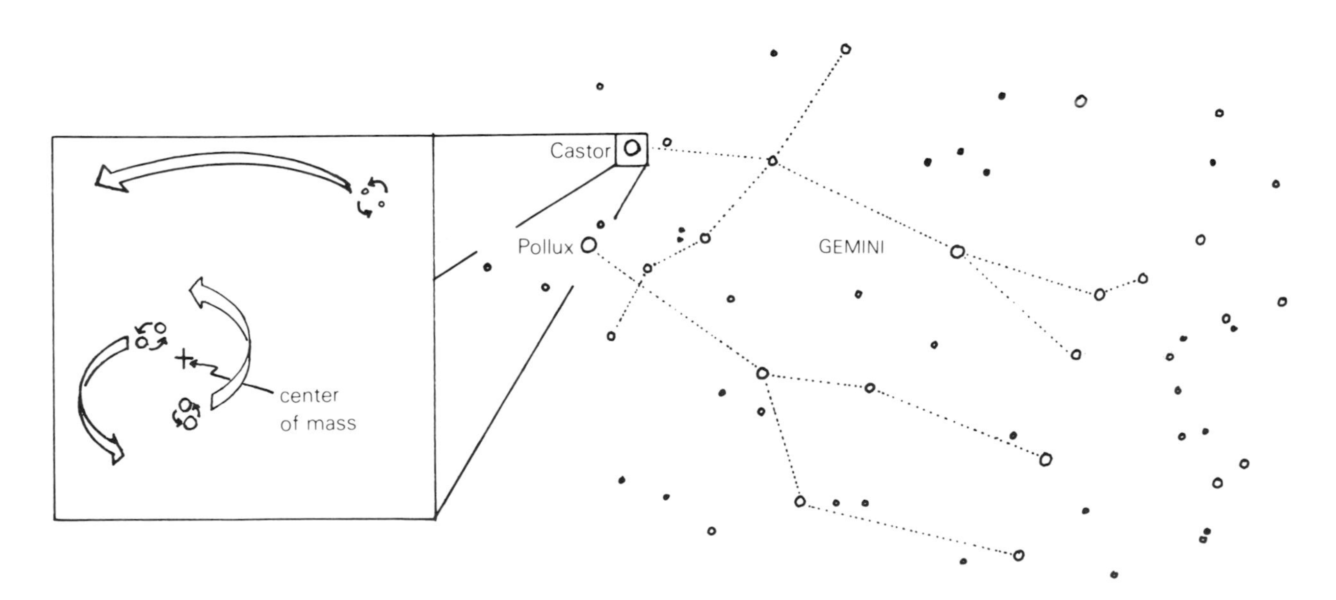

Figure 7–6. The star Castor actually consists of three pairs of stars, which all are in orbit around one another.

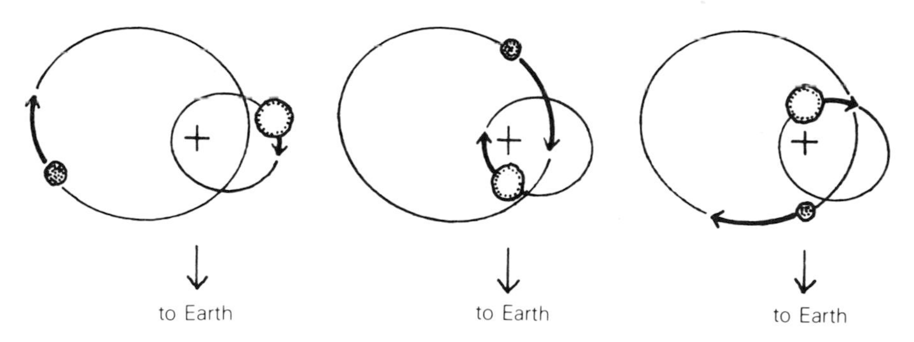

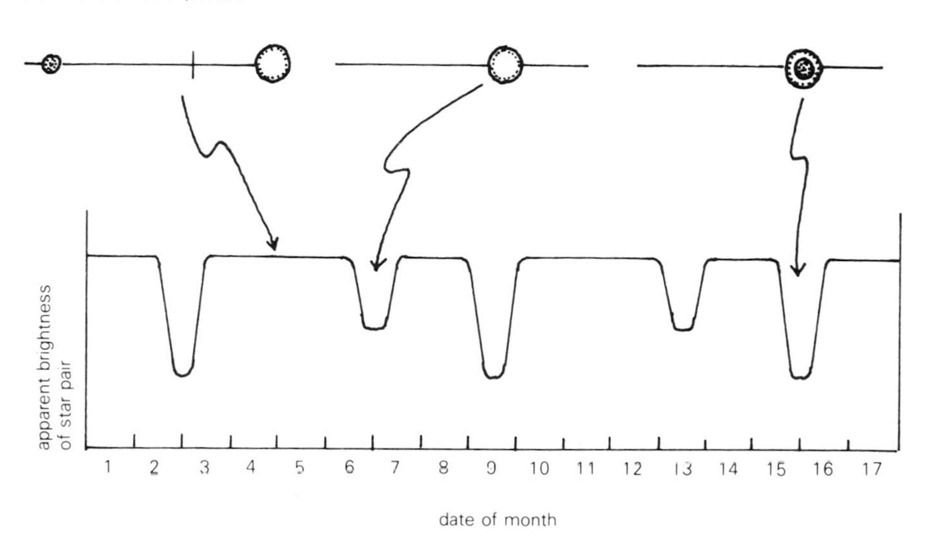

Figure 7–7. If we happen to be looking along the plane of two stars' orbits around their common center of mass, we can see one star eclipse the other periodically as the stars move along their orbits.

that nuclear reactions inside stars produce the energy that allows us to live, and that the same sort of nuclear reactions could transform life on Earth, for better or worse. Let us see how stars solved the energy problem billions of years ago, and how they can continue to produce energy for billions of years in the future.

Stars produce photons by converting energy of mass into energy of motion. Usually we don't enjoy remembering equations, but Einstein's best one has conquered the hearts of millions of nonscientists. $E = mc^2$ states that the energy of mass equals the mass times the speed of light squared, anytime, anywhere. Thus a greater mass will, if turned into energy of motion, liberate a greater amount of energy. So it has been, and so it will be.

When a particle meets its corresponding antiparticle, *all* of the energy of mass of the particle and antiparticle is converted into energy of motion. This would be the most efficient way for a star to produce energy of motion, but a star could never form from an even combination of matter and antimatter, because the two would annihilate each other long before any condensation shrank to the size of a star. The process of thermonuclear fusion that produces energy of motion inside stars has an efficiency of only about one percent: The particles involved in the thermonuclear fusion reactions lose about one percent of their mass, hence one percent of the energy of mass turns into the energy of motion liberated inside stars.

Why do thermonuclear reactions occur inside stars? And why don't they blow stars apart? After all, hydrogen bombs involve the same sort of thermonuclear reactions that power the stars, yet stars don't explode. We can think of the interiors of stars as fusion reactors, which liberate the same amount of energy per second over a time period of billions of years. The fusion reactions inside stars occur because the temperature in stellar interiors grows so large that the simple hydrogen nuclei (protons) fuse together into more complicated nuclei. In these fusion processes some energy of mass becomes energy of motion, but the star holds itself together by gravity well enough that even the outward rush of this newly liberated energy of motion does not destroy the star.

To understand how fusion reactions begin inside a star, we consider what happens to a contracting protostar. Suppose that a clump of gas has the mass and the density necessary to hold itself together by its own self-gravity. That is, the mutual gravitational attractions of the atoms in the clump can overcome their tendency to wander apart because of their random velocities. If the gravitational forces are strong enough, the clump will start to contract (Figure 7–1). As the clump condenses to a smaller and smaller size, the density of matter within the clump will grow larger and larger. This increase in density will be most noticeable at the center of the clump, where the matter will tend to concentrate. Each new chunk of matter pulled toward the center will increase the density there by bringing more material along with it. The constant arrival of more and more matter increases the gas pressure in the central regions of the clump. Thus the gas in the central regions will be squeezed together more and more, and it will resist this squeezing by increasing its temperature. Each atom in the gas must move around more and more rapidly, so that the gas at the center, although it is very dense, uses the random velocities of its individual atoms to avoid being squeezed into no space at all. The balance of the atoms' random motions with the growth of density at the center follows naturally as more and more material responds to the gravitational force and heads toward the center of the clump.

As the protostar continues to condense, the temperature inside it rises higher and higher, especially in the central regions. The initial cloud of gas had a temperature of ten or twenty degrees above absolute zero. As the cloud contracts under the influence of its own gravity, the temperature in its central regions rises to hundreds and thousands of degrees absolute. When the temperature reaches ten thousand degrees absolute, the atoms have enough energy of motion to ionize each other when they collide. Collisions among atoms then will knock loose the atoms' electrons, and if any electrons should recombine onto the ionized atoms, they will soon be knocked loose again by the high velocity of impact between the particles. This ionization process does not stop the condensation of the protostar, so the central regions keep getting denser and hotter. What eventually does stop the collapse of the star is the start of nuclear fusion reactions at its center. These reactions liberate energy of motion, which keeps the center so hot that it resists further contraction induced by the star's self-gravitation.

The fusion reactions involve protons, which constitute a major part of the star. Since the protostar formed from interstellar gas, we expect that the star and the gas will have a similar composition. Interstellar gas contains mostly hydrogen atoms (90% by number, 75% by mass) and helium atoms (10% by number, 25% by mass). All heavier atoms (carbon, nitrogen, oxygen, etc.) provide at most one percent of the mass of the protostar. Clearly the major part of the gas, both now and during the past billions of years, consists of hydrogen. The ionization of hydrogen inside a condensing star produces a mixture of protons and electrons. The ionization of helium atoms produces more electrons and helium nuclei. All other forms of nuclei are rare exceptions to the basic composition of protons, electrons, and helium nuclei. The protons constitute most of the protostar's mass (they are 1836 times more massive than electrons are), and fusion reactions among protons liberate most of the energy that a star produces.

The Proton-Proton Cycle

When the central regions of a collapsing protostar reach a temperature of ten million degrees absolute, pairs of protons begin to fuse together. At lower temperatures, the protons are moving too slowly to overcome their mutual repulsion through electromagnetic forces. Since protons all have a positive electric charge, they tend to repel one another, and as we discussed previously (Page 48), their mutual gravitational attraction is insufficient to overcome this repulsion. However, if two protons approach each other with a sufficient energy of motion, they can overcome their mutual electromagnetic repulsion enough to get into the distance range where strong forces act (Page 48). Strong forces hold protons and neutrons together in atomic nuclei. They always attract one

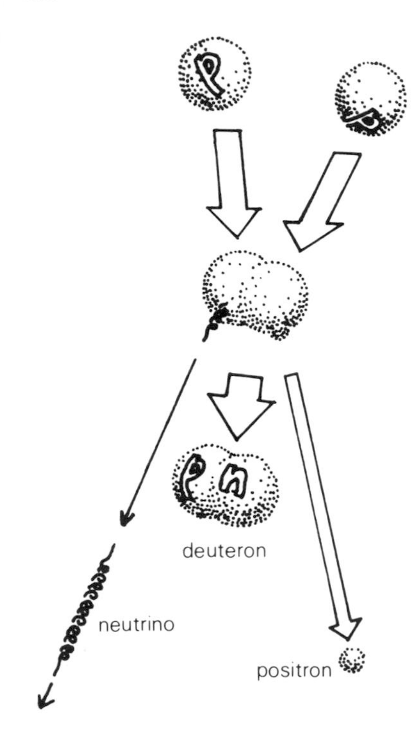

Figure 7–8. In the first step of the proton-proton cycle, two protons fuse to produce a deuteron, a positron, and a neutrino, plus additional energy of motion.

nucleon (proton or neutron) to another, but they are effective only over small distances, a few times 10^{-13} centimeter or less. Since protons at a temperature of ten million degrees absolute are moving at a speed of 10^{10} centimeters per second, this means that strong forces must act within a time span of 10^{-23} second, the time that the two protons are close enough to each other to feel the effect of the strong forces. Electrons do not feel the effect of strong forces in any situation, so the electrons simply repel each other by electromagnetic forces no matter how close they come to each other.

The fusion of two protons by strong forces forms the first step in the *proton-proton cycle*. This fusion reaction produces a deuteron, a neutrino, and a positron, as shown in Figure 7–8.[1] A deuteron consists of a proton and a neutron held together as one nucleus. We might expect that since a neutron has slightly more mass than a proton, a deuteron would have slightly more mass than two protons. But in fact a deuteron has *less* mass than two protons. This violates our intuition but is crucial to the liberation of energy. The fact that many nuclei have *less* total mass than the sum of the masses that went into them arises from the way that strong forces work. This decrease in mass does not occur for all nuclei: Some heavy nuclei, most notably uranium-235, have slightly *more* mass than the elementary particles from which they were formed. But all the lighter nuclei, such as helium, carbon, nitrogen, oxygen, neon, and even iron have slightly less mass than the sum of the masses of the protons and neutrons that form these nuclei.

Hence a deuteron, the nucleus that consists of a proton and a neutron, has less mass than two protons do. The deuteron, in fact, has less mass than the mass of two protons *plus* the mass of the positron that is produced by the fusion of two protons to form a deuteron (remember that a neutrino has no mass at all). As Table 7–1 shows, the two protons that fuse have a total mass of 3.3448×10^{-24} gram. The total of the deuteron's mass plus the positron's mass is only 3.3441×10^{-24} gram. The decrease in mass, and thus in the energy of mass, after the fusion reaction appears as energy of motion. That is, a total of 0.0007×10^{-24} gram of matter has disappeared, to be replaced by energy of motion, which adds to the energy of motion of the particles that emerge from the reaction. We can find the amount of energy of motion by using Einstein's equation, $E = mc^2$. If we multiply the mass difference, 0.0007×10^{-24} gram, by the speed of light squared, $c^2 = 9 \times 10^{20}$ centimeters squared per second squared, we find an energy increase of 6.3×10^{-7} erg. (One erg equals one gram times one centimeter squared per sec-

[1]This first step of the proton-proton cycle involves weak forces as well as strong forces. The weak forces signal their presence by the appearance of the neutrino, and although strong forces do the basic work in binding the deuteron together, weak forces also play an important role in making the reaction occur.

Table 7–1
Change in Mass During the First Step of the Proton-Proton Cycle

Before		After	
Mass of proton	$= 1.6724 \times 10^{-24}$ gram	Mass of deuteron	$= 3.3432 \times 10^{-24}$ gram
Mass of proton	$= 1.6724 \times 10^{-24}$ gram	Mass of positron	$= 0.0009 \times 10^{-24}$ gram
		Mass of neutrino	$= 0$ gram
Total mass before	$= 3.3448 \times 10^{-24}$ gram	Total mass after	$= 3.3441 \times 10^{-24}$ gram
Energy of mass before	$= (3.3448 \times 10^{-24}$ gram$) \times c^2$	Energy of mass after	$= (3.3441 \times 10^{-24}$ gram$) \times c^2$
[Energy of mass before] − [Energy of mass after] $= (0.0007 \times 10^{-24}$ gram$) \times c^2 = 6.3 \times 10^{-7}$ erg			

Note: The positron will annihilate with an electron to liberate another 54×10^{-7} erg of energy of motion.

ond squared, for reasons lovers of physics easily can understand.) Every fusion of two protons into a deuteron, a positron, and a neutrino converts 0.0007×10^{-24} gram of mass into 6.3×10^{-7} erg of energy of motion. This energy of motion, which was formerly a tiny part of the energy of mass of the colliding protons, appears as extra energy of motion of the deuteron, positron, and neutrino. In other words, if two protons collided with an energy of motion of 5×10^{-9} erg, so that the total energy of motion before the collision was 10^{-8} erg, then the total energy of motion after the collision would be $10^{-8} + 6.3 \times 10^{-7}$ erg. This new energy of motion can be shared in different proportions by the deuteron, the positron, and the neutrino, but their total energy of motion always will be 6.3×10^{-7} erg more than the total energy of motion of the two protons before the collision.

In this first step of the proton-proton cycle, about one five-thousandth of the protons' energy of mass turns into energy of motion. The new energy of motion, 6.3×10^{-7} erg, could not be felt by a human being if it tickled our tongues, but since about 10^{39} of these reactions occur each second inside a star, their total effect can be seen quite easily. To be thorough, we must examine the second and the third steps of the proton-proton cycle, the basic energy-liberating process inside stars.

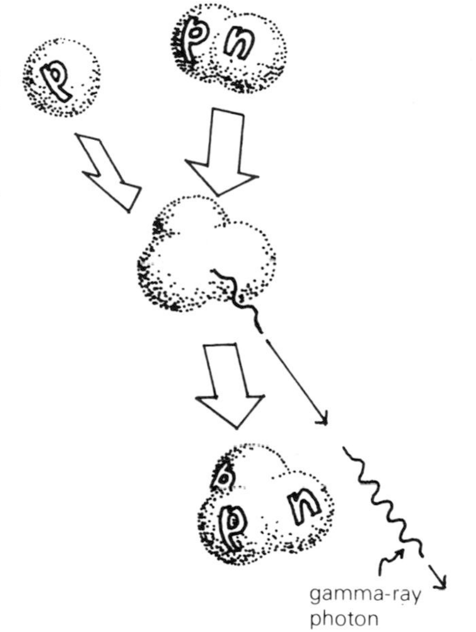

Figure 7–9. In the second step of the proton-proton cycle, a proton fuses with a deuteron, thereby producing a photon, a nucleus of helium-3, and additional energy of motion.

In the second step of the cycle, a proton collides with a deuteron at speeds high enough to produce a fusion reaction (Figure 7–9). The collision produces a nucleus of "helium-3" and a photon, plus some extra energy of motion. Helium-3 (He^3) has a nucleus of two protons and one neutron, whereas ordinary helium, called "helium-4" (He^4), has a nucleus of two protons and two neutrons. The mass of a helium-3 nucleus, 5.0059×10^{-24} gram, is slightly less than the mass of a proton plus the mass of a deuteron. Therefore the particles that appear after the collision (a photon and a nucleus of He^3) together have slightly more energy of motion, and slightly less energy of mass, than the particles that collided. Again some energy of motion, in this case 88×10^{-7} erg, has turned into additional energy of motion.

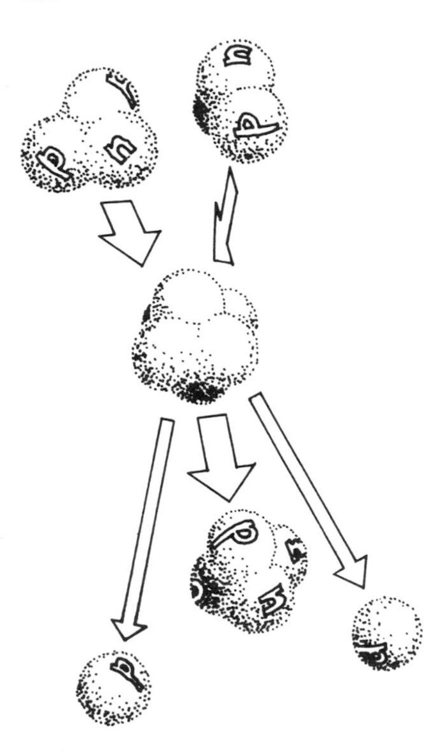

Figure 7–10. The third and final step of the proton-proton cycle fuses two nuclei of helium-3 to produce one nucleus of helium-4, two protons, and still more energy of motion.

The third step of the proton-proton cycle actually liberates most of the energy of motion. Once helium-3 nuclei have been made, two of them can collide (Figure 7–10). This collision will produce a nucleus of helium-4 and two protons, plus additional energy of motion. In this third step, the particles after the collision have a total mass that is 2.28×10^{-26} gram less than the mass of the two colliding helium-3 nuclei. The collision thus converts an energy of mass equal to 2.28×10^{-26} gram times c^2, or 205×10^{-7} erg, into additional energy of motion of the nuclei that appear after the collision.

The three steps of the proton-proton cycle, which are summarized in Figure 7–11, have the cumulative effect of taking six protons to start with and ending up with two protons, two photons, two positrons, two neutrinos, a nucleus of helium-4, *and* 4×10^{-5} erg of additional energy of motion. The reason for so many two's is that the third stage of the cycle requires the collision of *two* helium-3 nuclei. Each He^3 nucleus emerges from the first two steps, so we need two of Steps 1 and 2 to have one Step 3 in the proton-proton cycle. This leaves us with the end products listed above. The neutrinos escape directly from the star, taking a small part of the new energy of motion along with them. The positrons annihilate with electrons almost immediately, and the photons that result from this annihilation add a little energy to the energy of motion produced by the proton-proton cycle (we already have included this in the total of 4×10^{-5} erg given above.) These photons, plus the two photons from the second step of the cycle, tend to escape outwards from the center of the star, where they were produced, but they will undergo many collisions with electrons and with nuclei before they escape all the way into space. The protons and the helium-4 nucleus stay in about the same place as the original six protons that entered the cycle. Hence the net result of the proton-proton cycle is this: Four protons combine to make one nucleus of helium-4, plus photons and neutrinos, which carry away most of the energy of motion that the proton-proton cycle liberates from the conversion of energy of mass. The mass of the initial colliding particles is about 1% greater than the mass of the final particles, thus 1% of the original energy of mass turns into energy of motion, while the other 99% remains as energy of mass.

The photons produced in the proton-proton cycle carry off about 90% of the energy of motion liberated by these fusion reactions. As they try to escape, the photons constantly hit the elementary particles, mostly electrons and protons, that surround them. In these collisions, the photons yield some of their energy of motion to the protons and electrons immediately around them, thereby giving more energy of motion to the protons and electrons. These particles in turn collide with the protons and electrons immediately outside *them*, and the cascading result of the liberation of energy at the star's center is that *all* of the particles in the

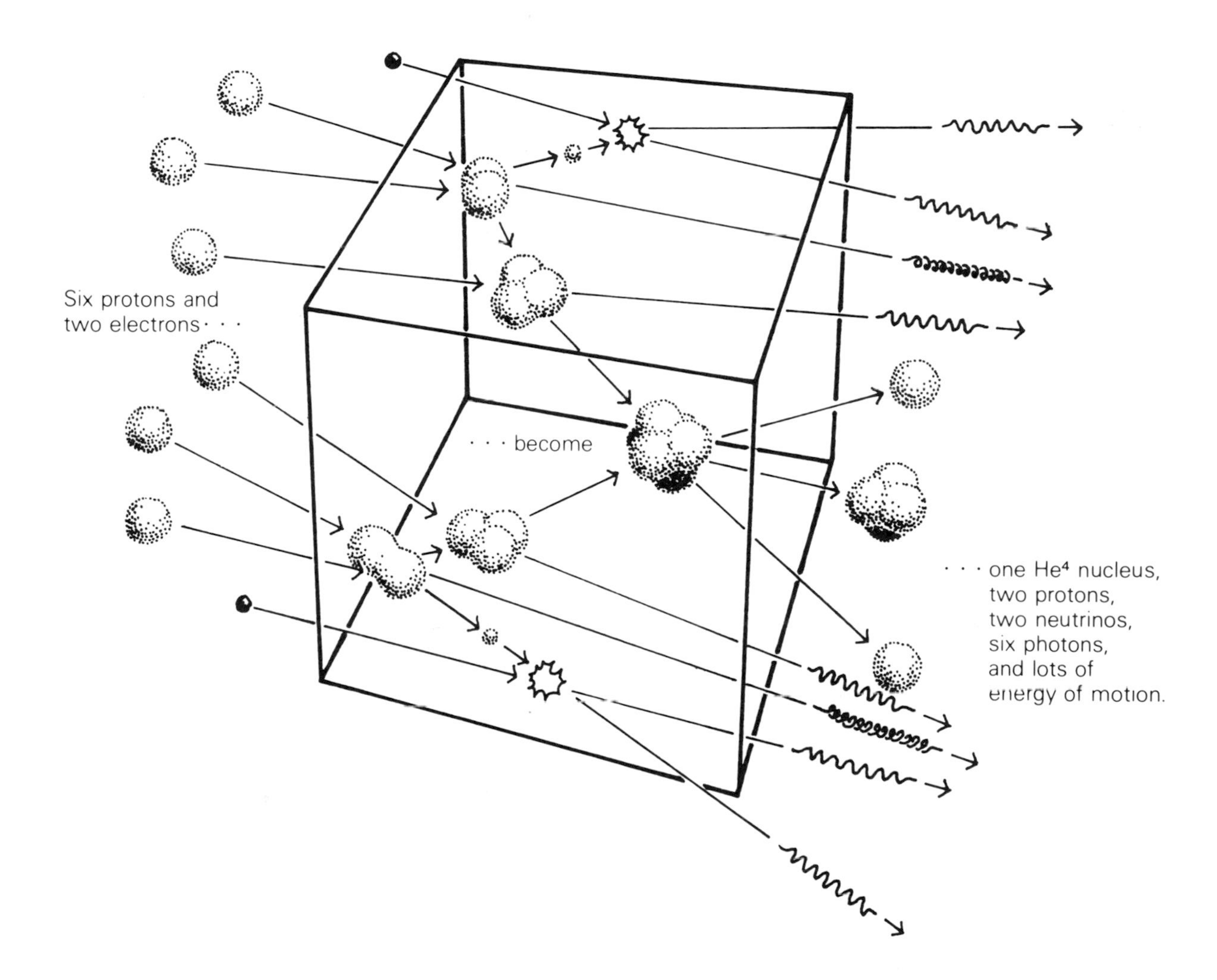

Figure 7–11. Together, the three steps of the proton-proton cycle turn six protons and two electrons into two protons, one nucleus of helium-4, two neutrinos, and 4×10^{-5} erg of energy of motion. The two positrons produced in the proton-proton cycle annihilate with two electrons, thereby converting all of their energy of mass into energy of motion.

star acquire some of the energy liberated at the center. The particles nearest the center acquire the most energy per particle, and the particles farthest out get the least. The energy of motion that each particle receives from its numerous collisions originates with the hydrogen-to-helium fusion reactions in the hottest, innermost parts of the star.

When a protostar condenses, its central regions grow hotter, and the peak temperature occurs at the center of the condensation. When the temperature reaches ten million degrees absolute, the proton-proton cycle of nuclear fusion begins in the star's interior. We say *star* because the onset of nuclear fusion reactions marks the transition from a condensing protostar to an energy-liberating star. The energy liberated by nuclear fusion, communicated outward to all the particles, gives the entire star the energy of motion that its constituent particles need to provide a counter-pressure against the inward pull of gravity. In every star, these opposing tendencies—gravitational self-compression and the outward flow of energy of motion (communicated by collisions among the particles)—have a chance to balance each other and hold the star at a constant size.

Stars shine by thermonuclear fusion reactions among their elementary particles, and the most important of these reactions are in the proton-proton cycle. Another series of reactions called the "carbon cycle" (Figure 7–12) uses a carbon nucleus as a catalyst for turning protons into helium nuclei, the same result as that achieved by the proton-proton cycle. The protons combine onto the heavier nucleus, which first turns into a nitrogen nucleus and then becomes an unstable oxygen nucleus, as protons fuse with it. The oxygen nucleus splits into a carbon nucleus and a helium nucleus, so the net result of the carbon cycle almost equals the result of the proton-proton cycle: Four protons combine to form one helium nucleus, plus some photons and neutrinos, which carry away the extra energy of motion made from energy of mass. The carbon nucleus can be used over and over again as a site for this process, because the carbon cycle begins with a carbon nucleus and ends with a carbon nucleus.

Most stars, including the sun, produce their energy of motion from the proton-proton cycle. Some young, bright stars, which contain a sufficient number of carbon nuclei and have a large enough central temperature (more than twenty million degrees), liberate most of their energy of motion through the carbon cycle.

Masses and Ages of Stars

The most fundamental differences among the various stars are their *masses* and *ages*. A star's mass determines how heavily its outer layers will press inward at the center. This pressure must be balanced by the energy of motion of each particle within the star. *Temperature simply measures the average energy of motion per particle.* In most stars, the

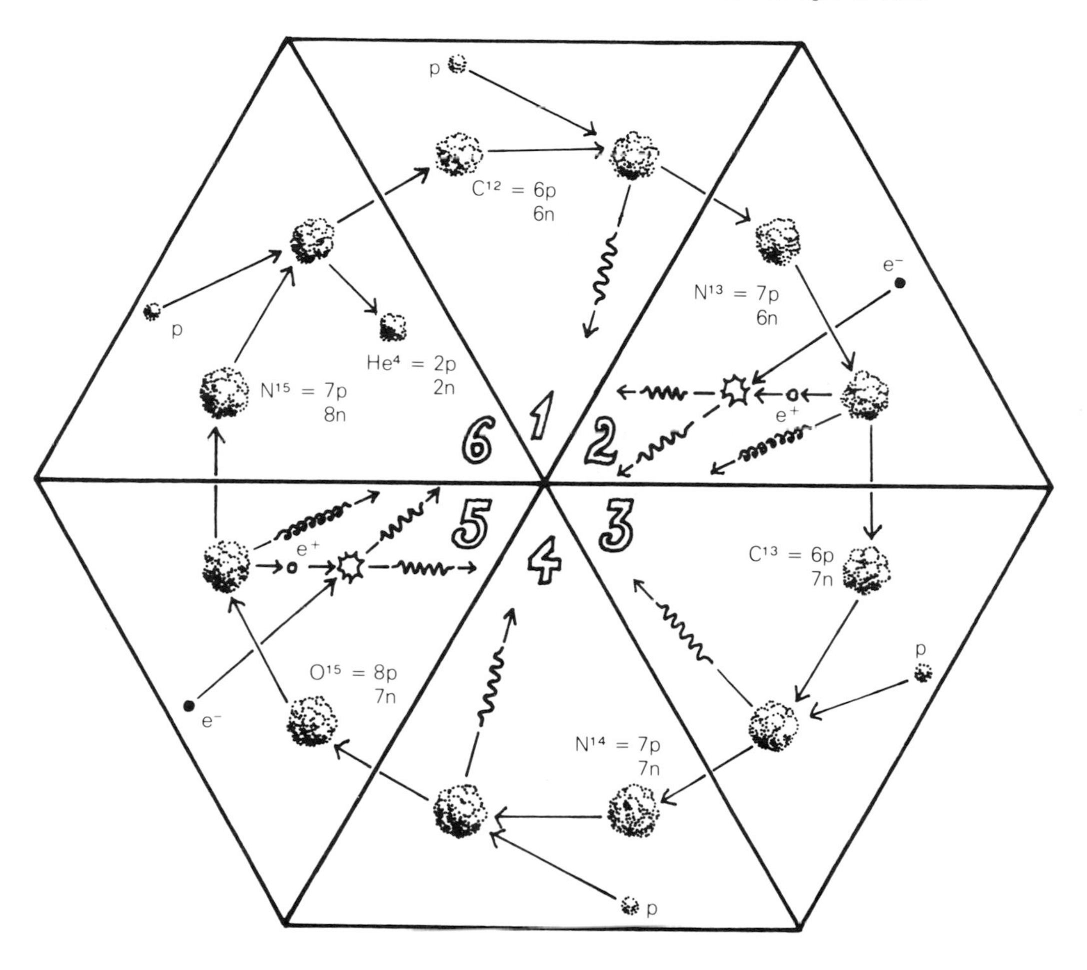

Figure 7–12. In the carbon cycle a nucleus of carbon-12 acts as the catalytic site for nuclear reactions to occur. First a proton fuses with a carbon-12 nucleus to make a nucleus of nitrogen-13, a photon, and additional energy of motion. Next, the nitrogen-13 nucleus changes into a nucleus of carbon-13 plus a positron and a neutrino. (This decay is analogous to the decay of the neutron mentioned on Page 49.) The carbon-13 nucleus will fuse with a proton to make a nucleus of nitrogen-14 plus a photon, and then the nitrogen-14 nucleus will fuse with yet another proton to produce a nucleus of oxygen-15 and a photon. The oxygen-15 nucleus will decay into a nucleus of nitrogen-15 plus a positron and a neutrino; this is another weak reaction similar to the decay of carbon-13. Finally, the nitrogen-15 nucleus will fuse with a proton to make a nucleus of carbon-12 and a nucleus of helium-4, plus additional energy of motion. Thus we recover our original carbon-12 nucleus, and the net effect of the carbon cycle, like that of the proton-proton cycle, is to convert four protons into a helium-4 nucleus, two positrons, and additional energy of motion. The positrons soon meet electrons, and their mutual annihilation liberates still more energy of motion.

temperature of the central regions varies between eight million and thirty million degrees absolute, and the more massive stars are hotter at their centers than the less massive stars are. Higher temperatures, which indicate more energy of motion per particle, allow the positively charged protons to overcome their mutual electromagnetic repulsion more easily. Thus higher temperatures lead to a more rapid release of energy of motion in nuclear fusion reactions. These relationships arise naturally from the influence of a star's total mass on the conditions at its center.

If a star's outer parts weigh more heavily on its center, this greater pressure will cause the particles there to move more rapidly, to fuse more readily, and to liberate more energy of motion each second. This increased rate of energy liberation appears as an increased energy of motion per particle, that is, as a higher temperature. Thus the entire process is self-sustaining: More massive stars *need* a higher central temperature to support their greater weight, and they get it from the greater rate of nuclear reactions that this very weight induces at the star's center.

As a natural consequence of this interrelationship between the temperature in the star's center and the pressure from the weight of the star's overlying layers, stars can keep their rate of energy liberation at a constant level. If the particles at the center should happen to push out too much, the star's interior will expand slightly, thereby causing a drop in the pressure, a decrease in the temperature, and a decline in the rate of nuclear reactions. The star then will release less energy of motion per second, so the star's center will contract slightly to regain its original size. In this way, the interiors of stars regulate their rate of liberating energy to have the outward push of gas pressure (from the particles' energy of motion) just in balance with the contracting forces of gravity. More massive stars must release more energy of motion per second to balance their larger self-gravity, and this energy of motion appears at the stars' surfaces to give the more massive stars much greater true brightnesses than the less massive stars have. Figure 7–13 shows the relationship between the masses and true brightnesses of stars, as determined for those stars (mostly in double and multiple star systems) for which the mass can be estimated from the way that stars move around one another.

A star's *age* deserves our attention for two reasons. First, the age of the star determines how large a proportion of the star will consist of elements heavier than helium (Page 245). Second, no star can continue to liberate energy of motion indefinitely. Every star eventually must run out of fuel for the nuclear reactions that cause it to shine. As this day approaches, the star can embark on the various courses of "late stellar evolution" described in the next chapter, all of which lead to catastrophe in terms of the liberation of photon energy.

When we observe stars, we cannot easily determine their masses or their ages. For most stars, these two quantities can be estimated only from what we know about how stars work. Observations of stars do give us two other stellar characteristics: the stars' surface temperatures and their apparent brightnesses. For this reason, most of the classification of stellar types is in terms of the observational data, surface temperatures and apparent brightnesses, but we seek to relate the classification of stars made on this basis to the more fundamental characteristics of masses and ages.

Surface Temperatures and Absolute Luminosities

Astronomers can determine the surface temperature of stars from the spectra of photons that they emit. As the photons produced in a star's interior pass outwards toward its surface, they constantly collide with elementary particles and with the atoms and ions that exist within the star. An individual photon produced in nuclear fusion reactions inside a star will change its energy, and hence its identity, millions of times as it bounces through the star, eventually to appear at the surface. As a result of the many scatterings, absorptions, and reemissions of photons, the inside of a star closely resembles an ideal radiator (Page 72). In most of the star's interior, *almost* the same amount of energy enters and leaves each small volume every second. The exceptions to this rule are the innermost regions where the energy of motion is produced, and the outermost layers where the photons escape directly into space. In the intermediate zones, which form most of the star, great numbers of pho-

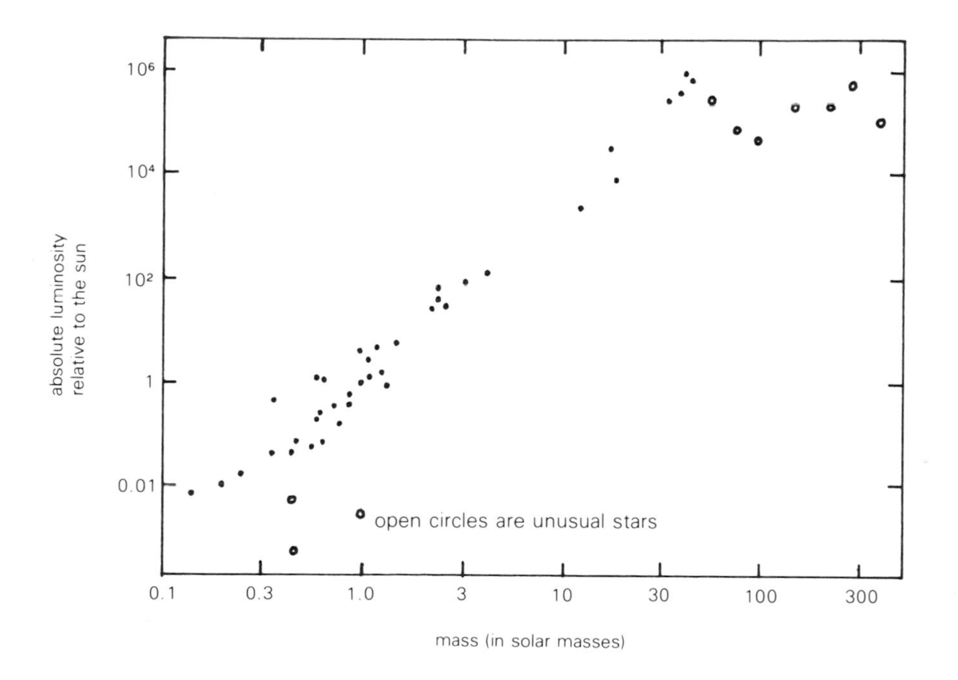

Figure 7–13. If we graph stars' absolute luminosities (true brightnesses) against their masses, we can see that there is a definite relationship: More massive stars have much larger absolute luminosities.

tons dart in every direction. The number of photons headed inward almost equals the number darting outward. The *net* flow is indeed outward, and this causes the energy produced at the center to escape eventually at the surface. However, in each small volume inside the star, this net outward flow comes from subtracting a huge inward flow from a slightly larger outward flow (Figure 7–14).

Because the inward and outward flows of photon energy are nearly equal inside a star, this interior volume is almost an ideal radiator. Inside each small volume, photons constantly collide with elementary particles and thus exchange energy back and forth with them. Thus the interior of a star has a general resemblance to the early universe, before the photons stopped interacting with the other particles (Page 74). The number of photons with various energies forms an energy spectrum close to that of an ideal radiator. The photons inside each small volume have an energy spectrum characterized by the ideal radiator *shape*, and the spectrum has its peak (greatest number of photons) at the energy that characterizes the temperature of the matter within this volume (Figure 7–15). The temperature inside the star decreases steadily outwards from the center, where temperatures are ten or twenty million degrees, to the surface, where temperatures are measured in thousands or tens of thousands of degrees above absolute zero.

When we observe the spectrum of photons from a particular star, we examine the number of photons at each energy (or at each wavelength) that leave the star's surface each second. The general distribution of

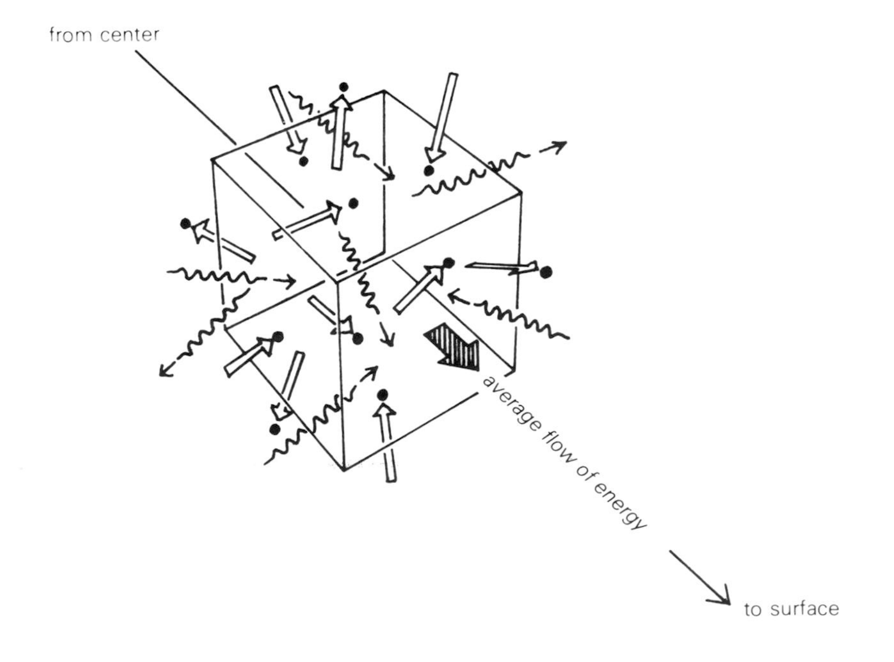

Figure 7–14. In each small volume inside a star, particles bring in energy of motion from all directions and carry off energy of motion in all directions. These particles include both photons and particles with mass, such as protons and electrons. When we add the energies coming in and going out in all directions, we find that the overall result is a flow of energy toward the outside of the star.

photon energies depends on the temperature in the star's surface layers. If this temperature is relatively high, say twenty thousand degrees absolute, then the largest number of photons have ultraviolet energies and wavelengths, because the peak of the ideal-radiator spectrum occurs at ultraviolet energies. If the star's surface temperature is low, say three thousand degrees absolute, like the star Antares in the heart of Scorpio, then the photon spectrum peaks at wavelengths and energies characteristic of red light. Stars with intermediate surface temperatures, such as our sun (six thousand degrees at the surface) or Sirius (ten thousand degrees), have a photon spectrum that peaks in the orange or yellow part of the spectrum. Various stellar spectra therefore differ first of all in the energy at which they emit the largest number of photons.

This peak energy in turn reflects the temperature of the surface layers of the star. Thus the star's *color*, which depends on the energy at which the largest number of photons appear, gives an estimate of the star's surface temperature. Astronomers have agreed on a series of standard wavelengths at which they measure a star's output of photons. From a comparison of the number of photons emitted at these various wavelengths we then can assign a "color index" to the star. The simplest color index measures simply the ratio of the star's brightness at a standard blue wavelength to its brightness at a standard yellow wavelength. To determine this basic color index, astronomers can photograph part of the sky with blue-sensitive photographic plates, and then rephotograph

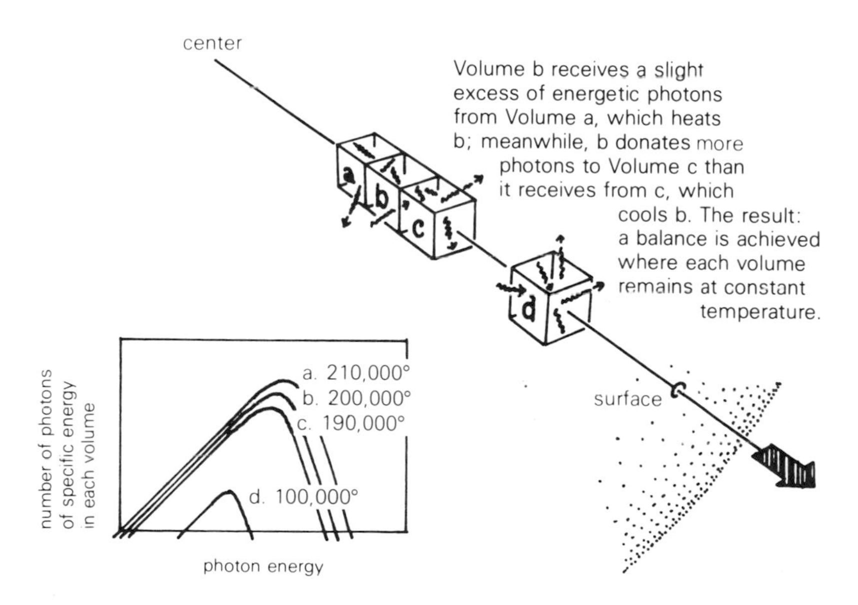

Figure 7–15. Within each small volume inside a star, the numbers of photons with different energies can be calculated. This photon energy spectrum within the volume closely resembles the energy spectrum of photons emitted by an ideal radiator. The temperature of the matter inside the volume determines where the *peak* number of photons will lie when we plot the spectrum of photon energies.

the same stars with yellow-sensitive plates. By measuring the brightness ratios of the stars' images on both plates, astronomers then can find the blue-to-yellow color index for all the stars. The nice part about this color index is that the star's distance doesn't matter, since the *ratio* of its output in blue light to its output in yellow light should be independent of the distance of the star. Thus the star's *apparent* blue-to-yellow light ratio should give the absolute blue-to-yellow ratio that reflects the star's particular surface temperature.

If the apparent brightness of a star (or a galaxy) is extremely faint, we may not be able to do more than to measure its brightness at a few standard wavelengths. But for many stars, we can study the spectrum of the photons in more detail than is given by simply naming the apparent brightness in certain standard colors and by making the appropriate color indices from their ratios. In addition to determining a star's color, the energy spectrum also shows the absorption of photons with certain particular energies that we described in Chapter 2. These photons are absorbed during the photons' final dash for freedom at the star's surface. Thus we can learn which kinds of atoms, ions, or molecules are present in the star's surface layers by studying the exact frequencies or wavelengths at which photons have been absorbed.

For example, in stars such as Sirius, hydrogen atoms near the surface absorb photons with a particular series of energies called the "Balmer series."[2] These energies are those that an electron in a hydrogen atom's second smallest orbit needs to jump to the third smallest, the fourth smallest, the fifth smallest, and so on in the orbital sequence (Page 34). The energy spectrum of the light from Sirius shows the Balmer series of absorption lines as its predominant features, and the presence of this series shows that Sirius's outer layers must contain many hydrogen atoms. Furthermore, since the Balmer series arises from electronic excitation from the *second* smallest orbit, we can conclude that at any moment, many of these hydrogen atoms have the electron not in the smallest possible orbit, but in the second smallest orbit. This excitation into a larger orbit comes from collisions among the hydrogen atoms, which can make an electron jump to a larger orbit. But if the collisions are too violent, they will destroy the hydrogen atoms completely by knocking the electrons loose.

The force of a collision depends on the average energy of motion that each atom possesses. This in turn depends on the temperature in the star's surface layers. At high temperatures the hydrogen atoms will collide so violently that they ionize themselves. At low temperatures, hydrogen atoms will exist, but only a few of them at any time will have

[2]It would be superfluous to point out that the man who first investigated this series thoroughly was named Balmer.

the electron in the second smallest orbit, ready to absorb a photon with one of the energies in the Balmer series. Therefore we expect that in neither the high-temperature nor the low-temperature surface layers will we find that the Balmer series of absorption lines stand out as an obvious feature. And indeed this proves to be true: Only those stars with surface temperatures between five thousand and twenty thousand degrees absolute show the Balmer lines in their energy spectra (see Figure 7–16). Stars with higher surface temperatures have few hydrogen *atoms* in their surface layers, and stars with lower surface temper-

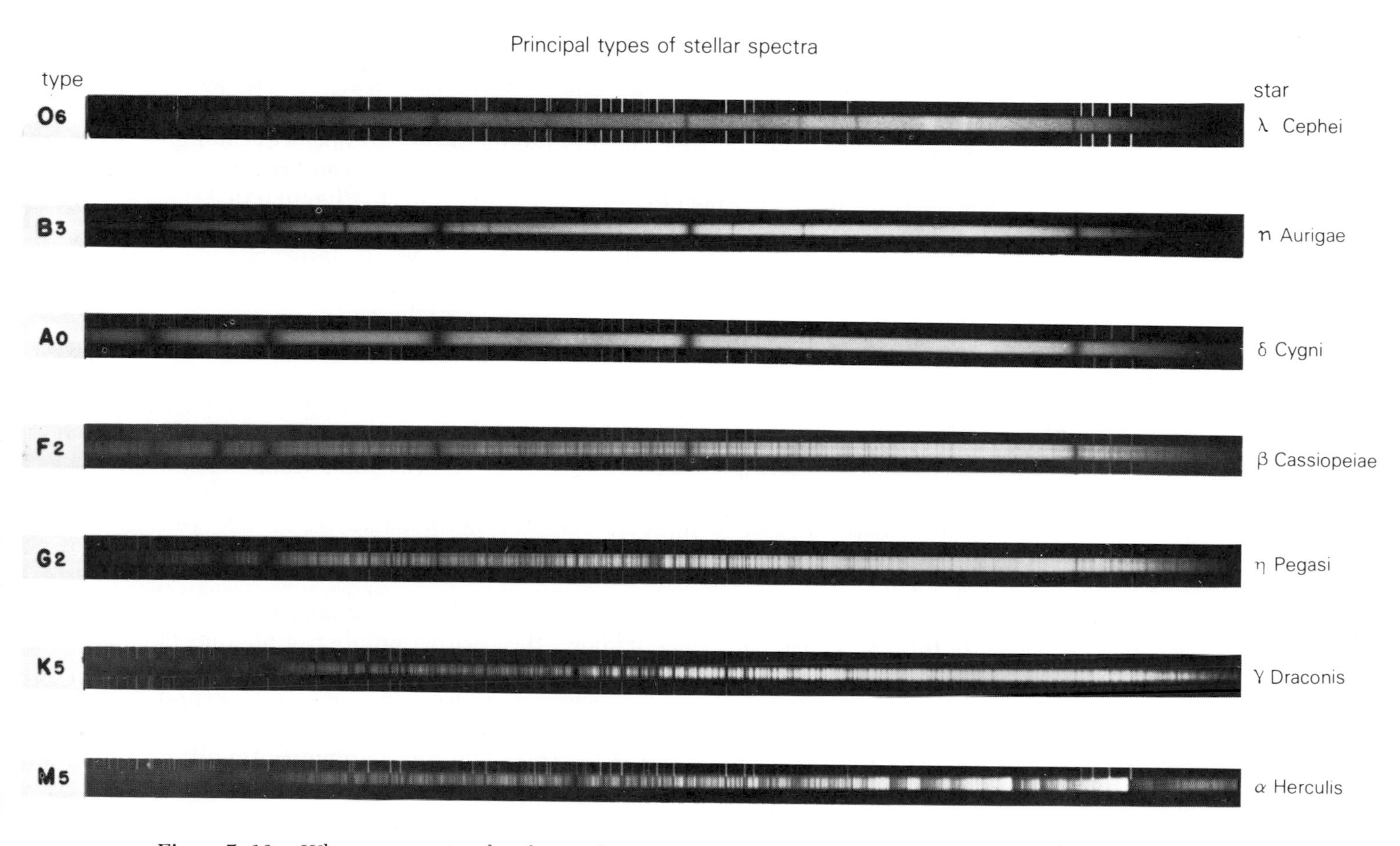

Figure 7–16. When we examine the photons from various types of stars, we find prominent Balmer lines (from absorption by hydrogen atoms) in those stars whose surface temperatures lie between 5000 and 20,000 degrees absolute. Notice that the Balmer lines are most prominent in the spectra of A0 stars such as Delta Cygni (third from the top), and lose their prominence for stars with surface temperatures much above or below the 10,000 degrees absolute that characterizes A0-type stars.

atures have almost no hydrogen atoms with electrons in the second smallest orbit.

When astronomers began to classify the spectra of stars, they used the prominence of the Balmer series as their basic criterion. This was natural because of the way that the Balmer-series lines stand out in the spectra of the visible light from stars. Had astronomers then been able to see the ultraviolet part of the spectrum, which our own atmosphere absorbs, they could have seen the Lyman series of absorption energies that come from photons being absorbed by hydrogen atoms with electrons in the smallest possible orbits, and they probably would have made an entirely different classification of stellar spectra on the basis of the still more intense Lyman lines. But eighty years ago, astronomers labeled the spectra with the most outstanding Balmer lines as "A," the spectra with the slightly less prominent Balmer lines as "B," and so on to class "O" and even beyond. Unfortunately, the reason *why* the Balmer lines stood out more or less prominently had not been understood. As we have seen, the absence of Balmer lines can arise from a star's surface layers being either too hot or too cold. By the time that the importance of this explanation became clear, the classification scheme had stuck (it was invented at Harvard), although some classes had been dropped or had been melded together. As a result of this history, the order of stars' spectral classification, from the highest to the lowest surface temperature, look like this:

O B A F G K M

(like the letters on an optometrist's chart), which can be understood only as an historical accident.[3] These spectral categories, or spectral types, each specify a range of surface temperatures and a characteristic appearance of the star's spectrum (Table 7–2). The types have been subdivided by placing numbers after the letters, like G2 or K6. The surface temperature of a G9 star barely exceeds that of a K0 star, and so forth.

The spectrum of a star's photons serves as the chief available indicator of the star's character, since it shows the relative number of photons of each energy that emerge from the star's surface layers. The *total* amount of energy that photons carry away from the star each second determines the star's absolute luminosity. If we can find the distance to a given star, we can determine the star's absolute luminosity from this

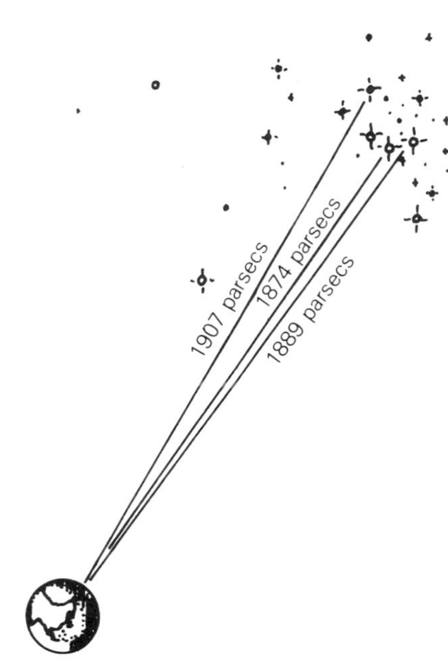

Figure 7–17. When we observe a cluster of stars, the differences in the distances to the individual stars are much less than the average distance to the stars in the cluster. Hence the stars' *apparent* brightnesses reflect the comparison of the stars' *absolute* brightnesses with acceptable accuracy.

[3]Some astronomers find the memory rule "O, Be A Fine Girl, Kiss Me" a helpful if sexist way to remember the spectral types in order of decreasing temperatures. Another such mnemonic is "O Buy A Fresh Green Kilo, Mister," which has capitalistic overtones.

Table 7–2
Surface Temperatures and Spectral Features of Different Types of Stars

Spectral type	Average surface temperature (degrees absolute)	Outstanding spectral features
O	30,000	Absorption lines from ionized helium; weak hydrogen absorption
B	20,000	Absorption lines from un-ionized helium; stronger hydrogen absorption features
A	10,000	Hydrogen absorption lines strongest; un-ionized helium atoms still show absorption features
F	7,000	Hydrogen lines still dominate spectrum; absorption features from "heavy" elements
G	5,500	Hydrogen lines weak; many absorption features from once-ionized and un-ionized heavy elements
K	4,000	Increasing number of absorption features from un-ionized atoms of "heavy" elements
M	3,000	Absorption features of un-ionized atoms and of simple molecules (in particular, titanium oxide) dominate the spectrum

distance and our measurement of the star's apparent luminosity. Together with the star's surface temperature, the star's absolute luminosity provides the most basic facts about the star that we can measure.[4] We can compare the surface temperatures and absolute luminosities of various stars to find out how stars are similar, and how they differ. If we have a cluster of stars that appear to be all at the same distance from us (Figure 7–17), we can compare the stars' absolute luminosities by comparing their apparent luminosities. Since the stars all are about the same distance from us, the ratios of their apparent brightnesses should be the same as the ratios of their true brightnesses. Hence even if we cannot measure the distance to a star cluster exactly, we can profit from our measurements of the stars' apparent brightnesses to *compare* the stars' true brightnesses with one another.

[4]We would consider the star's mass to be a still more basic fact, if we would measure it. However, only for a few double stars (Page 190) can we measure stellar masses.

The Temperature-Luminosity Diagram

When astronomers first turned their attention to the brightness and spectral types of stars in clusters, they found something interesting. If they made a diagram that showed the stars' apparent brightness in one direction and the surface temperature along the perpendicular direction (Figure 7–18), the stars did not scatter all over the graph. Instead, these "temperature-luminosity diagrams" of most star clusters showed a strong concentration close to one line in this graph of luminosity against surface temperature. Figure 7–18 shows the temperature-luminosity diagram of the Pleiades, an open cluster of relatively young stars. Notice that astronomers like to make these diagrams with the spectral type O (highest surface temperatures) at the left and spectral type M (lowest surface temperatures) at the right, which is just the opposite of what would be expected from the way most graphs are drawn. This astronomers' peculiarity, yet another historical accident, means that the surface temperatures *decrease* from left to right. In Figure 7–18 most of the stars fall on the "main sequence" of the temperature-luminosity diagram. That is, most stars show a definite relationship between their

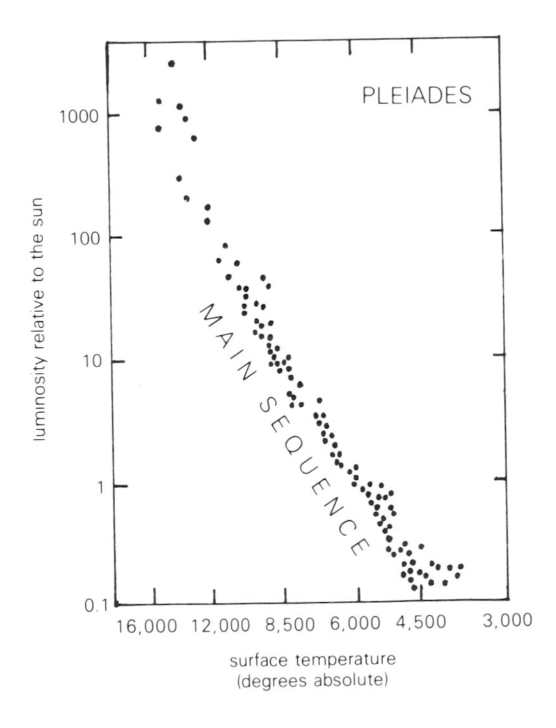

Figure 7–18. The diagram of luminosity (brightness) against surface temperature for the stars in the Pleiades shows that most of the stars fall near the "main sequence."

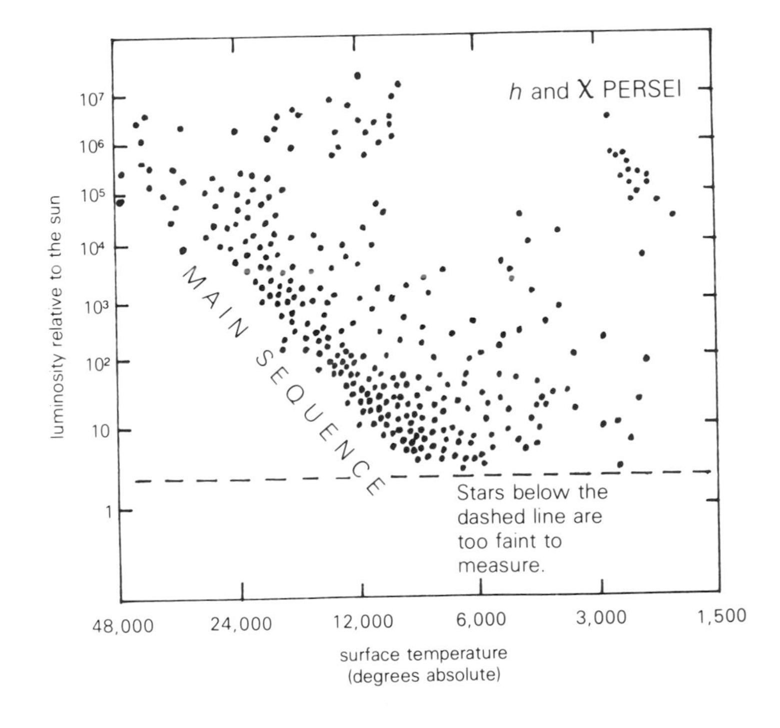

Figure 7–19. The temperature-luminosity diagram for stars in the open cluster *h* and X Persei shows that again most stars fall near the main sequence. In contrast to the Pleiades, this cluster also contains some red-giant stars.

surface temperature and their luminosity. The figure shows the apparent luminosities of the stars relative to the brightest star in the cluster, but if all the stars have about the same distance from us, the relationship also must be true for the stars' absolute luminosities. Larger surface temperatures go with larger absolute luminosities for all the stars on the main sequence. This pattern repeats for cluster after cluster (Figure 7–19), along with a minority of stars that do not fall on the main sequence in the temperature-luminosity diagrams.

The relationship between the surface temperature and the absolute luminosity of stars on the main sequence in the diagram eventually made sense to astronomers. Hotter stars turned out to be brighter, as we might expect. We now know that the hotter stars on the main sequence are the more massive ones, and the cooler, fainter stars have smaller masses, so that the masses, luminosities, and surface temperatures of stars all decrease as we go down the main sequence (Figure 7–20). Astronomers now have so much confidence in the main sequence that they have measured for cluster after cluster that if a star's spectrum

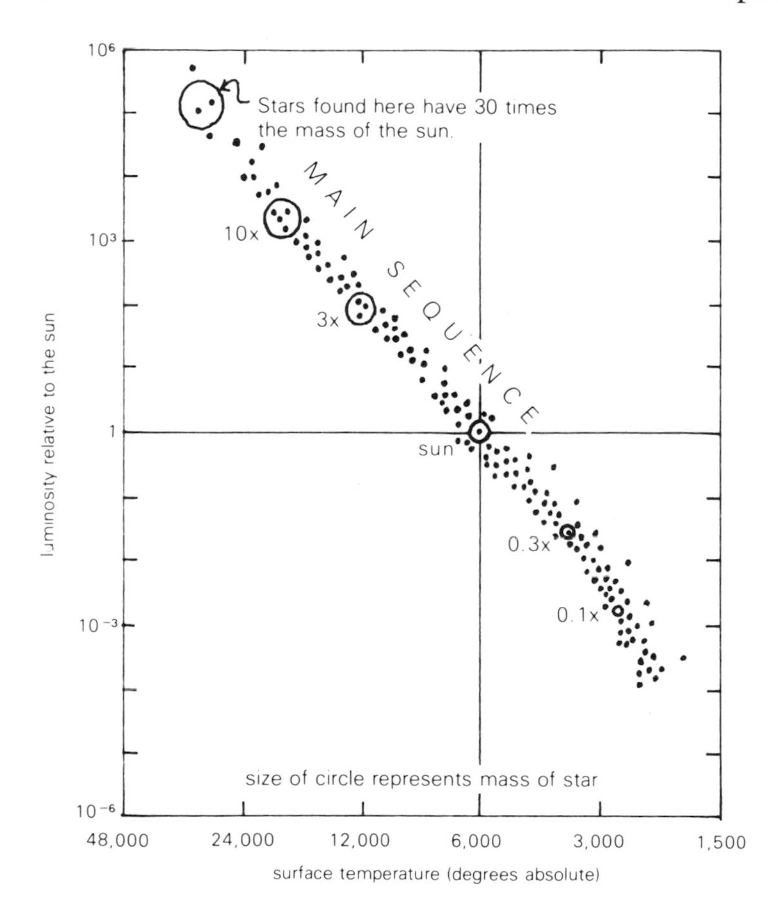

Figure 7–20. Stars lower down the main sequence (lower luminosity) all have progressively smaller masses and lower surface temperatures than the brighter stars on the main sequence.

looks like the spectrum of an ordinary main-sequence star, astronomers will assume that the star *is* a main-sequence star, since such stars form the vast majority of all stars. We then can assign the star a surface temperature from the appearance of its spectrum, and the assumption that the star lies on the main sequence tells us what the star's absolute luminosity should be. From this estimate of the star's absolute brightness and the measurement of the star's apparent brightness, astronomers then can estimate the star's distance. Such "spectroscopic" distance estimates, based on the idea that the star's spectrum tells us how bright it really must be, are not as reliable as the trigonometric methods discussed on Page 4, but they are useful in situations where no other method can be used. With careful cataloguing, astronomers have even been able to apply the spectroscopic method to the stars that do *not* fall on the main sequence in the temperature-luminosity diagram.

Red Giants and White Dwarfs

What are these stars that do not appear on the main sequence? These exceptions fall into two categories. Some stars have much larger true brightnesses than main-sequence stars with the same surface temperature, whereas other stars have far smaller true brightnesses (Figure 7–21). Astronomers realized some fifty years ago that these stars must be either extraordinarily large or extraordinarily small in comparison with normal stars. They knew this by applying the laws of physics that deal with ideal radiators, which stars closely resemble. Experiments had shown that the absolute luminosity of an ideal radiator varies in proportion to its surface area times the fourth power of its surface temperature. For stars with a particular surface temperature, the stars' absolute luminosities should vary in proportion to the stars' surface areas. The amount of surface area in turn varies as the square of the stars' radii. Twice as large a radius means four times as much surface area, and four times as much absolute brightness at the same temperature. Thus if two stars have the same surface temperature (as determined from the stars' spectra), but one star has a hundred times the absolute luminosity of the other, it must have ten times the radius of the other. What astronomers were finding fifty years ago was that one star with the same surface temperature as another had ten billion (10^{10}) times its absolute luminosity, and so had to be a hundred thousand (10^5) times larger!

Such stars of immense size were named red giants and red supergiants because of their colors and sizes. The red supergiant Antares has the same surface temperature as the nearby star Groombridge 34 A (the "A" means the brighter member of a double-star system),[5] but Antares'

[5]This double star was number 34 in Groombridge's catalogue. It is only four parsecs away, but nonetheless is far too faint to be visible without a telescope.

absolute luminosity is about a million times greater. Both of these stars have surface temperatures of about 3000 degrees absolute and belong to spectral class M, but Antares' radius must be the square root of one million, or one thousand times greater than the radius of Groombridge 34 A. It turns out that Antares' radius is about a hundred times larger than the sun's radius, which is almost as large as the Earth's orbit! Red supergiant stars such as Antares have tremendous sizes, but their masses are not that much greater (perhaps ten or twenty times larger) than the sun's mass. Therefore the average density of matter inside a red supergiant star is thousands of times less than the density of our atmosphere. Red supergiant stars consist of an energy-liberating core, even smaller than that in the sun, surrounded by a near-vacuum that can be hundreds of millions of kilometers thick.

What about the stars that lie below the main sequence, with far *less* absolute brightness than main-sequence stars at the same temperature?

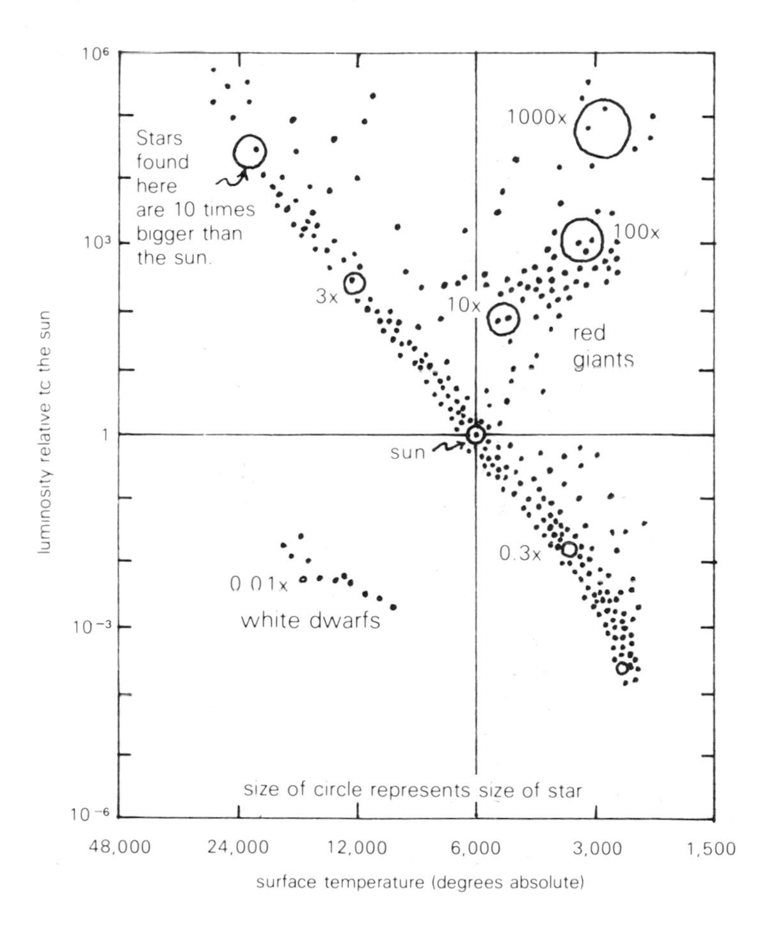

Figure 7–21. Some stars have much larger true brightnesses than main-sequence stars with the same surface temperature. These stars are much larger than the main-sequence stars of the same surface temperature and are called "red-giant" stars. The stars whose true brightnesses are far *less* than main-sequence stars with the same surface temperature comprise the "white-dwarf" stars.

These stars must be much smaller than main-sequence stars, and represent a numerous minority of stars called "white dwarfs." The term "white" means only that these stars are neither particularly red nor particularly blue. White-dwarf stars turn out to be relatively common, but they are hard to spot because of their small absolute luminosity. The best known white-dwarf star is a companion to Sirius; Sirius and its white-dwarf companion orbit around their common center of mass as they move around the galactic center. The bright star (Sirius A) is a normal main-sequence star and shines with an absolute luminosity ten thousand times greater than its faint companion (Sirius B), yet the two stars have almost the same surface temperature, ten thousand degrees absolute. Thus Sirius B must have a surface area about ten thousand times smaller than Sirius A, so its radius must be a hundred times smaller. We can conclude that the entire white-dwarf star, which has a mass almost equal to the sun's mass, has a size no larger than the Earth. Inside a white dwarf such as Sirius B, the average density of matter is enormous, a million times the density of water, hence thousands of times denser than any substance known on Earth. We shall see later (Page 231) how such tremendous densities occur as stars grow older and contract their insides.

Three major categories of stars—main-sequence stars, red giants and supergiants, and white dwarfs—include almost all of the stars found in the sky. Other names of star categories are either refinements of these three major groupings (like orange giants or yellow supergiants), or are entirely strange kinds of stars, such as the neutron stars we shall discuss in Chapter 12. We use the terms *red* giants and *red* supergiants to characterize the largest stars because it is at the red end of the temperature scale that giant stars differ most from main-sequence stars. So-called "blue supergiant" stars are simply the hottest, brightest stars on the main sequence (Figure 7–20).

The variation among different stars with the same surface temperature thus embraces a variation in the *sizes* of the stars. This difference in size appears as a change in the stars' average densities, because the variations in *size* far overshadow the changes in *mass* between one star and another. The star's average density, which equals the star's mass divided by its volume, depends mostly on the star's volume, which in turn depends on the cube of the star's radius. Astronomers have learned how to tell the spectrum of a giant or supergiant star from the spectrum of a main-sequence star *with the same surface temperature* by the subtle changes that the difference in density makes in the spectra. If two stars have the same surface temperature, the star with the higher density in its surface layers will show a greater tendency to have atoms rather than ions. The higher density means that ions and electrons are more likely to recombine into atoms, which then will be available to absorb a particular set of photon energies. Thus a main-sequence star of spectral type A1, such as Sirius, will show a stronger pattern of Balmer-series absorption lines than a giant star of almost the same spectral type, such as Deneb (type A2). The two stars have about the same surface temperature, but Deneb has an absolute brightness 2500 times that of Sirius, and so must be about fifty times larger in radius. Conversely, Sirius' companion, the white-dwarf star Sirius B, also has a type A spectrum, but the pattern of the Balmer lines in its spectrum is even more pronounced than in the bright star Sirius A, because the higher density leads to the more efficient recombination of electrons with protons to form hydrogen atoms.

To the absolute luminosity and surface temperature of stars, astronomers add "luminosity classes" to typify the stars' sizes. Thus they denote a main-sequence star by the Roman numeral V, and the sun's classification becomes G2V. The spectral type G2 indicates a surface temperature of six thousand degrees absolute, and the luminosity class V means that the sun has a size typical of main-sequence G2 stars. Luminosity Class I denotes the most extreme supergiants; thus Deneb is a type A2I star. Luminosity classes II, III, and IV are larger than main-sequence stars with the same surface temperature but smaller than stars of luminosity class I: Aldebaran, the brightest star in Taurus, is a type K2III star. White-dwarf stars have no Roman numerals; instead, astronomers, who never tire of inventing new ways to catalogue stars, put a special "w" *in front of* the spectral type; thus Sirius B is a wA5 star. Even these classifications have been further subdivided as the effort in stellar taxonomy continues, but we need not worry right now about the quirks of astronomers' notation. It is time to examine how stars evolve from one spectral type to another, from main-sequence stars to red giants to white dwarfs.

Summary

Stars seem to have condensed out of collapsing gas clouds billions of years ago, although in our own galaxy and in other spiral galaxies (and in irregulars), stars continue to form even today. In spiral galaxies, the passage of a "density wave" through a given region appears to trigger a burst of star formation in interstellar gas clouds. The Orion Nebula provides the nearest example of a cloud that has formed stars within the past few million years.

Stars shine by turning some energy of mass into energy of motion, in accord with Einstein's formula $E = mc^2$. To do this, most stars use a series of nuclear reactions called the "proton-proton cycle," whereas the more massive stars employ the "carbon cycle" for this purpose.

Both of these cycles of reactions have the effect of turning four protons into one helium nucleus, with the release of some energy of motion as the energy of mass decreases slightly. Huge numbers of these reactions occur in the hottest, innermost regions of stars each second. This energy, shared among all the particles inside the star through collisions, gives the particles enough energy of motion to resist the star's tendency to collapse under its own self-gravity. Because of this balance between the outward flow of liberated energy of motion and the inward pull of self-gravitation, most stars automatically can keep their rate of energy liberation at the center equal to the amount needed to oppose their tendency to collapse. Stars that achieve a steady rate of energy liberation in this way appear on the "main sequence" of a graph of stars' true brightnesses and surface temperatures. Stars with larger masses have larger surface temperatures and *much* greater true brightnesses than stars with smaller masses do.

Questions

1. Why were the clumps that formed globular clusters among the first units to condense within a protogalaxy?
2. How have other protostars formed? Is it unusual for a protostar to produce a pair or a triplet of stars? How can we tell?
3. How can we discover the existence of double stars without observing the two stars as separate points of light?
4. What is the difference between "spectroscopic" double stars and "eclipsing" double stars?
5. The most efficient way to turn energy of mass into energy of motion is to bring matter and antimatter together. Why are stars unlikely to consist of matter and antimatter in equal amounts?
6. How do stars turn energy of mass into energy of motion? Why are high temperatures needed for this process to work?
7. How could such high temperatures be produced inside contracting gas clouds?
8. Why did the liberation of energy of motion from energy of mass stop the contraction of protostars?
9. Why doesn't the steady liberation of energy of motion inside the sun and other stars make these stars expand?
10. What is an "ideal radiator"? Why does the inside of a star resemble such an ideal radiator?

11. How does the peak in a star's energy spectrum (the energy at which the largest number of photons are emitted) allow us to determine the star's surface temperature?

12. What is the "main sequence" in the temperature-luminosity diagram? Why do most stars in the temperature-luminosity diagram fall along this main sequence? What kinds of stars do not appear in the main sequence?

13. What happens to the positrons that appear in the insides of stars as part of the proton-proton cycle or the carbon cycle?

14. The amount of energy released by an ideal radiator each second varies as the object's size times the fourth power of its temperature. Suppose that two stars have the same size, but one is twice as hot as the other at the stars' surfaces. How many times more energy would the hotter star emit each second, according to the ideal-radiator law? What would this ratio be if one star's surface temperature were *three* times the other's?

Further Reading

G. Gamow, *A Star Called the Sun* (Mentor Books, New York, 1961).

W. Baade, *The Evolution of Stars and Galaxies* (Harvard University Press, Cambridge, Mass., 1963).

G. Herbig, "The Youngest Stars," *New Frontiers in Astronomy* (W. H. Freeman and Co., San Francisco, 1975).

Chapter 8 Stars—Maturity and Senescence

"My position
Up to this time
Has been
Quite frankly
Nobody
Ever
Told me
A damn bit
Of this."

R. M. Nixon

Astronomers once thought that stars begin life at their hottest and cool down thereafter. Reality turned out to be more complex, and although we don't yet know the full story of how a star passes through its life cycle, we do know enough to understand that the inside and the outside of a star may behave in quite different ways. To determine how stars evolve from observations of their outermost layers is difficult but not impossible. The process of deduction is something like figuring out what must be happening inside our bodies as we age, solely by making observations from the outside and (figuratively speaking, of course) never removing our clothes. Astronomers never have lacked hypotheses to explain their results from observations of stars, and they have rejected a large number of theories to arrive at their present ones, which they hope, naturally enough, are more correct than the ones they abandoned.

Evolution From the Main Sequence

We have seen that the nuclear reactions in both the proton-proton cycle and the carbon cycle convert four protons into one helium nucleus, two positrons, two neutrons, two photons, and extra energy of motion. (To be honest, *three* photons come out of the carbon cycle.) The positrons quickly annihilate with electrons to produce photons with still more energy of motion, while the neutrinos escape directly from the center of the star out into space. The bulk of the energy of motion liberated by nuclear reactions inside the star has to fight its way to the surface through millions of collisions among photons and elementary particles. The process of emerging takes about a million years in the sun or in similar stars, hence a change in the rate of energy production at the star's center would cause a corresponding change in the output at the star's surface only a million years later. However, as we discussed on Page 202, stars regulate their energy-production rate extremely well, and the sun, for example, has kept its energy output constant (to within a few percent or less) over at least the past ten or twenty million years, as we can tell from fossil records of life on Earth.

Thus most of a star's lifetime consists of a steady internal liberation of an energy flow, which diffuses outward and escapes into space. Although this tells the story of most of the star's lifetime, the initial and final stages of a star's life, like the start and end of biological life, command special attention on account of the changes they embrace and the future toward which they point.

To produce energy, stars constantly turn hydrogen nuclei (protons) into helium nuclei. When a star first begins to liberate energy in its interior, about ninety percent of the nuclei (with seventy-five percent of the mass) inside it are protons.[1] As time goes on, the star turns more and more of these protons into helium nuclei, until finally the star's basic fuel supply, protons for the proton-proton or carbon cycles, become scarce in the star's interior. This exhaustion of nuclear fuel follows from the laws of nature, which state explicitly that nothing comes free: If the star consumes protons, it eventually will run out of them. The time when the star has used up the supply of protons in its interior regions by making them into helium nuclei marks the star's passage from middle age, the main-sequence phase, into the glorious senility of the red-giant phase of stellar evolution.

How long does it take for a star to exhaust its supply of protons and become a red giant? The answer depends strongly on the star's *mass*. Strangely enough, more massive stars use up their nuclear fuel, and thus pass through their middle age on the main sequence, far more rapidly than less massive stars do. The potential of each star for the liberation of energy of motion varies in proportion to the star's initial mass, which consists mostly of protons. The *rate* at which stars liberate energy, however, varies as the cube of the mass or even (for the largest stars) as the mass raised to the fourth power. A star such as Rigel, in Orion's foot, has ten times the sun's mass, and thus ten times the sun's total potential for liberating energy. But Rigel, in fact, liberates energy at a rate ten thousand times the sun's; therefore Rigel's total lifetime as an energy-liberating star will be a thousand times *less* than the sun's. (Ten times the sun's energy potential divided by ten thousand times the sun's energy-liberation rate gives Rigel one thousandth of the sun's lifetime.) Stars stay on the main sequence in the temperature-luminosity diagram so long as they steadily are converting protons into helium

---

[1]Recall that the big bang (Page 66) already had fused some of the protons into helium nuclei to give the proportion mentioned above. Nuclei heavier than helium did not appear in abundances greater than one *millionth* of the proton abundance as the result of the big-bang fusion processes.

nuclei deep inside. The sun and other stars with the same mass as the sun will remain on the main sequence for about nine billion years, as nearly as we can calculate. In the case of the sun, half of this time already has gone by. A star with 1½ times the sun's mass will spend only two billion years on the main sequence, a star with five solar masses has a main-sequence lifetime of only eighty *million* years, and a star with fifteen solar masses lasts only ten million years on the main sequence, a mere instant in cosmic history. Stars with masses smaller than the sun's will remain on the main sequence longer than the sun—some twenty billion years for a star with 80% of the sun's mass (2×10^{33} grams).

What happens to a star as it runs out of protons in its interior? Why does it leave the main sequence in the temperature-luminosity diagram? A star running low on nuclear fuel finds itself in a difficult pickle. Its self-gravity keeps pulling its outer layers inward, and the energy released at the center must oppose this inward pressure or the star will collapse (Figure 8–1). As the star's center uses up its supply of protons, the star must continue to liberate the same amount of energy each second. Hence the star must process its dwindling store of protons into helium nuclei at a faster rate than before, so that the smaller number of protons can release the same amount of energy each second. To consume its nuclear fuel more rapidly, the star contracts, and thus heats, its central regions. The higher temperature (more energy of motion per particle) makes the protons fuse together more rapidly (Page 202). The star's outer layers are pulled toward the sun's center, and as the fuel supply runs low in the center, so that less energy would be released there each second, the pressure of the outer layers make the center contract and heat up slightly. Then the higher temperature makes for more nuclear reactions each second, and the extra energy released

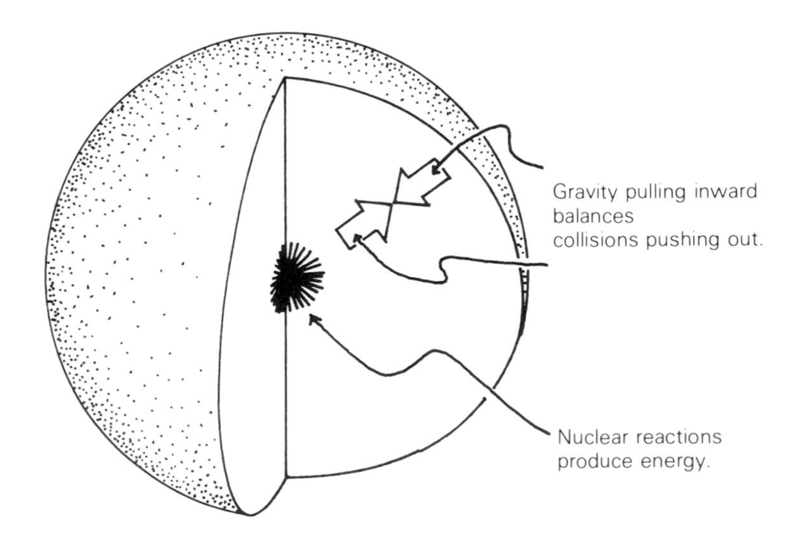

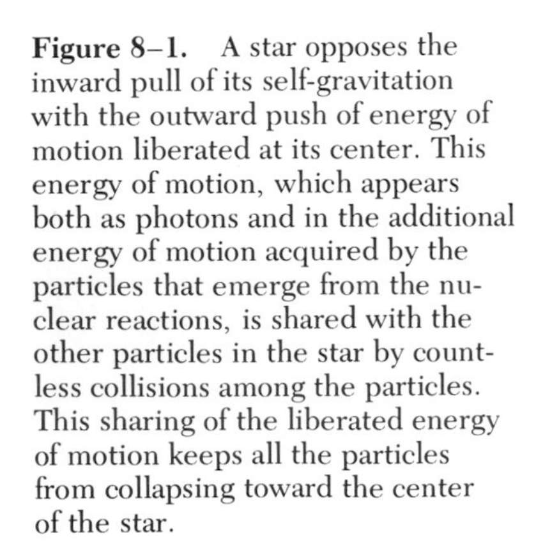

Figure 8–1. A star opposes the inward pull of its self-gravitation with the outward push of energy of motion liberated at its center. This energy of motion, which appears both as photons and in the additional energy of motion acquired by the particles that emerge from the nuclear reactions, is shared with the other particles in the star by countless collisions among the particles. This sharing of the liberated energy of motion keeps all the particles from collapsing toward the center of the star.

keeps the star from contracting a bit longer. The result is that as the star's supply of protons runs down, its central regions contract slowly but steadily, always increasing the central temperature to consume the star's fuel ever more rapidly, and in fact the total energy liberated each second continues to *increase*.

As the central core of the star contracts, the star's outer layers expand. Part of the extra energy liberated in the core goes into pushing outward in the star's outer layers, and part of the surplus escapes into space. The star's absolute luminosity increases, because more photon energy escapes into space each second. However, the star's surface temperature does not increase, and in fact may decrease, because the expansion of the star's outer layers cools these layers. In their newly expanded state, each square centimeter of the star's outermost surface will radiate no more energy, and perhaps even less energy, per second. But since there now are many more square centimeters in the total surface of the star, the total amount of energy radiated away each second increases dramatically. The star becomes a "red giant" of low surface temperature and large absolute luminosity. A typical red-giant star may have a radius a thousand times larger than the sun's, while its helium-rich core will be only a few times larger than the Earth's radius (Figure 8–2). This ratio of sizes is like having a core inside the Earth one kilometer across that contains half of the Earth's mass.

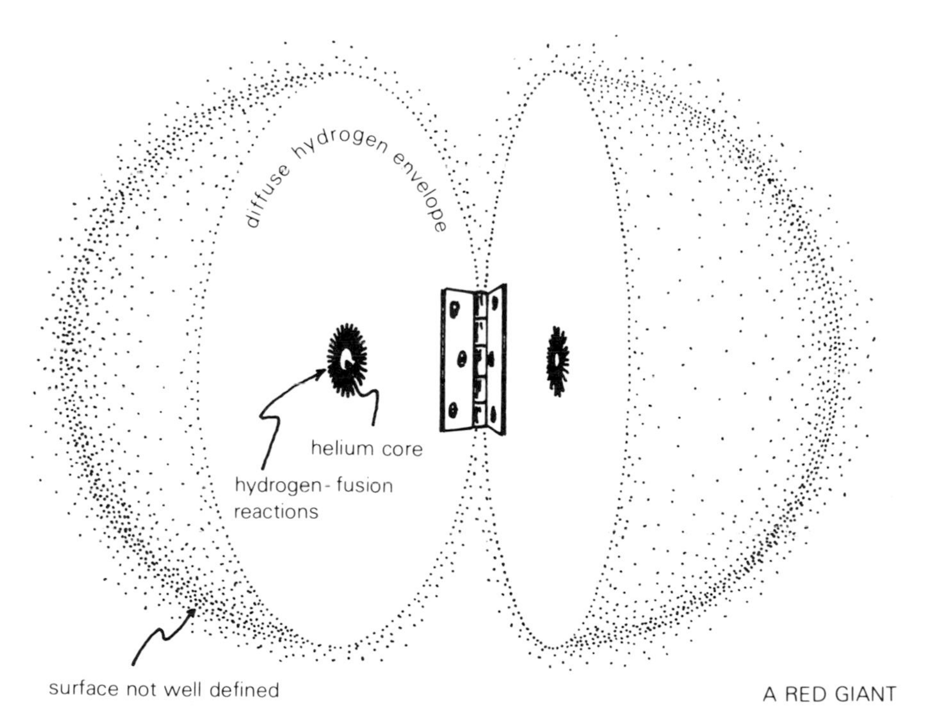

Figure 8–2. The helium-rich core of a red-giant star has a diameter of tens of thousands of kilometers, while the star itself may be hundreds of millions of kilometers across. Energy of motion is released from hydrogen-to-helium fusion reactions within a shell of material that surrounds the helium core.

Figure 8–3 shows how a star's surface temperature and absolute luminosity change as the star's core contracts and its outer layers expand. Stars of different masses, which pass most of their lives at various points on the main sequence, behave differently as their core regions exhaust their protons. Still, all of the stars expand and grow cooler in their outer parts as the stars' cores contract and grow hotter.

Suppose that we construct the temperature-luminosity diagram of the stars in a cluster. Those stars that lie above and to the right of the main sequence form the "red-giant branch" of the diagram. Not all of these stars are truly red (surface temperatures 1500 to 3000 degrees absolute), but they all are redder (lower surface temperature) than main-sequence stars with the same absolute luminosity. All of the stars in the red-giant branch of the temperature-luminosity diagram have contracting cores and a dwindling supply of protons for nuclear reactions.

Estimating the Ages of Star Clusters

If we recall that stars of large mass consume their nuclear fuel more rapidly than stars of small mass do, we can see that the brightest stars

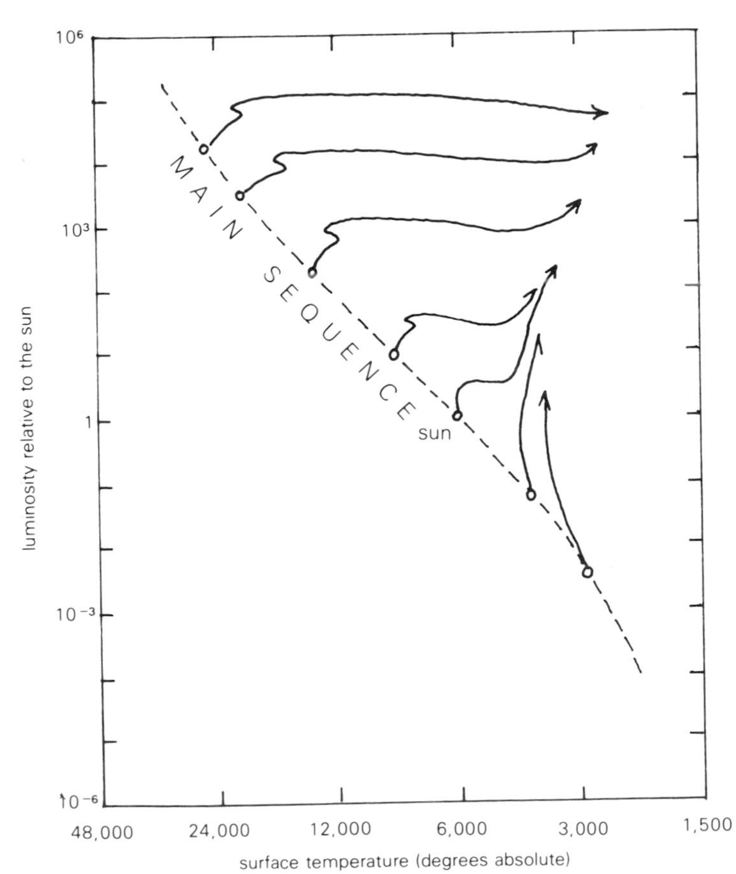

Figure 8–3. As a star leaves the main sequence of the temperature-luminosity diagram, the star's absolute brightness increases and its surface temperature decreases or remains about the same.

will spend the least time as main-sequence stars and will be the first to become red giants, assuming that the stars all began to shine at the same time. These bright main-sequence stars have the largest mass. Less massive stars have smaller absolute luminosities when they are main-sequence stars, and they take longer to use up their fuel. Because of this inverse relationship between a star's absolute brightness as a main-sequence star and its lifetime on the main sequence, we can estimate the age of a cluster of stars that we assume to have all formed at about the same time. We do this by seeing how far the cluster's main sequence on the temperature-luminosity diagram extends upward to greater surface temperatures and greater true brightnesses.

The youngest star clusters will still have stars of extremely great brightness and surface temperature at the top of the main sequence, and perhaps not even the most massive stars in the cluster will have had time to become red giants. In a young cluster of stars, the "turnoff" point from the main sequence to the red-giant branch lies at a high surface temperature and a large absolute luminosity (Figure 8–4). In an

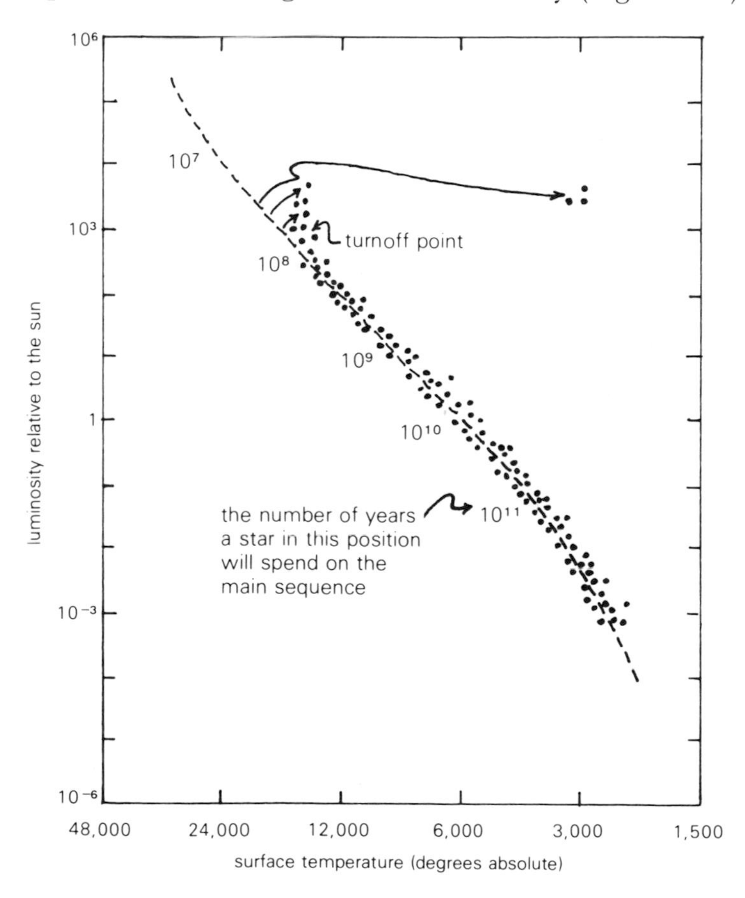

Figure 8–4. The point on the temperature-luminosity diagram where the red giants "turn off" from the main sequence depends on how long the cluster's stars have been liberating energy of motion. An older cluster will have a turnoff point farther down the main sequence, because it takes longer for the less massive stars down the main sequence to exhaust their supply of protons. The turnoff point therefore can tell us the age of the cluster since the time when the stars began to liberate energy.

older cluster, the turnoff point will occur at lower surface temperatures and smaller absolute luminosities, because more stars will have had time to become red giants. Thus we can find the age of the cluster's stars by measuring the location in the temperature-luminosity diagram of the turnoff point of the red-giant branch. Since we believe that the original main sequence for all clusters appears in the same location on the temperature-luminosity diagram, it is sufficient to measure the surface temperature of stars at the turnoff point from the main sequence. The lower this temperature is, the older is the star cluster.

Using this technique, astronomers have found that globular clusters all have ages measured in billions of years, often ten or fifteen billion. In contrast, open clusters have ages that range from five *million* years to ten billion years. A young open cluster such as *h* and χ Persei (Figure 8–5) shows a main-sequence turnoff at surface temperatures of 25,000 degrees absolute, indicating an age of ten million years, whereas an old open cluster such as M 67 (Figure 8–5) shows a turnoff at surface temperatures of 6500 degrees absolute, which implies an age of four billion years for the stars in this cluster. These ages refer to the time since the stars began to liberate energy at a steady rate, that is, since the time when they first appeared as main-sequence stars. This time, however, was not long (by cosmic standards) after the stars began to condense from protostars—only a few millions or tens of millions of years (Page 187).

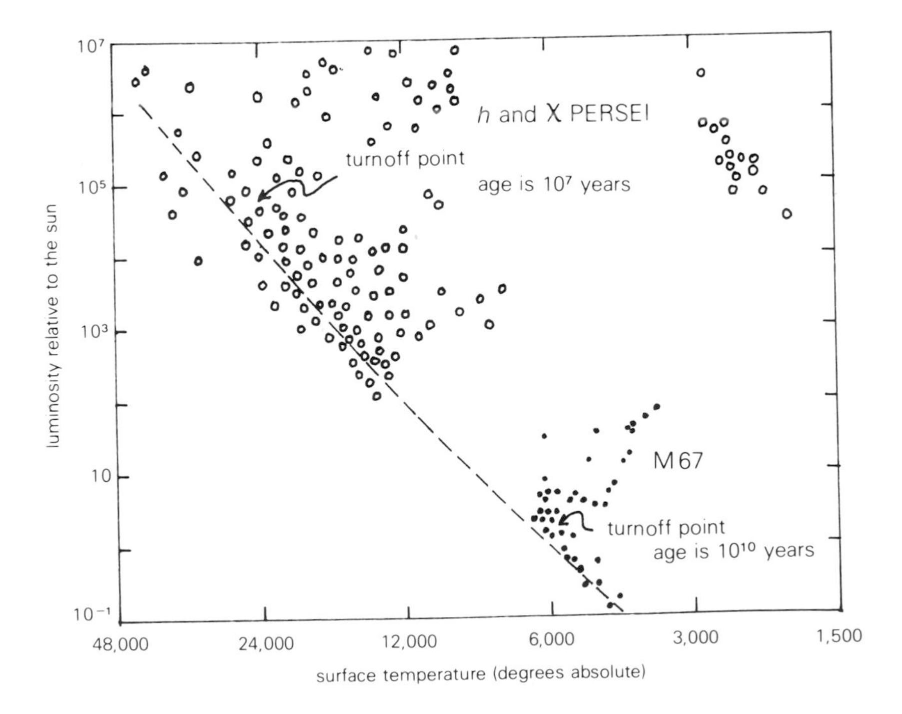

Figure 8–5. If we plot the temperature-luminosity diagrams for the two clusters M 67 and *h* and χ Persei, we find that the turnoff from the main sequence occurs at a much lower surface temperature in M 67 than it does in *h* and χ Persei. This indicates that M 67 is by far the older of the two clusters.

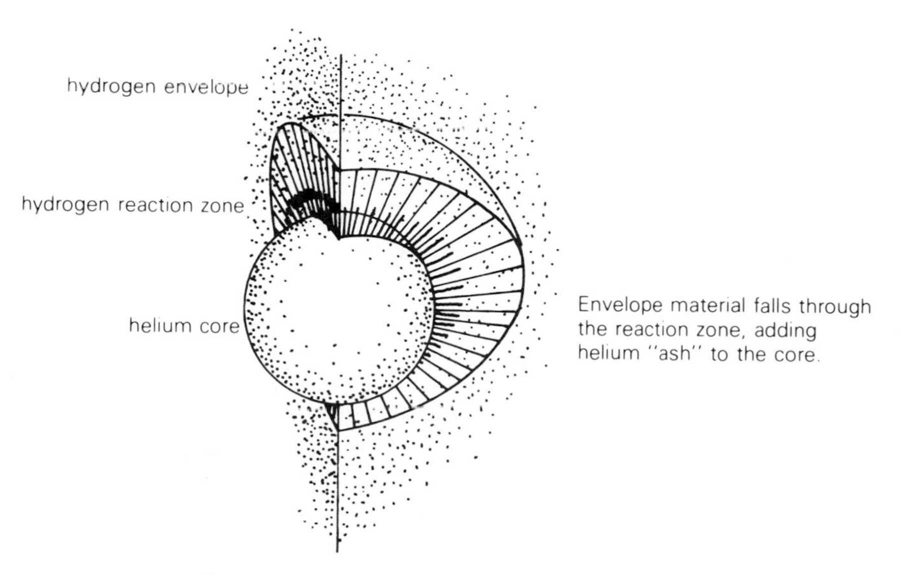

Figure 8–6. As a red-giant star grows older, more matter at its center becomes almost pure helium, and around this helium core a shell of material continues to fuse hydrogen nuclei into helium nuclei to release energy of motion. The helium core may be about as large as the Earth, but the hydrogen envelope around it can be larger than the Earth's orbit around the sun.

Further Evolution of Stars

What happens to a red-giant star as it continues to process its remaining stock of protons?[2] The star discovers several ingenious ways to keep on liberating energy of motion, each of which lasts a shorter time. First of all, when the star processes all of its central hydrogen into helium, the star's central regions consist of a shell where protons still fuse into helium nuclei that surrounds a pure helium core (Figure 8–6). As time passes, the hydrogen-consuming shell embraces more and more of the star's middle regions, but all of these regions must keep on contracting to raise their temperature and thus to process the protons more rapidly. Likewise, the helium core contracts under the pressure from the layers that overlie it, and the temperature there keeps increasing. Eventually, the helium nuclei in the core have a temperature so large that *they* can fuse together and liberate more energy of motion. Because helium nuclei each have twice the positive electric charge of hydrogen nuclei (protons), they repel each other more strongly than protons do by elec-

[2]We are specifically discussing here the evolution of a Population I star, which astronomers think they understand somewhat better than the evolution of Population II stars. However, in both stellar populations the general pattern of the stars' evolution should be the same.

tromagnetic forces. Therefore, although protons will fuse together at temperatures of ten to twenty million degrees absolute, the fusion of helium-4 nuclei requires a temperature near two hundred million degrees absolute.[3] Inside the helium core of a red-giant star, the temperature will reach this value at a time when the contraction of the core has increased the density of particles to ten thousand grams per cubic centimeter, a hundred times the density at the center of the sun. When the temperature grows large enough for helium nuclei to fuse together, the star's core suddenly liberates a burst of energy called the "helium flash." Nuclei of helium-4 fuse together to form nuclei of beryllium-8, and then another helium-4 nucleus fuses with each beryllium-8 nucleus to make nuclei of carbon-12, the common variety of carbon. The extra photon energy from the "helium flash," which comes from the decrease in the energy of mass during the fusion processes, deposits itself in the now-dense shell that surrounds the helium core, and these photons expand this shell to a much larger size. The expansion of the proton-consuming shell reduces its density and temperature, and thus decreases the rate of nuclear reactions in the shell. Thus the helium flash has the perverse result of lowering the total rate of energy output from the star, once the initial brief flash has passed.

Figure 8–7 shows the calculated change of a star's position in the temperature-luminosity diagram as it becomes a red giant and undergoes the helium flash. At some moments in this evolution, the star is likely to become somewhat unstable and to undergo periodic changes in its size and its absolute luminosity. These periodic changes make the star's surface area progressively larger and smaller (Figure 8–8). The star's absolute luminosity also fluctuates cyclically, growing largest in the middle of the star's contracting phase and smallest in the middle of the star's expanding phase (Figure 8–8). These periodic variations in some red-giant stars have a special importance for astronomers because they can be observed as fluctuations in a star's brightness, with a definite period and even a definite pattern in the shape of the curve of brightness plotted against time (Figure 8–8). Best of all, these "Cepheid variable" stars show a direct correlation between the period of one fluctuation cycle in their light and the star's true brightness (Page 153). The Cepheid variables have large absolute luminosities (Figure 8–9), so they can be seen far away, even in other galaxies. A subclass of Population II Cepheids called "RR Lyrae" variable stars differ somewhat from

---

[3]The fusion of helium-3 nuclei, the third step in the proton-proton cycle, represents an exception to the general rule that more highly charged nuclei require higher temperatures to fuse together. Helium-3 nuclei are especially eager to fuse with one another because of their nuclear structure, and this eagerness balances their greater electromagnetic repulsion.

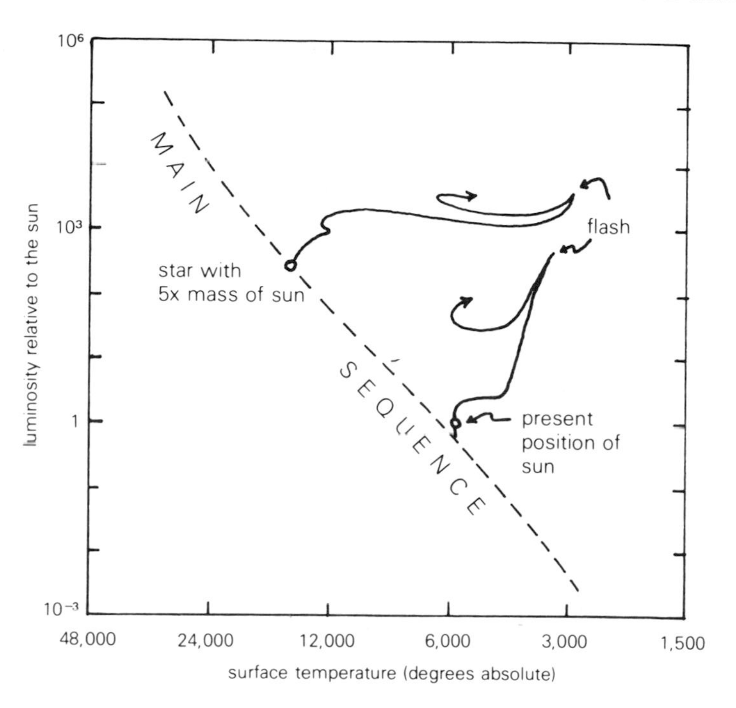

Figure 8–7. The moment of the "helium flash" marks the highest absolute luminosity (barring any sudden explosion later on) that a star achieves along its evolutionary path through the temperature-luminosity diagram. After the helium flash, the star's surface temperature increases (because we now can see deeper into the star after the helium flash loosens the surface layers), while the star's absolute brightness decreases. This is the time when the star is likely to start pulsating.

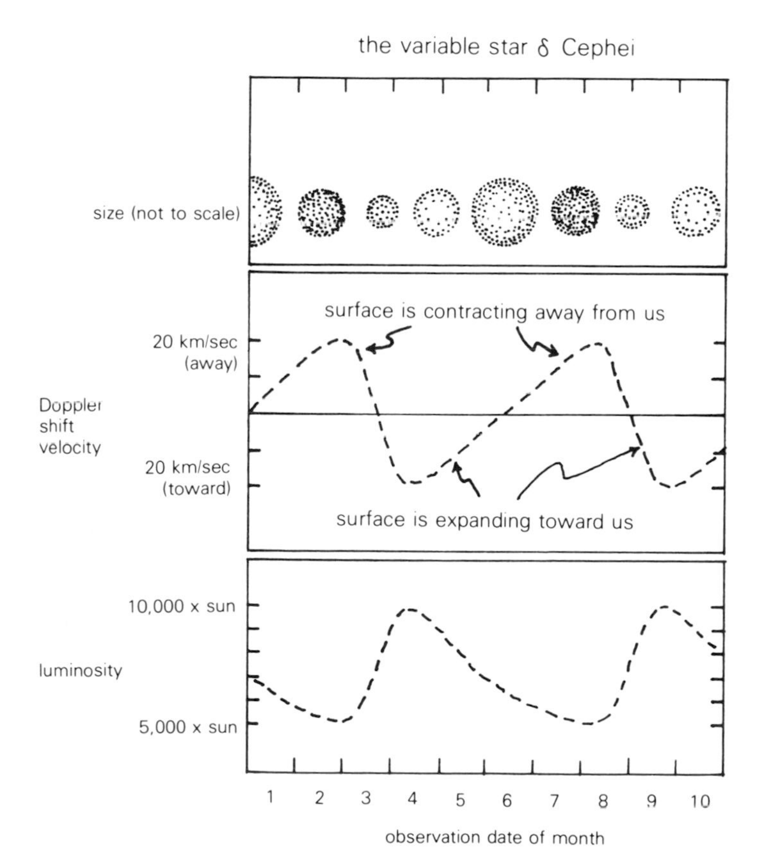

Figure 8–8. A pulsating star, such as the original Cepheid variable star Delta Cephei, undergoes cyclical changes in brightness that are related to changes in the star's size. The star is brightest during the middle of its contracting phase and faintest during the middle of its expanding phase. We know this from measuring the Doppler shifts of the absorption features in the star's spectrum and correlating these shifts with the star's changes in brightness.

Cepheids: The RR Lyrae variables have slightly less absolute luminosity and a different period of light variation. Both the Cepheid variables and the RR Lyrae variables show up in the temperature-luminosity diagram within certain well-defined regions called "instability strips" (Figure 8–9). *All* of the stars whose surface temperatures and absolute luminosities place them in these regions are "unstable," and they pulsate in and out, up and down in their energy output, during the time that their evolution carries them through this region in the temperature-luminosity diagram.

We have seen that stars convert hydrogen into helium, and then they process the helium nuclei into carbon nuclei. Both of these fusion processes release energy of motion, but the law of diminishing returns has set in. Each hydrogen-to-helium reaction releases fifteen times as much energy of motion as each helium-to-carbon fusion. Therefore the later stages of a red giant's life unfold far more rapidly than its earlier, main-sequence history did. A star with a mass equal to the sun's will spend nine billion years on the main sequence. The *entire* red-giant phase, including the stage when protons fuse in a shell around the helium core and the unstable stage after the helium flash, will take a total of three or four billion years, less than the star's main-sequence lifetime. As the red-giant phase ends, almost all of the star's protons have been changed

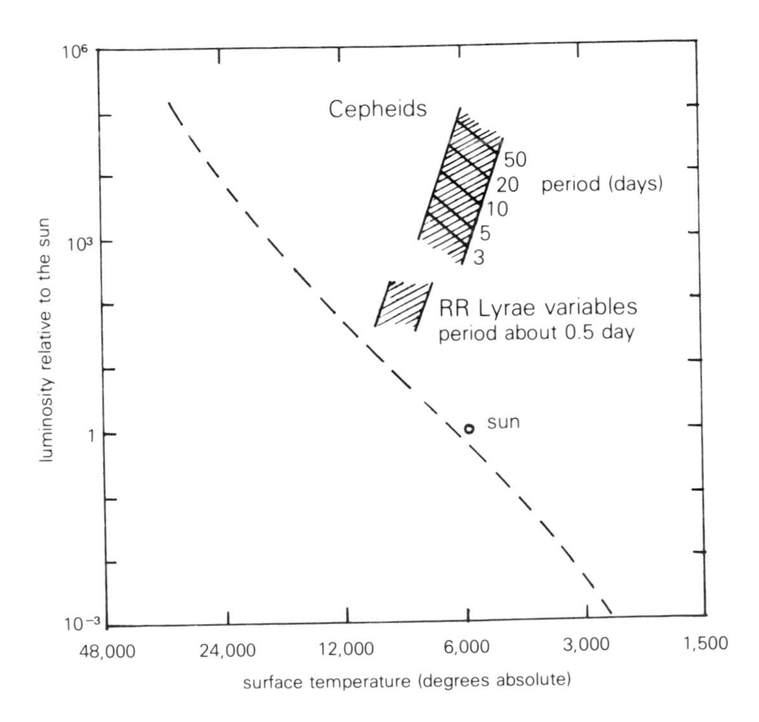

Figure 8–9. Stars tend to vary cyclically in their light output when they occupy regions of the temperature-luminosity diagram called "instability strips." The stars that appear in these regions of the diagram apparently have undergone their helium flash and are evolving toward the lower regions of the temperature-luminosity diagram, although they may have periods of large apparent brightness as they do so.

into helium nuclei, and the helium nuclei themselves have been changed into carbon nuclei.

What can the star do next to keep its core from collapsing under its own self-gravity? More nuclear fusion reactions indeed are available: carbon plus helium makes oxygen, oxygen plus helium makes neon, neon plus helium makes magnesium, carbon plus carbon also makes magnesium, and so on. Each of these fusion reactions releases some energy of motion, but the total energy released from these reactions remains far less than that released by the original hydrogen-to-helium cycles. Thus all of the later reactions must occur faster and faster, so that in a few hundred million years the star will have processed most of its mass into progressively heavier nuclei—oxygen, neon, magnesium, silicon—all the way to iron.

With iron the process stops working. Iron is not the heaviest nucleus, although it is the heaviest of the relatively abundant nuclei (those whose abundance is more than a millionth of the abundance of hydrogen). Most iron nuclei are iron-56, each of which has 26 protons and 30 neutrons. Nuclei of lead, silver, gold, and uranium are larger and heavier than iron nuclei, but with iron we encounter a problem: To fuse nuclei heavier than iron does not release energy of motion; rather, such fusion processes *require* energy of motion. When we fuse nuclei lighter than iron—for example, two helium nuclei—the products have less total mass than the original mass of the two nuclei. The decrease in mass, and thus in energy of mass, appears as new energy of motion released by the fusion process. But when we fuse an iron-56 nucleus and a helium-4 nucleus, the product nucleus (nickel-60) has *more* mass than the total mass of the fusing nuclei. The increase in mass, and in energy of mass, must come from the energy of motion in the colliding nuclei. Thus the fusion process decreases the energy of motion contained in the particles that undergo such reactions.

Hence iron marks the end of the line for releasing energy of motion through fusion reactions. A star with a central core that has steadily decreased its size and increased its temperature, that has fused helium nuclei into carbon, into magnesium, silicon, sulfur, argon · · · all the way into iron, has run out of ways to release the energy of motion that will continue to oppose its own self-gravity. Such a stellar core has reached the point of catastrophe, and catastrophe there will be. Before we examine this process, we must pause to determine which stars will reach this brink of collapse, and which stars will find some way to avoid this apparently inevitable calamity.

A star's mass once again turns out to be what determines whether the star can avoid collapse. Stars with large masses end in catastrophe; stars with small masses fade into an ever-dimming obscurity. Let us see why.

The Exclusion Principle

Stars with masses less than about 1½ times the sun's mass can avoid collapsing because of a strange resistance of elementary particles to being pushed together. All matter resists compression to some extent, as we know from our attempts to squeeze water from stones. But elementary particles such as electrons, protons, and neutrons have a special refusal to be packed tightly called the exclusion principle.

Physicists have found that if they try to pack together a bunch of electrons, the electrons avoid one another far more than would be expected from their mutual electromagnetic repulsion. The summary of experimental results, the exclusion principle, states that electrons refuse to be near each other when they have almost the same velocity. This rule has a precise mathematical form, which we need not examine, that amounts to this: The positions and velocities of the electrons in any group cannot all be the same. If we try to get each electron in a group to have the same position, their velocities will differ more and more, and if we try to make the velocities all the same, then the electrons' positions (locations) will differ more and more. This resistance of a group of electrons to all have the same positions and the same velocities seems to be absolute. That is, it goes beyond any repulsive force, such as the electromagnetic repulsion between electrons, which can be overcome by a stronger force pushing the electrons together. Instead, the exclusion principle keeps the electrons from all having almost the same position and the same velocity, period. The "almost" here includes the mathematics that we have left out. But all atoms show the effect of the exclusion principle in their electron orbits, because the exclusion principle underlies the impossibility of having more than two electrons in the smallest orbit, eight electrons in the next smallest orbit, and so on (Page 29). In this way, the exclusion principle is responsible for the chemical behavior of all the elements, since chemical interactions depend on the number of electrons in each orbit.

Powerful as its effects are, the exclusion principle seems to act selectively, only among certain kinds of identical particles. Electrons, protons, and neutrons obey the exclusion principle. Photons do not, nor do helium-4 nuclei and other nuclei made of an even number of nucleons (protons plus neutrons). Nature moves in mysterious ways, and it is through the action of the exclusion principle that the electrons inside a star reappear as the key determinants of the star's evolution, just at the time when the nuclei are running out of helpful things to do. We have concentrated on the heavier nuclei (protons, helium-4, carbon-12, and so on) rather than on the lighter electrons, because the nuclei fuse to liberate the energy of motion that makes stars shine. Electrons, which are not affected by strong forces (Page 48), do not fuse, so they do not play an important part in the star's main-sequence lifetime, although each star contains enough electrons (10^{57} or so) to balance the positive electric charge of its nuclei and thus to have a total electric charge of zero.

As an aging star's central core contracts, the nuclei and the electrons inside it are squeezed closer and closer together. Each cubic centimeter contains more and more electrons and nuclei, and the electromagnetic forces between the two kinds of electric charge assure that each cubic centimeter has a total charge of zero. Now the electrons always tend to move around more rapidly than the nuclei, because the temperature of the matter represents the average energy of motion per particle. For velocities much less than the speed of light, each particle's energy of motion varies as its mass times its velocity squared. Since each electron's mass is thousands of times less than the mass of a nucleus, the electrons have much larger velocities than the nuclei do, if all of them are in the same cubic centimeter with the same temperature. Thus the electrons, which have velocities perhaps a hundred times larger than their neighboring nuclei, feel the effects of the exclusion principle sooner than the nuclei do. The electrons refuse to be squeezed past a certain density, because the exclusion principle prevents them from all being in almost the same place with almost the same velocity. If the electrons all could be in almost the same place for an instant, their differing velocities would spread them over a much larger volume in the next instant.

The exclusion principle keeps the electrons from squeezing into an ever-shrinking volume as the star contracts through its own self-gravitation. What about the nuclei—helium, carbon, or whatever? The electrons hold *them* up by electromagnetic forces. If the nuclei were to try to filter through the electrons and pack toward the center, the star would have a large shell of negative charges (electrons) surrounding a sphere of positive charges (nuclei). The attractive electromagnetic forces between unlike charges then would pull the two kinds of charges back together. Since the electrons cannot pack more tightly (because of the exclusion principle), the positively charged nuclei would have to move outwards. In fact, the electromagnetic forces are so strong that the nuclei never do get far from the electrons, and the particles in each cubic centimeter continue to have a total electric charge of zero.

Thus the exclusion principle supports the entire star, electrons and nuclei, against its tendency to contract gravitationally. Matter in which the exclusion principle limits the bulk motions of particles, as it does inside these aged stars, receives the name "degenerate" from physicists. A "degenerate" star is one in which the exclusion principle, rather than the nuclear reactions, supports the star against gravitational collapse. Degenerate matter always must be extremely dense, far denser than any matter on Earth. The exclusion principle does not much affect the motions of electrons that are not attached to atoms *until* the density of matter reaches a million times the density of water, or fifty thousand times the density of gold. If matter this dense were brought to the Earth's surface, each cubic centimeter of it would weigh a thousand tons.

The cores of stars can and must stop contracting when they become degenerate. The exclusion principle does the trick: The electrons can't be compressed any farther, and the electrons hold the nuclei. The point in its lifetime at which a star's core becomes so dense that degeneracy occurs as the exclusion principle becomes effective will vary for stars of different masses. Lower-mass stars are more prone to develop degenerate cores. The reason is that the density in the centers of stars varies inversely as the stars' masses. The most massive stars have the smallest density of matter at their centers when they are main-sequence stars, because the temperature at the center of all main-sequence stars is about the same, namely the ten or twenty million degrees needed to make protons fuse. But a more massive star can achieve this central temperature at a lower central density, because the star has more mass pressing in on the center. The general rule that larger-mass stars have lower central densities holds true as the stars age, and this rule produces an important distinction between large-mass and small-mass stars. If a star has a mass less than about 1.4 solar masses, its core will become degenerate at about the time that all the protons there have become helium nuclei. When helium nuclei start to fuse inside the core, the energy released by the helium flash expands the core and removes the degeneracy. The core than starts shrinking again, and by the time the helium nuclei have fused into carbon nuclei, the core again will be degenerate, this time for good. Figure 8–10 shows these stages of a star's evolution on the temperature-luminosity diagram. After degeneracy sets in for the second time, the star's core contracts no farther. Almost no nuclear fusion reactions occur in the core, because the exclusion principle that supports the star also inhibits nuclei from colliding with each other.[4] The star's outer layers slowly expand out into space, forming a "planetary nebula" (Figure 8–11), and eventually only the naked core remains. These cores are the white-dwarf stars that fill the part of the temperature-luminosity diagram below and to the left of the main sequence.

Planetary Nebulae

Many aging stars eject their outer layers into a spherical shell that surrounds the star, some small fraction of a parsec away. This shell of matter does not contain much of the star's mass, but its appearance can be spectacular (see Figure 8–11). Stripped of its outer layers, the star's new surface shows a temperature of eighty or a hundred thousand degrees absolute, which is more than ten times hotter than the sun's sur-

[4]In degenerate matter, where the exclusion principle is important, the simple relationship between the temperature and the average energy of motion per particle that exists for nondegenerate matter must be replaced by a more complicated relationship.

face. The high-energy photons from the star then ionize many of the atoms in the gas. As these atoms recombine, they emit photons that make the entire shell gleam. When we observe one of these glowing shells in a small telescope, it looks something like the disk of a planet, so these shells of ejected material acquired the inappropriate name of "planetary nebulae," although they have nothing to do with planets. A

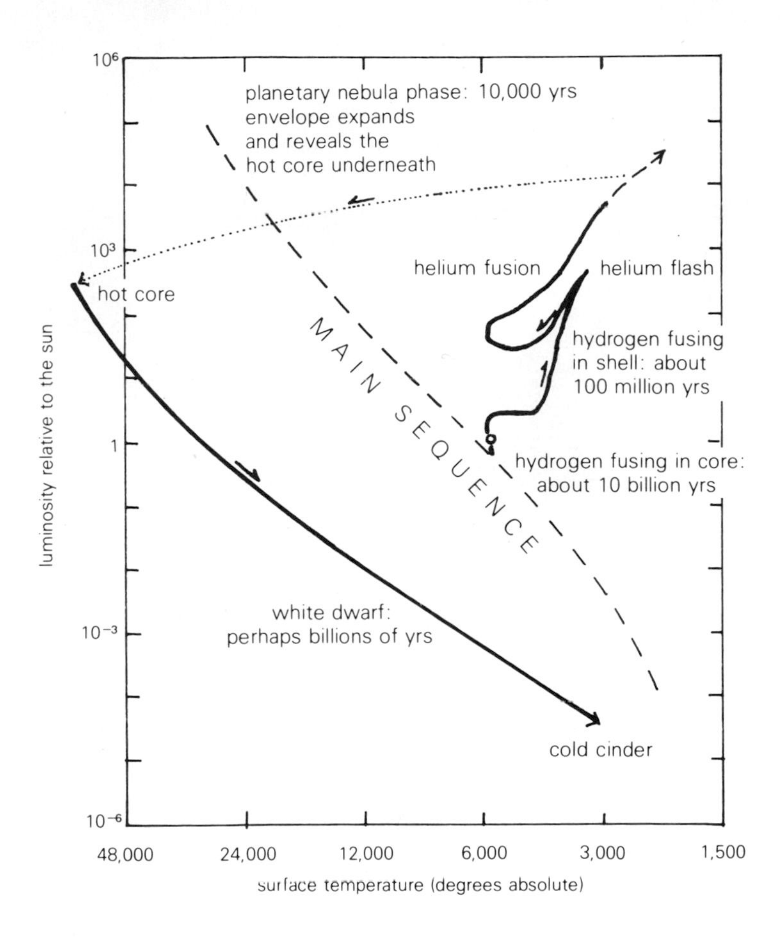

Figure 8–10. A representation of various stages in a star's evolution through the temperature-luminosity diagram, as calculated for a star with a mass equal to the sun's mass. These calculations are not as accurate as astronomers would like, but they are the best that we have now. Notice that a star in the "planetary nebula" phase has a high surface temperature and a large absolute luminosity, so in the temperature-luminosity diagram the star appears for a while nearly at the same place as the blue supergiant stars at the top of the main sequence. However, the planetary-nebula stars are near the end of their lives, whereas the blue supergiant stars are extremely young. White-dwarf stars are the remnant cores of stars that have already passed through the planetary nebula phase.

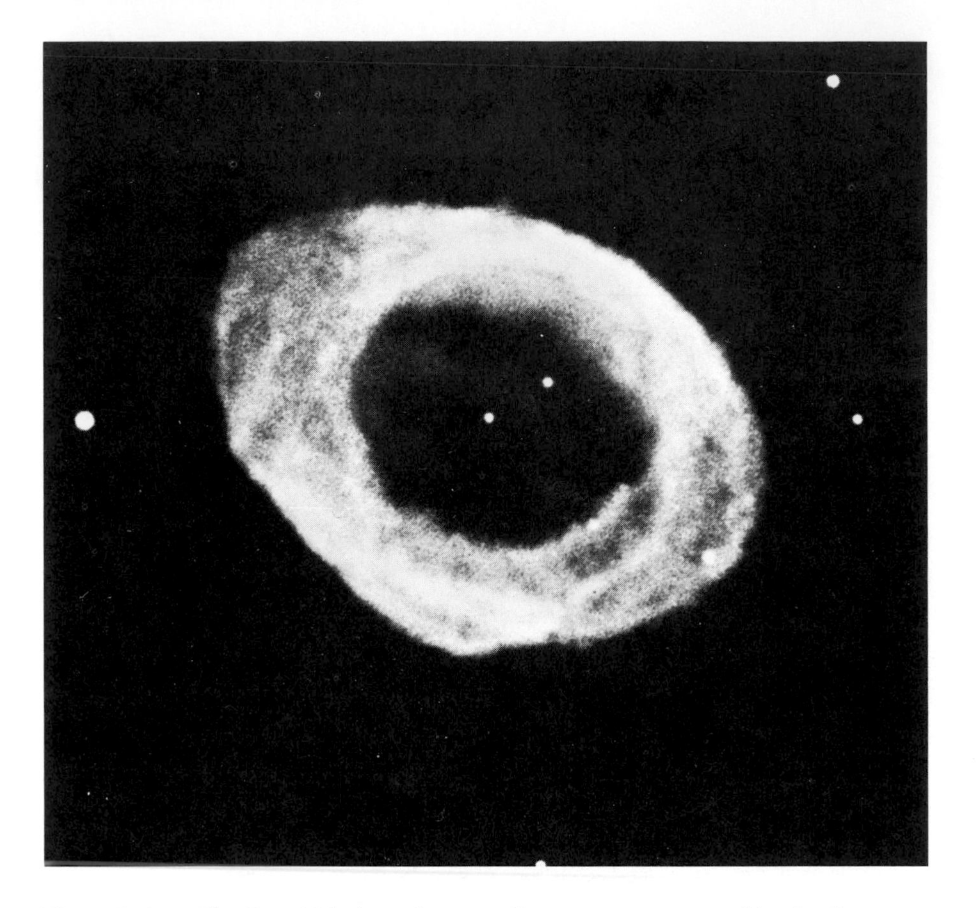

Figure 8–11. The Ring Nebula in the constellation Lyra is powered by the faint star at its center, which in fact produces huge numbers of ultraviolet photons that do not show up on visible-light photographs. These photons ionize many of the hydrogen atoms in the shell of gas, presumably ejected by the star, that surrounds the central source of photons. As the atoms recombine, they emit photons characteristic of the jumps between the possible orbits in hydrogen atoms, such as the Balmer series mentioned on Page 206.

large telescope will reveal the shell structure of a typical planetary nebula such as the "Ring Nebula" in the constellation Lyra (Figure 8–11). The central star that powers this nebula is so hot that most of its photons have ultraviolet frequencies rather than those of visible light, so the star itself appears misleadingly faint in visible light. In fact, the light from the entire nebula emerged originally from the hot central star as photons of ultraviolet energies.

Stars in the "planetary nebula" stage pass over the top of the main sequence (Figure 8–10) on the temperature-luminosity diagram. This phase lasts for a few million years, as the star sheds more and more mass until the core stands revealed as a white dwarf. The energy radiated by the central star of a planetary nebula comes from its continuously contracting core, which will cease to liberate energy of motion when it becomes degenerate.

Mass Limits on White-Dwarf Stars

If a star's core becomes so dense that degeneracy sets in, so that the exclusion principle prevents any further fusion reactions from liberating more energy of motion, then the stellar core is destined to become a white-dwarf star. Once the outer layers have expanded away into space, the remnant core must live on its small hoard of accumulated energy. White-dwarf stars shine because they slowly are radiating away energy of motion brought up from their central regions. Because white dwarfs radiate far less energy per second than main-sequence stars do, they can last a long time on their reserves of energy stored inside from their years of nuclear fusion. The white-dwarf companion of Sirius, called Sirius B, produces energy at a rate three hundred times less than the sun, and ten thousand times less than Sirius A. The white dwarf's surface temperature almost equals that of Sirius A, but its radius is about a hundred times less, about twice the Earth's radius. Yet Sirius B has a mass as large as the sun's, so the average density of matter inside it is several hundred thousand times the density of water.

Thus white-dwarf stars are the exposed cores of not-so-massive stars. They consist mostly of carbon nuclei and "degenerate" electrons, and they will take billions of years to leak away their remaining energy, slowly growing fainter but not smaller. The exclusion principle allows them to resist their own gravitation, as we described previously. All this can be well calculated, but a strange conclusion emerges from the calculations that astrophysicists have made. The exclusion principle *cannot* support a star against its own gravitational forces *if* the star has more than 1.4 solar masses. More massive stars can overwhelm the exclusion principle in a sudden rush that we will describe on Pages 237–243, and these stars will collapse into a state of matter far denser than even the incredible compactness of white-dwarf material. This upper limit on the mass of any white-dwarf star poses a problem to any star with more than 1.4 times the sun's mass. If the star's core ends up with a mass above this limit, the star will collapse.

Mass Loss From Stars

Observations of aging stars have shown that many of them, especially the more massive red giants and red supergiants, are losing mass from their outer layers. All stars eject some matter from their outermost regions (as the sun, for example, does in the "solar wind" described on Page 273), but for these red-giant stars the mass loss proceeds at a much greater rate, amounting to the loss of a sizable part of the star's mass in a few million years. Red-giant stars can lose mass more rapidly and easily than others because their tenuous outer layers are far from the core of the star, and this distance diminishes the gravitational force from the star. It is tempting to think of red giants' mass loss as the stars' attempt to pass below the upper limit allowed for a white-dwarf star, and although this tendency to assign a human viewpoint to stars cannot be quite correct, the result may be the same. The high rate of mass loss indeed may allow some red-giant stars to avoid catastrophic collapse and thus permit them to end their lives as serene cinders rather than fiery blowouts. Consider, for example, the star Sirius B, which presumably was born at the same time as Sirius A. Since Sirius B has evolved farther, all the way to the white-dwarf stage, we may easily believe that it began with more mass than Sirius A, which has 2.3 solar masses. Yet the present mass of Sirius B is just under one solar mass, which strongly suggests that Sirius B managed to lose more than half of its original mass on its way to becoming a white dwarf with less than the upper limit of 1.4 solar masses.

What happens to the mass expelled by aging stars? Some of it forms the "planetary nebula" shells described above, and some of it ends up dispersed through the interstellar gas, ready to be mixed into a new generation of stars. The interstellar dust particles described on Page 176 probably came from the outer atmospheres of red-giant stars, where the low temperature allows atoms to combine into molecules, and the molecules into long chains, without being broken apart by collisions. Some of the gas that red-giant stars blow into space emerges as individual atoms rather than as large dust grains. We can observe that the outer layers of red-giant stars are expanding because the atoms and molecules there show a Doppler shift in the characteristic frequencies of the photons they absorb. We can interpret this Doppler shift to slightly higher frequencies as arising from the motion of the stars' outer layers toward us.

We can expect that the sun will pass through the red-giant phase, shed a small fraction of its mass, and end as a white dwarf. Since the sun's mass already lies below the upper limit of 1.4 solar masses for any white-dwarf star, the sun's future mass loss is not essential for it to become a white dwarf. We probably shall not be around as a species to check up on what happens, though, because as the sun enters its red-giant phase, some four or five billion years from now, its absolute luminosity will increase a thousandfold, thereby causing the temperature at the Earth's surface to quadruple from its present 300 degrees to 1200 degrees absolute (2200 degrees Fahrenheit). By this time, humanity must have found a safer haven if it is not to perish along with the oceans and atmosphere of our original home, the Earth, whose origin and evolution we shall examine in the next chapter.

Supernovae

The mass that a star has as it approaches the end of its energy liberation will determine the star's final resting place. If a star has less than 1.4 solar masses, or if it manages to reduce its mass below this limit during its red-giant phase, then the star almost certainly will become a white dwarf, quietly and rigidly letting its remaining energy of motion seep outwards bit by bit. As a white dwarf cools, it leaks less and less energy per second out into space, so the star has more and more time to shine feebly. Only after tens of billions of years will the star cease to shine entirely, earning the name "black dwarf" as opposed to the "white dwarfs" that we can see. Stars of small mass turn out to be the quiet ones: They contract, begin nuclear reactions, release energy of motion, pass through the red-giant stage, become degenerate dwarfs, then fade away at an ever-lower luminosity. Each stage in such a star's life takes hundreds of millions or billions of years, and the longest stages—main-sequence energy release and white-dwarf senescence—each take nine billion years or more for a star with the sun's mass.

But what of the stars with large mass, which exhaust their nuclear fuel far more quickly, only to confront the possibility of catastrophic collapse? We left our description of these more massive stars on the brink of the abyss (Page 229) as they had used all possible nuclear fusion reactions to liberate energy of motion by fusing their nuclei all the way to iron-56. Since the fusion of iron nuclei uses up energy of motion rather than liberating energy of motion, the star no longer can oppose its own self-gravitation with energy from fusion reactions, and is threatened with collapse. And collapse it will, so long as its center is not dense enough to be degenerate when the nuclei have fused into iron. Deprived of the support from newly released energy of motion, the central regions all fall in toward the center of the star. This collapse occurs so quickly, in less than one *second*, that we may think of it as taking the exclusion principle by surprise. In this collapse, most of the iron nuclei come apart into helium nuclei and neutrons. (Each nucleus of iron-56 can yield thirteen nuclei of helium-4 plus four neutrons.) This process is the opposite of the fusion of helium nuclei to liberate energy of motion, and the star must *provide* energy of motion to make iron nuclei come apart. The necessary energy of motion comes from the motions of the collapsing layers rushing toward the center, and this is definitely a one-time affair. However, in this rush, the motions are violent enough to break apart the iron nuclei, and even to break the resultant helium nuclei into protons and neutrons.

This does not complete the tale by a long shot. The *electrons* now begin to fuse with protons, one by one, to form neutrons and neutrinos. This is the reverse of the neutron's decay into a proton, an electron, and an antineutrino that we discussed on Page 49. Since the electrons do not feel strong forces, we can be sure that weak forces are at work in this kind of fusion, as they are when the neutron decays into the three particles named above. As a result of the fusion of protons and electrons, the star's electrons disappear rapidly (in less than a second), so that when the exclusion principle wants to hold the electrons apart, there are no longer any electrons. Finally, the neutrons that soon constitute most of the star indeed will be held apart by the exclusion principle, but at a density far greater than that inside a white-dwarf star, because neutrons become degenerate at much greater densities than electrons do.[5]

What about the neutrinos that appear when protons and electrons combine? Amazingly enough, they blow off the outer layers of the collapsing star! Neutrinos pass right through ordinary matter with barely an interaction. Almost all of the neutrinos produced 700,000 kilometers inside the sun during the proton-proton cycle (Page 196) pass right through the sun and through the Earth if they come this way. The best detector that we can build on Earth to interact with neutrinos can stop one neutrino in every billion billion (10^{18}) that pass through it.[6] But when matter becomes immensely dense—not just white-dwarf dense—then neutrinos start to interact with other matter in an important way. A density of one million grams per cubic centimeter is like a vacuum to neutrinos; even a density of ten billion grams per cubic centimeter barely slows a group of neutrinos. At densities of one hundred billion grams per cubic centimeter, neutrinos finally notice that they are encountering real matter, and they begin to transfer some of their energy to it. At a density of one trillion grams per cubic centimeter, the matter will acquire most of the neutrinos' energy of motion as it becomes a true obstacle to their motion.

The collapse of the star's central regions provides the energy of motion to break down the nuclei into protons and neutrons. Then the electrons, dragged along by electromagnetic forces, combine with protons to form neutrons and neutrinos. This fusion reaction also needs

[5]As we mentioned in the section on electron degeneracy, the electrons feel the effects of the exclusion principle at far lower densities than the heavier nuclei—or neutrons—do because the electron's mass is so small. Hence the electrons always tend to have much higher velocities than the nucleons.

[6]The best detector is a tank of about a million liters of perchloroethylene (cleaning fluid), located a kilometer underground (to shield out other sorts of particles). Every once in a while, a high-energy neutrino, probably from the sun, hits an atom of chlorine-37 right in the nucleus and changes it into a nucleus of argon-37. This rare form of argon (most argon is argon-40) is radioactive and can be measured *atom by atom.* Oddly enough, the sun seems to be emitting only (at most) one half as many neutrinos per day as calculations say it ought to. Whether this discrepancy indicates something fundamentally new, rather than a calculational error, remains a subject of hot debate.

energy of motion, which comes from the one-shot infall of the star's center, and it occurs when the density has risen to about one trillion grams per cubic centimeter—the density at which the neutrinos no longer can pass through matter. The neutrinos, trying to escape in all directions, push the outer layers of the star out with them. This push returns part of the energy of motion from the collapse back to the star; in fact, it takes the last gasp of energy (the collapse) from the central regions and puts it into the more outward-lying layers of the star. These outer layers have less mass than the central parts, so the energy of motion that the neutrinos deposit in them blows them off into space at enormous velocities, tens of thousands of kilometers per second or even more.

We have described what astronomers have calculated to happen inside a collapsing star, and the visible result is a *supernova* explosion, which is a titanic burst of photons from the star's exploding outer layers. This outward burst masks the preceding (by a few seconds) catastrophic infall of the star's center, which we cannot see. As the neutrinos generated by the collapse sweep the outer layers into space, the newly acquired energy of motion of this matter produces all sorts of collisions among the nuclei and the electrons within them. New nuclei can be formed; existing nuclei briefly can capture electrons and lose them again; electrons can collide with each other or with nuclei, and not necessarily be captured into definite orbits. The supernova explosion process resembles a scaled-down version of the primeval fireball, with less energy per particle, although enough for many different reactions among particles to occur.

And what do we see from all this? A star undergoing a supernova explosion can be as bright as an entire galaxy for several months. Even in a giant galaxy, a supernova stands out as a noticeable fraction of the light output all by itself (Figure 8–12). This last gasp of energy bursts out suddenly where no star was seen before, and the remnants of a supernova explosion expand for thousands and millions of years (Figure 8–13). In a big galaxy such as our Milky Way, supernova outbursts occur about once a century, as another massive star resolves its problems in collapse and explosion. The last supernova within our galaxy seen from Earth appeared in 1604 A.D., and the one before that in 1572 A.D. Perhaps others since then have been hidden by interstellar obscuration (Page 177), although it takes a lot of obscuration to hide a supernova. The "new stars" of 1572 and 1604 were bright enough to be seen in the daytime for a month or so, and were visible at night for the best part of a year after that, even though they were thousands of parsecs away. In the year that follows its explosion, a supernova can radiate as much energy as the star that made it had radiated during the millions or billions of years of main-sequence life that preceded its explosion.

Figure 8–12. The supernova that appeared in the spiral galaxy NGC 4303 in 1959 (arrow), was briefly able to make a noticeable increase in the galaxy's light output. The other points of light that appear in the photograph are foreground stars in our own galaxy, about a million times closer to us than the supernova in NGC 4303. Hence these foreground stars are about a trillion times fainter (in absolute brightness) than the supernova in the faraway galaxy.

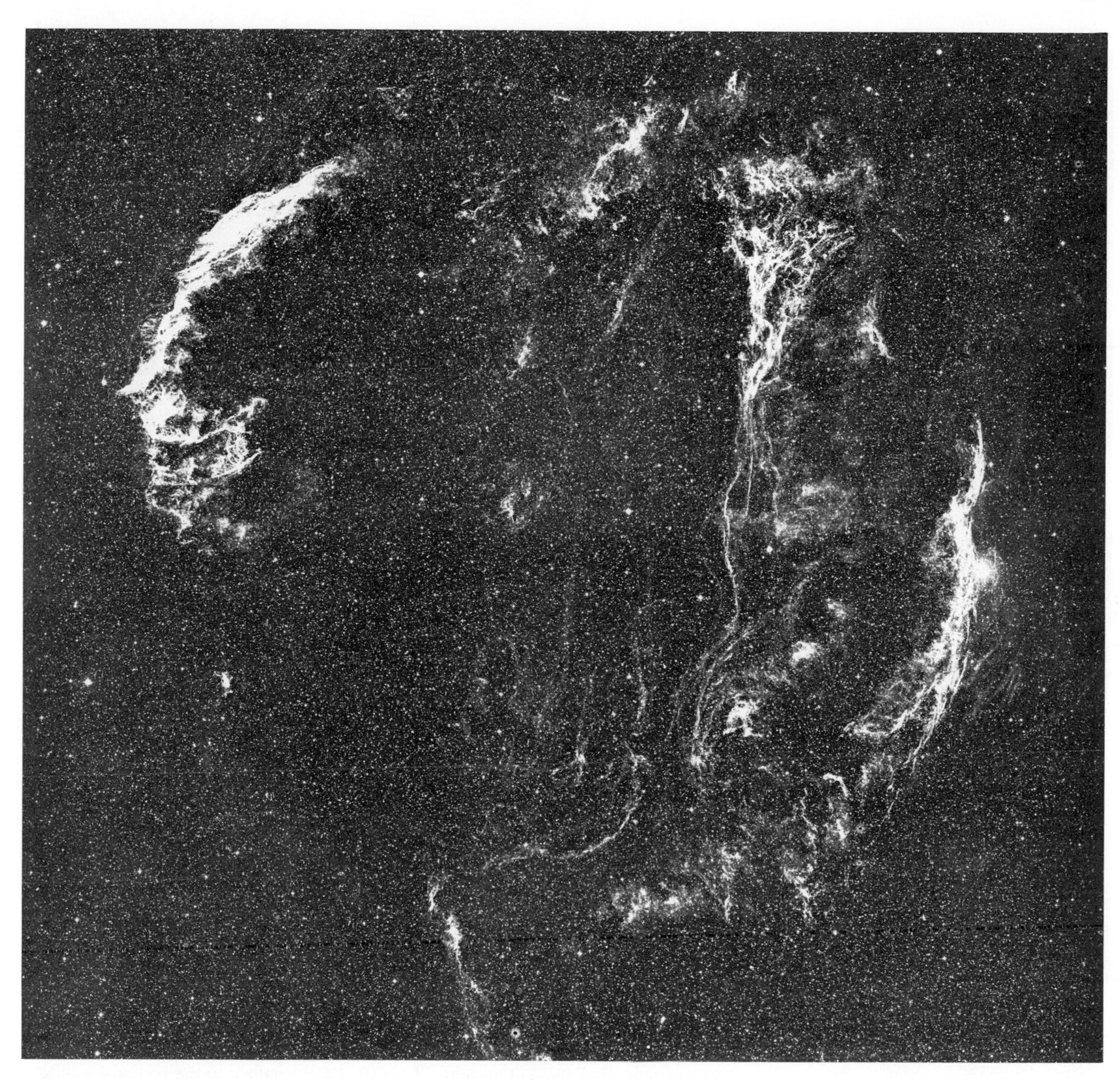

Figure 8–13. The "Veil Nebula" in Cygnus appears to be the remnant of a supernova that exploded about 30,000 years ago. The filaments of gas photographed here are still expanding from the point of explosion.

Figure 8–14. A photograph of the Crab Nebula taken with a filter to show only the photons emitted by one of the Balmer-series lines, namely the photons emitted when the electron in a hydrogen atom jumps from the third-smallest to the second-smallest orbit. This filtering process reveals gaseous "filaments" in the supernova remnant that owe much of their light to exactly this sort of electron jump, as we can see by comparing this photograph with Figure 2–17, which shows the Crab Nebula in the entire band of visible light.

The best-known supernova produced the remnant we call the Crab Nebula (Figure 8–14), which we shall describe in Chapter 12. This explosion was seen in the year 1054 A.D., though it apparently went unrecorded by European and Islamic scientists. However, Chinese astronomers made extensive observations of the new star's brightness changes, as indeed they had of dozens of other supernovae during the previous 2500 years, and the star-conscious Americans of that time left behind the record shown in Figure 8–15. We should note that since the supernova's distance from us was about a thousand parsecs (Figure 8–16), the explosion seen on Earth in 1054 A.D. actually must have occurred about three thousand years earlier. Similarly, the supernova that appeared in the Andromeda galaxy in 1885 A.D. must have exploded about two million years previously. We observe the past history of the universe, always a little out of date by the time we study it. The next supernova that we will observe blazing forth in our galaxy almost certainly has exploded by now, and the news is traveling toward us at the speed of light.

Figure 8–16. The supernova of 1054 A.D. occupied a galactic position slightly farther out from the center than the sun does.

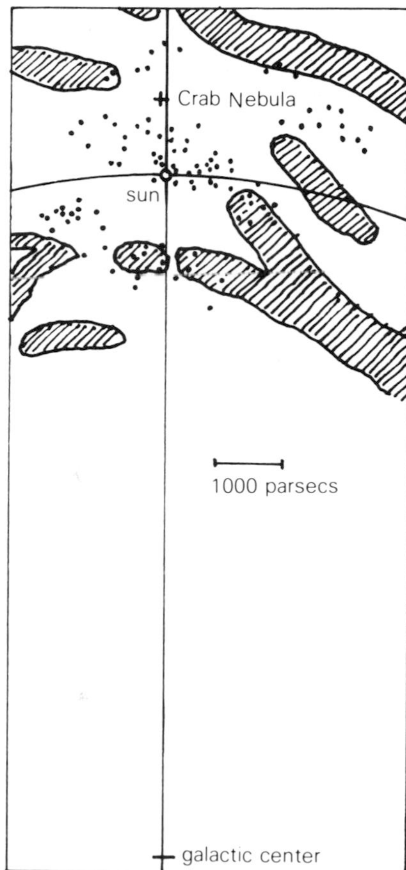

Figure 8–15. This pictograph carved on a rock in northern Arizona may represent how the supernova of 1054 A.D. looked one morning when the supernova was near its brightest and the moon happened to appear close to it on the sky. The reversal of the crescent moon from its correct appearance may have occurred because the record carver was looking over his or her shoulder at the rare event.

Table 8–1
Average Abundance of Certain Elements Relative to Hydrogen in the Solar System

Atomic number (number of protons in nucleus)	Element	Number of atoms of element for every 10^{12} hydrogen atoms
1	Hydrogen	10^{12}
2	Helium	8×10^{10}
3	Lithium	1500
4	Beryllium	25
5	Boron	10^{4}
6	Carbon	4×10^{8}
7	Nitrogen	1.2×10^{8}
8	Oxygen	7×10^{8}
9	Fluorine	7.5×10^{4}
10	Neon	1.1×10^{8}
11	Sodium	2×10^{6}
12	Magnesium	3×10^{7}
13	Aluminum	2.5×10^{6}
14	Silicon	3×10^{7}
15	Phosphorus	3×10^{5}
16	Sulfur	1.5×10^{7}
17	Chlorine	1.8×10^{5}
18	Argon	3.5×10^{6}
26	Iron	2.5×10^{7}
29	Copper	7×10^{4}
47	Silver	15
79	Gold	6
82	Lead	125
92	Uranium	0.8

So much for the glorious appearance of supernovae; now for their real importance. Supernovae seed the universe with nuclei heavier than helium, nuclei that appear in succeeding generations of stars and planets that condense in interstellar space. Recall that the original big bang (Page 65) produced most of the helium nuclei found in stars, but did not form appreciable amounts of carbon, nitrogen, oxygen, or any heavier nuclei. All of these elements, which are essential to living organisms (and for that matter form the bulk of the Earth), came from fusion processes inside stars. We believe this to be the best explanation now available for the existence of these heavier nuclei, and astronomers are almost at the point where their calculations can explain the relative numbers of these heavier elements and their abundance relative to hydrogen (Table 8–1). Although stars belch forth some heavier elements during their red-giant phases, the amount of this material falls far short of the one percent or so that we find for elements heavier than helium in stars such as the sun. The supernova explosions apparently produced enough energy of motion to manufacture even the heaviest nuclei, such as lead, gold, and uranium. About every nucleus heavier than hydrogen and helium formed inside a massive, rapidly evolving star, shot out into space in a supernova explosion, and found itself billions of years later incorporated in matter of quite different appearance, such as interstellar gas, Population I stars, or human bodies.

The distinction between Population I (younger) and Population II stars (older) is mainly in their different abundances of elements heavier than helium. Population I stars typically have one or two percent of their mass in the form of these "heavy" elements. This amount of heavy elements doesn't seem like much, but it is ten or a hundred times the relative abundance of heavy elements in Population II stars. It now appears likely that when galaxies began to form, an initial burst of star formation produced many massive stars that became supernovae and seeded the galaxies with enough heavy elements to explain the Population I abundances. The less massive Population II stars took longer to evolve and are still around today. The distinction between Population II and Population I, although helpful, cannot be completely sharp, because supernovae have continued to spread more and more heavy nuclei through the galaxies for the next batch of protostars to use.

Supernovae, then, made almost all the nuclei heavier than helium. Anything else? Indeed there is, because supernovae also are believed to be the source of "cosmic rays," which are protons, electrons, and other elementary particles that fly through space at almost the speed of light (Figure 8–17). These particles, which each have a huge energy of motion (for a proton or an electron), appear to fill our galaxy fairly uniformly, and thus seem to be the result of many individual sources rather than a few gigantic explosions. Given the scenario for an exploding supernova that we have outlined, we can suggest that cosmic-ray particles come from the outermost layers of supernovae, which acquire the largest velocities as the star explodes. The continual outbursts of supernovae in various parts of our galaxy, and of other galaxies, would have filled all of space with a small number of these cosmic-ray protons and electrons, each moving at speeds greater than 99% of the speed of light. Curious, but are they much use to us? Well, yes, they are, since cosmic rays are thought to be a reasonable explanation for the sudden changes or "mutations" in living organisms from one generation to the next. And mutations? They play a key role in the evolutionary process (Page 341) that carries us on to new experiences in the cycle of life.

Figure 8–17. Cosmic rays passing through these "lexan" plastic sheets left behind the tracks seen here (each a few millimeters long) because of the cosmic-ray particles' large energies of motion. Once the particles slow down from almost the speed of light, they are indistinguishable from ordinary electrons, protons, helium nuclei, and heavier nuclei that exist in abundance on the Earth.

So supernovae do it all. Caught on our lump of cosmic debris, itself the accreted remnant of countless supernova explosions,[7] some of the heavier nuclei have formed into "organic" molecules. The continuing inward flux of cosmic rays bombard these molecules and change the genetic material in some of them as the molecules replicate, helping evolution on its way. The hand that wrote this book, the paper on which it appears, the eye that reads it, and the mind that interprets it all owe their reality to the stars that exploded during the past five or ten billion years. We are living on the product, as the product, and by the product of supernova explosions. Cosmic catastrophe? It depends on the point of view; from ours, the collapse of these stars represents good fortune

[7]Although we see the Crab Nebula as a small cloud of gas in the constellation Taurus (Figures 8–14 and 8–18), the supernova also ejected particles that have now spread throughout the region between ourselves and the Crab Nebula.

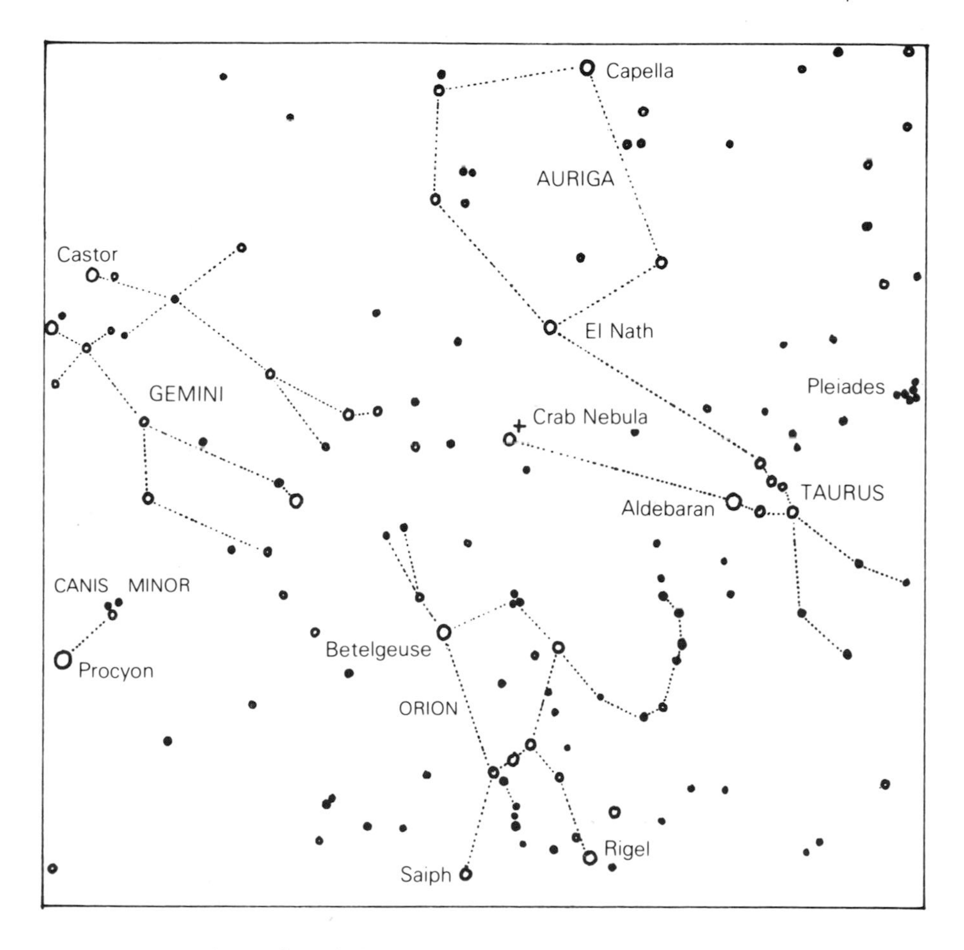

Figure 8–18. The Crab Nebula appears on the sky among the constellations Orion, Auriga, Gemini, and Taurus. These constellations can be seen rising in the east about 9 p.m. on December 21, or about 11 p.m. on November 21. During the winter months they appear toward the south and southwest early in the evening.

for the Earth. Without these catastrophes that spread "evolved" nuclei through space, each star would hoard its own output of heavy nuclei, and new stars would form pale copies of the old ones. Because of these explosions, new stellar material contains an increasing fraction of nuclei heavier than helium, and cosmic evolution really occurs. Some say that you can "see a world in a grain of sand" (William Blake). The grain of sand and the ability to see it are linked to the past history of the universe as surely as the cosmic background radiation or the farthest galaxies that we can see.

Summary

As stars grow older, they start to exhaust their supply of protons for fusion into helium nuclei at their centers. This deprives the star of the ability to achieve a steady rate of energy liberation. For a while, the star can compensate for the shortage of protons by contracting its central regions, thereby raising the temperature there to fuse the remaining protons at an ever-increasing rate. The additional energy of motion liberated in this way expands the star's outer layers, cooling them slightly in the process and producing a "red-giant" star that has a huge outer surface but a small, fast-burning inner core. More massive stars will run out of protons more rapidly than less massive ones, so in a cluster of stars all of the same age, the more massive stars will leave the main sequence of the temperature-luminosity diagram to become red giants before the less massive stars do. A red-giant star eventually will contract its core enough to start fusing the helium nuclei into carbon nuclei. This "helium flash" temporarily expands the core and sets the star into a period of "instability," during which it is likely to pulsate in size and brightness, as Cepheid variable stars do.

Smaller stars will become "degenerate" when they have fused helium nuclei into carbon nuclei: The star's interior will be so dense at this point that the exclusion principle keeps the electrons from packing any tighter. The electrons hold the nuclei from contracting through electromagnetic forces, thus the exclusion principle supports the star against its self-gravity. After the star's outer layers evaporate, the inner remnant persists as a "white-dwarf" star capable of a long, slow fade into dimness. However, no white dwarf can exist if its mass exceeds about 1.4 times the mass of the sun. Stars with more than 1.4 solar masses are likely to produce "supernova" explosions as they age. Supernovae arise from the fact that the more massive stars will not be "degenerate" at the point when they have fused helium nuclei into carbon nuclei. Rather, these stars will fuse carbon into heavier and heavier nuclei to keep on liberating energy of motion, but at the stage of the production of iron nuclei, no more energy-liberating fusions are possible, so the star's core collapses. The collapse produces huge numbers of neutrinos, which blow the outer layers of the star out into space at enormous speeds, making the star exceedingly bright for a while in a last gasp of defeat. Such supernova explosions seed heavier nuclei throughout the galaxy, and they probably produce most of the "cosmic rays," which are nuclei and electrons traveling at almost the speed of light, that we have detected in the solar system.

Questions

1. How long has the sun been converting protons into helium nuclei? How long will it take for the sun to run out of protons?
2. As the sun starts to exhaust its supply of protons, why will its central regions contract? What effect will this contraction have on the sun's outer layers?
3. Do we expect all stars to pass through a red-giant phase?
4. In a cluster of stars that all began to liberate energy of motion at the same time, which stars will become red giants first? How can we use this fact to find the ages of all the stars in such a cluster?
5. The Pleiades and the Hyades both are open clusters that contain about a hundred stars each. In the Pleiades the hottest main-sequence stars are of the type B5, whereas in the Hyades the hottest main-sequence stars are of type A1. Which cluster is the older?
6. What happens to a star when its "helium flash" occurs? Does this make the star's interior hotter or colder?
7. Consider two Cepheid variable stars, one of which fluctuates in light output over a period of one day, while the other has a cycle of light fluctuation that takes half a day. Which star has the greater true brightness? Why are the true brightnesses of each of these stars greater than the true brightnesses of main-sequence stars with the same surface temperature?
8. A star with ten solar masses liberates energy of motion about four thousand times more rapidly than the sun does. If the sun's supply of protons will allow it to liberate energy of motion at its present rate for a total of ten billion years, how long will the star with five solar masses last at its rate of energy liberation?
9. Why does the fusion of helium nuclei into carbon nuclei fail to keep a star going for as long a time as the fusion of protons into helium nuclei?
10. How does a white-dwarf star differ from a main-sequence star with the same surface temperature? What characteristic of a star determines whether the star can become a white dwarf?
11. What effect does the exclusion principle have on electrons? Does the exclusion principle affect helium nuclei? How do the electrons interact with the nuclei inside a white-dwarf star?
12. Which stars are likely to eject matter into space over a time span of several million years? Which stars are likely to eject much of their mass into space in a period of a few minutes?

13. Is the sun likely to produce a "planetary nebula" at some point in its lifetime? If a typical planetary nebula has a diameter of a tenth of a parsec, would the Earth's orbit lie well inside or well outside such a nebula, if the sun were at the center?
14. Why do the centers of some stars collapse as they age? How does this collapse lead to an outward explosion?
15. Why are supernova explosions important in the evolution of other stars in the same galaxy as these supernovae?

Further Reading

T. Page (Editor), *The Evolution of Stars* (Macmillan Co., New York, 1968).

R. Jastrow, *Red Giants and White Dwarfs* (New American Library, New York, 1969).

J. Greenstein, "Dying Stars," *Frontiers in Astronomy* (W. H. Freeman and Co., San Francisco, 1970).

J. Oort, "The Crab Nebula," *Frontiers in Astronomy* (W. H. Freeman and Co., San Francisco, 1970).

Planets—Formation and Orbits

Chapter 9

"But the Solar System!" I protested.

"What the deuce is it to me?" [Sherlock Holmes] interrupted impatiently: "You say that we go round the sun. If we went round the moon it would not make a pennyworth of difference to me . . . ."

Arthur Conan Doyle,
A Study in Scarlet

We have waited until now to discuss the Earth and the sun's other planets, because the planets' small sizes and masses make them tiny dust specks in the universe of stars and galaxies. Therefore we first have surveyed the structure and the history of the universe of larger distances, to gain a general impression of our cosmic environment. Yet we would be mistaken if we thought that planets' small sizes make them insignificant. As we have seen, the tiniest elementary particles still govern the behavior of stars, and we eventually may be surprised to find that planets have an impact on the universe far greater than their small sizes would imply. If living organisms tend to originate on the surfaces of planets—as life on Earth apparently did—and if advanced forms of life can reshape planetary systems and even entire galaxies—as has been suggested—then the planets could be the origin of an evolutionary force in the universe stronger than the basic patterns that we call the laws of nature, namely the forces of intelligence. But before we can speculate about the powers of intelligence on other planets or in other galaxies (Chapter 11), we must see how the one planetary system that we know in some detail came into existence, for better or worse, temporarily perhaps, but flourishing with life as this book is written.

The Formation of the Solar System

Astronomers now have good evidence that the sun's planets formed at the same time as the protosun contracted toward its present size. When the sun condensed from interstellar gas and dust, the condensation process went faster and faster as the protosun grew smaller and smaller. The protosun and the planets around it have condensed from a cloud of gas and dust much larger than the present size of the solar system. The original cloud must have had a diameter comparable to the distance to the next nearest star, which equals about ten thousand times the size of the sun's planetary system.

This planetary system includes four small planets—Mercury, Venus, Earth, and Mars—and four much larger and more distant ones—Jupiter,

Figure 9–1. A schematic drawing of the solar system gives an idea of the ordering of the planets out from the sun: Mercury, Venus, Earth, Mars, Jupiter, and so on. ▶

Saturn, Uranus, and Neptune. The ninth and most distant planet, Pluto, is an exception to the pattern of four small inner planets and four large outer planets (Figure 9–1). Between Mars and Jupiter lie a host of minor planets, or asteroids, and throughout the solar system we find meteors, which are rocks and dust in orbits around the sun like the planets and asteroids. Finally, a host of comets also orbit the sun, spending most of their time in orbit far beyond the outermost planet, Pluto.

In comparison to the distances between stars, even Pluto nestles close to the sun. If we make a model of the solar system in which we represent the sun with a light bulb, the planets would be tiny gnats circling the light bulb at distances between half a meter and one hundred meters, so the entire solar system (except for the comets) would fit easily inside an Olympic stadium. But the next closest *star* would be another light bulb five hundred kilometers away, the distance between Montreal and Buffalo (Figure 9–2). In this model the Milky Way galaxy would consist of a hundred billion light bulbs spanning the distance from the Earth to the sun at intervals of about a hundred kilometers.

When astronomers try to figure out how the solar system formed, they look for clues in the present arrangement of the sun and planets. The planets do not move around the sun in random orbits. Instead, their motions show a strong correlation with each other and with the sun's own rotation. First, all of the planets orbit the sun in almost the same plane, so that the solar system is as flat as our own galaxy (Figure 9–3). Second, the planets all orbit in the same direction around the sun. Third, the sun rotates once a month, in the same direction as all the planets orbit around it (Figure 9–4). From this overall behavior of the solar system, we can reconstruct the following scenario for the formation of the solar system. More than five billion years ago, a cloud of gas and dust began to contract. This cloud, which became the sun, the planets, asteroids, meteors, and comets, originally was thousands of times larger than today's solar system. As the cloud contracted, it tended to flatten from its original shape, which presumably was almost spherical, into a squashed disk. This flattening occurred because the cloud was rotating, slowly at first, more rapidly as it contracted. Thus the contraction of the protosun recapitulated, on a scale a million times smaller, the contraction of protogalaxies billions of years earlier (Chapter 4). Like the protogalaxies, the protosun and other protostars must have acquired some tendency to spin as the clouds that formed them detached from the general background of gas and dust. The clouds' spin tended to support them against contraction in the plane perpendicular to the spin axis (Figure 9–5). Therefore, although the clouds contracted in all directions, they contracted the fastest along their spin axes, so the clouds flattened into disks rotating around their axes of

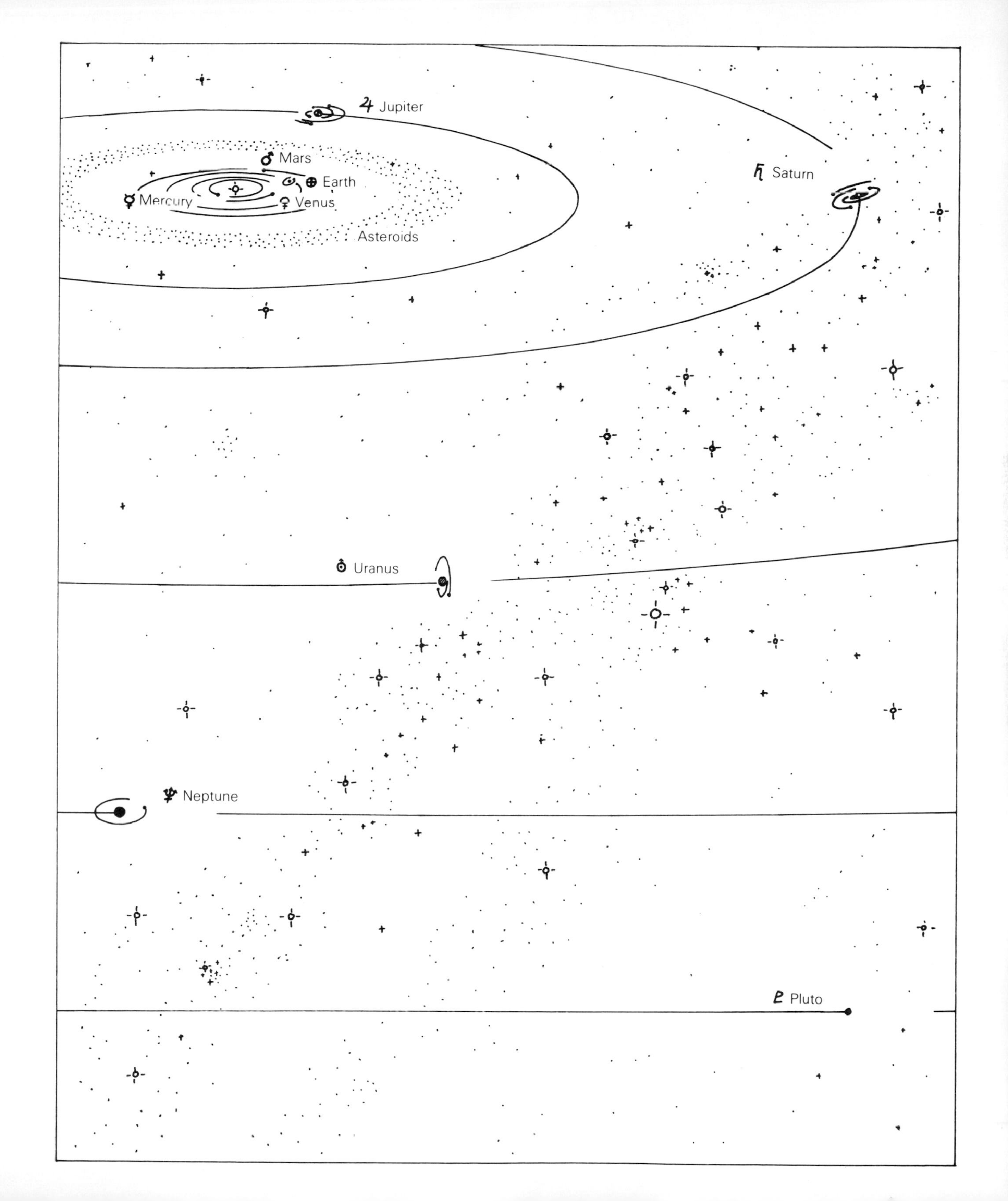
Jupiter
Mars
Earth
Mercury
Venus
Asteroids
Saturn
Uranus
Neptune
Pluto

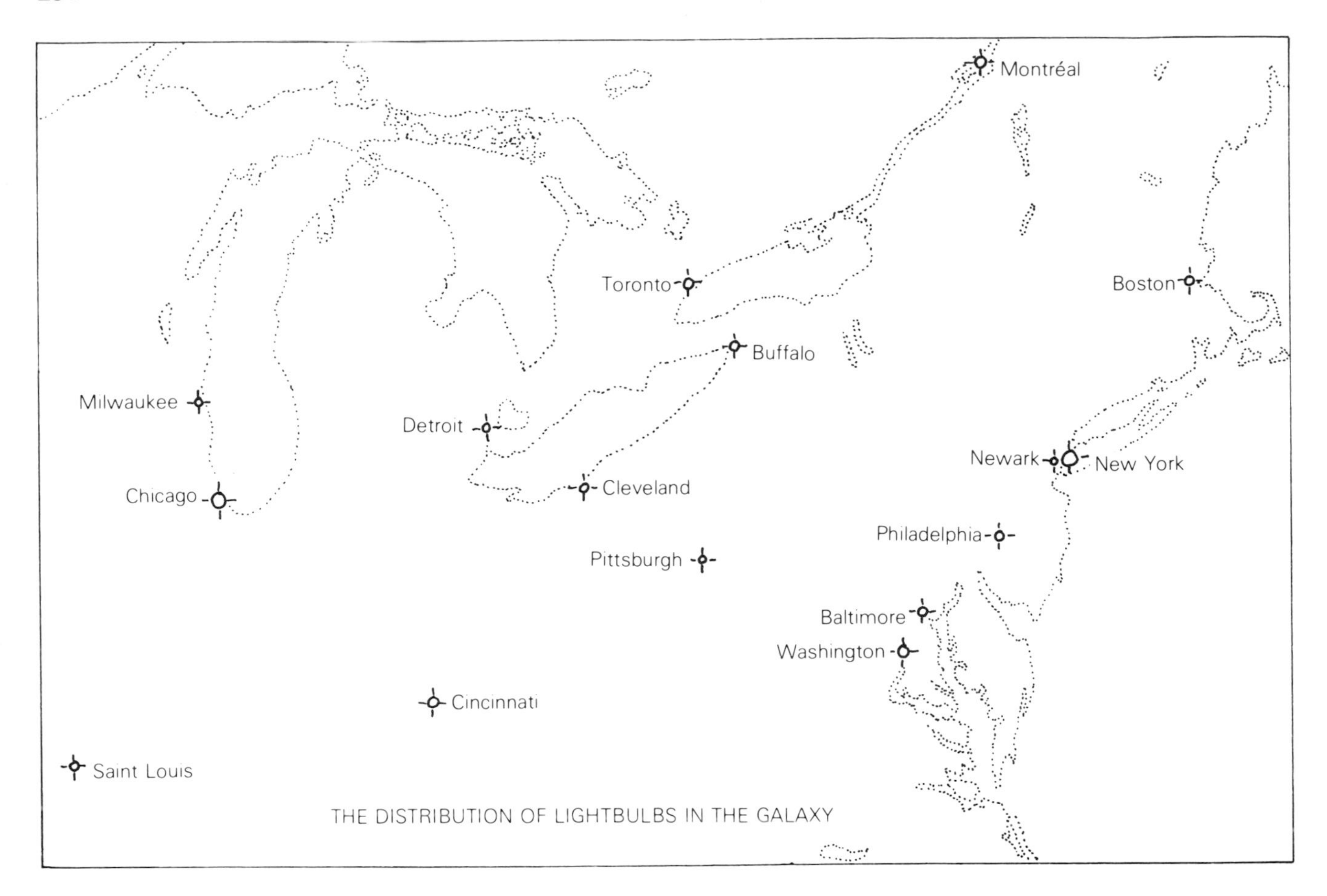

Figure 9–2. If the sun were the size of a light bulb, nearby stars would be spaced a few hundred kilometers apart. On this scale, the planets orbiting the sun could be represented as gnats flying around the light bulb at distances of half a meter to one hundred meters.

spin. The clouds' original rate of spin at the time the contraction began must have been far less than the rates of spin that we see now. For example, the cloud that formed the solar system might have been rotating once every hundred million years when it began to contract. As the protosun condensed, its continuing contraction made it spin faster (Page 92), so the spin rate steadily increased.

When the individual planets condensed out of the disklike, rotating cloud, their motions reflected the cloud's spin rate at that time and at the places where the planets formed. At any time, the inner parts of the cloud were rotating more rapidly than the outer parts. Today, the var-

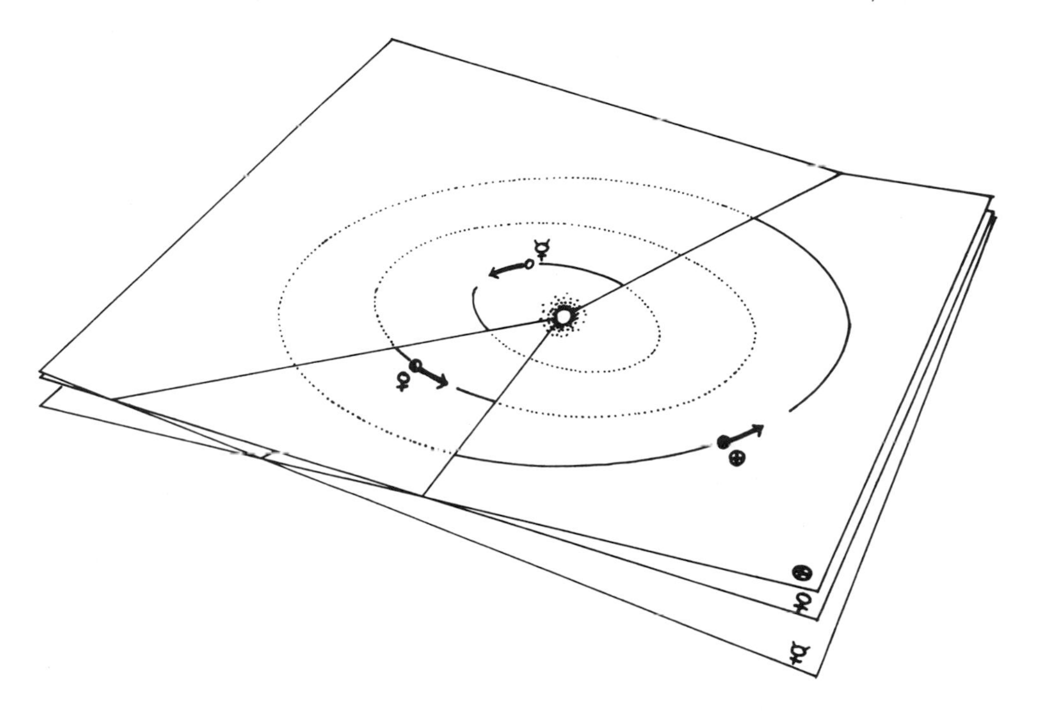

Figure 9–3. All of the planets orbit the sun in nearly the same plane and in the same direction.

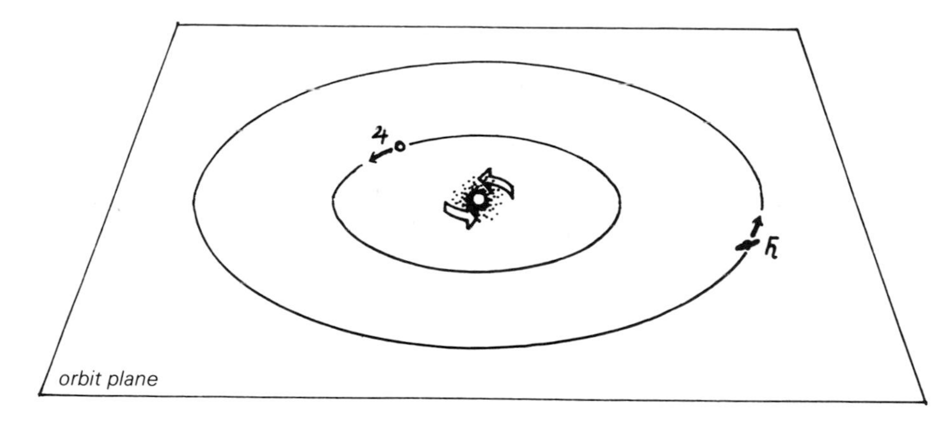

Figure 9–4. The sun rotates in the same direction as the planets orbit around the sun.

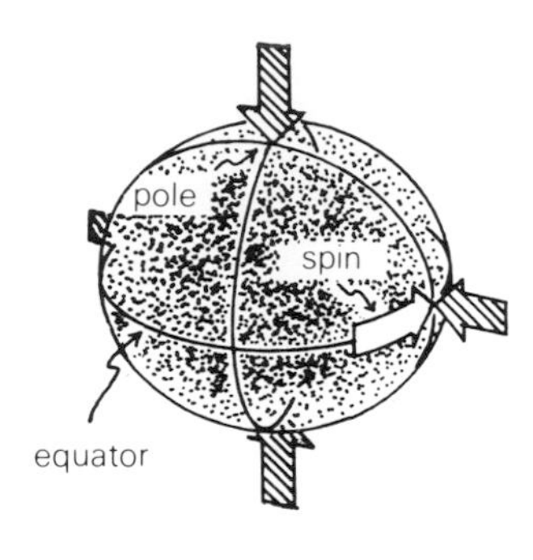

Spin resists gravity at the equator, but not at the poles. The cloud flattens to a disk.

Figure 9–5. If the contracting clouds of gas and dust that became protostars were rotating, their spin would tend to make them form disks.

ious planets complete their orbits around the sun in times that range from three months (for Mercury) to 250 years (for Pluto). Each successively farther-out planet moves more slowly along its orbit than the planets with smaller orbits do, which was also the case in the rotating cloud that produced the planets and the sun.

The planets' orbital motion gives each of them a certain amount of angular momentum.[1] The sun's own rotation, carried over from the rotation of the protosun, provides the sun with some angular momentum of its own, but compared to the angular momentum carried by the planets in their orbits, the sun's own angular momentum is quite small. In other words, the angular momentum of the original contracting cloud ended up preferentially in the planets instead of staying with the sun. The sun contains more than 99% of the mass in the solar system, but it has only two percent of the system's total angular momentum. The contracting protosun shed almost all of its original angular momentum as it left behind the clouds that formed the planets.

When we use the Doppler effect (Page 21) to measure how fast other stars are rotating, we find that the bright, hot stars at the top of the main sequence rotate much more rapidly than the cooler main-sequence stars such as the sun. Although it is tempting to speculate that the hotter stars did not produce planets while the cooler stars did form planets, this is almost certainly not the correct explanation for the differences in the stars' rotation speeds. Instead, the difference comes from the internal structure of the stars, and the way that the energy produced in their centers flows outward to their surfaces.

Planetary Orbits

A planet's angular momentum balances the gravitational attraction from the sun. As we mentioned on Page 46, the gravitational force between two bodies varies as the product of their masses, divided by the square of the distance between their centers. The willingness of an object to move in response to a force (such as gravity) varies in inverse proportion to the object's mass. Thus, although you pull on the Earth with the same amount of gravitational force as the Earth pulls on you, if you leap from a tall building, the Earth will not come up to meet you half way. In fact, it will move only about 10^{-18} centimeter, a distance much smaller than the size of an electron, and completely unmeasurable.

The planets and the sun all exert gravitational forces on one another. Since the planets have masses thousands of times less than the sun's mass, they tend to move toward the sun much more than the sun tends to move toward them. If the planets were motionless, they would start to fall into the sun as soon as they could, and would take only a few years

[1]An object's angular momentum in orbit is proportional to its mass times its velocity in orbit times its distance from the center of the orbit.

to do so. But instead of falling into the sun, the planets are falling *around* the sun. Their angular momentum keeps them moving in nearly circular orbits instead of heading straight toward the sun. A similar situation arises for objects in orbit around the Earth, such as the moon or our artificial satellites. If we launch a rocket with the right speed and direction, instead of falling back to Earth when its fuel runs out, it will fall around the Earth (Figure 9–6). This way of thinking about objects in orbit first was suggested by Isaac Newton when he considered the Earth's natural satellite, the moon. The gravitational force from the Earth keeps trying to pull the moon toward the Earth, but the moon's motion in orbit, its angular momentum, keeps it falling around the Earth in a roughly circular orbit. The same is true for the planets orbiting the sun. If the sun suddenly disappeared, the planets' momentum would carry them out of their circular orbits in straight lines (Figure 9–7). In contrast, if the planets were stopped in their tracks for a second, they would start to fall directly toward the sun. The tendency to move in a straight line works against the gravitational force from the

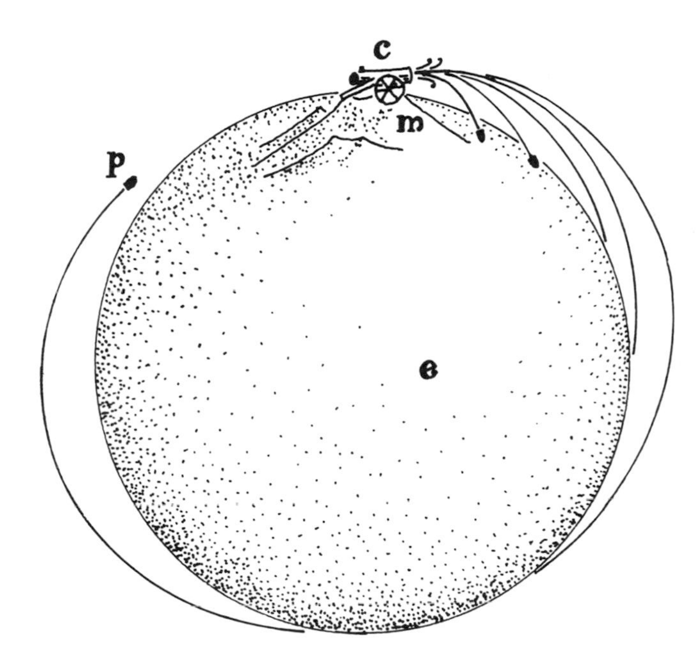

Figure 9–6. If we fire a projectile (p) from the Earth with a modest velocity, it will travel part way around the Earth before falling back. As Newton realized when he drew the diagram on which this figure is based, if we should fire the projectile with a large enough velocity, the projectile will "fall" into orbit around the Earth.

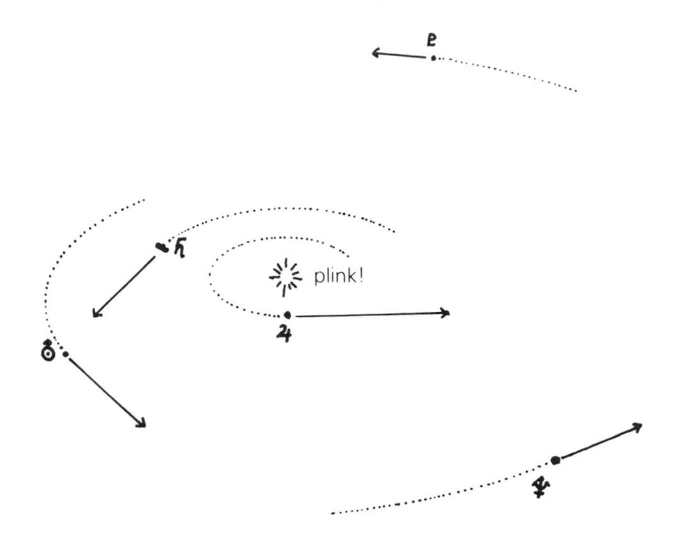

Figure 9–7. If the sun suddenly disappeared, the planets' paths would cease to be nearly circular orbits and would become straight lines. The momentum that each planet has would tend to carry the planet onward in the same direction, and with the same velocity, that it had at the moment that the sun disappeared.

sun, and the struggle produces an almost circular orbit in which each planet falls around the sun. If a planet's orbit is somewhat elongated, when the planet comes a little closer to the sun, it moves faster. When the planet is farther away, it moves more slowly in its orbit. The more distant planets move much more slowly in their orbits around the sun, but each planet has just the speed required to balance the sun's gravitational force. This force decreases when the planet-to-sun distance increases (in fact, the force decreases as the *square* of this distance).

The Earth moves around the sun in an almost circular orbit at a speed of 30 kilometers per second, which is enormously fast by our standards but just average in cosmic terms. Since the Earth's orbit has a radius of 150 million kilometers, the orbit's circumference totals 940 million kilometers, and the Earth takes 32 million seconds to complete each orbit. On Earth, we call this one year (actually, the Earth completes an orbit in 31,556,926 seconds, approximately). By comparison, the planet Jupiter is five times farther from the sun than the Earth is. Jupiter also covers an almost circular orbit around the sun, but it has five times farther to go than the Earth does. Furthermore, Jupiter's speed in orbit is only 13 kilometers per second compared to 30 kilometers per second for the Earth. Therefore Jupiter takes 12 earth-years to orbit the sun once.

All of the planets' orbits follow a general rule that their orbital periods (time to complete one orbit) vary in proportion to the planets' distances from the sun raised to the $3/2$ power. This intriguing relationship was discovered by Johannes Kepler in 1610 A.D. Kepler was also the first to

Table 9–1
Characteristics of the Planets and Their Orbits

Planet	Diameter (Earth = 1)	Mass (Earth = 1)	Average density (grams per cm^3)	Orbital period (years)	Distance from sun (Earth = 1)	Gap between planetary orbits	Predicted gap
Mercury	0.38	0.054	5.5	0.241	0.387		
Venus	0.95	0.82	5.1	0.615	0.723	0.336	0.3
Earth	1.00	1.00	5.5	1.00	1.00	0.277	0.3
Mars	0.53	0.11	4.0	1.88	1.52	0.52	0.6
[Asteroids]					2.77	1.25	1.2
Jupiter	11.04	318.0	1.3	11.86	5.20	2.43	2.4
Saturn	9.17	95.1	0.7	29.46	9.54	4.34	4.8
Uranus	3.79	14.5	1.6	84.01	19.18	9.64	9.6
Neptune	3.85	17.3	2.2	164.79	30.06	10.88	19.2
Pluto	0.5	0.1	5	247.69	39.44	9.38	38.4

establish that the planets' orbits, although nearly circular, are in fact elliptical.[2] Half a century after Kepler, Isaac Newton showed that Kepler's conclusions about our solar system follow from the law of gravity: *Any* object orbiting around a much more massive body will move in an elliptical orbit, and the time for it to complete an orbit always will be proportional to the $3/2$ power of its average distance from the more massive object. Newton showed that these conclusions follow naturally from the fact that gravitational forces decrease as the *square* of the distance.

Table 9–1 lists the planets' sizes, masses, densities, orbital periods, and distances from the sun. Most planets' orbits have "eccentricities" near zero (Figure 9–8) and thus are almost circular. Even the most "eccentric" or elongated orbits, those of Mercury and Pluto, look almost circular to the untrained eye (Figure 9–9), but the noncircularity of planets' orbits already was known four centuries ago.

As far as we can tell, the planets have followed their elliptical orbits around the sun for about $4\frac{1}{2}$ billion years. When the protosun had contracted to about the size of the present solar system, things started to happen relatively quickly. The formation of the planets, as well as the shrinking of the sun to its present size of 1.4 million kilometers from a protosun a trillion kilometers across, took only a few million years. Compared to the initial contraction of the protosun from an interstellar cloud of gas and dust, which took tens or hundreds of millions of years, this final birth process was almost instantaneous. As the density of gas

[2]In addition, Kepler recognized that a planet orbits more slowly when it is farther from the sun. For example, the Earth's speed in orbit varies between 29.3 and 30.3 kilometers per second.

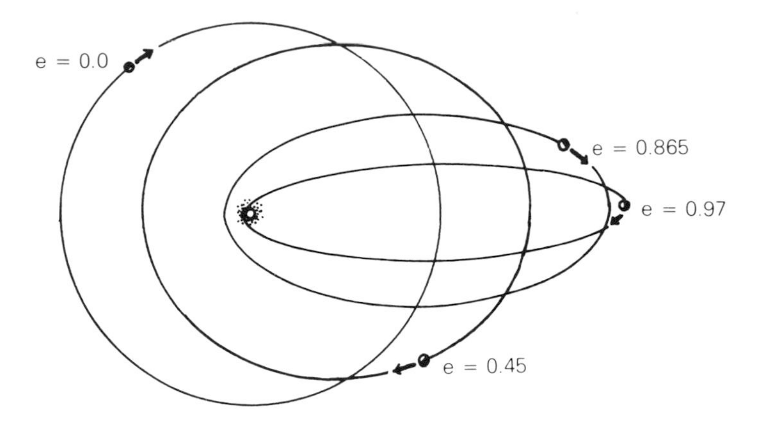

Figure 9–8. The "eccentricity" of an orbit measures the amount of elongation. A circular orbit has an eccentricity of zero, whereas an eccentricity of 0.97 means that the orbit is tremendously elongated. For a precise definition of the eccentricity, read further: Consider an object in orbit around the sun, and divide the object's greatest distance from the sun by its smallest distance from the sun. If we call this ratio R, the orbital eccentricity equals $(R - 1)/(R + 1)$. Hence if $R = 1$, the eccentricity is zero; as R becomes much larger than unity, the eccentricity grows nearly equal to unity. Orbits with an eccentricity greater than 1.0 are not closed paths.

and dust increased within the contracting disk, individual lumps began to form from random swirls and eddies. Dust particles that collided would stick together, thereby forming a larger target for other dust particles. As a lump grew toward planetary size, its own gravitational forces would provide significant help in capturing atoms and molecules that came near. In this way, the interstellar gas and dust of many billions of years ago became the solar system of today. We should never forget, though, that this interstellar matter itself had changed in the fiery furnaces of the stars, to be sown by supernova explosions among a new generation of protostars such as our sun.

Our solar system now contains nine planets, six of which are known to have satellites (Table 9–2). The Earth has one moon, Mars has two, Jupiter thirteen, Saturn ten, Uranus five, and Neptune two. Before the invention of telescopes, people knew only of the Earth's satellite.[3] Before telescopes were invented, some people refused to believe in the "double motion" of the moon around the Earth as the Earth orbited around the sun. These skeptics claimed that such a double motion as

[3]The planets Uranus, Neptune, and Pluto themselves remained undiscovered before telescopes revealed them in 1781, 1846, and 1930, respectively.

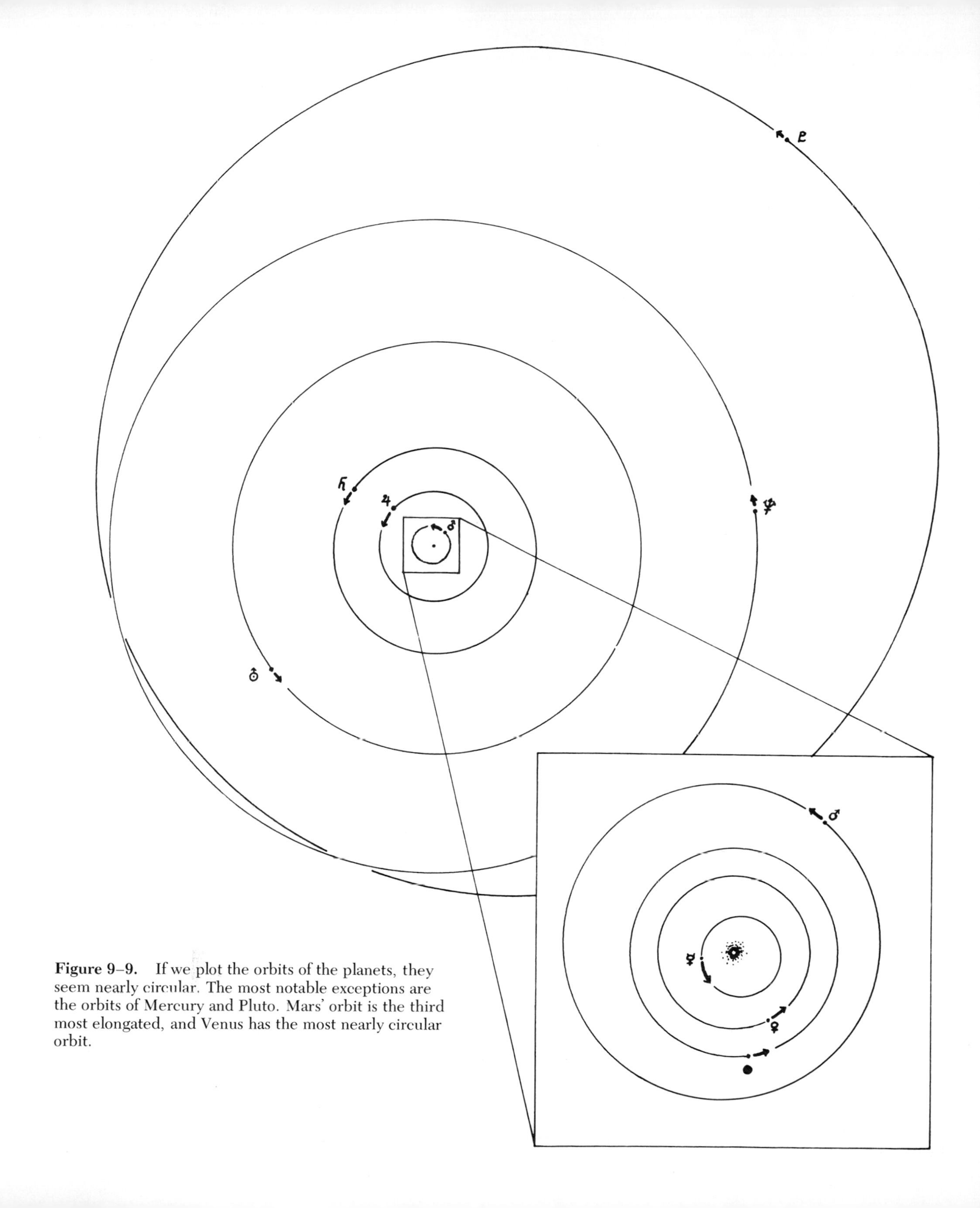

Figure 9–9. If we plot the orbits of the planets, they seem nearly circular. The most notable exceptions are the orbits of Mercury and Pluto. Mars' orbit is the third most elongated, and Venus has the most nearly circular orbit.

Table 9–2
Satellites in the Solar System

Planet and satellite	Average distance from planet (kilometers)	Period of revolution (days)	Diameter (kilometers)	Orbital eccentricity
Earth				
Moon	384,000	27.32	3475	0.055
Mars				
Phobos	9,000	0.32	20	0.021
Deimos	23,000	1.26	11	0.003
Jupiter				
Amalthea	181,000	0.50	140	0.003
Io	422,000	1.77	3640	0.000
Europa	671,000	3.55	3050	0.000
Ganymede	1,070,000	7.15	5250	0.002
Callisto	1,883,000	16.69	4900	0.008
Leda	11,100,000	239	8	0.147
Himalia	11,470,000	250.6	100	0.158
Elara	11,740,000	259.6	20	0.207
Lysithea	11,850,000	263.5	15	0.130
Ananke	21,200,000	625[a]	12	0.169
Carme	22,560,000	714[a]	16	0.207
Pasiphae	23,500,000	735[a]	20	0.378
Sinope	23,700,000	758[a]	16	0.275
Saturn				
Janus	159,000	0.75	320	
Mimas	186,000	0.94	600	0.020
Enceladus	238,000	1.37	600	0.004
Tethys	295,000	1.89	1000	0.000
Dione	377,000	2.74	800	0.002
Rhea	527,000	4.52	1500	0.001
Titan	1,222,000	15.94	5500	0.029
Hyperion	1,481,000	21.26	400	0.104
Iapetus	3,560,000	79.32	1500	0.028
Phoebe	12,950,000	550.37[a]	200	0.163
Uranus				
Miranda	128,000	1.41	200	0.000
Ariel	192,000	2.52	600	0.003
Umbriel	267,000	4.14	400	0.004
Titania	438,000	8.71	1000	0.002
Oberon	586,000	13.46	800	0.001
Neptune				
Triton	353,000	5.88[a]	3600	0.000
Nereid	5,600,000	359.42	200	0.750

[a]Denotes satellite orbiting in "retrograde" motion; that is, the satellite orbits the planet in the opposite sense to the planet's rotation.

Copernicus proposed for the moon in his sun-centered solar system must be too complex to conform to the divine simplicity that they sought in a universe centered on the Earth. In fact, the universe has turned out to be immensely complicated; as Einstein said, "God may be complex, but not malicious." When Galileo first looked at Jupiter with a telescope in 1609 A.D., he saw four moons orbiting the planet, clearly performing a "double motion" as Jupiter moved in its own orbit. Once this discovery penetrated the minds of other scientists, the Earth-centered model of the universe passed away forever. Today, almost 370 years after Galileo discovered them, these four "Galilean" moons of Jupiter provide the likeliest landing sites for any future expedition to the largest planet.

The "double motion" of planetary satellites deserves a closer look. The moon orbits around the Earth once a month, but in fact both the Earth and the moon orbit around their common center of mass. Because the Earth's mass is 81 times the moon's mass, the Earth-moon center of mass lies 81 times closer to the Earth's center than to the moon's center. This places the center of mass inside the Earth, although not at the Earth's center (Figure 9–10). As the Earth orbits around the sun, it also swings around the Earth-moon center of mass, making about twelve such swings during each orbit around the sun. Thus the Earth too has a "double motion," but neither its orbiting of the sun nor its monthly swing around the Earth-moon center of mass disturbs our tranquil lives on its outer surface.

Because the moon's orbit around the Earth-moon center of mass is not quite circular, the moon's apparent size on the sky varies from time to time. At its nearest to us, the moon has an angular size of 33 minutes of arc, whereas at the farthest point of its orbit, the moon's angular size is only 29 minutes of arc. Our memories are not good enough for us to notice the changes in the moon's apparent angular diameter, but the effect can be measured easily.

Just as the Earth and the moon orbit around their common center of mass as this center of mass orbits around the sun, so too the other satellites in the solar system all orbit around the planet-satellite centers of mass, while the centers of mass themselves make elliptical orbits around the sun. Because none of the other satellites has a mass as much as one one-thousandth of the mass of its parent planet, these centers of mass also all lie inside the planets. Each system of satellites, moving in elliptical orbits, forms a miniature "solar system," in which the outer satellites orbit more slowly than the inner ones, just as the outer planets move in their orbits more slowly than the inner planets do.

We can speculate easily that the planets' satellites were made during the later stages of protoplanet formation. As the planets themselves

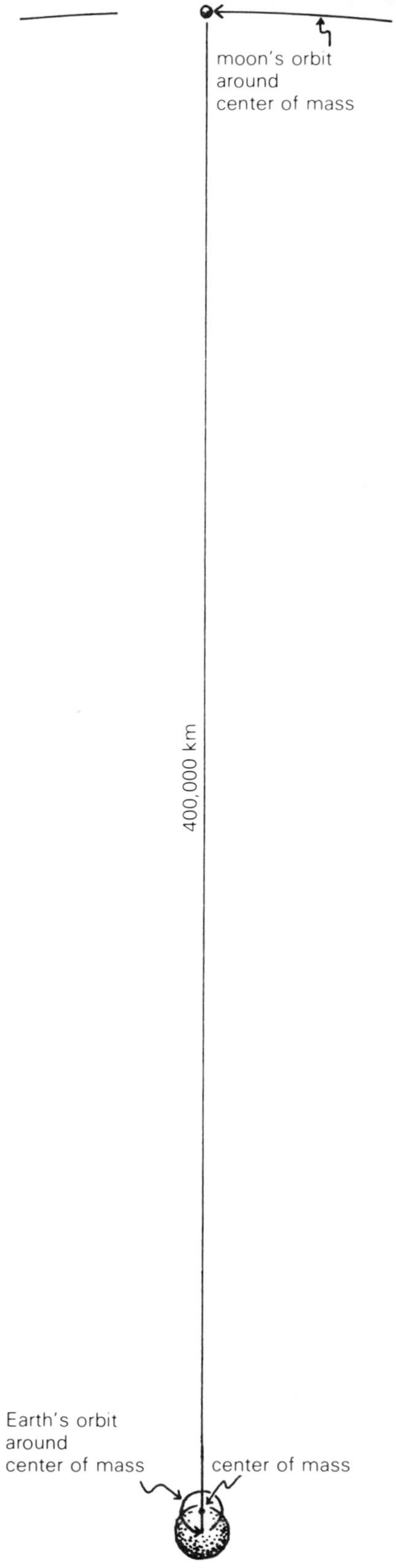

Figure 9–10. The center of mass of the Earth-moon system lies 81 times closer to the Earth's center than to the moon's center, because the Earth has 81 times the moon's mass. This places the center of mass inside the Earth, about 5000 kilometers from the Earth's center in the direction of the moon at any given time.

acquired more and more mass, some of a protoplanet's mass might have formed subclumps around the planet. This process seems especially likely for the four giant outer planets—Jupiter, Saturn, Uranus, and Neptune—which together have 29 of the 32 satellites in our solar system. These four planets are so massive (15 to 318 times the Earth's mass) that their own gravitational forces must have helped to attract more and more material as they formed. The inner planets—Mercury, Venus, Earth, and Mars—owe their formation more to the tendency of particles to stick together when they collide, although the planets' masses did grow large enough to hold on to some atmospheric gases.

Most of the original gases, however, have escaped from the planets. The reason that the four giant planets are so much more massive than the inner four rests in their ability to hold on to hydrogen and helium. The material from which the planets formed consisted, like the rest of the universe, mostly of hydrogen and helium. Because the temperature in the contracting disk was fairly low before the sun began to shine, atoms did not collide with enough energy of motion to ionize themselves. When the protosun had contracted to the point where it began to release photon energy (Page 195), this energy began to warm the protoplanets. We shall consider this life-giving radiation in Chapter 11, but for now the important point is that the inner planets got warmer than the outer planets.

At any temperature, each of the atoms had about the same energy of motion as its neighbors. This means that the less massive atoms had greater velocities than the more massive atoms, because each atom's energy of motion was proportional to its mass times the square of its velocity. As each atom in a protoplanet moved about randomly, it had a chance of flying off into space. This chance depended on the atom's velocity and on the force of gravity that the planet could exert. The lightest atoms, hydrogen and helium, always have the greatest velocities and hence the best chances to escape. Because the four inner planets are warmer than the giant planets, the hydrogen and helium that formed most of the contracting gas cloud has escaped from their vicinity. The inner planets now have masses too small to hold on to hydrogen and helium atoms even if the atoms were still present in their atmospheres. In contrast, the four giant planets still consist mostly of hydrogen and helium, which gives them their large masses—and the large masses in turn help them hang on to the hydrogen and helium atoms. If Jupiter were moved to the Earth's distance from the sun, the planet still could hold on to its hydrogen molecules (pairs of hydrogen atoms) because of its huge gravitational forces, which are ten times stronger than the Earth's at the planet's surface.

Spacing of the Planets' Orbits

If we consider the various distances of the planets from the sun, we notice a certain pattern. Going outward from the sun, we find that the spacing between the planets' orbits steadily increases. The gap between successive planets approximately doubles each time we go to the next outward planet.

Let us write the planets' distances from the sun as a percentage of the Earth's distance from the sun, 150 million kilometers. Astronomers call this distance one "astronomical unit," reflecting our belief in the importance of the Earth. If we express all the planet-sun distances in astronomical units, we have the results shown in the sixth column of Table 9–1. The next column in this table shows the gaps between successive planetary orbits, as measured in astronomical units. With some fudging, we can see that the size of these gaps follows a regular progression. The Mercury-Venus and Venus-Earth orbital separations each are 0.3 astronomical units, but then we get successive gaps (measured in astronomical units) of about 0.6, 1.2, 2.4, 4.8, 9.6, and so on, with each gap twice as large as the preceding one.[4] To be brutally frank, the actual sizes of the spaces between orbits do not quite follow this direct doubling pattern, but instead are 0.52, 1.25, 2.43, 4.34, 9.64, and so forth—not to mention the complete failure of the rule at the outer edge of the solar system. However, if we do not insist on complete mathematical accuracy, we see that there is a noticeable tendency for each gap to be twice as large as the preceding one.

The doubling of the sizes of the gaps between planets' orbits holds as a general rule, but there are some exceptions. First of all, there seems to be a planet missing between Mars (1.52 astronomical units from the sun) and Jupiter (5.2 astronomical units). Second, the outermost planets, Neptune and Pluto, don't seem to fit the rule at all.

Astronomers have explanations for these exceptions. Filling the gap between Mars and Jupiter, they note, are the asteroids, or minor planets. The largest asteroid, Ceres, has a diameter of 700 kilometers, which is five times less than our moon's. Although the total mass of the thousands of known asteroids adds up to far less than a planet's mass, astronomers believe that the asteroids represent the remains of a protoplanet that failed to coalesce properly, but got only as far as a host of subplanetary bodies. These subplanets must have been unable to cohere into one planet, perhaps because Jupiter already had formed and kept disrupting any attempts at further coalescence by the gravitational forces it exerted on the subplanets. In any case, we now find the asteroids moving in nearly circular orbits (although some asteroids have

[4]This rule for orbital spacing is called the "Bode-Titius Law" after its two discoverers.

highly elongated trajectories) whose average distance from the sun is 2.77 astronomical units, which almost equals the distance predicted by the rule of doubling the size of the gap between successive planets.

A few asteroids do have highly elongated orbits, with eccentricities as high as 0.83 for the minor planet Icarus. The distances of the asteroids with elongated orbits from the sun can vary from as much as 10 astronomical units to as little as 0.3 astronomical units for Icarus, which at the point of its orbit closest to the sun goes inside the orbit of Mercury. A few asteroids (Icarus, Geographos, Apollo, Adonis, Hermes, and Eros, for example) have orbits elongated enough and small enough to bring them across the Earth's orbit from time to time. Thus there is a small but finite chance that the Earth someday might collide with one of these asteroids, which are at most a few kilometers across but still capable of making quite a splash upon collision.

Astronomers continue to discover asteroids at the rate of several per year, to add to the thousands already found, although by now we are down to finding asteroids mostly less than a kilometer across. The astronomer who discovers a new asteroid gets to name it, and in the past, impecunious astronomers sold this right, while others named asteroids for their favorite wives, children, cities, or fantasies. Thus we have asteroids named Chicago, Adelaide, Limburgia, Fanatica, Hela, Ivar, and Marlene.

The other exception to the rule of regular doubling of the gaps between successive planets appears in the outermost planets, Neptune and Pluto. Astronomers discovered Pluto, the farthest planet, by means of the slight effect that Pluto's gravity has on Neptune's orbit. Pluto's gravitational force deflects Neptune slightly from the orbit that it would have if Pluto did not exist, and by an arduous series of observations and calculations, astronomers were able to predict about where Pluto ought to be. A long search through pairs of photographs taken a few days apart revealed Pluto, in 1930, by its small apparent motion against the background of stars.[5] Neptune had been discovered in 1846 by its gravitational effect on the orbit of Uranus. Uranus, however, was discovered by accident, or rather by William Herschel, who noticed one night, in 1781, that what others had recorded as a star actually showed a disk like a planet in his telescope.

Although Uranus and Neptune both are giant planets, Pluto turned out to have a far smaller size and mass than the other outer planets. In addition, the distances of Pluto and Neptune from the sun do not fit

[5]Even though Pluto's existence had been suggested by calculations, these calculations were inexact enough for us to say truthfully that Pluto really was discovered by this search alone.

the pattern of regular spacing established by the other planets' orbits. Finally, in this list of odd facts about Pluto, astronomers found that Pluto has the most elongated orbit of any planet (Mercury is second), and its orbit actually passes inside the orbit of Neptune (Figure 9–9). Because of Pluto's peculiar orbit, and because Pluto is only about 70% larger than Neptune's larger satellite, Triton, some astronomers have suggested that Pluto originally was a satellite of Neptune, perhaps ejected into its own orbit around the sun by a near collision with Triton. This theory may not necessarily be true, but it offers an interesting possible explanation of Pluto's small size, and the collision hypothesis might explain why Neptune's and Pluto's distances from the sun do not fit the general pattern of orbits in the solar system.[6] Additional evidence for a past disturbance in Neptune's satellite system comes from the fact that Neptune's larger moon, Triton, orbits the planet in the opposite sense to the planet's rotation. Triton is the only large satellite in the solar system to have such a "retrograde" orbit. Neptune's smaller satellite, Nereid (only 200 kilometers in diameter as compared to Triton's 3600) has by far the most elongated orbit of any planet or satellite. Nereid's orbit has an eccentricity of 0.75, and its distance from Neptune along this orbit varies from 1.4 million kilometers to almost 10 million kilometers.

Pluto takes almost a week to rotate, in contrast to the giant planets, which all rotate in less than half a day. This difference again speaks in favor of the suggestion that Pluto was once a satellite of Neptune, but the question remains undecided, and probably will stay that way until we get a close look at both Neptune and Pluto. These planets are so distant that earth-based telescopes cannot reveal anything about the features on their surfaces.

Meteors and Comets

In addition to the nine planets, their satellites, and the asteroids between Mars and Jupiter, our solar system contains swarms of dust and rocks called "meteors," and a host of larger meteorlike objects called "comets." Meteors orbit the sun along somewhat elongated trajectories that often cross the planets' almost circular orbits (Figure 9–11). At different times during its yearly orbit around the sun, the Earth intersects the orbits of different swarms of meteoritic dust particles. When this happens, the large velocity of the meteor swarm relative to the Earth (perhaps 10 to 30 kilometers per second) means that any meteoritic particle that enters the Earth's atmosphere will glow from the friction produced by its pas-

[6]To be more honest, the collision hypothesis might explain why Pluto's orbit does not fit the pattern of orbital spacings, but it is harder to see how this would explain Neptune's failure to conform to the pattern.

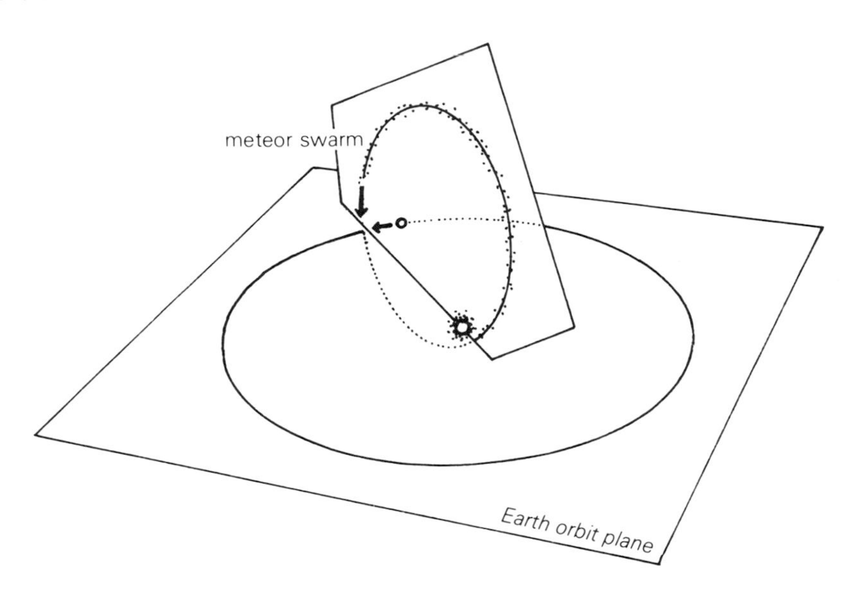

Figure 9–11. Meteor swarms move around the sun in somewhat elongated orbits that often cross the Earth's orbit. Because the meteors are not moving in circular orbits, they do not have the same velocity in orbit as the Earth does at the time when they cross the Earth's orbit, so the meteors have a sizable *relative* velocity with respect to the Earth. This relative velocity, which is increased by the pull from the Earth's gravity and by the Earth's rotation, causes the meteors that enter the Earth's atmosphere to have a high velocity through it. The friction that arises from this high velocity consumes most of the meteors that enter our atmosphere.

sage through the atmosphere (Figure 9–12). This friction burns up the meteors to make "shooting stars," as we can see for ourselves on any clear night if we stand outside for a few minutes. On those nights when the Earth's orbit intersects a large swarm of meteors, we can see several shooting stars each minute. The average meteor that we see shooting across the sky weighs less than a gram, and a meteor that weighs a kilogram will make an especially bright trail as friction burns it away. Still larger meteors occasionally survive (in part) their frictional burning in our atmosphere and strike the Earth's surface. The meteor crater in Arizona (Figure 9–13), which is more than a kilometer across, bears witness to the impact of a gigantic meteor about 50,000 years ago. Thanks to our atmosphere, most meteors that reach the Earth never strike its surface, but some do arrive each year. Twenty years ago, a woman in Alabama was hit on the leg by such a meteor—luckily on the first bounce. She suffered only minor tribulations as a result of this encounter with an extraterrestrial object.

Figure 9–12. Photographs of stars often capture "meteor trails" by accident. The brightest meteors can leave behind a trail of ablated matter and shock-heated gas that lasts a few minutes before disappearing.

Figure 9–13. The great meteor crater near Winslow, Arizona shows the effects of a large-mass impact. Much of the meteor must have vaporized as it struck the Earth, because test borings in the crater floor have failed to come up with more than small meteoritic lumps.

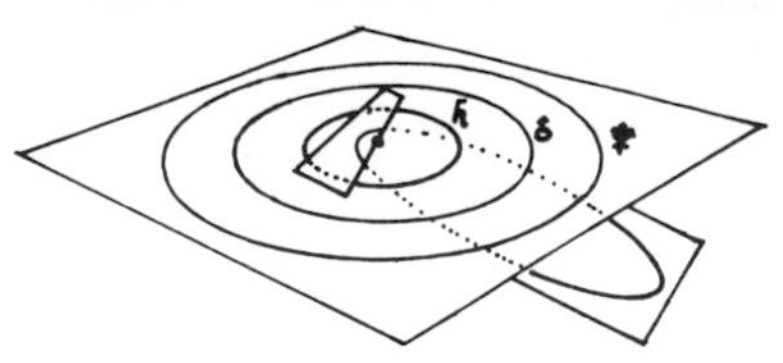

More exciting than meteors, comets move around the sun in extremely elongated orbits. Halley's comet, the best known, has an orbit with an eccentricity of 0.97 (Figure 9–14), yet this orbit is *less* elongated than most comets' orbits, which tend to have eccentricities near 0.9999. The comets with these tremendously elongated orbits spend most of their time far past the orbit of Pluto, reaching distances from the sun of 50,000 or 100,000 astronomical units—half way to the nearest star. Halley's comet has an orbit that takes it outside the orbit of Neptune and back inside the orbit of the Earth. The comet completes an orbit about every 76 years: It passed close by the Earth in 1910, and will be back in 1986 (even now it has crossed Uranus' orbit and is picking up speed as its orbit takes it closer to the sun). A small minority of comets have orbits similar to the orbits of asteroids, but more elongated. That is, these comets have an average distance from the sun of three or four astronomical units, and orbital eccentricities of about 0.5 to 0.7. Apparently the planet Jupiter has captured these comets into orbits much smaller than their original ones, through the power of its gravitational forces at a time when the comets passed close by the planet. Because these captured comets take only five to eight years to orbit the sun, they are called "short-period comets" in contrast to the vast majority of comets, which take anywhere from one hundred to one million years to orbit around the sun. Halley's comet, with an orbital period of about 76 years, stands between the short-period and long-period comets in the size of its orbit and the time that it takes to orbit the sun once. People had observed Halley's comet dozens of times on its successive passes by the Earth before Edmund Halley realized that they must have been seeing the same comet over and over again. Halley died before his realization could be verified by the comet's next return, but we honor him today as the man who first understood cometary orbits. Before Halley made his prediction in 1682, people tended to be terrified of comets, with their unusual "hairy" appearance and their unpredictability. Thus, for example, the Bayeux tapestry shows King Harold of England being warned by Halley's comet (in 1066 A.D.) that catastrophe lay in store for him, and indeed before the year was out William the Conqueror's army had slain him at Hastings.

Comets consist of frozen ammonia, frozen water, and dust particles, which together form a solid lump for most of the comet's journey around the sun. During the time that a comet comes relatively close to the sun, ultraviolet photons from the sun vaporize part of the comet to

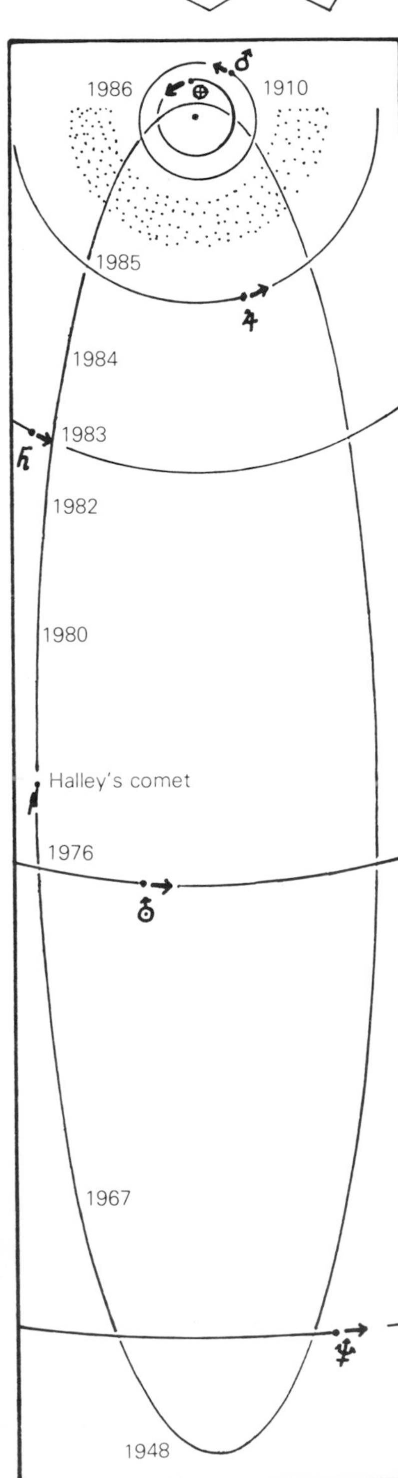

Figure 9–14. Halley's comet has an orbital eccentricity of 0.97, which makes the comet's orbit far more elongated than that of any planet, although less elongated than most comets' orbits are. Halley's comet moves around its orbit in about 76 years, passing inside the planet Mercury at its closest approach to the sun and outside the orbit of Neptune at its most distant. Perturbations in the comet's orbit that are produced by the gravitational forces from the planets (especially Jupiter and Saturn) make the orbital period vary from 74 to 78 years. At the present time, Halley's comet is between the orbits of Uranus and Saturn, heading in toward the sun for its next close approach, in 1986.

produce a long, rarefied "tail" behind the main part, or "nucleus," of the comet. We shall discuss what comets are made of in more detail in the next chapter, but for now we can point out that comets seem to represent some of the original chunks of matter out of which the protoplanets formed, chunks that have spent most of their time in a celestial deep freeze between the stars. Thus to study a comet close up would greatly please those astronomers who seek to understand how our solar system condensed around the contracting protosun. Someday we may dissect a comet, which would have perhaps a billionth of the Earth's mass, to uncover the past history of the planets and the sun.

The sun, which is at the focus of the solar system, holds it together and keeps it warm. Important as the planets are to us, we should not forget that even the largest, Jupiter, has only a thousandth of the sun's mass. If other stars have planetary systems, we could not hope to see them even with our largest telescopes. The planets' small sizes, plus their relative nearness to their parent stars, plus the fact that they shine only by reflected light, leads to the planets' being lost in the glare from the stars they orbit. In fact, astronomers believe that a large fraction of the stars around us probably do have planetary systems. They base this belief on the idea that many protostars should have contracted in a manner similar to the condensation of our protosun. Most contracting protostars also could have formed a rotating disk that later condensed into planets. The regular spacing of the planets' distances from the sun seems to imply that the planets formed naturally as a result of the protosun's condensation, and not, for example, by random capture of passing bodies. The nearly identical chemical composition of the sun and the giant planets also implies that they arose from the same original material.[7] Since the sun seems to be an average star, we can speculate that many of the other hundred billion stars in the Milky Way galaxy also have planets orbiting around them. The exceptions to this line of reasoning may be the double and multiple star systems. As these protostars contracted, any planet-sized condensations might have been pulled apart by the competing gravitational forces from the various nearby stars. But even these multiple stars might have planets in orbits around one of the stars, or around all of them. Such planets would not have a simple day or night that arises from the planets' rotation, but could have a mixture of day or night that arises from having two or four or six suns in their skies, each of a different color and each providing a different amount of light.

---

[7]Once we allow for the differences in gravitational holding power among the planets and the sun, we find that the relative abundance of the various elements is almost the same in all of them. Almost, but not quite; for example, the minor differences between the relative abundances in the moon and in the Earth (Page 295) seem to show that the moon was never a part of the Earth.

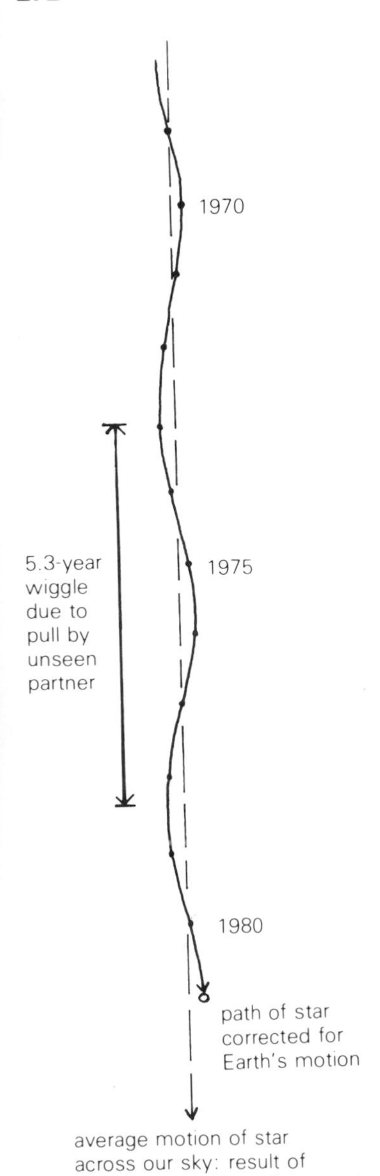

Figure 9–15. If we plot the apparent position on the sky of a nearby star relative to the background of much more distant stars, we may find that even after we allow for the back-and-forth "parallax shift" (see Page 3), the nearby star appears to move with respect to the background. This "proper motion" depends on the distance to the star and on the star's velocity. From any small wiggles in the straight-line proper motion that we measure, we can deduce that the star may have planets in orbit around it that pull the star first one way and then another by a small amount.

Even if double and multiple stars do not have planets, and even if the (relatively scarce) young and hot stars that we observe to be rotating rapidly also do not have planets (Page 256), still the remaining stars would provide billions of planetary systems in our own galaxy, if most of them developed planets as they contracted. How could we detect any other such planetary systems, if we can't see them in telescopes? Another method of detecting planets relies on the gravitational forces between the planets and their parent star. Even though the planets have far less mass than the star, their gravitational pull can produce small effects in the star's motion through space. When we observe the stars closest to us over a period of several years, we can detect the changes in their position that arise from the stars' motion relative to the sun. This motion appears as a small displacement in the stars' positions relative to the background of more distant stars, which are too far away for their motions to be detectable in only a few years' time. We do have to allow for the parallax effect produced by the Earth's motion around the sun (Page 4), but once we do this, we can plot the trajectory of a nearby star as shown in Figure 9–15. The star's path through space that we measure looks basically like a straight line. However, if we can detect wiggles in the star's straight-line path, we may assume that the star is responding to the gravitational pull from one or more planets orbiting around it. A nearby star, called "Barnard's star" after a famous American astronomer (Barnard), seems to show such deviations from a straight line in its motion through space. Barnard's star, which is only four parsecs away, is a good deal smaller and cooler than our sun, so if it does have planets, they are likely to be much colder than the sun's. Because the Earth's atmosphere blurs our attempts at accurate measurements of stars' positions on the sky, this search for other planetary systems through the planets' effect on their stars' motions could work much better from above our atmosphere. Thus one of the results that astronomers hope to obtain from the Large Space Telescope could be the first definitive detection of other planetary systems. The best evidence that we *now* have for the existence of such systems consists of the fact that the sun appears to be an average star.

The Sun

The golden appearance of the sun masks a seething exterior that itself conceals the thermonuclear fusion reactions in the sun's internal regions. Inside the sun the temperature reaches 13 million degrees absolute, which is sufficient to make protons fuse together to form helium nuclei. As the energy liberated in these reactions spreads outwards from the sun's center through many collisions, the photons that carry the energy pass through layers of steadily decreasing temperature and density. By the time the photons reach the sun's surface, the temperature has fallen from 13 million degrees to six thousand degrees absolute, which is hotter than a blast furnace on Earth. We define the sun's "surface" as the depth to which we can see before the sun's gases become opaque to visible light. Above the "surface" we find (by definition) the sun's atmosphere.

The sun has an atmosphere that extends far into space. In truth, this solar atmosphere engulfs the Earth and reaches beyond it, in the form of a "solar wind" of electrons, protons, and heavier nuclei that are expelled constantly outwards from the sun. This solar wind carries off a tiny fraction of the sun's atmosphere, which is at once replenished from below. Occasional outbursts called "solar flares" can produce still more energetic particles moving at almost the speed of light, which may pose a constant hazard to any space travelers in our part of the solar system. When the charged particles in the solar wind reach the Earth, the Earth's magnetic field deflects them by electromagnetic forces. The particles that penetrate our upper atmosphere produce the photons in the "aurorae" or "northern lights" that are especially prominent near the north or south magnetic poles of the Earth (Figure 9–16). Sometimes the sun will emit an especially sharp burst of high-energy charged particles that will distort the layer of charged particles, called the ionosphere, in the Earth's upper atmosphere. This ionospheric layer reflects long-wavelength radio photons. When solar bursts distort the ionosphere, long-distance radio communications become more difficult, because we have a harder time bouncing radio waves off the ionosphere. In contrast, auroral displays become more frequent at times when the sun emits greater numbers of fast-moving charged particles. Somewhere past the orbit of Mars, the solar wind merges with the general background of energetic charged particles (cosmic rays) that permeate the galaxy.

A strange fact about the sun's atmosphere is its high temperature, often much greater than the sun's surface temperature. The outermost part of the sun's atmosphere, called the corona, has a temperature of millions of degrees absolute. This corona can be seen during a total eclipse of the sun extending several times the sun's radius, or millions of kilometers, outwards into space (Figure 9–17). In fact, we can regard the solar wind as the outermost part of the corona, and say that the corona reaches hundreds of millions of kilometers into space, constantly expanding outwards away from the sun. How does the corona get so hot? This is a good place to remember the difference between heat and temperature, as physicists use the words. Heat measures the amount of

Figure 9–16. The aurora or "northern lights" arises from the interaction of high-energy particles in the solar wind with the particles in the Earth's upper atmosphere. The Earth's magnetic field influences the motion of the charged particles in the solar wind, and as a result the aurora appears more frequently in regions close to the Earth's magnetic poles. Likewise, auroral displays grow more intense after a solar outburst, when more high-energy particles reach the Earth in the solar wind.

Figure 9–17. The sun's corona flashes into view during a total eclipse of the sun. At other times the corona, which has a total brightness about equal to the brightness of the full moon, remains completely hidden by the glare from the sun's surface, which has a total brightness fifty thousand times greater than the corona's.

energy of motion in a given volume, whereas temperature measures the average energy of motion per particle. If we have very few particles in some volume, they can have an enormous temperature without representing a great amount of heat. We can see this difference at home if we compare a pot of water at its boiling point, 373 degrees absolute, with a pot of steam at the same temperature. The *total* amount of energy of motion in the pot of steam is far less than that in the same pot filled with boiling water, because there are far fewer particles in the pot when it is filled with steam than when it is filled with water. If we dive into a large pot of boiling water, the higher density of particles means that many more molecules will strike our bodies each second than would be the case in a vat of steam. The water has more heat (total energy of motion) than the same volume of steam at the same temperature.

The small density of the gas in the sun's outer atmosphere allows it to have an enormous temperature with a relatively small total amount of energy of motion. All of this energy arises from the turbulent roiling and boiling of the outer layers of the sun, which pass some of their energy of motion on to the sun's atmosphere. As the density of the atmosphere decreases going outwards, the number of particles to be shaken around decreases. Thus each particle's average energy of motion (temperature) increases, even though the total energy of motion does not. The outer corona has a temperature of two million degrees absolute, which is 300 times the temperature at the sun's surface. This high temperature makes the sun's corona emit x-ray and ultraviolet photons

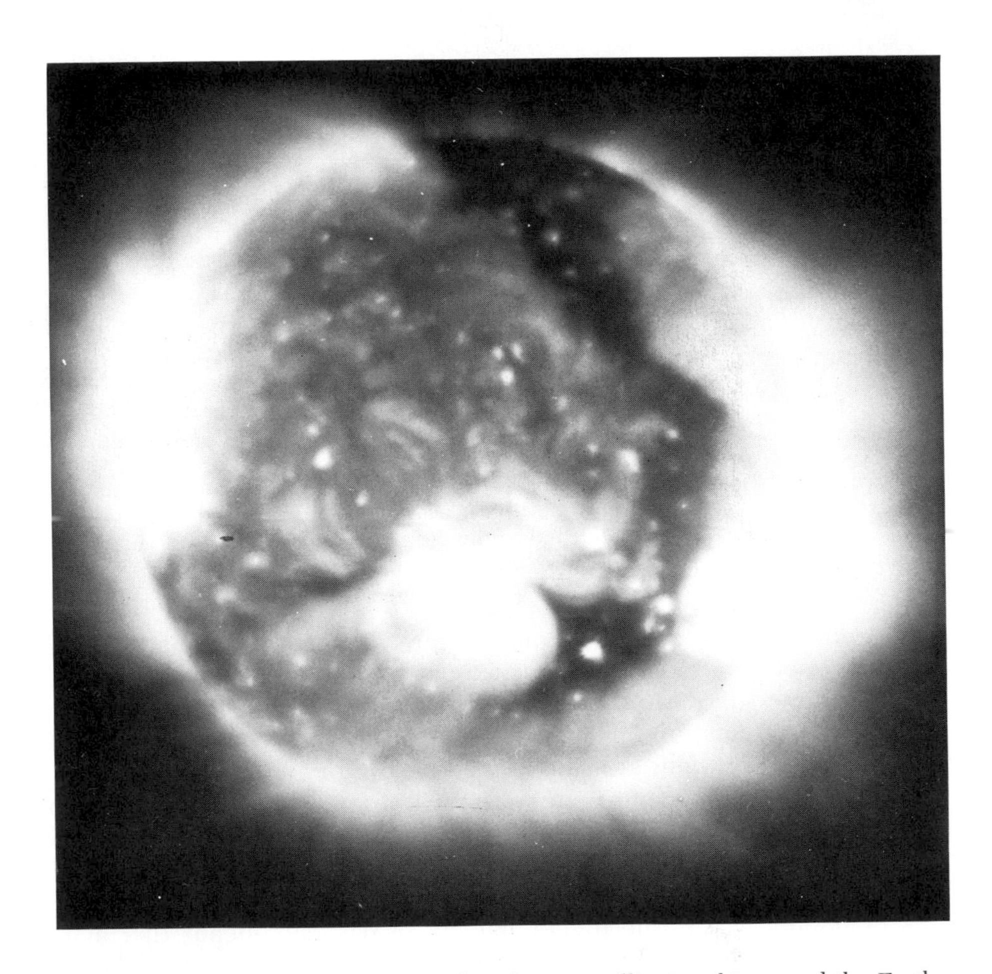

Figure 9–18. This photograph was taken from a satellite in orbit around the Earth, using detectors sensitive only to x-ray photons. The high temperatures in the sun's chromosphere and corona lead to the production of x-ray photons, and the brightest regions on the photograph show x-ray photons from "active regions" where violent motion occurs.

as well as visible-light photons. Figure 9–18 shows a picture of the sun in x-ray light, taken high above the Earth's atmosphere, which prevents us from observing the sun's x-ray photons when we are below the atmosphere.

The transition regions from the sun's surface to its corona form the sun's "lower atmosphere" and "chromosphere." In these regions we observe fantastic arches and filaments of hot gas called prominences (Figure 9–19). These loops of gas appear and disappear over a period of several hours or days. The prominences are localized condensations

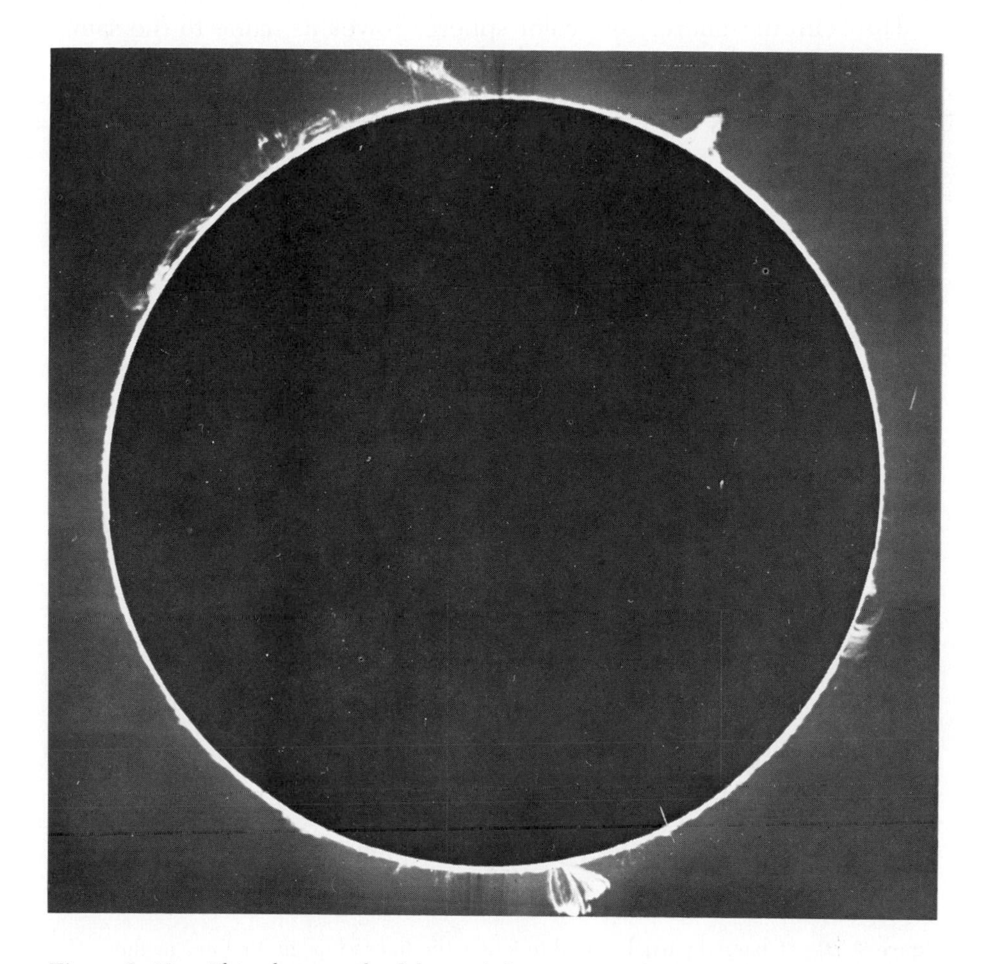

Figure 9–19. This photograph of the sun's lower atmosphere was taken using a machine, called a "coronagraph," that artificially blocks out the light from the solar disk to simulate the effects of a total eclipse of the sun. We can see the loops and surges of solar "prominences" pushing out through the lower chromosphere; each of these prominences is larger than the Earth.

within the sun's atmosphere, and their complicated structure shows the presence of a magnetic field at the sun's surface. This magnetic field affects the motions of charged particles, so they do not move in straight lines but follow the curvature of the magnetic lines of force (Figure 9–20). Most of the visible light that prominences emit comes from hydrogen atoms that have just recombined from protons and electrons. The most intense photon frequency in the visible-light frequency region comes from newly captured electrons making the jump from the third-smallest to the second-smallest orbits of the hydrogen atoms (Page 32). This jump produces photons of red light, so the prominences look red to us. The "chromosphere," or "color sphere," owes its name to the same hydrogen-atom electron jumps, which make the chromosphere appear red. Usually we cannot see the chromosphere, because like the corona it is far fainter than the light from the sun's surface, but it does appear

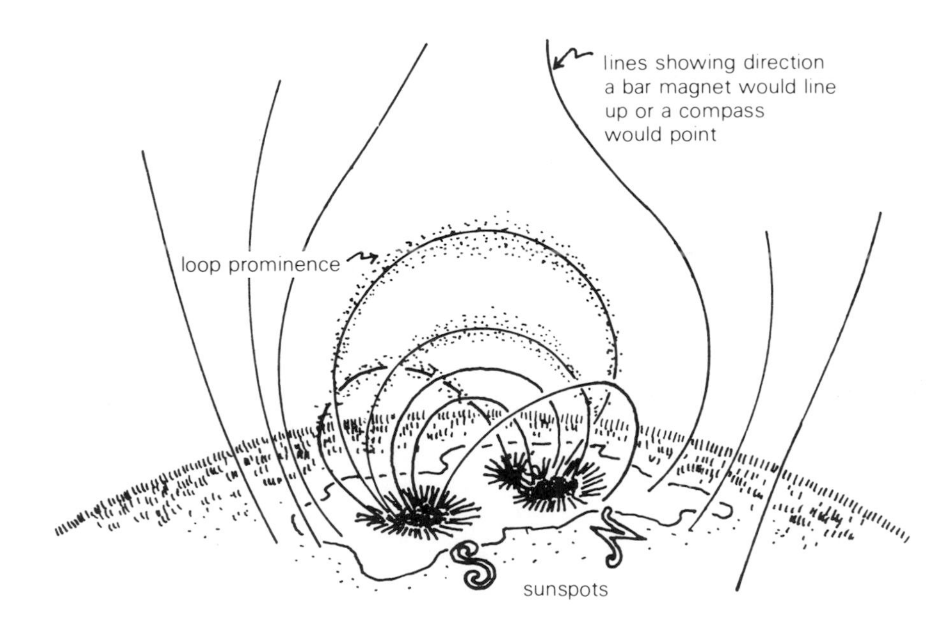

Figure 9–20. Charged particles tend to follow the lines of magnetic force as they move. The material in solar prominences, which has condensed from the gases in the sun's upper atmosphere, consists of electrons, protons, and other (positively charged) nuclei. Therefore the motion and location of the particles in prominences (Figure 9–19) reflect the structure of the magnetic field in that region of the sun's surface and atmosphere.

during a total solar eclipse and thus was evident long ago to admiring astronomers who gave it its peculiar name.

The most notable features on the sun's surface are the cooler, darker regions called sunspots (Figure 9–21). Sunspots appear almost black on photographs, but this blackness arises from the contrast effect of the cooler sunspots against the hotter, brighter background of the ordinary solar surface. Sunspots have temperatures one or two thousand degrees less than the 6000 degrees absolute of the rest of the sun's surface. This lower temperature apparently comes from the effect of the sun's magnetic field, which is much stronger inside a sunspot than outside.

Sunspots first appear as small dark spots, grow larger, and then last for a few days or weeks before fading away. A large sunspot group can cover an area a hundred times the size of the Earth, but still a small fraction of the sun's surface. Astronomers can use their observations of

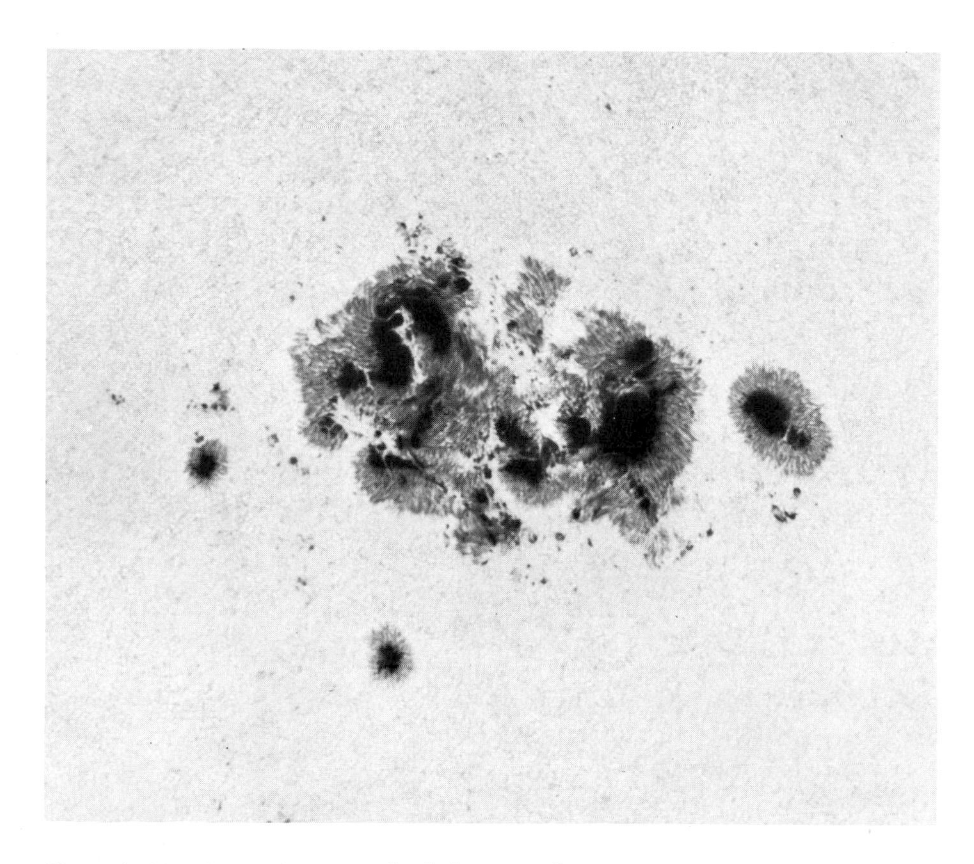

Figure 9–21. Sunspots consist of a darker central region called the "umbra" and a surrounding, not-so-dark area called the "penumbra." Together they may span fifty thousand kilometers or more.

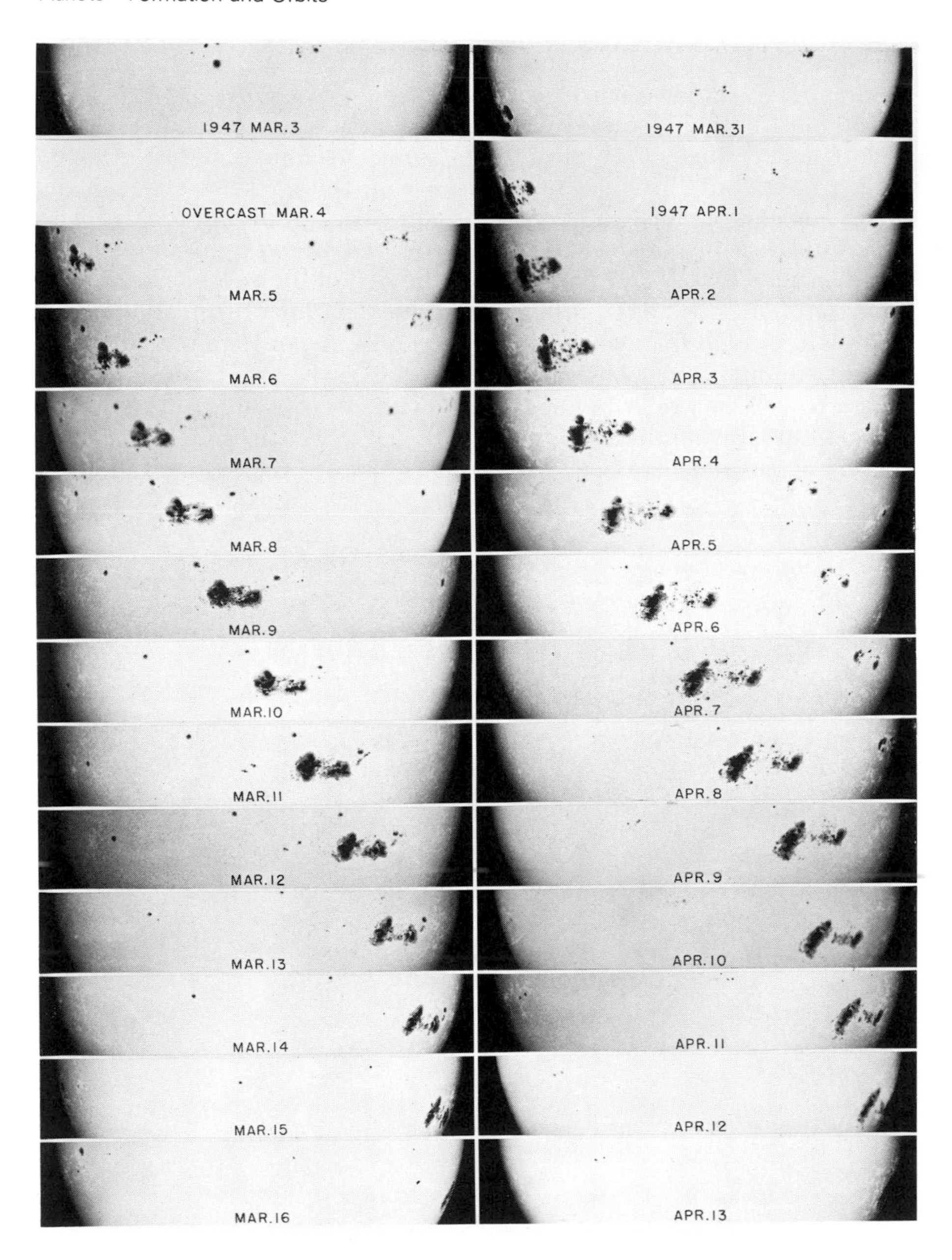

Figure 9–22. The large spot group of March and April, 1947, lasted long enough for the sun to rotate 1½ times. Sunspots can be used to measure the length of time that the sun takes to rotate. However, because sunspots do not appear near the poles of the sun (they are seen only up to solar latitudes of about 30°), they can be used to find only the rotational period near the solar equator. To find the length of time that regions near the poles take to rotate, we must measure the Doppler shift of the light from small portions of the sun.

sunspots to determine the rate of the sun's rotation (Figure 9–22). They have found that the sun takes about 27 days to rotate once at its equator, but because the sun is not a solid body, the regions away from its equator rotate more slowly than the equatorial parts do. The regions of the sun at 45 degrees latitude away from the equator take four or five days longer to rotate once than the equatorial regions do.

Sunspots appear on the sun more often every eleven years or so. The next period of "sunspot maximum" will occur around 1980. This sunspot cycle affects the Earth, because during the periods of sunspot maximum, the sun emits more outbursts of high-energy solar-wind particles, which have a small but noticeable effect on the Earth's atmosphere (Page 273). Many theories have been spun connecting cycles of terrestrial events to the repeating sunspot cycle of eleven years' duration. One favorite theory suggests that sunspots affect the weather, which in turn affects our well-being, and that people think best during times of sunspot maximum. Opponents of this theory suggest that it was conceived during a deep sunspot minimum.

So much for the sun, our own typical star. If we have passed lightly over its features, let us realize that we can do this because the sun functions so steadily and so simply inside. A host of astronomers now await the next generation of satellites that will help them understand the *details* of how the sun produces energy of motion and passes it outwards, but the sun's basic process—hydrogen-to-helium fusion—works for us now and (not quite) forever.

Summary

The nine planets in our solar system orbit the sun in nearly circular paths, all in the same direction and in almost the same plane. The planets apparently owe their orbital regularity to the way that they formed from the contracting cloud of gas and dust that included the protosun. In addition to the regularity of orbital *direction*, the approximately regular *spacing* between the planets' orbits—each successive gap between planets' orbits about doubles the preceding one as we go outward from the sun—seems to reflect the planets' formation process. In this pattern of regular spacing, the asteroids collectively fill the gap between Mars and Jupiter, while Neptune and Pluto fail to conform to the regularity established by the inner planets.

The four "terrestrial" planets—Mercury, Venus, Earth, and Mars—lie closest to the sun. These planets are smaller and denser than the four giant planets—Jupiter, Saturn, Uranus, and Neptune. The giant planets consist mostly of hydrogen and helium, while the terrestrial planets contain little of these lightest elements, but consist mainly of silicon, oxygen, aluminum, magnesium, carbon, and iron.

Outside the solar system (as defined by the planetary orbits) a swarm of comets move along highly elliptical orbits. Comets contain the primordial material from which the solar system formed—methane, ammonia, and water ice, frozen around lumps of rock and smaller dust particles. At the solar system's center we find the sun, which has 98% of the system's total mass. Above the sun's surface, which occasionally is pocked by a slightly cooler "sunspot," loops and filaments of hot gas called "prominences" poke through the lower atmosphere or "chromosphere." The sun's outermost atmosphere, the "corona," forms an ever-expanding shell that must be replenished constantly from below. The rarefied gas in the corona has the surprisingly high temperature of one or two million degrees absolute (the sun's surface is six thousand degrees absolute), and the outwardly moving corona sweeps past the Earth as a "solar wind" of electrons, protons, and heavier nuclei.

Questions

1. The sun's mass is a thousand times the mass of Jupiter and 300,000 times the mass of the Earth. If we made a model of the solar system with an elephant (6000 kilograms, or six million grams) for the sun, how massive would Jupiter and the Earth be?
2. In the model described in Question 1, if we let the width of the elephant (2.8 meters) represent the size of the sun (which in fact is 1.4 million kilometers), how far away from the elephant should the Earth and Jupiter be, if we round off their actual distances from the sun to 140 million kilometers and 700 million kilometers, respectively?
3. Why do the planets all orbit around the sun in the same direction and in almost the same plane?
4. Why doesn't the moon fall into the Earth? How does this explanation relate to the planets' orbits around the sun? How does it relate to the launching of artificial satellites from the Earth?
5. Johannes Kepler discovered that the time that a planet takes to orbit the sun varies in proportion to the $^3/_2$ power of the planet's distance from the sun. How long would it take a planet nine times the Earth's distance from the sun to complete one orbit? How long for a planet 25 times the Earth's distance from the sun? Why doesn't the time for one orbital period vary in *direct* proportion to the planets' distances from the sun?
6. Which planets have the most moons? Is it an accident that these are the most massive planets?
7. Why does Jupiter consist mostly of hydrogen and helium, whereas the Earth contains only a small percentage of these elements in its mass?

8. What are asteroids? Are they the remnants of an exploded planet? How are meteors similar to asteroids?
9. Why do comets move more rapidly in their orbits when they are closer to the sun? Why do they change their appearance when they approach the sun?
10. How can we hope to detect planets in orbit around other stars, if we can't hope to see the planets' reflected light because of the glare from their parent stars?
11. Why is the sun hotter at its center than at its surface? Why is the sun's outermost atmosphere (the corona) also much hotter than the sun's surface layers?
12. Why do sunspots look dark? What do sunspots have in common with solar prominences?

Further Reading

F. Whipple, *Earth, Moon, and Planets* (Harvard University Press, Cambridge, Mass., 3rd ed., 1968).

D. Menzel, *Our Sun* (Harvard University Press, Cambridge, Mass., 1959).

R. Richardson, "The Discovery of Icarus," *Frontiers in Astronomy* (W. H. Freeman and Co., San Francisco, 1970).

O. Gingerich, "The Solar System Beyond Neptune," *Frontiers in Astronomy* (W. H. Freeman and Co., San Francisco, 1970).

Chapter **10**

A Tour Through the Solar System

"The Space Wanderer had been thinking about moving on. There didn't seem to be much here for him. The Shaltoonians did not even have a word for philosophy, let alone such as ontology, epistemology, and cosmology. Their interests were elsewhere. He could understand why they thought only of the narrow and the secular, or, to be exact, eating, drinking, and copulating. But understanding did not make him wish to participate. His main lust was for the big answers."

Kilgore Trout,
Venus on the Half Shell

Small as the solar system may be on a stellar scale of distances, it is our home, and—for the time being at least—our back yard. During the past ten years, we have begun to send cameras and other instruments past the four planets closest to us, and men have explored our own moon. These voyages have revealed much that is new about the solar system, and such exploration may, before long, resolve key questions about the origin and history of the planets. Before we had space probes like the Pioneer and Mariner vehicles, astronomers spent years in observing, for example, the planet Mars, without being able to see surface "details" a hundred kilometers across. Today a single picture from a spacecraft flying around Mars can show a hundred times more detail than any picture of the planet taken on Earth, and we even can land automated probes, such as the Viking spacecraft, on the planet's surface, to sample its soil and to determine if any primitive life forms exist there.

Let us anticipate the space vehicles of a few years from now and imagine ourselves on a tour of the solar system, starting near the sun and passing outward past Pluto. Of course, we must limit our description to what we know now, so our tour guide cannot reveal many fascinating items that still await our discovery once such a trip becomes possible. But without more apologies, let us be off.

A few million kilometers away from the sun, we find ourselves in the sun's outer corona, which generates the outward stream of charged particles we call the solar wind. Future astronauts might find a way to deploy gigantic sails, perhaps made of electromagnetic fields, against which the solar-wind particles could push to propel their spacecraft outward. In the nineteenth century, some astronomers claimed to have seen a planet inside the orbit of Mercury, tentatively named "Vulcan."

Such a planet almost surely does not exist, but our journey would give us a good chance to make another search. Otherwise we pass only an occasional meteor on the part of its highly elliptical orbit that brings it close to the sun, or perhaps a comet on its point of closest approach to the sun, called its "perihelion."

The Inner Planets

Continuing on, at a distance of 60 million kilometers from the sun we reach the innermost planet, Mercury (Figure 10–1). Mariner-10 photographs taken close by the planet show a remarkable resemblance between Mercury's surface and our moon's surface. We see a heavily cratered skin of the planet that shows little sign of erosion, despite the enormous range of surface temperatures (500 degrees absolute on the day side, 150 degrees absolute on the dark side) that slowly alternate for any given point on the planet's surface. Mercury spins only 1½ times as fast as it revolves around the sun, so it rotates three full turns every time it makes two complete orbits around the sun. This exact relationship between Mercury's rotation and its orbital period means that the planet's rotation has been "locked in" to its orbital motion. Apparently Mercury's mass is not distributed in an entirely symmetric way throughout the planet. Since Mercury has a rather elongated orbit around the sun, it comes significantly closer (46 million kilometers) to the sun at its point of closest approach than it does on the average (58 million kilometers) or at the farthest point along its orbit (70 million kilometers). Every second closest approach of Mercury to the sun brings the *same* part of Mercury, presumably the denser part, closest to the sun and to its force of gravity. The sun's gravity has captured this denser part of Mercury and forced the planet to rotate in such a way as to bring the denser part directly toward the sun at every second time of closest approach. Such an interlocking of rotation and orbital periods appears again for Venus (where the Earth, surprisingly, exerts the key gravitational force in producing the "locked in" phenomenon) and for our moon (where the Earth has locked in the moon's rotation completely, so the moon rotates with the same period as it takes to orbit the Earth).

Mercury's diameter of 4900 kilometers is about ⅜ of the Earth's diameter and 1.4 times our moon's diameter. Thus Mercury has a size intermediate between the Earth's and the moon's. Because Mercury's mass is one six-millionth of the sun's mass, or one eighteenth of the Earth's mass, we can calculate that Mercury's average density is 5½ times the density of water. Thus the density of matter in Mercury almost exactly equals the average density of matter in the Earth. This density also describes fairly well both Venus (5.1 times the density of water) and Mars (4 times the density of water). In contrast, the outer giant planets, Jupiter, Saturn, Uranus, and Neptune, have average densities ranging from only 0.7 times to 2¼ times the density of water. The

Figure 10–1. A photograph of Mercury's surface taken from the Mariner-10 spacecraft in 1974, when the Mariner was about 200,000 kilometers away from Mercury. The largest craters are about 200 kilometers across.

inner four planets clearly are made preferentially of elements such as oxygen, carbon, silicon, and iron, at the expense of hydrogen and helium. Furthermore, the structure of the four inner planets, if we could slice them open to have a look, seems to be differentiated into a denser "core" and a less dense "mantle" around the core (Figure 10–2). The denser constituents of the inner planets ended up closer to the centers of the planets as they formed some 4½ billion years ago.

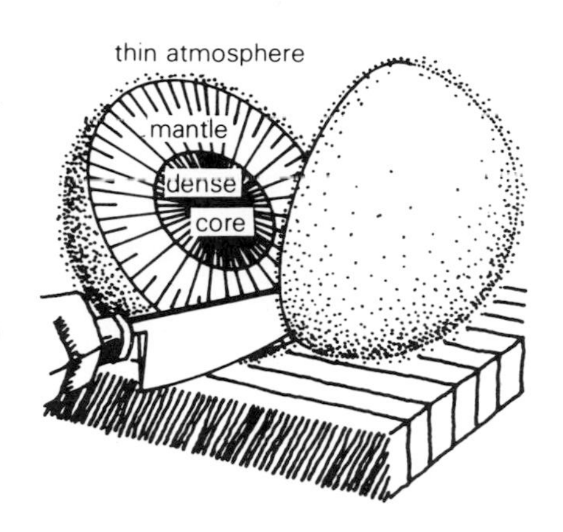

Figure 10–2. The inner planets have an internal structure that is "differentiated" into a dense core and a less-dense mantle that surrounds the core.

Mercury does have a thin atmospheric layer around it, although this atmosphere amounts to only a few billionths of the Earth's atmospheric pressure. The Mercurian atmosphere consists mostly of inert gases such as helium, neon, argon, and xenon. Such atoms are not destroyed easily by the intense solar radiation that falls on Mercury. Because Mercury is 2½ times closer to the sun than we are, each square centimeter of its surface receives 6¼ (2½ squared) times as much photon energy per second from the sun as a square centimeter on Earth does.

The outstanding puzzles about Mercury are its high density and the fact that the planet has a detectable magnetic field. We have mentioned why we believe the inner planets have larger densities than the giant planets, namely the escape of hydrogen and helium from the inner parts of the cloud that formed the protoplanets. However, the explanation of how Mercury came to have a density as great as the Earth's, the largest of any planet, with a mass only one eighteenth as large as the Earth's mass, still eludes the pursuit of those astrophysicists who speculate about how the solar system formed.

Mercury's magnetic field is difficult to explain in view of the fact that the planet rotates so slowly. Although we know little about the ways in which a planet (or a star) can produce a magnetic field, what we do know seems to suggest that the more rapidly rotating objects should have the largest magnetic field strengths. Thus, comfortingly enough, Jupiter has by far the strongest magnetic field of any planet, and Jupiter rotates the most rapidly (especially for its size). In contrast, Mercury rotates more slowly than any other planet except Venus; its rotation period is 120 times greater than Jupiter's and 56 times greater than the Earth's. Yet Mercury does have a magnetic field, which, although weaker than the Earth's (and far weaker than Jupiter's), remains far stronger than the magnetic field "should" be according to our most informed guesses.

Mysteries or no mysteries, Mercury presents an inhospitable environment for life, broiling in its daytime and freezing at night, with "day" and "night" each several months long. "Freezing" only mildly conveys the night-time temperature of 150 degrees absolute (−123 degrees Centigrade) that we would find on Mercury if we faced away from the sun, probably digging for shelter in the fractured landscape. Unaccompanied by any satellite, dense little Mercury speeds swiftly along its orbit, offering little comfort to us as we proceed on our way to Venus.

"Twin of the Earth," Venus has a size, mass, and density that are each a trifle smaller than the Earth's. To be numerically accurate, Venus has 95% of the Earth's diameter, 82% of its mass, and 93% of its density. In contrast to Mercury's tenuous atmosphere, Venus hides beneath eternal veils of cloud, and no one has yet seen the planet's surface.

Figure 10–3, also photographed by Mariner-10, shows a typical set of cloud patterns on Venus. The photograph cheats a little, because the cameras recorded ultraviolet light but the picture was printed in blue light. On Venus, even more than on the Earth, the atmosphere affects

Figure 10–3. This photograph of the cloud patterns in Venus' atmosphere was taken by Mariner-10 when it was about 700,000 kilometers from Venus.

the heat balance at the planet's surface. Some photons from the sun with frequencies of ultraviolet and blue light can penetrate deep into Venus' atmosphere, but when they interact with the atoms and molecules that they strike in the atmosphere or at the planet's surface, these photons change into photons with smaller frequencies that cannot escape through the atmosphere. That is, the atmosphere tends to let photons in, but does not let out the photons that they change into once they interact with atoms on or near the planet's surface. A similar process occurs on Earth, where higher-frequency photons penetrate our atmosphere more easily than lower-frequency photons can. This effect is fairly subtle, because the highest-frequency photons (x rays and gamma rays) do not penetrate the atmosphere at all, but for both Venus and the Earth, the average incoming photon has a higher frequency than the photons that can escape through the atmosphere. This difference in frequency means that less photon energy will escape as reflected light from the surface and lower atmosphere than will reach us as solar photons. Hence our atmosphere, and Venus' atmosphere even more so, can trap and hold some of the sun's photon energy. This atmospheric energy trap, which makes the planets' surfaces warmer than they would be if the planets had no atmospheres, is called the "greenhouse effect" because it reminds astronomers of what happens inside Earth-bound greenhouses.

The greenhouse effect caused by the *Earth's* atmosphere gives us a surface temperature of 290 degrees absolute (17 degrees Centigrade) rather than the 250 degrees absolute (−23 degrees Centigrade) we would have without an atmosphere. This minor change represents for us the difference between life and death. On Venus the atmospheric trapping of photon energy produces a far more spectacular change in temperature. Instead of the surface temperature of 300 degrees absolute that Venus would have without an atmosphere (300 rather than 250 because Venus orbits closer to the sun), Venus' actual surface temperature reaches 750 degrees absolute! To produce such an enormous effect, Venus' atmosphere must be far denser than the Earth's. In truth, the atmospheric pressure at Venus' surface exceeds by 90 times the pressure at the Earth's surface. On Earth we would have to descend a kilometer into the ocean to find an outside pressure equivalent to that at Venus' surface. The atmosphere on Venus consists mainly of carbon dioxide molecules (two oxygen atoms linked with one carbon atom). Astronomers also have detected small amounts of water, carbon monoxide, and hydrofluoric acid, but almost none of the free oxygen molecules that form 20% of the Earth's atmosphere. The clouds in Venus' atmosphere that constantly conceal the planet's surface from us may consist largely of sulfuric acid!

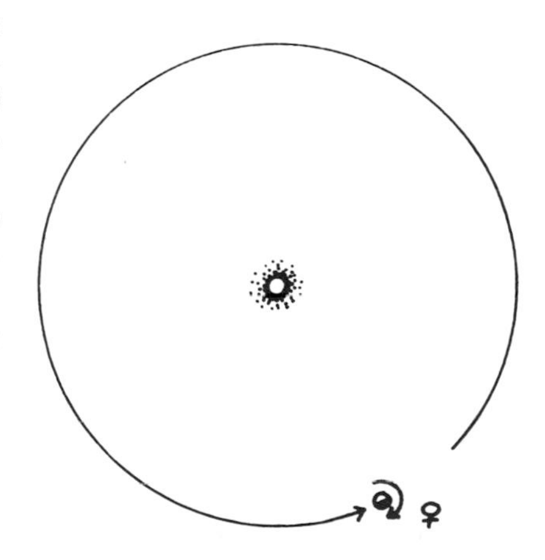

Figure 10–4. Venus rotates once every 243 days in the *opposite* sense to the planet's orbital motion around the sun.

Like Mercury, Venus rotates slowly compared to the other planets. Venus rotates once in 243 days, and the sense of this rotation, unlike

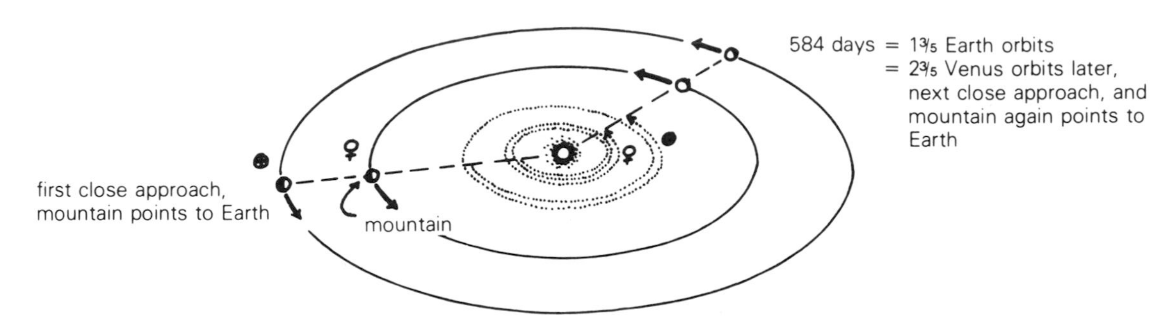

Figure 10–5. Because of the relationship between the time that Venus takes to rotate and the times that Venus and the Earth take to orbit the sun, every time that Venus comes closest to the Earth, the same part of Venus faces the Earth. If this part of Venus is somewhat denser than the rest, the Earth may have "locked on" to this part and forced Venus' rotation into its present period.

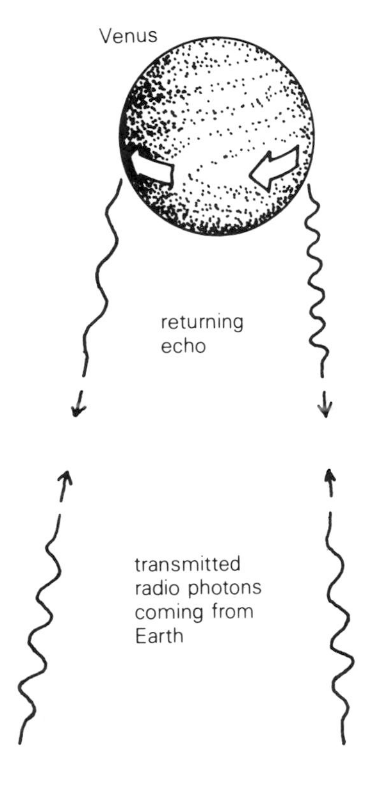

Figure 10–6. If we bounce radio waves off Venus, the radar echo from the side of Venus approaching the Earth (because of the planet's rotation) will return photons with higher energies than the side of Venus receding from the Earth. By combining this Doppler-shift information with the detailed timing of the radar echoes, we can discover how fast Venus is rotating and how rugged its surface is.

that of Mercury or the Earth, is backwards compared to Venus' motion around the sun (Figure 10–4). That is, if we trace the orbit of Venus on a chart, then Venus' rotation turns in the opposite direction to its orbital motion. For all the other planets (except Uranus), the circle of orbital motion and the planet's rotational spin are in the same direction. Venus' rotation period has locked on to its gravitational relationship with the Earth. Because Venus takes 224 days (almost 8 months) to orbit around the sun, while the Earth takes 365¼ days for each orbit, the two planets recover the same position relative to one another only after a period of 584 days (Figure 10–5). Every time that the Earth and Venus are closest to one another, the same part of Venus is closest to the Earth. Apparently Venus has some sort of bulge or mass concentration that the Earth has locked onto. Despite the fact that Venus feels the sun's force of gravity much more than the Earth's, the small gravitational force from the Earth seems gradually to have "captured" the rotation of Venus and made it what it is today.

Venus' surface remains hidden from us by clouds, but radar waves can penetrate these clouds, bounce off the surface, and return to the Earth. With a precision that seems almost unbelievable, astronomers can measure the time of transit of the photons in the radar waves to better than one part in a billion, and thus have determined the distance to Venus at any time to within a few meters. Furthermore, these "radar astronomers" have used the Doppler shift in the frequency of the radar echoes to measure the rotation of Venus, because the edge of the planet approaching us in rotation will return a higher-frequency echo than the edge receding from us (Figure 10–6). Finally, by computer analysis of thousands of echoes, we can determine even the ruggedness of Venus' surface and how well it reflects radar waves. Thus astronomers have proven (as already was suspected) that there are no huge basins or oceans on Venus, because its surface does not reflect as large, flat areas would. Instead, the Venus surface "terrain" (to use a word that ought to apply only to "Terra," the Earth) has an average ruggedness only a little less than the surfaces of the moon and Mercury. Radar astronomy now is becoming so accurate that we can be almost sure that Venus, like the moon, Mercury, and Mars, has craters on its (always invisible) surface.

In 1975, the Soviet spacecraft Venera-9 and Venera-10 sent the first pictures of Venus' surface to Earth. These pictures revealed a rugged terrain complete with many rocks and boulders. The high surface temperatures and enormous atmospheric density on Venus clearly have not melted the rocks or smoothed away the small-scale features of the landscape.

Venus thus has turned out to be a fantastically hot, stuffy planet overlain with an atmosphere pressing down like an ocean. In size and mass close to the Earth, this nearest of all planets, on account of its enormous surface temperatures, looks like a loser as far as human-oriented living conditions go. Even during the Venus "night" of more than a hundred earth-days, temperatures greater than 700 degrees absolute fill the opaque atmosphere that would deny us any view of the cooler worlds outside. It is time to sail on and admire our life-giving Earth.

The Earth and The Moon

Upon approaching the Earth but still a few million kilometers away, we first would notice how fine the Earth and the moon look together, with the moon's pitted brown surface setting off the swirling clouds above the aquamarine, green, and brown of the Earth's surface. No other planet has a satellite so large in comparison to itself as our moon is. More than a quarter of the Earth's diameter, with one fiftieth of the Earth's volume and one eightieth of its mass (hence with a density 3⅓ times the density of water), the moon ranks far above all other satellites in its relative size. None of the thirty-two other satellites has a mass even one thousandth the mass of the planet it orbits, or a diameter one tenth as large as its planet. Even though six of these satellites are larger than our moon (Table 9–2), they orbit planets so much larger than the Earth that they shrink to almost nothing in comparison. But our moon, although far smaller than the Earth, provides the exception among satellites by being so large compared to its planet.

During the past few years, men have removed several kilograms of rocks from the moon, spending about a billion dollars per kilogram to do so. The moon's rocks are interesting because they are fossil rocks from the early years of our solar system. The moon has no detectable atmosphere, and not much erosion has occurred during the past few billion years, thus the moon carries on its surface a record of conditions long ago. At present, the major force for erosion on the moon comes from the alternate heating and cooling of the moon as it rotates, once every 27⅓ days, presenting first one side and then the other to the sun. Lunar "daytime" temperatures reach 400 degrees absolute, while the two-week lunar "night" lowers the surface temperature to 150 degrees absolute (−123 degrees Centigrade). Only at the lunar poles does the temperature remain fairly constant.

Now that we have collected lunar rock samples, we have acquired detailed knowledge of the moon's surface layers to help reveal its origin and history. To take the large view first, the moon's cratered and mountainous surface shows a great difference between its two sides. The moon rotates every 27⅓ days, exactly the right amount of time to keep one side (the "man in the moon") always pointed toward the Earth. The interval between one full moon and the next one is not 27⅓ days but 29½ days, because the Earth is moving around the sun, so the moon must cover a little more than one complete orbit to regain the "full moon" position behind the Earth (Figure 10–7). The Earth has locked to the moon's rotation with its force of gravity and now holds the bulge in the moon's mass always toward the Earth. If the moon's period of rotation were not exactly locked to the moon's period of orbital motion around the Earth, we would not always see the same face of the moon (Figure 10–8), and we would not have had to send cameras behind the moon to photograph its other side. This back side does not show an assortment of huge lava-filled basins, bordered by mountain chains and large craters, that form the features of the front side's "man in the moon." Instead, we see a single lava basin ringed by circles of craters, and there are no tall mountain chains. If this other side of the moon had been the side always facing the Earth, instead of a "man in the moon," we would see a sort of giant "eye in the sky" that might have affected our ancestors' religious beliefs.

The mass of the moon must have a nonsymmetrical distribution, with some extra mass in the half facing the Earth. The features on this side are great plains of lava, incorrectly called seas ("maria" in Latin), that solidified billions of years ago soon after the moon formed. In those early years, an intense bombardment by meteors and other rocky debris in the solar system made the huge craters (up to hundreds of kilometers across) and the even larger lava basins. This bombardment decreased in intensity once most of the debris left over from forming the solar system had been used up. Because of the effect of the Earth's force of gravity, the meteors falling on the moon apparently hit preferentially on the moon's "leading edge" in terms of its orbit around the Earth (Figure 10–9). This infall of meteors might have been what altered the moon's distribution of mass, and eventually the Earth's gravity "captured" the moon's rotation until the area of greatest impact

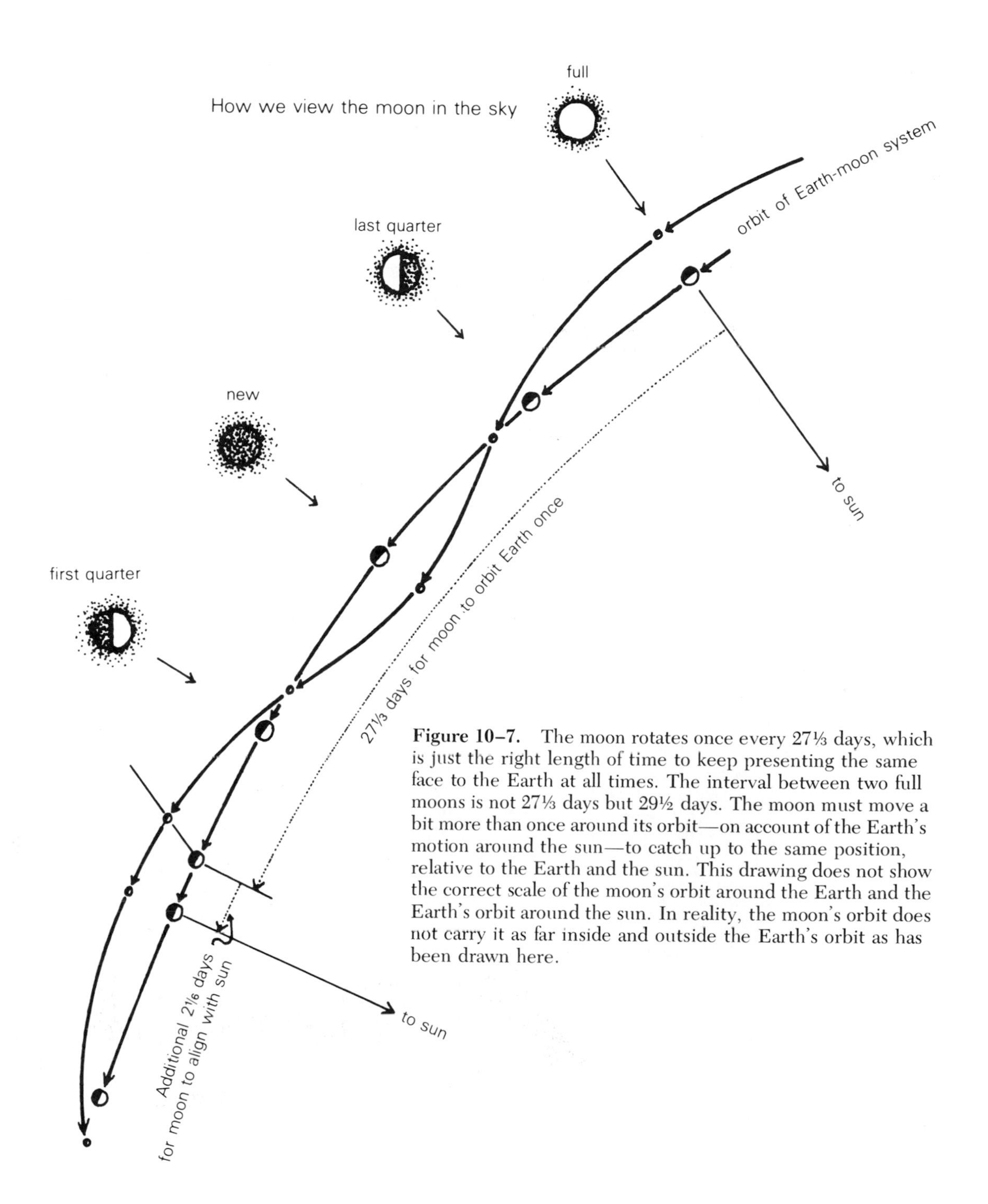

Figure 10–7. The moon rotates once every 27⅓ days, which is just the right length of time to keep presenting the same face to the Earth at all times. The interval between two full moons is not 27⅓ days but 29½ days. The moon must move a bit more than once around its orbit—on account of the Earth's motion around the sun—to catch up to the same position, relative to the Earth and the sun. This drawing does not show the correct scale of the moon's orbit around the Earth and the Earth's orbit around the sun. In reality, the moon's orbit does not carry it as far inside and outside the Earth's orbit as has been drawn here.

was always closest to the Earth (Figure 10–10). During this time, three or four billion years ago, lunar volcanoes and "moonquakes" must have been common. Since then volcanic activity on the moon seems to have been a rare occurrence, and moonquakes are minor tremors compared to the upheavals of old.

Thus the moon became a mass of craters and lava basins, while the Earth—thanks to its atmosphere and oceans—suffered fewer impacts and eroded the craters that did appear. Until the first rocks were

Figure 10–8. The side of the moon facing the Earth shows a combination of the roughly circular, dark lava beds ("seas" or "maria") with the lighter-colored lunar highlands and craters.

brought back from the moon, it was tempting to speculate that the moon was formed from part of the Earth, perhaps from the Pacific Ocean basin. However, analysis of the moon rocks has shown that the ratios of the abundances of the elements in them differ from the abundance ratios of Earth rocks, not enormously but enough to imply a separate origin for the Earth and the moon. In particular, many moon rocks are enriched in potassium, phosphorus, and "rare earth" elements compared to rocks on Earth. These elements, collectively called "kreep" (after the chemical symbols for potassium, etc.), appear in the lunar highlands, and not in the lunar basins, where lava flows have buried the oldest rocks. The record for the most ancient rock found by man now stands at 4½ billion years for a sample returned by the Apollo 14 mission. The oldest known Earth rock has an age of about 4¼ billion years. Current theories of the formation of the solar system assign an age of just over 4½ billion years to the Earth and the moon.

However the moon formed, it apparently used to orbit closer to the Earth. The moon now is receding slowly from the Earth, because of the tidal forces that the moon produces, by about one centimeter per year. That is, the moon's orbit grows larger by one meter per century, so if this rate has held steady, the moon was three percent closer to the Earth a billion years ago than it is now. Doubtless this rate of recession has changed, but it seems certain that the moon originally was much closer to the Earth than it is now.

When the moon had a smaller orbit, its greater gravitational force on the Earth must have raised far larger tides than the moon does now. The moon attracts all of the Earth, but it attracts the near side more strongly than the far side. All of the Earth responds to the moon's gravitational attraction, but the oceans, which are fluid, can react more easily than the land, which is solid. The tide raising effect at any point varies as the *difference* between the gravitational force from the moon at that point and the moon's gravitational force at the Earth's center (Figure 10–11). On the average, the moon attracts the Earth as if its mass all were located at the Earth's center, and as a result of this attraction, the Earth actually orbits around the Earth-moon center of mass (Page 263). The stronger force of gravity directly under the moon causes the oceans to bulge toward the moon on that side, and the weaker force of gravity on the opposite side from the moon makes the oceans bulge there too: The moon's gravity attracts the center of the Earth more than it attracts the seas on the side of the Earth away from the moon. The results of the moon's tide-raising effect are two high tides each day, which correspond roughly to the bulge directly underneath the moon and the bulge directly opposite the moon. The irregularity of ocean coastlines makes the timing inexact in terms of the moon's location, but there is a tendency to have high tide when the moon is either directly overhead or

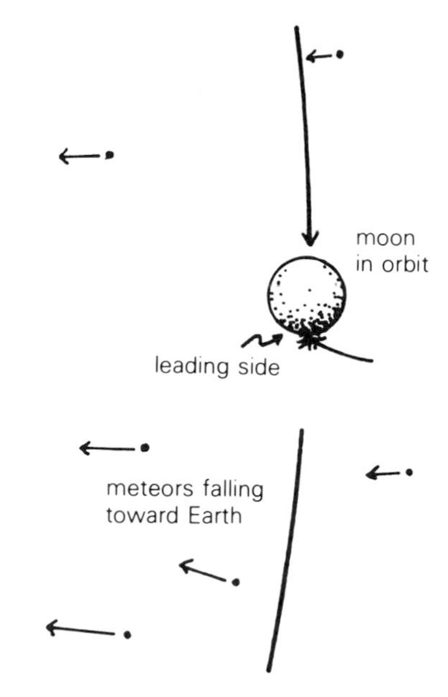

Figure 10–9. When meteors fell abundantly on the Earth and the moon, the Earth's force of gravity might have pulled many of the meteors toward the "leading edge" of the moon (in terms of its orbit around the Earth).

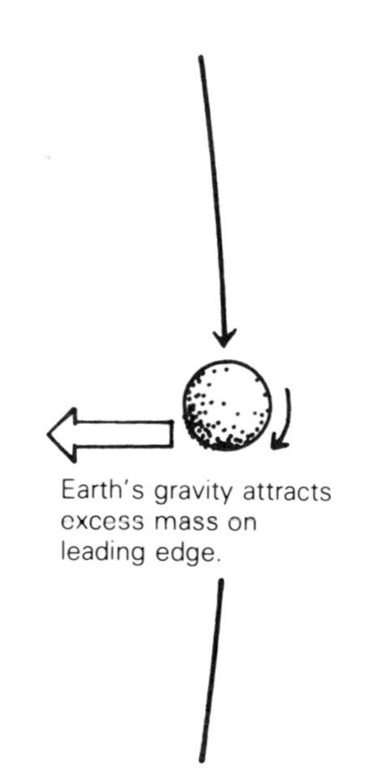

Figure 10–10. Once the moon had acquired additional mass from the infall of meteors on its leading edge, the Earth's gravity might have pulled on this mass until the relatively minor bulge faced directly toward the Earth as the moon rotated.

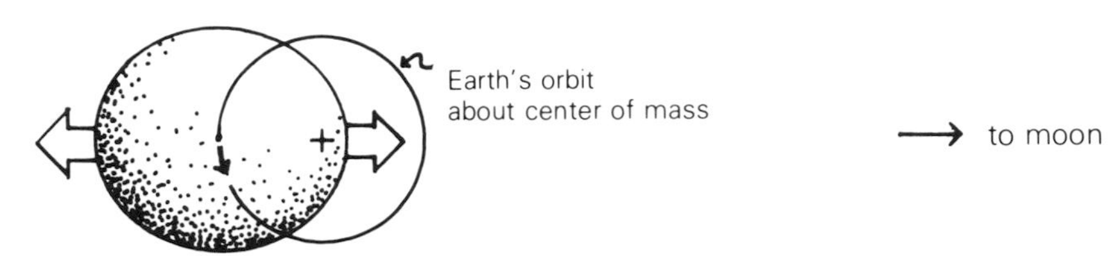

Figure 10–11. The moon attracts the near side of the Earth more than it attracts the Earth's center, and it attracts the center more than it attracts the far side. As a result of these *differences* in gravitational attraction, the Earth tends to bulge in directions directly toward the moon and directly away from the moon. The Earth's oceans respond more readily to this tendency to bulge than the land does, so the ocean tides slide up and down on the land.

exactly underfoot. The *size* of the rise between low and high tides depends on the shape and the slope of the coastline, so this rise can vary from a small amount on the gentle Pacific slopes to ten meters or more in the narrowing Bay of Fundy.

The sun also exerts a tide-raising effect on the Earth, but far smaller than the moon's. This seems odd when we remember that the sun's gravitational force on the Earth far exceeds the moon's (hence we orbit around the sun, not the moon). But the tide-raising effect depends on the *difference* between the gravitational force at a given point and the gravitational force at the Earth's center. This difference in gravitational force varies in proportion to the mass of the object exerting the force divided by the *cube* of the object's distance, not by the square of the distance that determines the gravitational force itself. Therefore, although the sun's gravitational force on the Earth exceeds the moon's by a factor of 150, the moon's tide-raising effect on Earth exceeds the sun's by a factor of 2⅔. When the sun and the moon almost line up, as they do twice a month at new moon and full moon (Figure 10–12), their tide-raising effects work together to produce especially large tides. Some marine organisms need these largest "spring" tides to survive, and barely live through the lower "neap" tides near the first and last quarters of the moon.

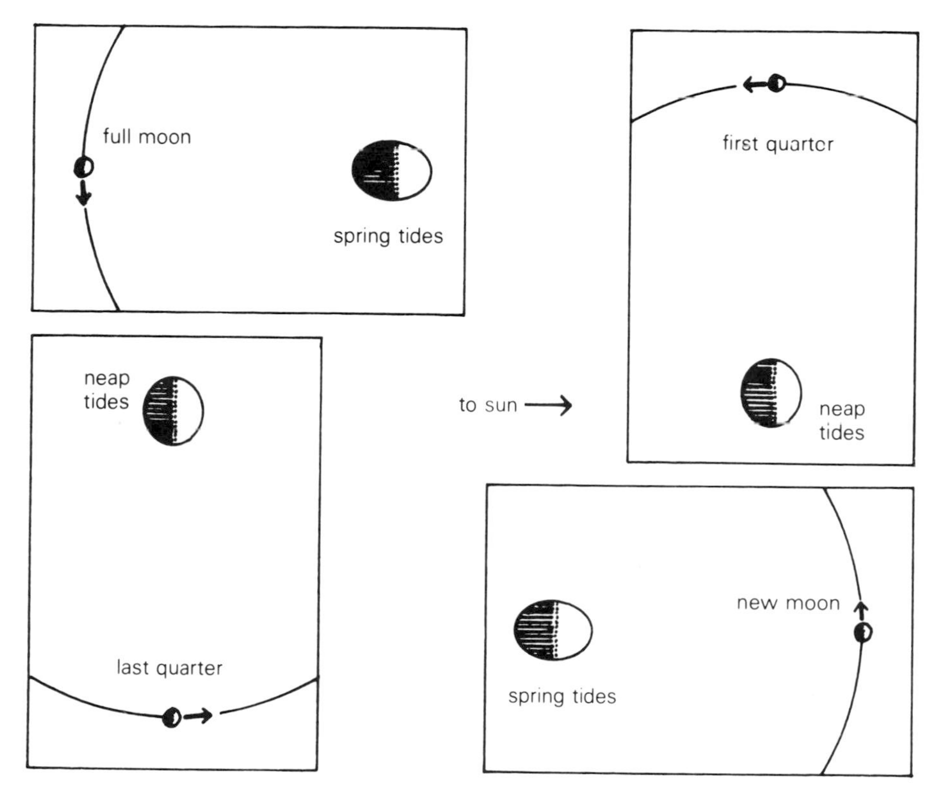

Figure 10–12. The sun also exerts a tide-raising effect on the Earth, but this effect is smaller than the moon's because the *difference* in gravitational attraction for the near and far sides of the Earth is less for the sun's force of gravity than it is for the moon's force of gravity. When the sun and the moon are in line with respect to the Earth, we have especially large "spring" tides; when the sun and moon are more nearly at right angles as seen from the Earth, we have relatively mild "neap" tides.

The sloshing back and forth of the tides, especially in shallow water such as the Bering and Irish Seas, leads to friction along the ocean bottoms that decreases the Earth's rate of spin by a tiny amount each year. The decrease in the Earth's rate of spin lengthens our day by a thousandth of a second every sixty years, so at this rate each day will be an hour longer in another 200 million years.[1]

In addition to their tide-raising activities, the sun and the moon also combine to produce the spectacle of solar eclipses. Although the sun's diameter exceeds the moon's by about 400 times, the sun is just about 400 times farther away from us than the moon. This odd coincidence allows the moon's disk to cover the sun's almost exactly if the moon's orbit brings it directly in front of the sun. If the moon comes in front of the sun at a time when its slightly elongated orbit brings it relatively close to the Earth (Page 275), the moon's disk appears large enough to

[1]The conservation of angular momentum implies that the slowing down of the Earth's rotation will be accompanied by an increase in the distance from the Earth to the moon.

cover the sun completely. However, if the moon is at the more distant points in its orbit when it eclipses the sun, the moon's disk is not large enough to eclipse the entire disk, and a thin ring of light remains around the moon during the eclipse. These "annular" eclipses (Figure 10–13) occur more often, but are not as much fun to watch as the spectacular total eclipses of the sun (Figure 9–17) when we can see the sun's corona and even some of the prominences rising above the sun's surface.

A total eclipse of the sun can be seen from only a small fraction of the Earth's surface. The moon's shadow barely can reach all the way to the Earth's surface (for an "annular" eclipse, it fails to reach even this far), and when it does reach here, it is only about 300 kilometers across (Figure 10–14). As the Earth rotates, the moon's shadow on the Earth seems to sweep across its surface at a speed of almost 1500 kilometers per second, so a total solar eclipse can cover a region 10,000 kilometers long but only 300 kilometers wide. Underneath the moon's shadow, observers can see the sun totally eclipsed for as long as eight minutes, if the moon's orbit puts it at its closest point to us at the time of the

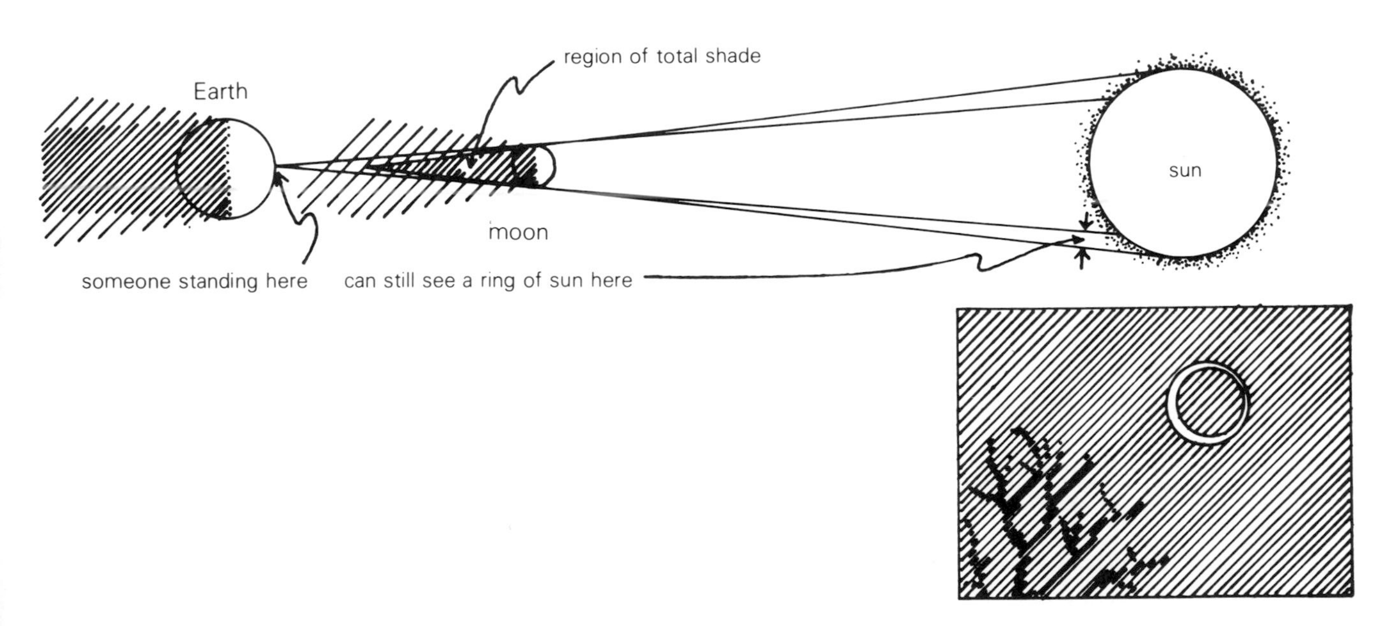

Figure 10–13. During an annular eclipse of the sun, the moon's apparent size is not quite enough to cover the sun's disk completely, because the moon is not at a relatively near part of its slightly elongated orbit around the Earth. Although the moon passes right across the face of the sun, some of the sun's light continues to reach us. The ring of light or "annulus" remains bright enough to prevent us from seeing the solar corona.

The like Eclipse having not for many Ages been seen in the Southern Parts of Great Britain, I thought it not improper to give the Publick an Account thereof, that the suddain darkness, wherein the Starrs will be visible about the Sun, may give no Surprize to the People, who would, if unadvertized, be apt to look upon it as Ominous, and to Interpret it as portending evill to our Sovereign Lord King George and his Government, which God preserve. Hereby they will see that there is nothing in it more than Natural, and nomore than the necessary result of the Motions of the Sun and Moon; And how well those are understood will appear by this Eclipse.

According to what has been formerly Observed, compared wth our best Tables, we conclude ye Center of ye Moons shade will be very near ye Lizard *point, when it is about 5 min: past Nine at* London; *and that from thence in Eleven minutes of time, it will traverse ye whole Kingdom, passing by* Plymouth, Bristol, Glocester, *Daventry,* Peterborough *& Boston, near wch it will leave ye Island: On each side of ye Tract for about 75 Miles, the Sun will be Totally darkned, but for less & less Time, as you are nearer those limits, wch are represented in ye Scheme, passing on ye one side near* Chester, *Leeds, and* York; *and on ye other by* Chichester, *Gravesend, and Harwich.*

At London *we Compute the Middle to fall at 13 min: past 9 in ye Morning, when tis dubious whether it will be a Total Eclips or no,* London *being so near ye Southern limit. The first beginning will be there at 7 min: past Eight, and ye end at 24 min: past Ten. The Ovall figure shews ye space ye Shadow will take up at ye Time of the Middle at* London; *And its Center will pass on to ye Eastwards, with a Velocity of nearly 30 Geographical Miles in a min: of Time.*

NB. *The Curious are desired to Observe it, and especially the duration of Total Darkness, with all the care they can; for therby the Situation and dimensions of the Shadow will be nicely determind; and by means therof we may be enabled to Predict the like Appearances for ye future, to a greater degree of certainty than can be pretended to at present, for want of such Observations.*

By their humble Servant Edmund Halley.

Sold by I. Senex, at the Globe in Salisbury Court, near Fleetstreet; who makes, and sells ye Newest and Correctest Maps, and Globes of 3, 9, 12, and 16 Inches Diameter, at moderate Prises.

Sold also by William Taylor at the Ship in Paternoster Row — *Entered in the Hall Book*

Figure 10–14. A "broadside" or public message that Edmund Halley, then the Astronomer Royal of England, prepared for observations of the total solar eclipse in 1715. Halley's calculations were accurate enough to predict where the moon's shadow would fall as the Earth rotated beneath it, but he was glad to use the actual observations to improve future calculations. The moon's shadow on England is elliptical rather than circular because the Earth's surface is not perpendicular to the light rays from the sun (because of the curvature of the Earth). Notice the reference to King George, who paid the bills.

eclipse. The full sweep of the moon's shadow across the Earth takes several hours, so if you have a fast airplane, you can fly along under the shadow and see several hours of eclipse. On the average, an observer who lives in the same place on the Earth will see a total solar eclipse every 300 years.

Although a total eclipse of the sun can be seen from only a tiny part of the Earth's surface, a total eclipse of the moon can be seen from fully half the Earth (Figure 10–15). In contrast to the moon's tiny shadow on the Earth, the Earth's shadow easily can cover the entire moon, because the Earth's diameter is four times larger than the moon's. A total eclipse of the moon can last more than an hour, the time it takes for the moon to move from one side of the shadow to the other in its orbit (Figure 10–15). During this time the moon would be completely blacked out from any sunlight, except for the fact that the Earth's atmosphere, like a lens, bends some sunlight into its shadow and onto the moon. Since red light bends more easily than blue light, the moon appears deep red or copper-colored during its eclipse, when the light that reaches it (to be reflected back to Earth) has been bent through our atmosphere.

Why don't we have eclipses of the sun at every new moon, and eclipses of the moon at every full moon? The reason is that the plane of the moon's orbit around the Earth is tilted with respect to the plane of

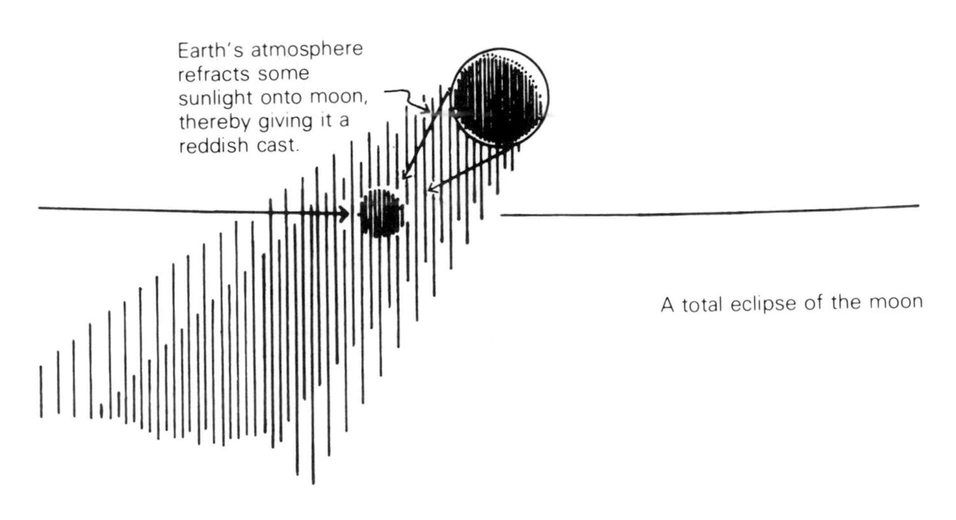

Figure 10–15. The Earth's shadow is about 10,000 kilometers wide at the distance of the moon from the Earth (400,000 kilometers). This is about three times the moon's diameter, so if the moon's orbit carries it directly through the Earth's shadow, the resulting total eclipse of the moon can last more than an hour and a half. During the total phase of a lunar eclipse, the Earth's atmosphere bends some sunlight into the shadow region. Therefore the moon remains dimly visible even when totally eclipsed.

the Earth's orbit around the sun. Thus at the time of new moon or full moon, the line from the sun to the Earth usually misses the moon (Figure 10–16). Only about one time out of six do the sun, moon, and Earth all fall in a close enough line for us to have an eclipse. Figure 10–17 shows the total eclipses of the sun that will occur before the year 1995, for those who wish to plan ahead.

We might pause here to reflect that since the moon's distance from the Earth is increasing constantly, total solar eclipses will become impossible in another few hundred million years. We are thus lucky to

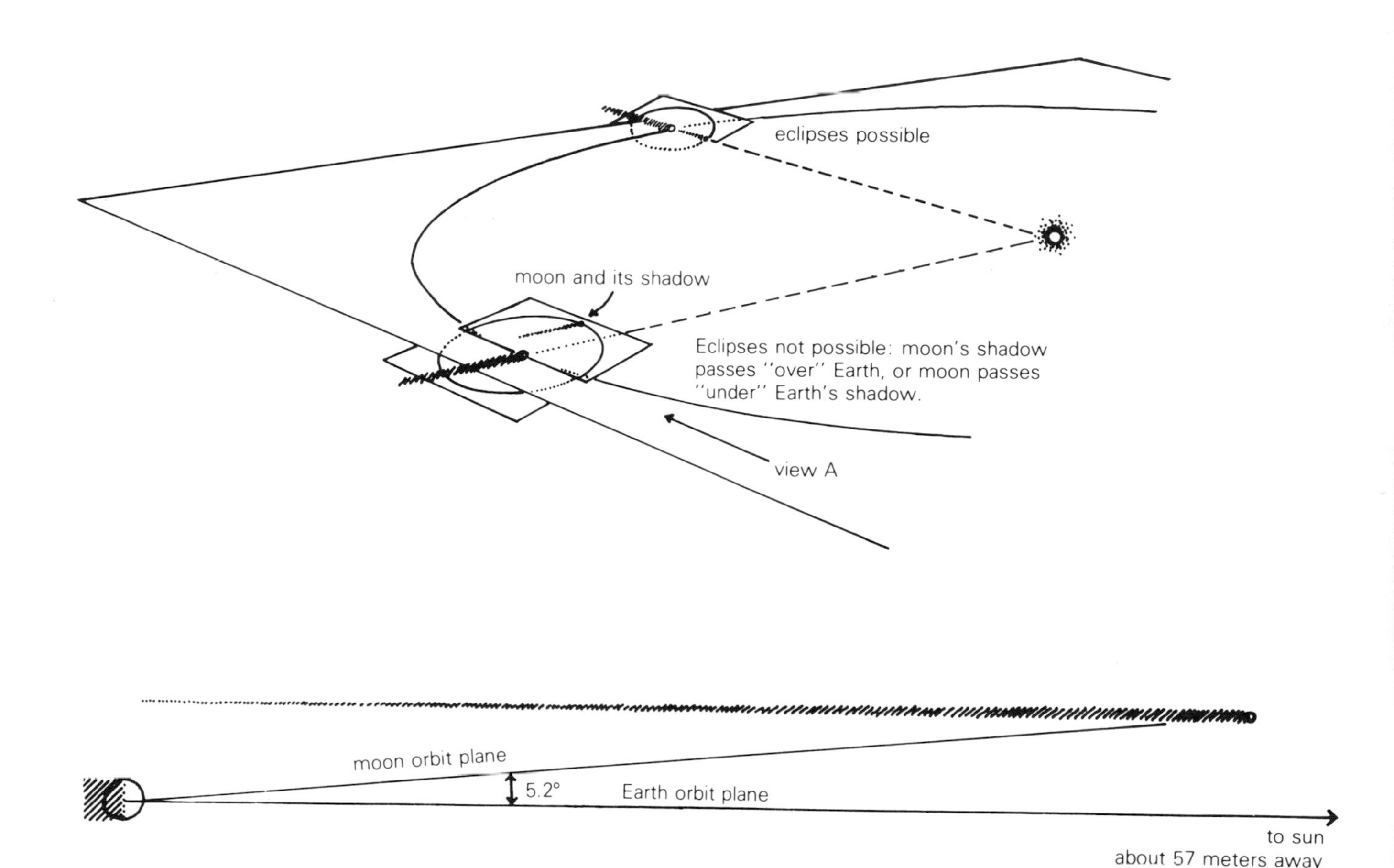

Figure 10–16. The plane of the moon's orbit around the Earth is tilted with respect to the plane of the Earth's orbit around the sun. As a result of this tilt, the sun, the moon, and the Earth usually do not line up exactly each time we have full moon or new moon.

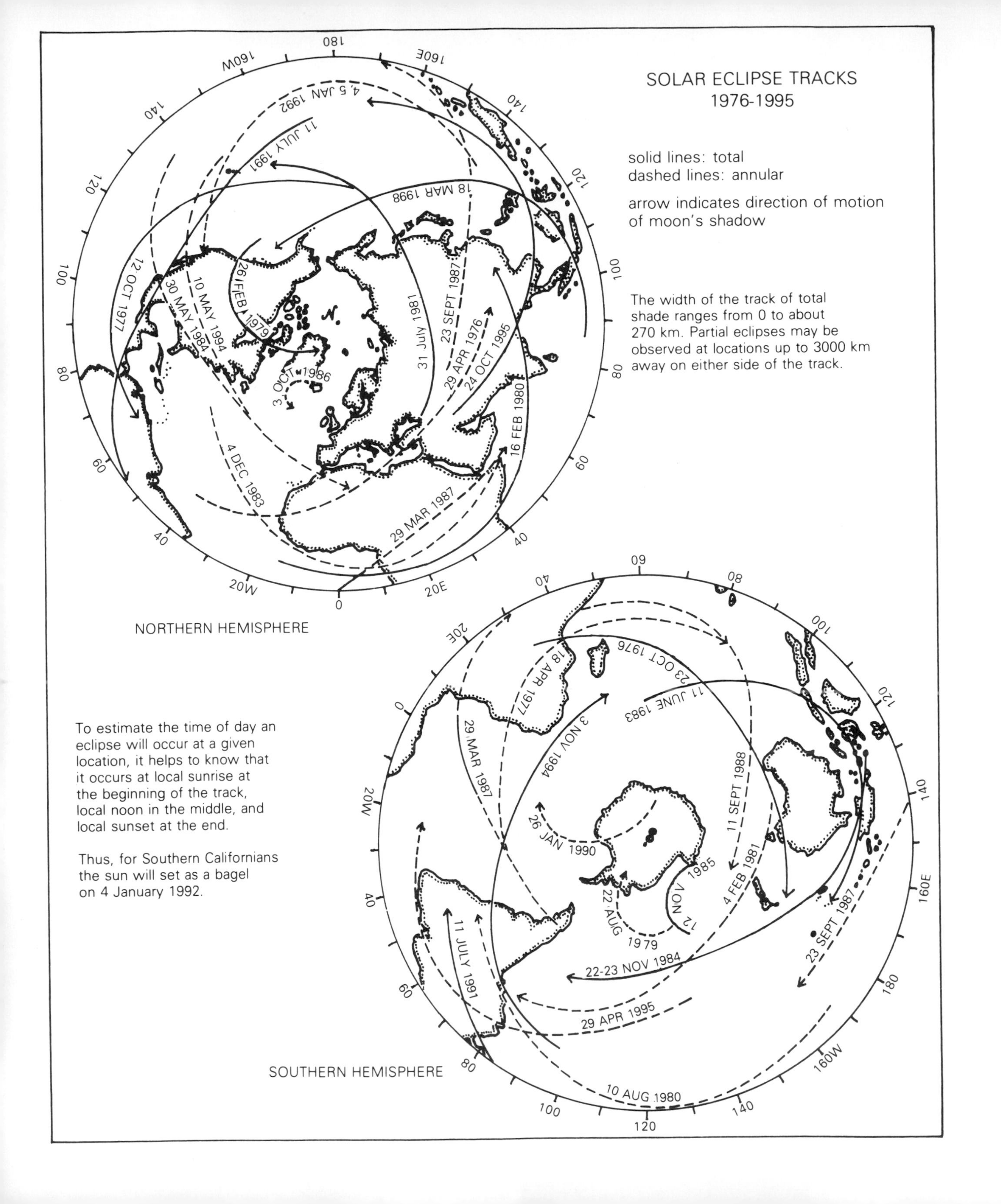

SOLAR ECLIPSE TRACKS
1976-1995
solid lines: total
dashed lines: annular
arrow indicates direction of motion of moon's shadow
The width of the track of total shade ranges from 0 to about 270 km. Partial eclipses may be observed at locations up to 3000 km away on either side of the track.
NORTHERN HEMISPHERE
4, 5 JAN 1992
11 JULY 1991
18 MAR 1988
12 OCT 1977
30 MAY 1984
10 MAY 1994
26 FEB 1979
31 July 1981
23 SEPT 1987
29 APR 1976
24 OCT 1995
16 FEB 1980
3 OCT 1986
4 DEC 1983
29 MAR 1987
SOUTHERN HEMISPHERE
18 APR 1977
23 OCT 1976
11 JUNE 1983
3 NOV 1994
29 MAR 1987
11 SEPT 1988
26 JAN 1990
12 NOV 1985
4 FEB 1981
22 AUG 1979
23 SEPT 1987
11 JULY 1991
22-23 NOV 1984
29 APR 1995
10 AUG 1980
To estimate the time of day an eclipse will occur at a given location, it helps to know that it occurs at local sunrise at the beginning of the track, local noon in the middle, and local sunset at the end.
Thus, for Southern Californians the sun will set as a bagel on 4 January 1992.

◀ **Figure 10–17.** This map shows the paths of total solar eclipses through the year 1995.

be around just at the time (in cosmic terms) when the moon has just the right distance to cover the sun exactly during some eclipses. Is this a sign of cosmic favoritism? Or is it, more likely, a happy coincidence to enjoy for its own glory?

The moon gives the Earth tides, eclipses, a nighttime beacon, and a source of wonder. Without the moon, life on Earth would be much duller. If early societies had not begun to watch the moon's phases and to speculate about the frightening eclipses it engaged in, astronomy might have taken far longer to develop. Even today, when we send space probes hundreds of times farther than the moon, we use a moon-based calendar to count "moonths," and set Passover and Easter on the first week after the first full moon after the beginning of spring. The ages of the Biblical patriarchs become more believable if we assume that they were reckoned in terms of "moons" rather than in years, as might well have occurred in a society ordered by the moon's changing phases.

Before we speed outward on our exploration of the solar system, let's have a quick look at the Earth, the seat of our billions of conflicting desires. Until quite recently, all human activity had produced nothing that could be identified as human-made from the distances at which Plate 7 was taken. Now we have city lights at night, and by day one could, with an effort, see the scars of the great highways and oil pipelines, or the great irrigated circles of farmland in arid countryside. Of course, what we can't see on the photograph may kill us eventually, if we destroy the Earth's ozone layer, poison ourselves with photochemical smog, or alter the Earth's climate by changing—even slightly—the composition of our atmosphere. The Earth's surface supports millions of species of living organisms, most of them quite small by human standards, while others are extremely large (ten thousand times a human mass). None of them can equal humanity as a threat to the Earth's surface and atmospheric environment.

Until now, the most significant kind of life on Earth has been plants. Plants use a series of simple chemical reactions called photosynthesis to turn the energy of motion of photons from the sun into bigger plants. Through photosynthesis, plants capture about one percent of the solar energy that reaches the Earth each second. Animals feed on plants (or on other animals that do), though they use sunlight directly to stay warm. We shall discuss the ways in which life has developed and maintained itself more fully in the next chapter, but for now we should notice that we tamper with the Earth's ecological balance at our own risk. We have the power to make the Earth colder or warmer, to spin it faster or slower, to change the one percent of solar energy used by plants to half a percent. To use this power simply because it exists, or because some immediate "benefit" will appear, ignores the interwoven pattern of life that has taken billions of years to evolve on Earth.

The Earth itself will abide, whether we destroy ourselves or not. The surface of the Earth doubtless will change, whatever we may do, since we do not yet have the power to stop the continents from drifting. Like Venus, Mars, and Mercury, the Earth has a dense core (largely iron) surrounded by a less dense mantle (Figure 10–18). Most of the iron in the Earth has collected in the center because of its greater density than the material in the mantle. Geologists believe that radioactive elements (mostly uranium) released enough energy of motion early in the Earth's history to *melt* most of the iron and thus to allow it to concentrate near the Earth's center while the lighter elements formed the outside mantle. Above the mantle, we live on the crust, which extends about 100 kilometers deep into the Earth. At this depth, the temperature inside

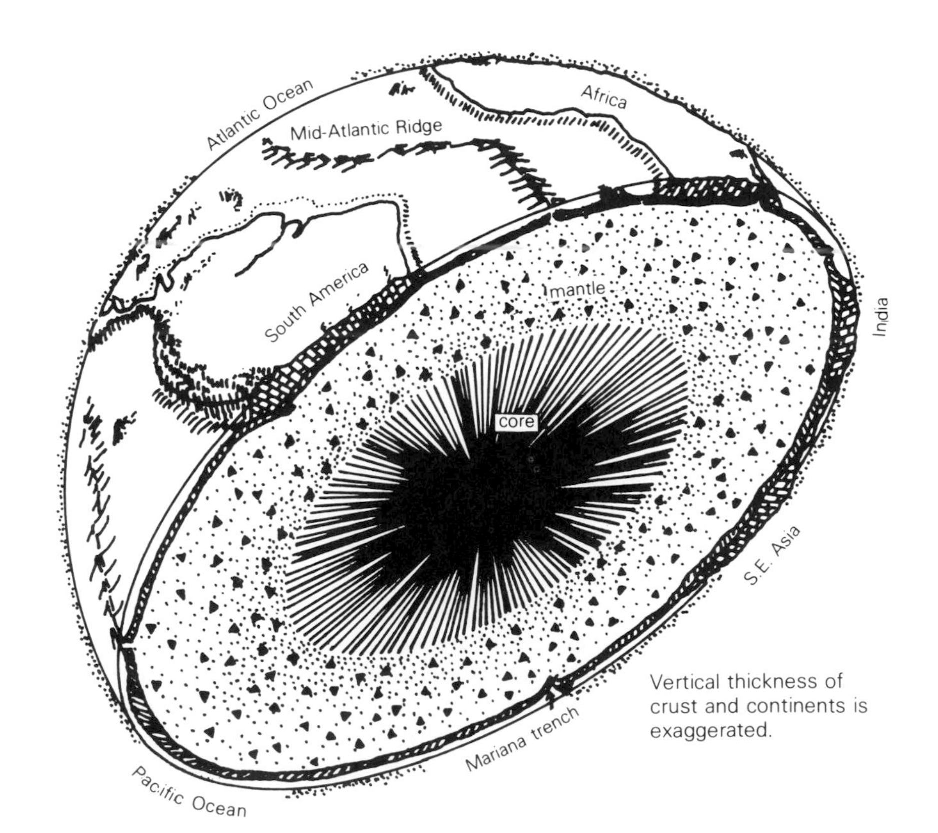

Figure 10–18. This schematic view of what we believe to be the Earth's internal consitution shows the iron-rich core, which has a diameter equal to about half of the Earth's diameter; the mantle, which fills about seven eighths of the Earth's total volume; and the outer crust, on the top of which we find the continental plates.

the Earth reaches the point at which the rocks are just ready to melt, like solder under a hot iron. As a result, pieces of the crust called "plates" actually can slide and collide like icebergs floating in an ocean slightly denser than they are. And like icebergs, the features of the Earth's crust are only the outermost tips of the plates.

When two plates collide, one tends to slide over the other. New mountain ranges appear, along with the volcanic activity and earthquakes that mark the regions of most recent crustal movement. Continents or subcontinents borne on the crustal plates can move toward or away from one another. For example, the Pacific Ocean plate, which includes all of California west of the San Andreas fault, is moving northward relative to the plate that carries the remainder of the United States. In ten million years, Los Angeles should be offshore of Berkeley. From reconstructing fossil and geological evidence, we now can be fairly sure that all the land masses were bunched together in one supercontinent about 200 million years ago, during the age of dinosaurs. Since then the continents have drifted, with a notable effect on the evolution of different types of animal species. The Himalaya mountains rose when the plate carrying the Indian subcontinent rammed into the Eurasian plate about 100 million years ago. This fascinating story now seems well established, but it doesn't hurt to remember that the man who thought of it first, Alfred Wegener, was considered an imbecile by professional geologists for many years, the more so because he was a meteorologist by trade.

Mars

Let us sail on in our spacecraft away from these vicissitudes of human fortune toward the red planet Mars. From the midpoint of the distance between the Earth and Mars (70 million kilometers when they are closest to one another), some similarities appear between the two planets. First, Mars has an atmosphere, and we can see its surface (Figure 10–19). Mars' atmosphere, like Venus', consists mostly of carbon dioxide molecules. The total atmospheric pressure at Mars' surface amounts to only 0.5% of the pressure on Earth, so Mars' atmosphere is two hundred times "thinner" than the Earth's. (The Earth's atmosphere is ninety times thinner than Venus' atmosphere.) Mars also resembles the Earth in possessing seasonal temperature changes. On both planets the seasons arise from the tilt of the planet's rotational axis. Like the Earth, Mars rotates in about one day (24 hours and 37 minutes for Mars), and like the Earth, Mars does not have its rotational axis lined up perpendicular to the plane of the planet's orbit around the sun (Figure 10–20). The angle of tilt from the perpendicular for the Earth (23½ degrees) almost equals that of Mars' rotation axis (24 degrees). The rotation axis always points along the same direction in space, so as the planet orbits

Figure 10–19. A mosaic of three photographs of Mars taken by Mariner-9 shows the planet's northern hemisphere. The north polar cap (top) was shrinking during the Martian summer at the time that this picture was taken in 1972. At the left in the bottom panel the great volcano Olympus Mons can be seen, and at the extreme bottom right we can glimpse the great equatorial canyon (see Figures 10–22 and 10–23).

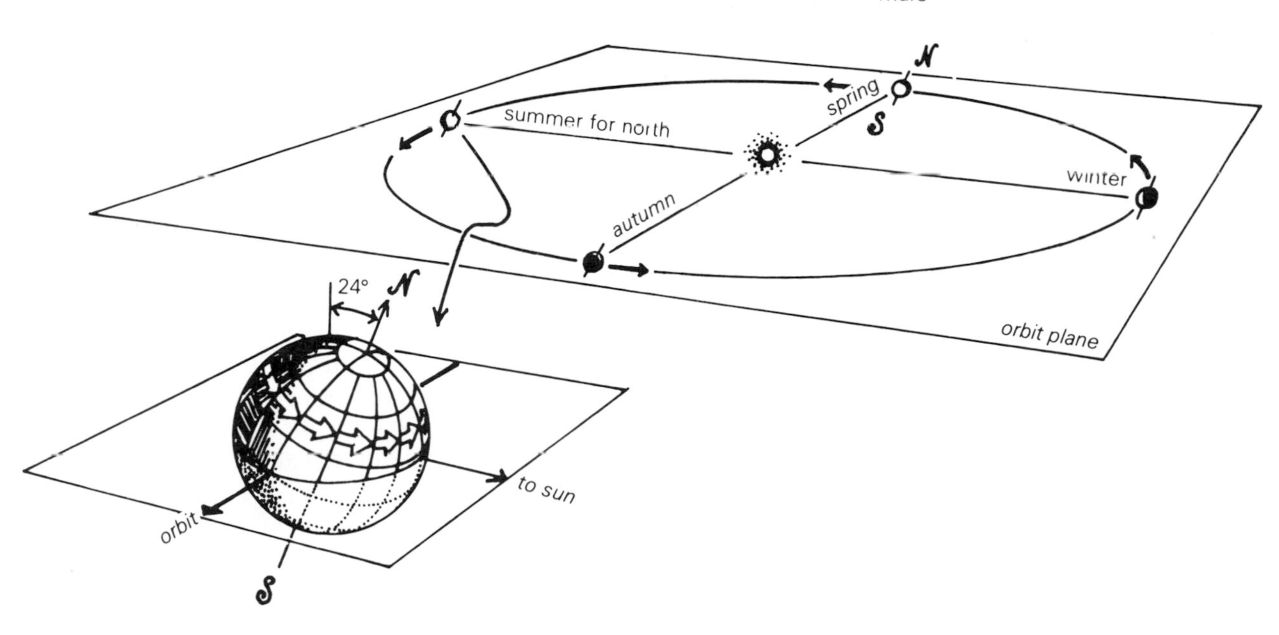

Figure 10–20. Mars' rotational axis, like the Earth's, tilts at an angle of about 24 degrees from being perpendicular to the plane of the planet's orbit around the sun. Therefore Mars has seasons like the Earth's: First the northern hemisphere, then the southern, receives the sun's rays more directly, because the planet's rotational axis always points toward the same direction in space as the planet orbits the sun.

the sun, first the northern hemisphere and then the southern hemisphere are more directly oriented toward the sun's photons (Figure 10–20). Thus at any given time we have winter in one hemisphere and summer in the other, or spring in one and fall in the other. Another similarity between Mars and the Earth is their polar caps, which grow larger in the winter of that hemisphere and smaller during the summer. The Martian polar caps seem to be frozen carbon dioxide, but there may be frozen water beneath this carbon dioxide layer.

Like the Earth, Mars shows seasonal changes in its surface appearance, but the Martian changes do not come from changes in vegetation, rather from the wind patterns that lift fine dust from the surface and deposit it elsewhere.[2] The variations in the wind patterns with the seasons thus are responsible for the varying appearance of Mars' surface. Sometimes Martian dust storms hide the planet's surface from our view for days on end.

---

[2]The Earth's seasonal changes in appearance are not as pronounced as those seen on Mars, despite the fact that they seem far more "real" to us.

So much for the similarities between Mars and the Earth. The differences between them deserve equal attention. To start with, Mars' diameter, which is twice the diameter of the moon, barely exceeds one half the Earth's 12,800 kilometers. Far smaller than Venus or the Earth, Mars has a mass only 11% of the Earth's and cannot compete with the Earth or Venus in holding on to a dense atmosphere. Mars' greater distance from the sun, as well as its thin atmosphere, lead to low surface temperatures, with a greater variation between night and day temperatures than on Earth. At the Martian equator noontime temperatures reach 290 degrees absolute (17 degrees Centigrade), but at night the surface cools to 200 degrees absolute (–73 degrees Centigrade) or even less. The thin Martian atmosphere contains traces of oxygen, carbon monoxide, water vapor, and ozone, along with its carbon dioxide, but none of these trace constituents provides more than a tenth of a percent of the (already extremely thin) Martian atmosphere. In fact, the Earth turns out to be the only planet where free oxygen molecules form a major constituent of the atmosphere.

Mars has no liquid water on its surface, and almost no water vapor in its atmosphere. The polar caps of Mars, which appear to be frozen carbon dioxide ("Dry Ice"), testify to the low temperature there, because carbon dioxide freezes only at temperatures below 200 degrees absolute. As on Earth, carbon dioxide frozen on Mars' surface will not become liquid, but rather will vaporize directly from the solid state into the gaseous state. The most striking aspect of Mars' surface, which was discovered by Mariner-7's photographs from an orbit around Mars, is the evidence of erosion by liquids some time in the relatively recent past. This evidence consists of sinuous channels highly reminiscent of river valleys on Earth (Figure 10–21). However, the largest channel system on Mars does not appear to be such a valley, but maps out some gigantic fault line for several thousand kilometers (Figures 10–22 and 10–23). Mars also has volcanoes (Figure 10–24) far taller and wider than those on Earth (hence, in comparison to the planet's size, immensely larger than the Earth's volcanoes), and Mars' heavily cratered surface testifies to the weak protection against meteors that its thin atmosphere offers (Figure 10–25). But the winding channels raise the question of how they possibly could have been made without water, since no other liquid comes to mind as a possibility. Some astronomers have said that there must be water somewhere on Mars, and the likeliest place for it now is inside the polar caps. These caps seem to consist of frozen carbon dioxide on their outside, but this could be only an outer covering that conceals large amounts of frozen water.

According to speculations based on the evidence of the Martian canyons, the water assumed to be frozen in the Martian polar caps must once have been liquid water on the planet's surface. How can we explain

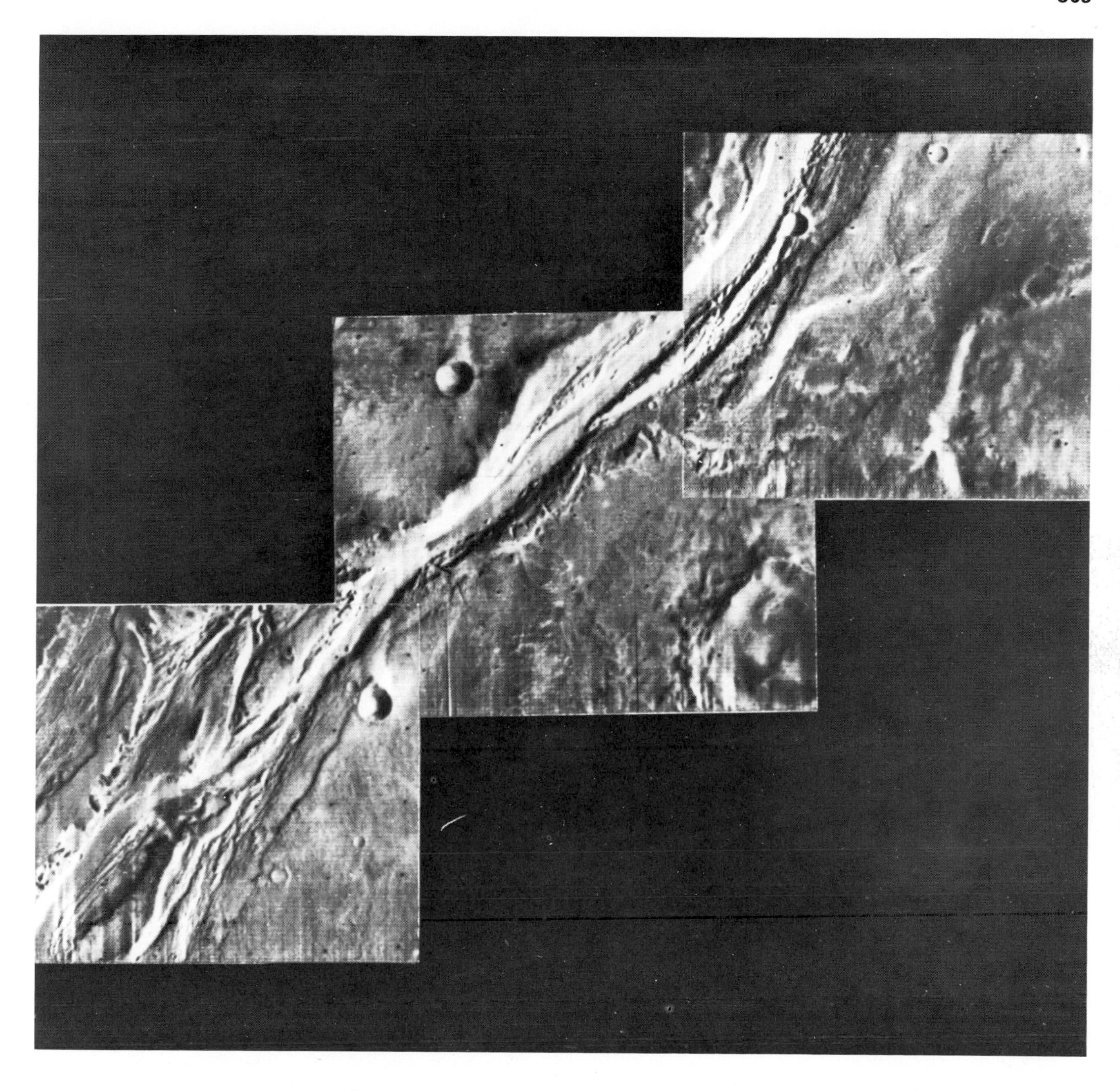

Figure 10–21. A channel thought to have been formed by running water lies near the Martian equator.

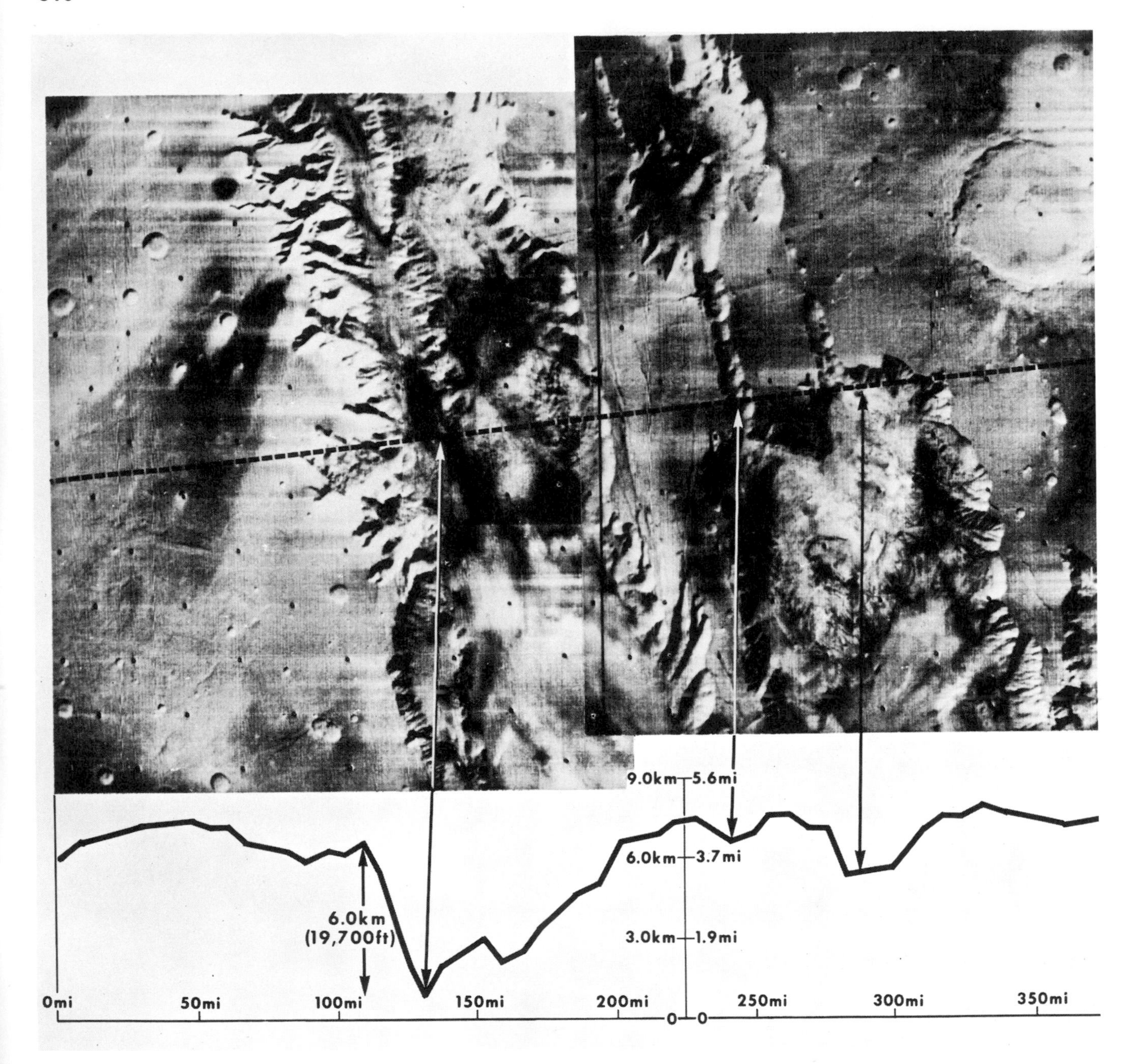

Figure 10–22. This mosaic of photographs taken by Mariner-9 shows the great equatorial canyon on Mars, estimated to be six kilometers deep (the Grand Canyon in Arizona is 1½ kilometers deep) and about 120 kilometers wide.

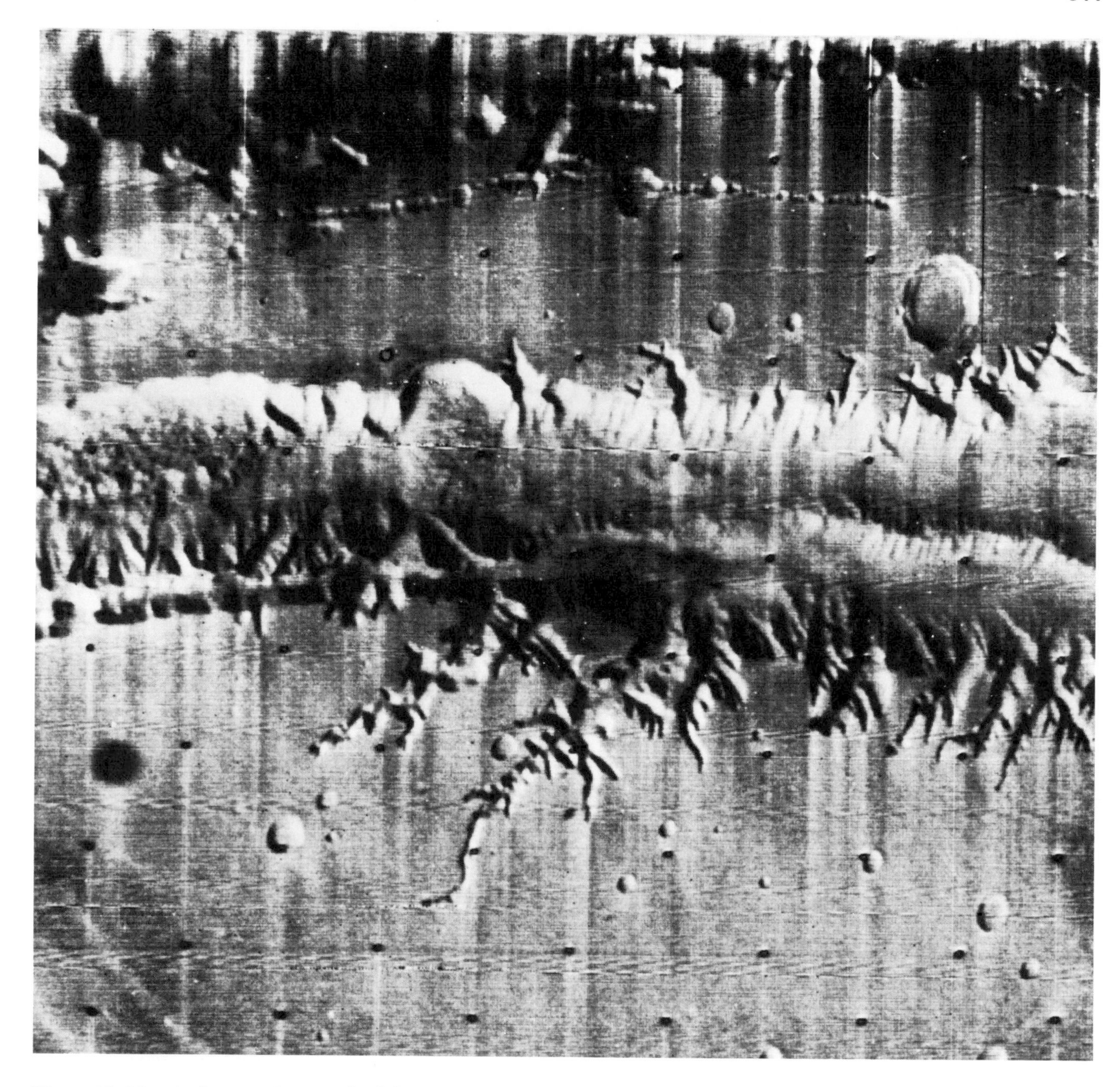

Figure 10–23. A close-up photograph of the canyon suggests that subsidence along lines of weakness in the Martian crust has contributed to its formation. This picture was taken in January 1972 when Mariner-9 was 2000 kilometers above Mars' surface.

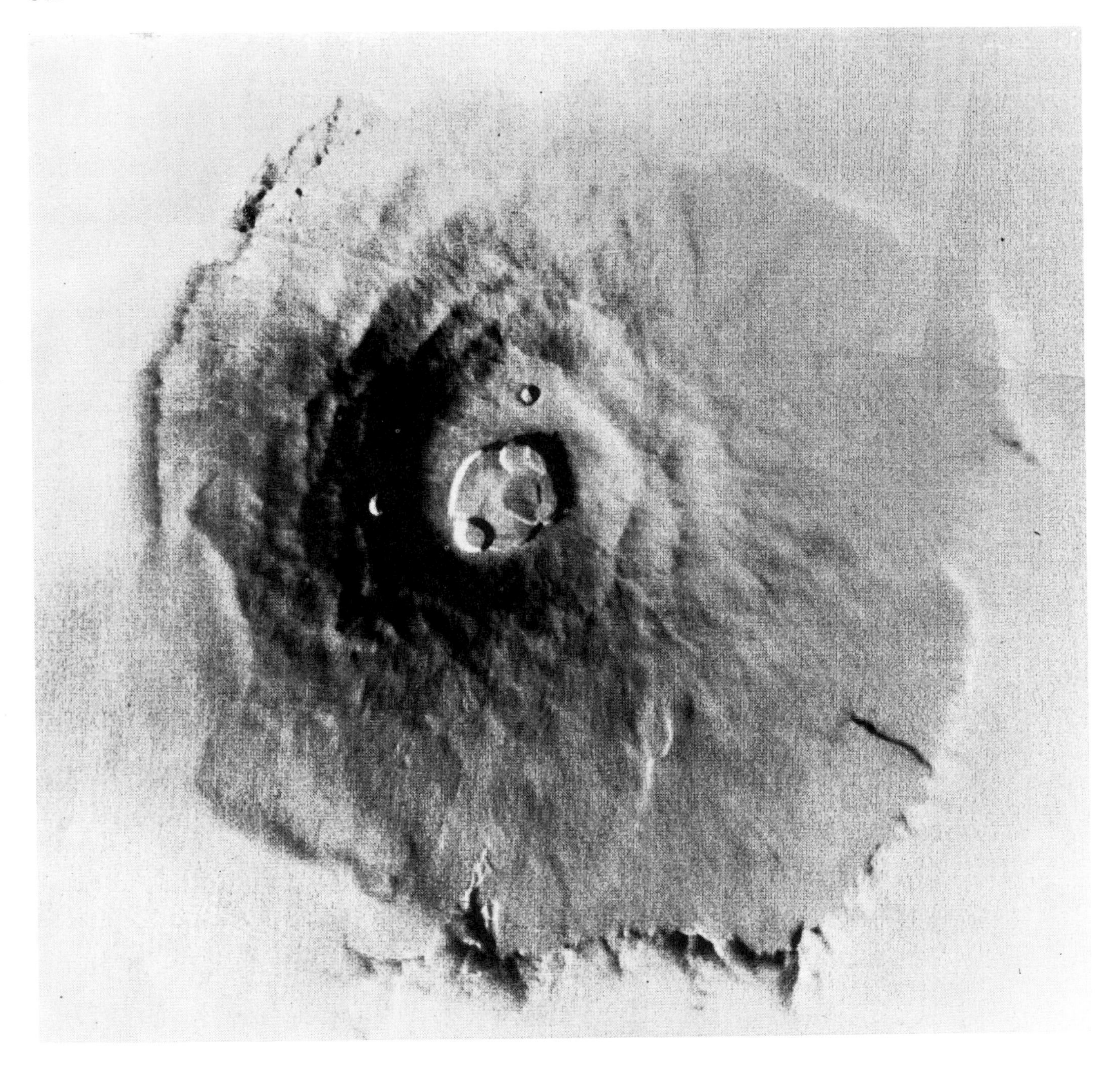

Figure 10–24. The volcano Olympus Mons stands 23 kilometers above the surrounding surface, and covers 600 kilometers in diameter at its base. (The volcanoes that form the island of Hawaii are 225 kilometers across and nine kilometers high.)

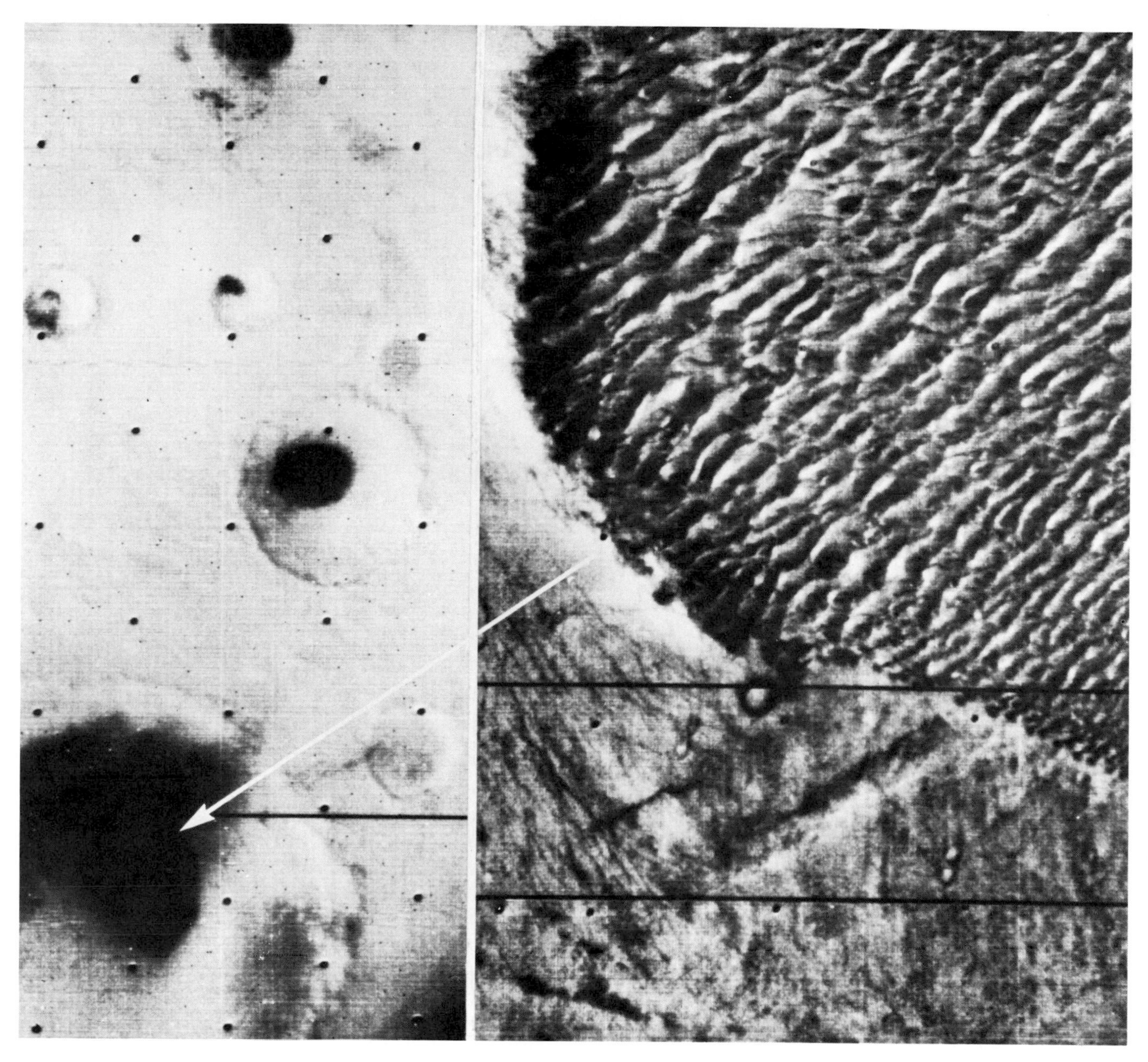

Figure 10–25. The sands of Mars appear in the enlargement (right panel) of a photograph of a Martian crater 150 kilometers in diameter (left). The dune field seems to be made of numerous long sand dunes about a kilometer apart. As in dune fields on Earth, the dunes at the margin of the field are smaller than those near its center. It is easy to ascribe the appearance of these sand dunes to the action of Martian winds.

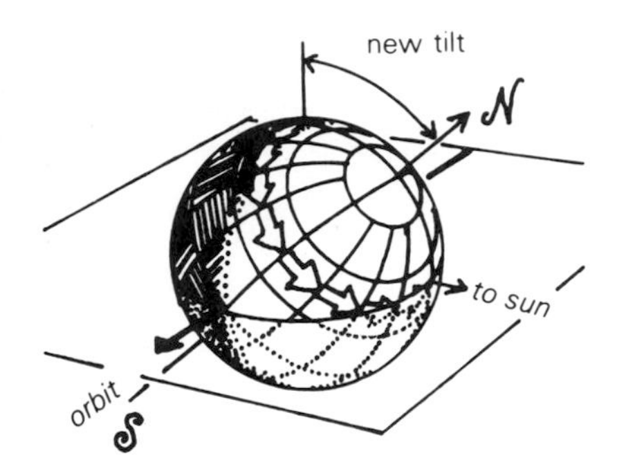

Figure 10–26. If the angle of tilt of Mars' rotational axis away from perpendicular to Mars' orbit around the sun were larger than it is now, sunlight would shine more directly on the summer hemisphere (and less directly on the winter hemisphere) than it does now. This would make the Martian summers warmer and the winters colder than they are now. Such increased changes in the seasonal temperatures on Mars might have caused water in the polar caps to melt from beneath the frozen carbon dioxide, according to some astronomers' speculations.

this change? One suggestion, among many made by Carl Sagan, is that the tilt of Mars' rotational axis has changed over the years. Mars now rotates around an axis that tilts at an angle of 24 degrees from the perpendicular as the planet orbits the sun. However, the angle of tilt changes by as much as 20 degrees over a cycle of about a million years. If this angle of tilt were greater, the Martian seasons would be more pronounced, because the sun's rays would shine still more directly on the hemisphere tilted toward the sun in summer (Figure 10–26). If the angle of tilt once had been large enough, say 45 degrees instead of 24 degrees, the sun's photons could melt the carbon dioxide and the ice (if it is there) in the polar caps during the summer in that hemisphere. Then the water could, at least briefly, pass through the cycle of evaporation, condensation, and rain familiar to us on Earth.[3] Any channels caused by Mars' rains would last a long time once the water again had been locked in the polar caps as the angle of tilt decreased. The thin Martian atmosphere and the present absence of water on the surface imply a slow rate of weathering of any features on Mars. This interesting speculation could be investigated further if we actually could land on Mars to examine the channels close up and to find the true composition of Mars' polar caps.

If liquid water has existed on Mars at some time in the past, then the chances of finding life on Mars improve enormously (see Chapter 11). Mars might have developed organisms that dry out completely for millions of years, until the planet's tilt again becomes large enough to release the water frozen at the poles. (Some terrestrial species of mosses can survive years of dehydration and then start to grow again when they obtain water.)

Mars has two tiny moons named Phobos and Deimos after the war-god's chariot horses. These two satellites are only 10 and 20 kilometers in diameter. The smaller one, Deimos, takes only 1¼ Earth days to orbit around Mars, while the inner satellite, Phobos, has an even smaller and faster orbit. Phobos' orbit around Mars lies only 7000 kilometers above the planet (our own moon orbits 400,000 kilometers above us), and the satellite moves so rapidly to stay in orbit that it completes each trip around the planet in just eight hours. Thus Phobos orbits around Mars in one third the time that Mars takes to rotate, so in the Martian skies Phobos moves from west to east, while all the other celestial bodies seem to move from east to west as Mars rotates. Phobos would make an ideal platform to observe Mars from close by, but otherwise this lump of rock about the size of Manhattan has little to offer the space traveler (Figure 10–27).

---

[3]The melting of the carbon dioxide as well as of the water ice is necessary to increase the atmospheric pressure enough to produce the evaporation-and-rain cycle for the water.

Figure 10–27. The larger satellite of Mars, called Phobos, looks like a lump of rock that has been scarred by repeated meteor impacts.

Mars slowly has begun to lose the mystery that came from astronomers' reports of "canals" on its surface. These straight lines, so carefully mapped by earth-bound observers in the early twentieth century, turned out to be optical illusions that arise from the tendency of the human eye to link together any prominent features as a straight line. Despite the fact that the "canals" are not real, Mars remains our best hope for finding life in the solar system outside the Earth.

As we travel outwards from Mars toward Jupiter, we embark on the most dangerous part of our journey, the passage through the asteroid belt. Thousands of known asteroids, and perhaps millions of other, smaller, uncharted rocks, orbit the sun in this region. Most of the asteroids' orbits lie near the plane of the solar system that contains the planets' orbits.[4] However, the orbits of some exceptional asteroids carry

---

[4]Only the orbits of Mercury and Pluto are tilted significantly (by 7 degrees and 17 degrees, respectively) out of the plane defined by the other planets' orbits. The other seven planets orbit the sun in almost the same general plane: The maximum deviation among the inclinations of the orbital planes of these seven planets does not exceed four degrees. Our choice of the Earth's plane of orbital motion as the fundamental plane of reference, the "ecliptic," of course is based on our happening to inhabit the Earth.

them far out of the plane defined by the Earth's orbit around the sun, and therefore no matter how we choose to sail above or below this "ecliptic" plane (so named because eclipses occur when the moon crosses this plane at full moon or new moon), we cannot count ourselves absolutely safe from the danger of colliding with an asteroid. To take a more positive viewpoint, if we choose to land on one of the larger asteroids, such as one called Psyche (300 kilometers in diameter), we would be standing on a specimen of relatively unchanged, original solar-system material. A chemical analysis of this matter could be even more valuable than an examination of rocks brought back from the moon in helping to unravel how the solar system formed. Also, it would not be hard to leave. Since the force of gravity on Psyche is only one percent of the force of gravity at the Earth's surface, we could launch ourselves into space by reaching a velocity of 200 meters per second, rather than the 11 kilometers per second needed to escape from the Earth's surface. On an asteroid one kilometer across, the escape velocity would be only about one meter per second, so that an easy leap would launch us off the asteroid's surface and back on our outward journey.

The Giant Planets

Eight hundred million kilometers out from the sun, we meet the planet Jupiter, giant among the sun's followers, with a mass 318 times the Earth's mass (but still just one thousandth of the sun's), and a diameter ten times larger than the Earth's. Jupiter has more mass than the rest of the planets and the asteroids combined, but its large size (140,000 kilometers across) gives it an average density of only 1.3 times the density of water, which is typical of the four giant outer planets. Pictures of Jupiter taken from Earth or from the Pioneer-10 spacecraft (see Plate 8) show the characteristic bands that run around the planet parallel to its equator. These bands, as well as the "Great Red Spot" that interrupts some of them, are atmospheric features rather than evidence of a solid surface on Jupiter. Although Jupiter is the largest planet, it also rotates the most rapidly, rotating its huge bulk in less than 10 hours at its equator. The middle latitudes take 10½ hours to rotate, as measured from the features that we see. The difference in the rotation period of the equatorial- and the middle-latitude regions provides definite evidence that we are observing atmospheric gases rather than a solid surface, which would all rotate with the same period. The observed lag of the middle-latitude regions behind the equatorial regions as the planet rotates helps to drive the circulation patterns that produce the bands of various colors. Similarly, the "Great Red Spot," which has shown a tendency to wander about a bit during the past two hundred years, probably is a semipermanent hurricane in Jupiter's atmosphere. Since Jupiter's diameter is ten times the Earth's and the planet spins 2½ times as fast as

Earth, the equatorial velocity on Jupiter that arises from the planet's rotation equals 25 times that on Earth. Our own rotation produces a velocity of 1600 kilometers per hour (0.4 kilometers per second) at the equator, but on Jupiter this velocity reaches 10 kilometers per second, which is almost equal to the escape velocity for Earth! Because the force of gravity on Jupiter exceeds that on Earth (by 2½ times), we would need a far larger velocity to escape from Jupiter than from Earth (the precise figure is 61 kilometers per second). Therefore the gases in Jupiter's atmosphere are not flung off the planet even though they are being wheeled around at speeds up to 10 kilometers per second. Such velocities do serve to produce a complex series of atmospheric disturbances whose topmost layer appears as the familiar bands and the Great Red Spot. Jupiter's rapid rotation also makes the entire planet bulge noticeably at the equator. Even our solid Earth has an equatorial diameter 0.3% larger than its polar diameter as the result of its rotation, but on Jupiter the diameter between the poles is 6% less than the diameter measured across the planet's equator.

Does Jupiter have a solid surface beneath its atmosphere? The answer seems to be no, the gases become denser and denser for thousands of kilometers going inward. These gases are mostly hydrogen and helium, together with large amounts of methane and ammonia, which are two simple molecular compounds of hydrogen with either carbon (for methane) or nitrogen (for ammonia). Jupiter's atmosphere also contains a small amount of water vapor, as well as other compounds of carbon and hydrogen such as ethane, which has two carbon atoms and six hydrogen atoms (as opposed to one carbon atom and four hydrogen atoms for methane).

Jupiter and the other three giant planets still cling to much of their original hydrogen and helium, thanks to their large masses and large gravitational forces. At the top of Jupiter's atmosphere, the temperature is about 140 degrees absolute (−133 degrees Centigrade), reflecting the great distances of the giant planets from the sun. At temperatures this low, we would expect that most of the water vapor, and indeed most of the ammonia vapor, would be frozen out of the atmosphere, although methane requires even lower temperatures (as do hydrogen and helium) to make it freeze. Certainly most of the water and ammonia at the top of the atmosphere must be frozen, and the fact that astronomers have observed gaseous water vapor and gaseous ammonia (by the characteristic energies of the photons that they absorb in the sunlight reflected from Jupiter) shows that they must be looking at higher temperatures somewhat deeper in the atmosphere.

Were we to descend into Jupiter's ever-thickening atmosphere, we would find that the temperature increases. Part of the reason for this increase is the fact that Jupiter has an internal source of heat. Some other planets also have such a heat source; for example, the Earth's radioactive rocks release energy of motion and help to heat its interior. But Jupiter's sources of energy of motion actually overshadow the energy received from the sun. Each second, the planet's insides release about 2½ times the amount of energy that reaches Jupiter from the sun! (Of course, at five times the Earth's distance from the sun, Jupiter receives only 1/25 as much energy per square centimeter of area as the Earth does.) How does Jupiter produce so much energy on its own? Not by nuclear reactions, because Jupiter's interior is not hot enough to make protons fuse, but rather by the slow contraction of its interior under its own self-gravitational pull.

If we recall how stars began to form (Chapter 7), we may remember that as the massive protostars condensed, the pressure of a star's overlying layers pushed on the interior regions, squeezing them to a greater density and making the particles there move faster and faster, until eventually the particles had enough energy of motion to begin to fuse. Jupiter's mass, which is far less than any star's, does not produce enough self-gravitation to pull the star together with pressure sufficient to make protons fuse. However, the energy of motion of all the particles in Jupiter's interior reaches a significant amount when we consider the enormous bulk of the planet. Jupiter in fact slowly is contracting its insides, and as it does so the contraction releases energy of motion. (The opposite process, making Jupiter expand slightly, would *require* the input of additional energy of motion.) About half of the energy of motion that Jupiter's contraction releases appears as extra energy of motion of the particles that form the planet, and about half is radiated into space in the form of photons.

If Jupiter had ten times its actual mass (thus one percent of a solar mass), nuclear fusion reactions probably would have begun by now as a result of the still larger energy of motion per particle (that is, the still larger temperature) that the contraction would provide. Then Jupiter would be a tiny star rather than a planet. At one hundred times its present mass (one tenth of the sun's mass), we know for sure that nuclear fusion reactions would have begun, because we can observe stars with one tenth or even one twentieth of the sun's mass, 2×10^{33} grams.

Before scientists knew about nuclear fusion, they thought that the *sun's* energy of motion might arise from a slow contraction of the solar interior like that inside Jupiter. However, they could calculate that if the sun released its energy of motion through this process alone, then a few hundred million years (less than a tenth of the sun's true lifetime) would cause the sun to contract into an unrecognizable form. Thus the scientists eighty years ago had hit upon the right theory for the wrong object. Protostars do contract and thus raise the temperature in their interiors, but when the central temperature reaches several million degrees absolute, nuclear fusion reactions begin and stop the star from contracting any more (Page 195). Jupiter may be thought of as a protostar that will never manage to become a star at all. Since Jupiter's internal energy source releases about one millionth as much energy of motion per second as the sun's does, Jupiter can continue to liberate this small amount of energy for a long time even though it has only one thousandth of the sun's mass.

Looking outside the planet once again, we can admire Jupiter's family of thirteen moons, which is the largest in the solar system. The four largest of these moons, first discovered by Galileo, all equal or exceed the Earth's moon in size and mass. (To be frank, the smallest of these four, Europa, is actually a bit smaller than the moon.) In fact, the two largest moons, Ganymede and Callisto, have diameters of about 5000 kilometers, and thus are as large as the planet Mercury. The smaller pair of the four major satellites, Io and Europa, have diameters of about 3200 kilometers. These four "Galilean" satellites carry the names of four of Jupiter's mythological lovers (three female and one male). The inner two, Io and Europa, are denser than the outer two, Ganymede and Callisto (3.5 times the density of water as compared to 2 times the density of water). This arrangement of densities imitates the solar system itself, in which the four inner planets have densities of 4 to 5½ times the density of water while the four giant outer planets have densities averaging 1½ times that of water.

The similarity in the densities, and in the decrease in densities outward from the orbital center, between the planets in the solar system and the four largest (Galilean) satellites of Jupiter does suggest that the satellites and the planets formed in the same series of processes. This similarity also implies that the Galilean satellites should have a chemical composition intermediate between that of the four inner planets and that of the four outer ones. Ganymede and Callisto should consist largely of frozen hydrogen, methane, and ammonia, while the denser, inner satellites, Io and Europa, should have a larger amount of heavier elements such as oxygen, magnesium, aluminum, sulfur, silicon, sodium, and calcium. We cannot test this conclusion thoroughly until we land a space probe on the Galilean satellites to analyze their composition. We do have definite proof that Io, the innermost large satellite, has sodium atoms on its surface, because some of these atoms absorb photons of a characteristic energy when sunlight reflects from the surface of Io. Additional evidence for the resemblance between the sun's planets and the Galilean satellites comes from the fact that the *orbits* of the four largest Jovian satellites follow a spacing that imitates the spacings between the planets' orbits. Once again, the gap between a satellite's orbit and its next outer neighbor is about twice the gap between that satellite and its next inner neighbor (Figure 10–28).

Jupiter's other nine satellites, which are much smaller than the four Galilean satellites, range from 8 to 140 kilometers in diameter. The outer four of these, and perhaps the outer eight, all seem to be captured asteroids because they have noticeably elongated orbits, which in four cases carry the satellites around Jupiter in the opposite sense to the orbits of the four major satellites and to the planet's rotation. Other moons of Jupiter may be as yet undiscovered, since the smallest known moon was found as recently as 1974.

Jupiter also controls two groups of asteroids that move in the same orbit and in the same direction as Jupiter, either one sixth of the way

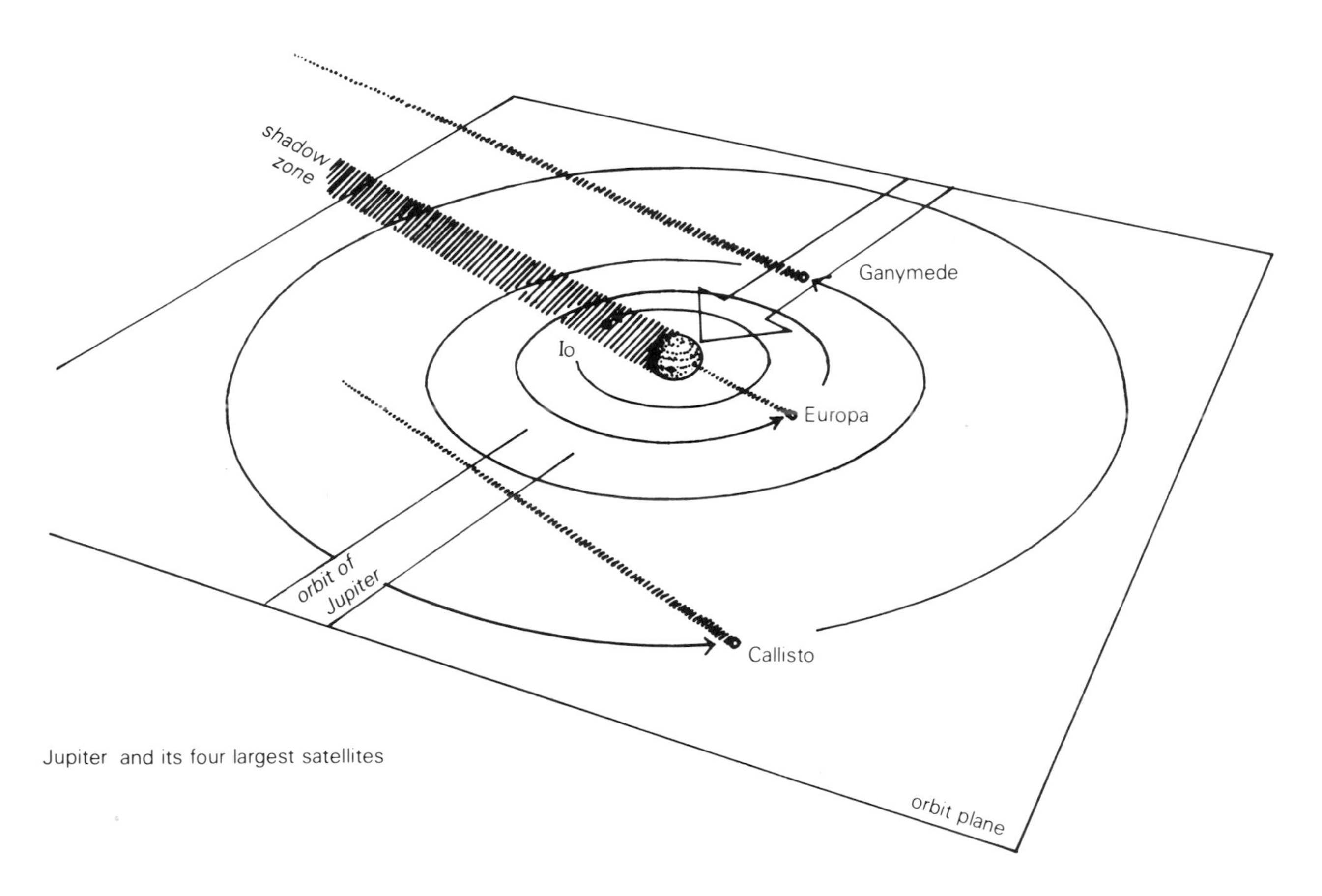

Figure 10–28. The spacing of Jupiter's four largest satellites' orbits around the planet is reminiscent, with a little imagination, of the spacing of the planets' orbits around the sun. Each successive gap between orbits is about twice as large as the next gap in.

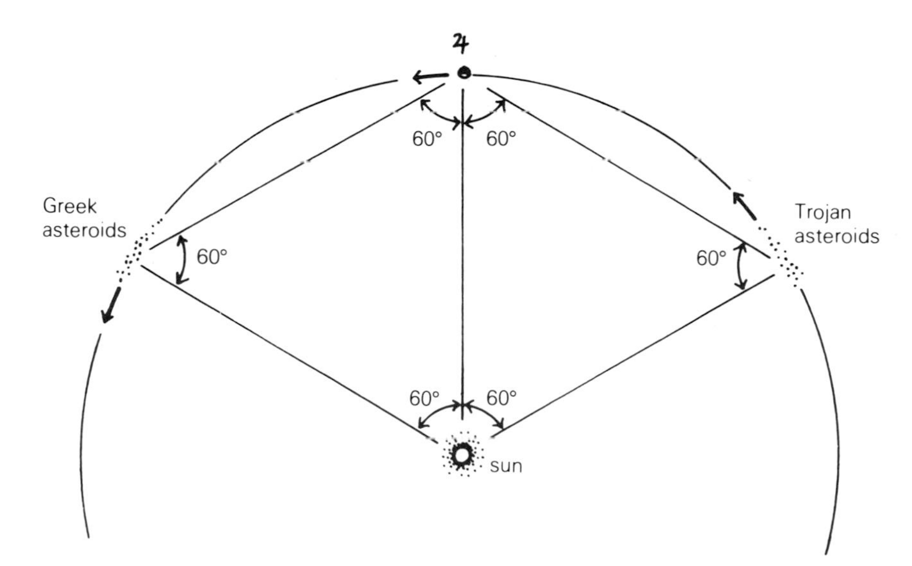

Figure 10–29. The "Trojan" and "Greek" asteroids orbit around the sun at the same distance from the sun that Jupiter does, but the asteroids are either one sixth of the way ahead of, or one sixth of the way behind, the planet Jupiter in its orbit. (Actually, the asteroids tend to wander a bit around these one-sixth points, but on the average their location corresponds to the equilateral triangles that are drawn here.)

behind Jupiter in its orbit (these are called the "Trojan" asteroids) or one sixth of the way ahead (the "Greeks").[5] Thus the sun, Jupiter, and either of these two groups of asteroids always form an equilateral triangle in their locations in space (Figure 10–29). Calculations show that once an asteroid reaches either the Greek or the Trojan position, the asteroid will be kept there by the interplay of the gravitational forces from Jupiter and the sun. Jupiter's strong gravitational forces also have affected the orbits of most of the ordinary asteroids, which lie between Jupiter and the sun. Jupiter's gravity perturbs each asteroid slightly from the orbit around the sun that it would have if Jupiter did not exist. The cumulative effect of these perturbations have been to move the asteroids' orbits into groups. The gaps separating these groups of asteroids represent orbits with periods that are a simple fraction of Jupiter's orbital period, such as ⅔ or ⅗ of the 12 years that Jupiter takes to orbit the sun. Any asteroid that had an orbital period that was a simple fraction of Jupiter's would be subject to recurrent tugs in the same direction every third or fifth orbit (for example) when it found itself closest to Jupiter once again, being tugged in the same direction each time. Figure 10–30 shows the gaps between the asteroids' orbits that Jupiter has produced by repeatedly tugging at any asteroids that happened to occupy these orbits.

---

[5]Owing to an early mistake, one of the "Trojans" is actually a Greek (Patroclus), and one of the "Greeks" is in fact a Trojan (Hector).

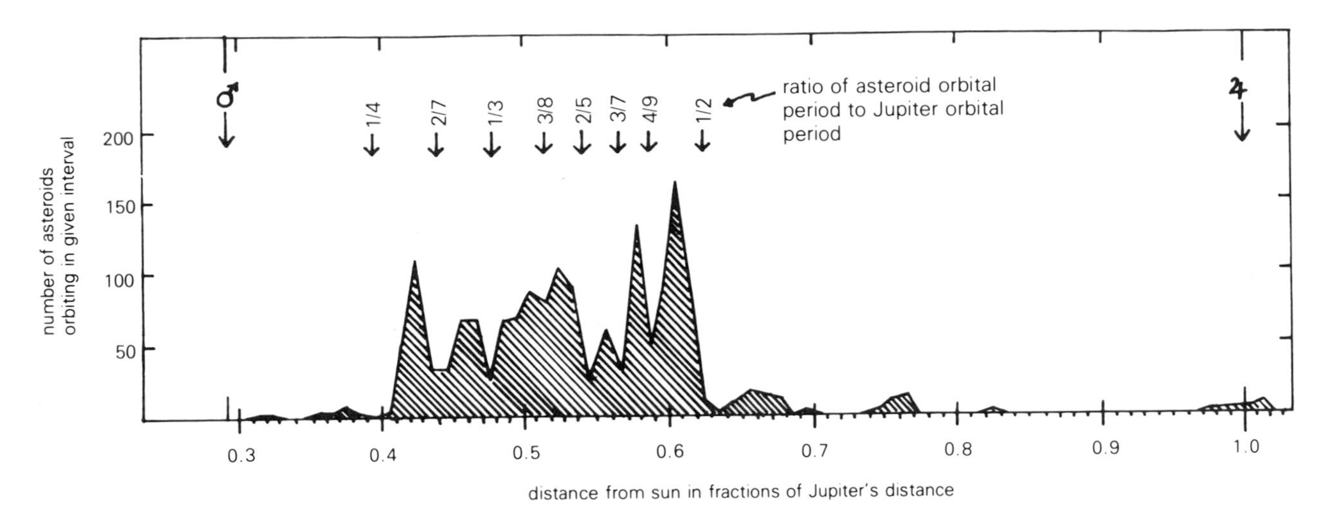

Figure 10–30. Jupiter's gravitational pull on the asteroids orbiting in nearly circular orbits has caused them to avoid those orbits whose periods are simple fractions of Jupiter's orbital period. This graph shows the number of asteroids whose orbits have various average distances from the sun. We can see a lack of asteroids with orbits that would make the asteroid's period of revolution equal to a simple fraction (¼, ⅓, ⅖, etc.) of Jupiter's orbital period.

Before we leave Jupiter for Saturn, we ought to notice the planet's large magnetic field, since if we had any delicate electronic instruments aboard they would suffer as a result of the field's existence. Like the Earth, Jupiter has a "dipole" (two poles—like a bar magnet) magnetic field that is roughly (but not exactly) aligned with its rotational axis. Thus the entire planet resembles a rotating bar magnet, and the effects of the magnetic field reach around the planet to affect the motions of any charged particles nearby through the electromagnetic forces that the field exerts. The Earth's magnetic field traps some of the incoming charged particles in the solar wind and keeps them in "Van Allen belts" for a while, until the particles leak through the magnetic field. Similarly, Jupiter's magnetic field (a hundred times stronger than the Earth's) captures charged particles, which can emit radio-wave photons as they accelerate or decelerate in the magnetic field (Page 39). Therefore Jupiter emits radio waves, often in sudden bursts. Some of these bursts correlate with the orbital motion of the innermost large satellite, Io. Apparently Io can disrupt the structure of the magnetic field as it revolves around Jupiter at a distance three times the planet's diameter.

Moving onward a long way—almost twice as far from the sun now as Jupiter is—we find Saturn, most impressive among planets, which is light enough to float in a bathtub were there a bathtub large enough to try. At this distance, 9½ times the Earth's distance from the sun, we find that the sun's apparent brightness has faded to one ninetieth of its apparent brightness as seen from the Earth. Saturn has a composition like Jupiter's—for that matter, all four giant planets appear to be quite similar in composition—but the light gases that compose it are spread still more diffusely than those in Jupiter. Saturn's diameter of 120,000 kilometers, 8½ times the Earth's, almost equals Jupiter's, but Saturn's mass hardly exceeds a quarter of Jupiter's, giving Saturn the lowest density of any planet (less than 70% of the density of water). Like Jupiter, Saturn rotates every 10 hours, and with the planet's low density this rotation gives Saturn the greatest flattening of any planet: Its polar diameter is 10% less than its equatorial diameter (Figure 10–31). Saturn also has a thick atmosphere—it may well be all atmosphere—made of hydrogen, helium, methane, and ammonia, and showing the banded pattern familiar to us from our studies of Jupiter. Unique to Saturn are its rings, which consist of millions of pebble-sized and dust-sized particles, each in orbit around the planet as a tiny satellite. These rings are efficient at reflecting light, because any particle's ratio of surface area to total volume increases as its volume decreases. (Applied to air

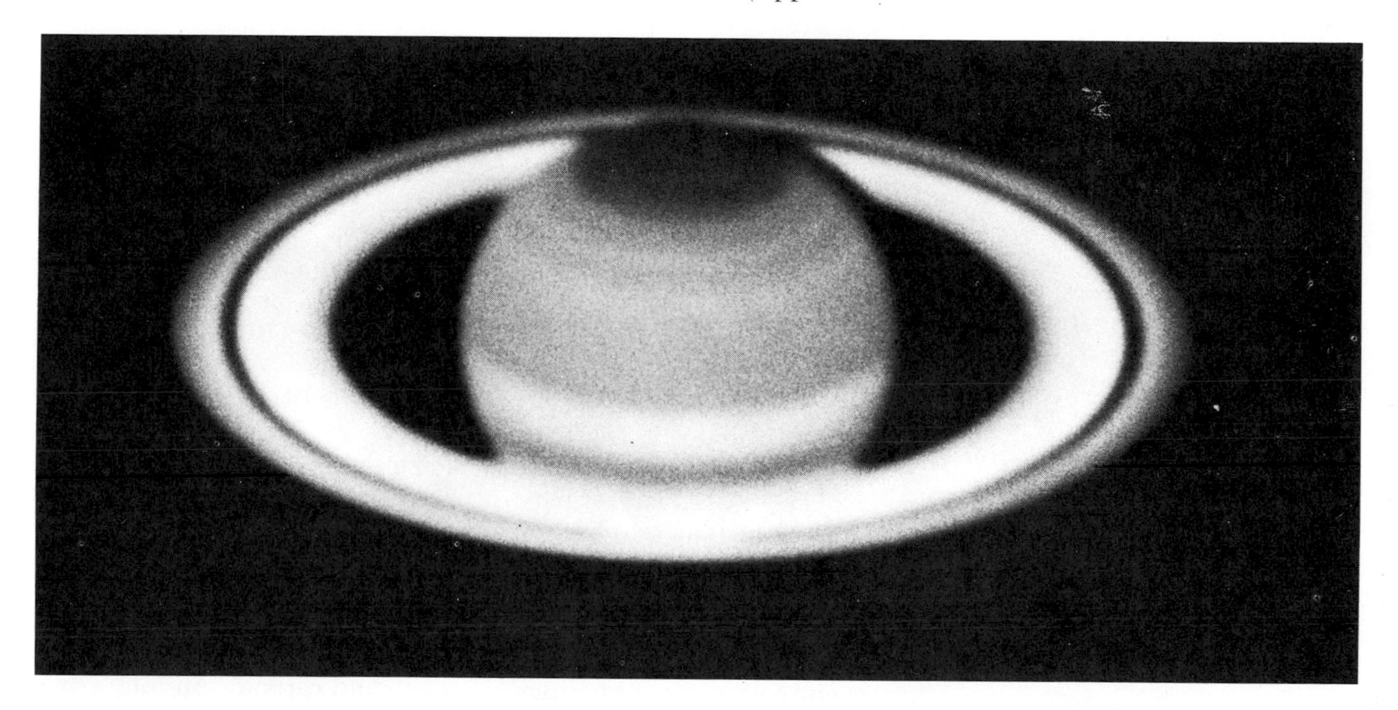

Figure 10–31. Saturn is the most flattened, or oblate, among the planets.

Figure 10–32. The planet Uranus and three of its moons (arrows). The spikes around the bright image of the planet are caused by struts in the telescope and do not reflect the true appearance of the planet.

resistance on falling bodies, this fact allows mice to drop distances on Earth that would kill a man.) Although the mass in the rings of Saturn totals less than one ten-thousandth of the planet's mass, these particles manage to reflect as much light as Saturn does by distributing their mass in a thin layer instead of inside a sphere.

The particles in Saturn's rings cannot coalesce into one good-sized satellite, because the tidal effect from Saturn (Page 296) would pull them apart. Any object held together by its own self-gravity, and in orbit around a larger body of the same density, will be disrupted by tidal effects from the primary body should it approach to a distance less than 2½ times the primary's radius. This calculation, which assumes that gravity (and not the electromagnetic forces that hold rocks and artificial satellites together) binds the satellite together, appears to be neatly verified by Saturn's rings. All of the ring material lies closer to Saturn than 2½ times its radius (the outer edge of the rings have a distance 2¼ times the planet's radius), while all of the true satellites of Saturn orbit more than this distance away. Similarly, all of the other satellites in the solar system, such as our moon, orbit around their planets at distances greater than 2½ times the planet's radius. Finally, the planets themselves orbit the sun at comfortable distances ranging from 80 to 8000 times the sun's radius.

We may speculate that Saturn's rings might represent the remains of a satellite that strayed too close to the planet and was broken apart by Saturn's tidal effect, as the planet pulled on the near side of the satellite more strongly than on the far side. Or the rings could be simply some debris left over from the protoplanet's contraction that went into orbit instead of falling onto the planet. In any case, the rings of Saturn, which are 270,000 kilometers from rim to rim but only a few kilometers thick, form one of the loveliest and thinnest systems known to us. The rings' thinness arise from the fact that if any particle's orbit should rise much above or below the plane of the rings, the combined gravitational pull from the rings, weak as it is, will tug the particle back into that plane. An analogous effect occurs if stars try to leave the plane of a spiral galaxy such as our Milky Way (Page 107).

Outside its rings Saturn has ten known moons, one of them (Titan) larger than Jupiter's largest moons (Ganymede and Callisto). Titan was the first satellite found to have an atmosphere of its own; more recently, similar atmospheres, mostly made of methane, have been detected around Jupiter's largest satellites. These giant satellites would make good reconnaisance bases for our expeditions to Jupiter and Saturn, especially if we carry along a methane converter to turn this noxious gas into useful forms of energy (plus hydrogen and carbon). An interesting fact is that Ganymede, Jupiter's largest satellite, has some frozen water on its surface, and might have a small amount of oxygen in its atmosphere. Radar echoes have shown Ganymede to have craters like our own moon's. Were it not for the low temperatures on satellites such as Ganymede and Titan (100 to 150 degrees absolute), we could speculate that these satellites might have some sort of life forms on their surfaces. The atmospheric pressure on Titan, the satellite with the thickest atmosphere, seems to equal half of that on Earth, and Titan's atmosphere is much thicker than Mars' atmosphere, a reminder that Titan (three fourths as large as Mars) has held on to far more gas than Mars has because of the much lower temperatures in the outer reaches of the solar system.

Aside from Titan, Saturn's other nine satellites have diameters of 200 to 1500 kilometers, not inconsiderable but far smaller than our own moon. The inner five of Saturn's ten satellites show spacings between their orbits that once again recalls the spacings of the planets' orbits around the sun, and in fact the general rule of doubling the gap each time we go out to the next satellite holds fairly well for all of Saturn's satellites.

Figure 10–33. In this photograph of the planet Neptune the arrow points to Neptune's larger satellite, Triton.

This rule holds again for the five satellites of Uranus. As Venus and Earth are nearly twins in their sizes and masses, so are Uranus and Neptune, the seventh and eighth planets. Thus we can compare these two planets easily, but we should not forget that the distance from Uranus to Neptune is 11 times the distance from the Earth to the sun, or more than the distance from the sun to Saturn. Uranus and Neptune both have diameters of about 45,000 kilometers, or four times the Earth's diameter, but Neptune's larger mass (17 times the Earth's, compared to 15 times for Uranus) gives Neptune a higher density (2¼ times the density of water, compared to 1.6 times for Uranus). Although not as large as Jupiter or Saturn, Uranus and Neptune (Figures 10–32 and 10–33) definitely are giant planets, and we can see the familiar (though muted) banded appearance that characterizes their thick atmospheres reacting to the planets' rapid rotation (11 hours for Uranus, 15 hours for Neptune). Like Jupiter's and Saturn's, the atmospheres of Uranus and Neptune are mostly methane, ammonia, hydrogen, and helium, but at the planets' great distances from the sun (3 billion and 4½ billion kilometers, respectively), the planets are so cold that ammonia freezes almost completely out of the atmosphere. All of the giant planets may have internal sources of heat, as Jupiter definitely does and Saturn appears to, that liberate as much energy inside them as they receive from the sun.

Uranus has five moons in almost circular orbits around its equator, but the planet's own rotational axis is tipped almost into the plane of its orbit around the sun, so the planet rolls on like a barrel with its satellites circling above. Given the coldness of the planet and the noxious composition of its atmosphere, we hardly can be surprised that no one on Uranus has complained yet about the planet's odd rotation.

Neptune's rotational axis has a much more normal tilt (29 degrees from the perpendicular to its orbital plane), but its satellite orbits are far more irregular than Uranus'. As we discussed in the preceding chapter, Neptune has one extremely large satellite, Triton, and one small satellite, Nereid. Uranus' five satellites all have diameters less than 1000 kilometers, but Triton has a diameter of 4000 kilometers and a mass that ranks with Ganymede and Titan (equal to half the mass of Mercury). Aside from the suggestion that Triton's retrograde orbit might be the result of a near-collision with Pluto (Page 267), there isn't much to say about Neptune and its satellites now because of the difficulty in observing them at billions of kilometers away from us.

When we sail outward to Pluto (Figure 10–34) we encounter the chance of resolving a long-standing mystery. Even supposing that Pluto is merely an escaped satellite of Neptune, the planet has something peculiar about it. Pluto was discovered because of the perturbations that its gravitational force causes in Neptune's orbit around the sun, as Neptune had been found by its gravitational perturbations of Uranus' orbit. From an analysis of these perturbations, astronomers could guess

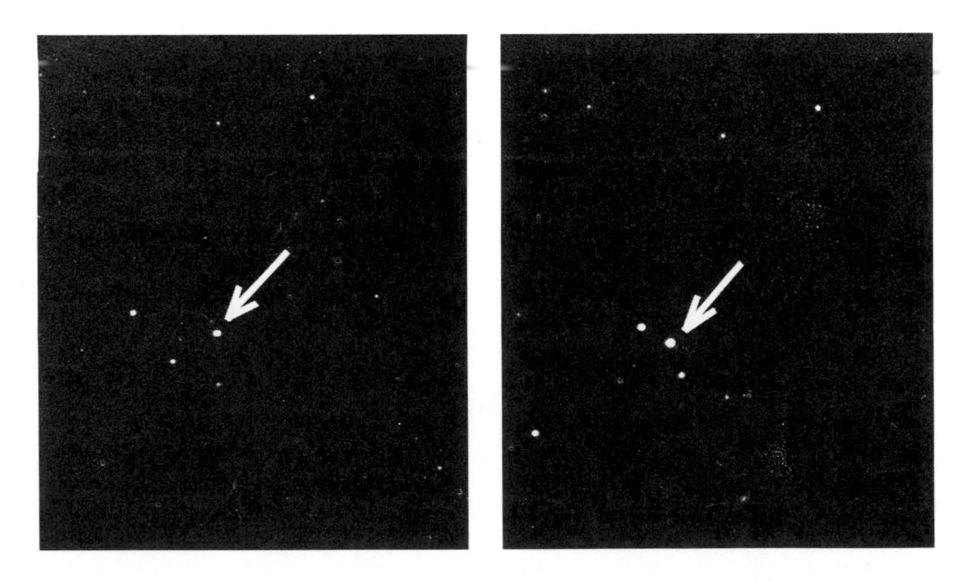

Figure 10–34. Two photographs of the planet Pluto taken several days apart show that Pluto indeed has moved slightly against the background of stars during that time.

both the location and the mass of the perturbing planet, Pluto. The best guess for the mass based on this method is one tenth the Earth's mass, or about twice the mass of Mercury. Pluto's diameter, however, seems to be at most barely larger than Mercury's (4800 kilometers). We cannot measure Pluto's diameter accurately, because the planet's distance from us is so great (eight times the distance to Jupiter) and the diameter is so small. But if the measured upper limit of Pluto's diameter were correct, Pluto would have at least five times the density of water, making it by far the densest object in the outer parts of the solar system. Can Pluto be made partly of lead? This is unlikely. Can Pluto's estimated mass or diameter be in error? This is more likely. Still, astronomers would relish the chance to visit Pluto to find out what the resolution of this puzzle might be. If we were to sail out past the orbit of Pluto, we could see whether Pluto's mass might have been overestimated because (so we may speculate) another planet exists beyond Pluto that also perturbs Neptune in its orbit. Astronomers now consider such a planet unlikely, but still we cannot rule out this possibility entirely. Our space voyage could allow us to make a close search for a tenth planet, "Planet X," if we like. If such a planet exists, then it, like Pluto, is surely a cold, distant, and forbidding place.

Beyond the Planets

Some astronomers would like most of all to visit a comet. Past the orbit of Pluto, tens of thousands of astronomical units from the sun, there are apparently billions of comets, whose total mass is less than a planet's mass. This cloud of comets occupies a celestial deep freeze that has preserved them much unchanged for billions of years. Only occasionally do comets approach the sun, because they move in tremendously elongated orbits that reach 50,000 to 100,000 astronomical units from the sun and take millions of years to complete. Thus an average specimen of such a comet would have made only a thousand or so orbits of the sun during its lifetime. We may speculate that every star has such a cometary family, numbering in the billions, that orbit slowly around the star, spending most of their orbital periods at distances from that star almost as great as the distance to the next star.

What are comets? We may surmise that they are nothing but the remnants of the original cloud from which the solar system formed. In fact, the material in comets should be older than the planets, since the cometary matter was left behind as the protosun's contraction got well underway. Meteors are solar-system material that failed to condense into planets *after* the protosun had shrunk to the size of the present planetary system. Comets, however, should contain the original chunks of matter from which the contracting solar cloud began to form. Thus the astronomers who wonder about the origin of the solar system, and of other stellar systems, would dearly love to get their hands on a piece of a comet (Figure 10–35), to see how its composition resembles and differs from the material in the inner solar system. Indeed, it would not be difficult to send a space probe into a comet during the comet's passage close (just a few tens of millions of kilometers) by the Earth at a time when a comet's orbit brings it relatively near the sun (Figure 10–36). Most of the presumed billions of comets far past the orbit of Pluto never will approach the sun even as close as this farthest planet, because their orbits are only somewhat elongated. Some of them, though, do have elongated orbits that bring them even inside the Earth's orbit. When

Figure 10–35. The comet Ikeya-Seki passed close by the Earth in 1967, and later split into two as it rounded the sun.

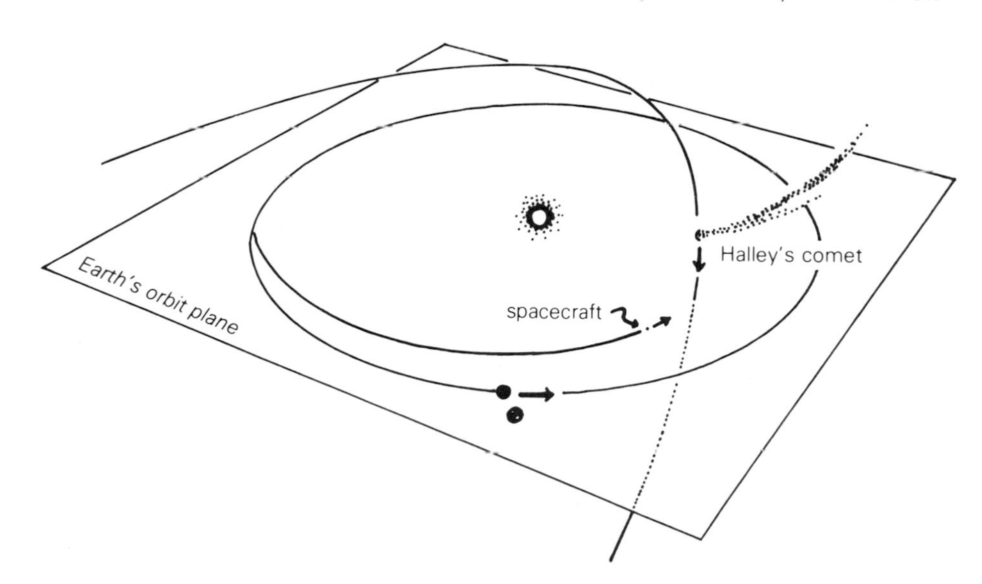

Figure 10–36. We could arrange rather easily to send a spacecraft to a comet if the comet's orbit passed within a few tens of millions of kilometers of the Earth. Halley's comet would be a natural target for such a flight in 1986.

we see such a comet, we are using our one chance to observe that particular comet for the next few hundred thousand or few million years.

Still, one comet seems much like another, so we would be delighted to examine even a single comet, especially one that had spent the last few million years in the cosmic deep-freeze between stars where a comet spends most of its orbit. Judging from the spectra of comets passing close to the sun, they consist of frozen water molecules and frozen ammonia molecules (with smaller amounts of other frozen substances) that form a large "snowball" only a few kilometers across. Perhaps rocks of one sort or another are embedded in this snowball; this is the sort of question we could answer if we landed on a comet. Such rocks would not evaporate when the comet passed close by the sun, so we could not detect them by studying the spectra of light from comets. For most of its orbit a comet consists merely of this primordial slush, interesting to the expert but not much to think about. However, when a comet comes relatively close to the sun (inside the orbit of Pluto, roughly) significant things start to happen.

The sun's photons heat the comet and begin to vaporize some of the matter in the snowball. This vaporization, which takes the material directly from the solid to the gaseous state, technically is known as sublimation. Because some of the material vaporizes, the comet loses a small fraction of its mass on each close approach to the sun. This small loss to the comet turns into a lovely gain for ourselves, because the matter vaporized from the comet's snowball or "nucleus" trails out for millions of kilometers as a cometary "tail." The matter in the tail reflects

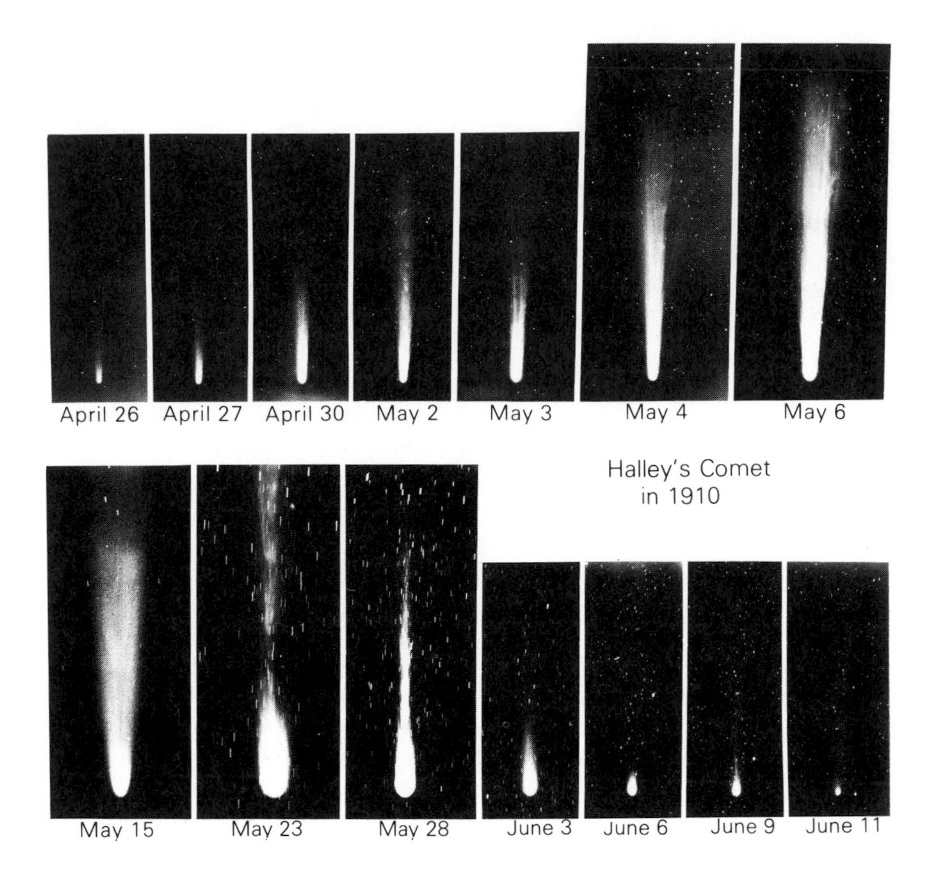

Figure 10–37. These fourteen photographs of Halley's comet, taken in 1910 (during the most recent orbit of the comet around the sun), show how the comet's tail grew longer and brighter as the comet approached the sun, then shortened as the comet sped around the sun and began its long journey past the orbit of Neptune.

sunlight efficiently, and we see the tail grow longer and longer as the comet comes closer and closer to the sun (Figure 10–37). At the same time, vaporized material close to the snowball forms a "coma," thousands of kilometers across, around the comet's nucleus. (The name "comet" comes from the Latin word for "hairy," which describes the fuzzy coma and tail of a comet when it is close to the sun and the Earth.) Some comets can produce a "coma" as wide as the sun's diameter (more than a million kilometers).

Astronomers have calculated the orbits of the few hundred comets that take less than a hundred years to orbit the sun, but for the vast majority of comets, which take thousands or millions of years to complete a single orbit, there is no predicting when they will appear and what their orbit will be. Most such comets are discovered by amateur astronomers who take the time to study the sky with medium-sized telescopes night after night. Professional astronomers tend to be too busy with more organized research to take the time to make probably fruitless searches for new comets. Once a new comet is found, however, the professionals become quite interested, even to the point (as in the case of Kohoutek's comet) of overestimating how bright and spectacular the comet will become.

Let us not imagine, though, that Kohoutek's comet was a total failure. Astronomers did observe certain kinds of molecules in the comet that they had not detected previously in other comets. Among these molecules were hydrogen cyanide (HCN) and water molecules, as well as the more complex molecules of methyl cyanide (CH_3CN). In addition, molecules of methane, ammonia, and hydrogen appeared in Kohoutek's comet, as they had in previous comets that had not been so well studied.

The reason why a comet's tail always points away from the sun comes from the sun itself. The solar-wind particles, and the photons from the sun, both push on the material freed from the comet as it warms up upon approaching the sun. The gas that is vaporized from the nucleus bends with the flow of the solar wind, while the dust particles are more susceptible to the radiation pressure from photons that we described on Page 15. As a result, a comet can have two sorts of tails, one made primarily of gas and the other primarily of dust particles, bending in different directions.

Thus the solar system: one sun, nine planets, thirty-three known satellites, many thousand asteroids, billions of comets, billions of meteors, countless particles in the solar wind. And on the third planet out, the best part of a billion different species of life forms have appeared. Whether this happened by chance or by nature receives our attention in the following chapter, as we head closer to the focus of the universe—ourselves.

Summary

The nine planets in the solar system show a variety of differences and some basic similarities. Mercury, like Mars (and our moon), shows craters produced by meteor impact, which is testimony to the small amount of protection against meteors that the thin atmosphere (for Mars) or almost nonexistent atmosphere (for Mercury) provides. Venus and the Earth have much denser atmospheres, mostly nitrogen and oxygen molecules in the case of Earth, while Venus' atmosphere (ninety times thicker than the Earth's) consists mainly of carbon dioxide molecules.

The four inner planets, denser by far than the four giant ones, have their internal structure "differentiated" into a denser core with a less dense surrounding mantle. The four giant planets, although they too are densest at their centers, have no such division into core and mantle; their atmospheres simply keep on thickening as we descend toward the center, providing an overall density for the planet about equal to the density of water (and the average density in the sun).

Mars, which shows seasonal changes in appearance on its surface (although not because of changes in vegetation) and in its polar caps, seems more like the Earth than any other planet. Venus, which is closer to the Earth in size, has an atmosphere so thick that the planet's surface temperature remains above 700 degrees absolute even during the months-long Venerean night. If Mars did have liquid water once, as the side channels of its giant gorge seem to indicate, the chances for finding life on its surface improve enormously.

The satellites of the giant planets would be better landing sites than the planets themselves, since the satellites have solid surfaces. The larger satellites are closer to the size of the terrestrial planets than the giant planets are.

Questions

1. How does Mercury's surface resemble the moon's? How can we explain this resemblance?
2. How do the four inner planets resemble one another? How do the four giant planets resemble one another? How can we explain these resemblances?
3. Why are the surfaces of Venus, Earth, and Mars hotter than they would be if these planets did not have atmospheres?
4. Why do we think that the moon was never part of the Earth?
5. Why can we see eclipses of the sun far less often than eclipses of the moon from a given point on the Earth's surface?
6. Could the channels on Mars have been made by liquid water? Where would the water have gone since the channels were made, given that there is now no liquid water on Mars' surface?
7. How does Jupiter generate heat? Do the other planets release heat (energy of motion) in the same way?
8. Is it an accident that the rings of Saturn lie closer to the planet than the orbits of all of Saturn's moons? Why?
9. Why does a comet's tail point away from the sun?

Further Reading

K. Weaver, "Mariner Unveils Venus and Mercury," *National Geographic* (June 1975).

W. Hartmann and O. Raper, *The New Mars: The Discoveries of Mariner-9* (National Aeronautics and Space Administration, No. SP-337, U.S. Government Printing Office, 1974).

F. Hoyle, *Highlights in Astronomy* (W. H. Freeman and Co., San Francisco, 1975).

R. Fimmel, W. Swindell, and E. Burgess, *Pioneer Odyssey: Encounter with a Giant* (National Aeronautics and Space Administration, No. SP-349, U.S. Government Printing Office, 1974).

"The Solar System" (twelve articles), *Scientific American* (September, 1975).

C. Sagan, *The Cosmic Connection: An Extraterrestrial Perspective* (Dell Publishing Co., New York, 1975).

Chapter 11 Life

"Before anything had a soul,
While life was a heave of Matter, half inanimate,
This little bit chipped off in brilliance
And went whizzing through the slow, vast, succulent stems."

D. H. Lawrence

What we call life has its origins with the rest of the universe, fifteen billion years ago in the big bang. From the big bang came hydrogen and helium, and from the insides of stars came more complex elements such as carbon, nitrogen, oxygen, and heavier elements. The right densities and the proper input of energy then led to the origins of life on Earth. Scientists now think that they are close to understanding the chemistry of living organisms, and thus to finding out how these organisms got started on the Earth billions of years ago. As far as we can tell, there is no reason to believe that life should not have evolved in other places that resemble the Earth, and the correspondence need not be exact.

What is life? A precise definition cannot please everyone, but we must try. We shall define a living organism as an object that can reproduce itself with an expenditure of energy far less than its energy of mass (its mass times the speed of light squared). Thus protons are not alive, even though two protons colliding at high energies can produce many more protons, because the energy involved is many times a proton's energy of mass. Living organisms characteristically reproduce, not by making new energy of mass out of energy of motion, but by rearranging the mass in atoms and molecules close at hand, expending relatively little energy (compared with their energy of mass) in doing so. Thus our definition makes some sense, but even so, there are problems. For example, are viruses alive? A virus contains thousands of atoms grouped together in molecules much like the molecules that we find in plants and animals. By themselves, viruses do not reproduce, because they lack the machinery through which more complex organisms make genetic copies. However, viruses that enter living "cells" can use the machinery that they find there to make more viruses like themselves. Taking a broad view, we might put viruses in the category of living things, but not everyone would agree with us. The virus question may seem a trivial problem of categories, but when we start to think about possible life forms outside the Earth, we shall see that defining what is and is not alive may prove a tricky business.

The Origins of Life

Terrestrial life forms began to appear more than three billion years ago, according to the fossil record, and life may have started on Earth as long as 4¼ billion years ago, almost immediately (just a few hundred million years) after the Earth formed. Thus life on Earth began sooner rather than later: Of the 4½-billion-year history of the Earth as a planet, more than two thirds, maybe even nine tenths, has included the history of life. In its early years the Earth was far different than the Earth now (even before humans started to work on it), and life itself has had much to do with changing our terrestrial environment.

When the Earth first condensed from the original cloud that formed the solar system, the atmosphere had a different mixture of atoms and molecules than it does now. The early atmosphere probably contained mostly molecules of hydrogen (two hydrogen atoms bound together), methane (four hydrogen atoms bound to one carbon atom), ammonia (three hydrogen atoms bound to one nitrogen atom), and water (two hydrogen atoms bound to one oxygen atom). That is, there was enough hydrogen available that almost all of the atmosphere's carbon atoms were in methane molecules, all the nitrogen atoms in ammonia molecules, and all the oxygen atoms in water molecules. This situation still exists in the atmospheres of the four giant outer planets.

On Earth the proximity of the sun made the difference; just as the sun supports life, so too the sun began the process that allowed life to develop. The early atmosphere of the Earth did not shield the surface against ultraviolet photons as our atmosphere does today. Our protection against these photons consists of the ozone (three oxygen atoms bound into molecules) and oxygen molecules (pairs of oxygen atoms) in the upper layers of the atmosphere. Ultraviolet photons from the sun constantly destroy some of these molecules, thereby becoming unable to penetrate deep into the atmosphere (Page 41). Other ozone and oxygen molecules constantly re-form at the rate needed to keep their relative abundance constant at the amount that has been established for billions of years. Three billion years ago, the Earth's atmosphere had almost no ozone or oxygen molecules, so ultraviolet photons penetrated the atmosphere much more easily. This penetration had important effects.

First of all, the ultraviolet photons helped to make "organic" molecules. The term "organic" gets thrown around rather loosely, and here we use it to describe the large, carbon-based molecules found in living matter on Earth. (One definition assigns the term "organic" to any molecule that contains at least one carbon atom.) The basic kinds of molecules in living matter contain mostly carbon and oxygen atoms, plus plenty of hydrogen atoms and other light atoms (nitrogen, phosphorus, sulfur), with carbon-oxygen double bonds a characteristic chemical feature. Carbon atoms like to participate in long chains of molecules, a property that makes them ideal for forming complex living systems, which consist of long molecular chains. The ability of carbon atoms to join in long molecular strings arises because carbon atoms have only four electrons in their outer "shell," which has room for eight electrons. Each carbon atom can join with as many as four other atoms in forming molecules.

Most organic matter today consists (by mass) of about 69% oxygen, 19% carbon, 10% hydrogen, and 2% other types of atoms, but these abundances were different in living systems when life began three billion years or more ago. Of the various molecules formed from this mixture, living organisms contain two especially important kinds: *enzymes* and *nucleic acids*. Enzymes are specialized molecules whose presence allows the basic chemical reactions in living organisms to proceed; that is, the enzymes "catalyze" their reactions. Enzymes are particular kinds of proteins, and protein molecules consist of simpler molecules called "amino acids" linked together in various ways. Each amino acid molecule contains ten to thirty atoms, each of which has the characteristic "amino" group (two hydrogen atoms bound to a nitrogen atom) along with the carbon and oxygen typical of almost every organic molecule.

Nucleic acids, which are more complex than enzymes, are the key to the reproductive process: They encode information to tell the next generation how to grow and how to function. The nucleic acid molecules called DNA (deoxyribonucleic acid), each of which contains millions of atoms, do this work for organisms on Earth. DNA molecules have the ability to duplicate themselves, because of their shape and structure. Each molecule of DNA has two spiral chains facing each other like intertwined spiral staircases (Figure 11–1). These chains depend on carbon atoms' ability to form long molecular strings, and they consist of carbon, nitrogen, oxygen, hydrogen, and phosphorus. Connecting the two chains in a DNA molecule are four kinds of molecules somewhat similar to amino acids. Each of these four kinds ("purines" and "pyrimidines") contains about fifteen hydrogen atoms and a

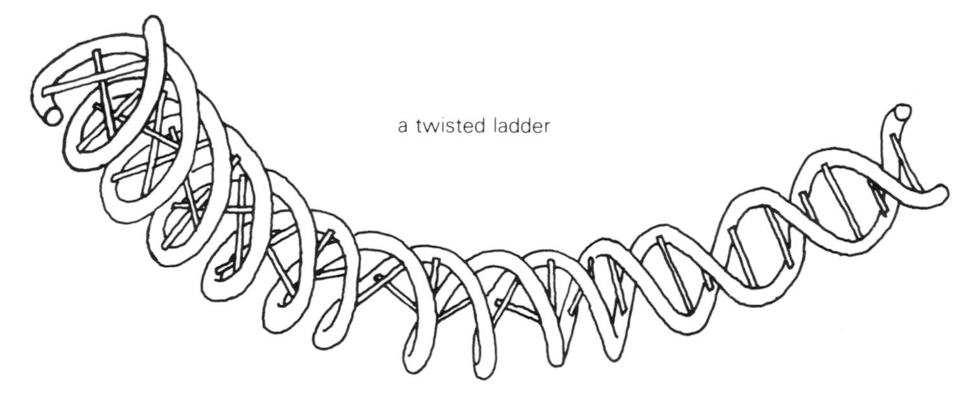

Figure 11–1. Each DNA molecule consists of two interwoven spiral chains of smaller molecules, connected by cross-links that consist of four different molecular varieties. Each DNA molecule contains about a million atoms.

dozen other atoms, mostly carbon, nitrogen, and oxygen. The cross-links between the two strands of a DNA molecule form from *pairs* of these molecules (Figure 11–2). But the crucial fact is that each of the four molecular types that form the cross-links can form a molecular bond with just one other type. This result comes simply from the space

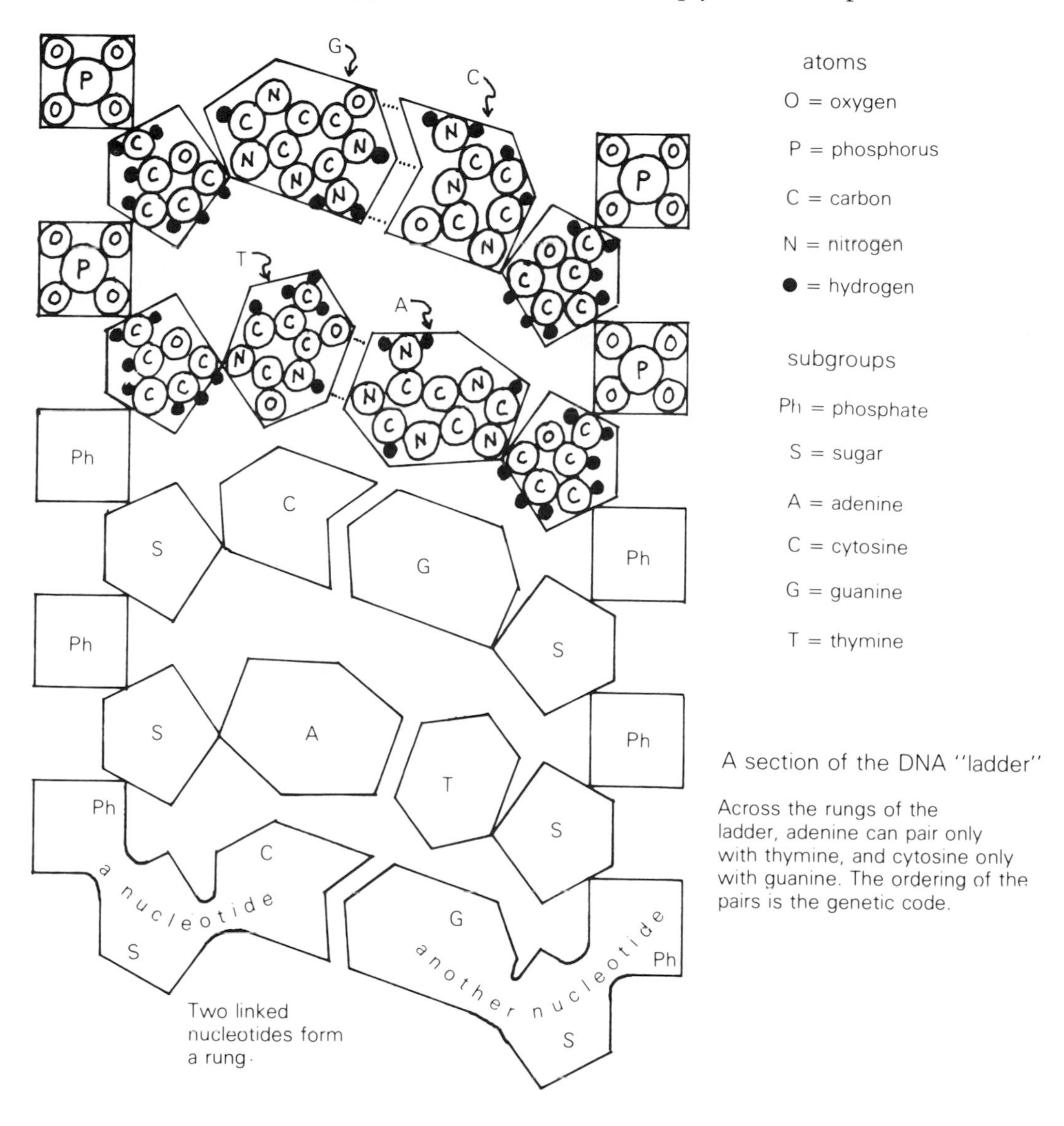

Figure 11–2. This drawing shows a section across the DNA spiral chains. Of the four types of molecules that form the cross-linked pairs in DNA molecules, each particular type can link with only one other type.

available between the double spirals, and the discovery of this fact was the big step in understanding how DNA works.

If a DNA molecule separates lengthwise down its middle, each half has the ability to form an exact duplicate of the original molecule. The single strand can do this because, if the strand acquires the new molecules to grow whole again, only the cross-linking molecular types that complement the original half can properly fill out the double-stranded DNA (Figure 11–3). These molecular cross-links carry the information that tells the next generation of organisms what to do. For example, twenty types of amino acids appear in living beings on Earth. Each successive triplet of cross-linked molecules encodes the message that a

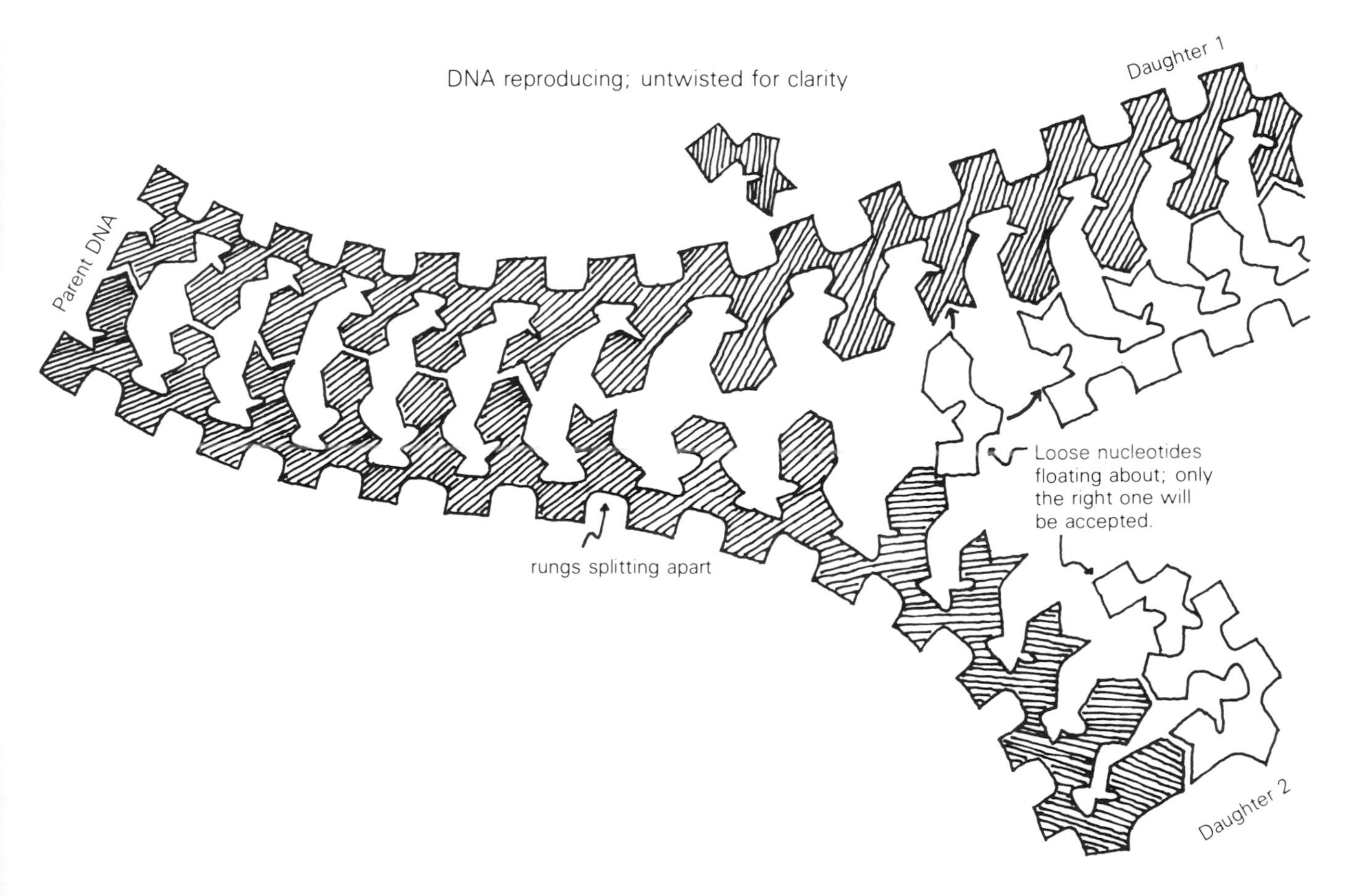

Figure 11–3. Because each cross-linking molecule can join with only one other type of molecule in a DNA strand, if we separate a DNA molecule into two halves and let the halves each reassemble the rest of the molecule, then *each* half will reduplicate the original molecule. The molecules that link the two strands are called "nucleotides."

particular one of these amino acids should be inserted into the new molecular chain. Reading along the strands of a DNA molecule, each successive triplet of cross-linked pairs specifies a particular amino acid; groups of amino acids make proteins, some of which are the enzymes needed to catalyze a specific reaction in the web of life. Other triplets specify when the reactions will occur, thus a single DNA molecule (which, however, may contain hundreds of thousands of cross-links) can prescribe the growth and function of an entire plant or animal.

Laboratory experiments have shown that if we take a mixture of hydrogen, methane, ammonia, and water vapor, and shine intense ultraviolet light on the mixture, or release electrical discharges through it, we find that after a while the simple molecules form many more complex molecules, including some of the amino acids *and* some of the four kinds of molecules that make the DNA cross-links. The amino acids that appear include not only the twenty kinds found in living organisms today, but other sorts as well that apparently have been excluded as unnecessary by evolutionary processes since life began. The results of these experiments go a long way toward indicating that amino acids appeared naturally in the early history of the Earth. Ultraviolet photons are not essential (although we think the sun provided plenty of those). We could have had lightning (electrical discharges) or even high-energy particles such as those in cosmic rays, because the experiments done with the same mixture but subjected to a bombardment of high-energy particles also yielded amino acids.

Today astronomers have detected ammonia, methane, hydrogen, and water vapor in dense interstellar clouds such as the one in the center of the Orion Nebula (Figure 11–4). Furthermore, astronomers have discovered more complex molecules, although as yet they have not found any amino acids inside these clouds. The basic mixture of gases found in the early atmosphere, which can produce amino acids after we add energy in various ways, seems to appear over and over again throughout our galaxy.

The problem in making first proteins and then DNA out of amino acids and other simple molecules comes from the fact that the same energy input that made these molecules can easily destroy them. Ultraviolet photons capable of breaking apart the four original types of molecules (methane, ammonia, water vapor, and hydrogen molecules) for their reassembly into larger molecules soon will break apart these larger molecules. Therefore, unless the molecules could find protection, they quickly would vanish from the early Earth. Without a layer of ozone and oxygen molecules in its atmosphere, the Earth billions of years ago offered protection from the sun's ultraviolet photons only in its oceans. Perhaps rain washed the molecules made near the Earth's surface into the protective bath of the seas, where only a few centimeters

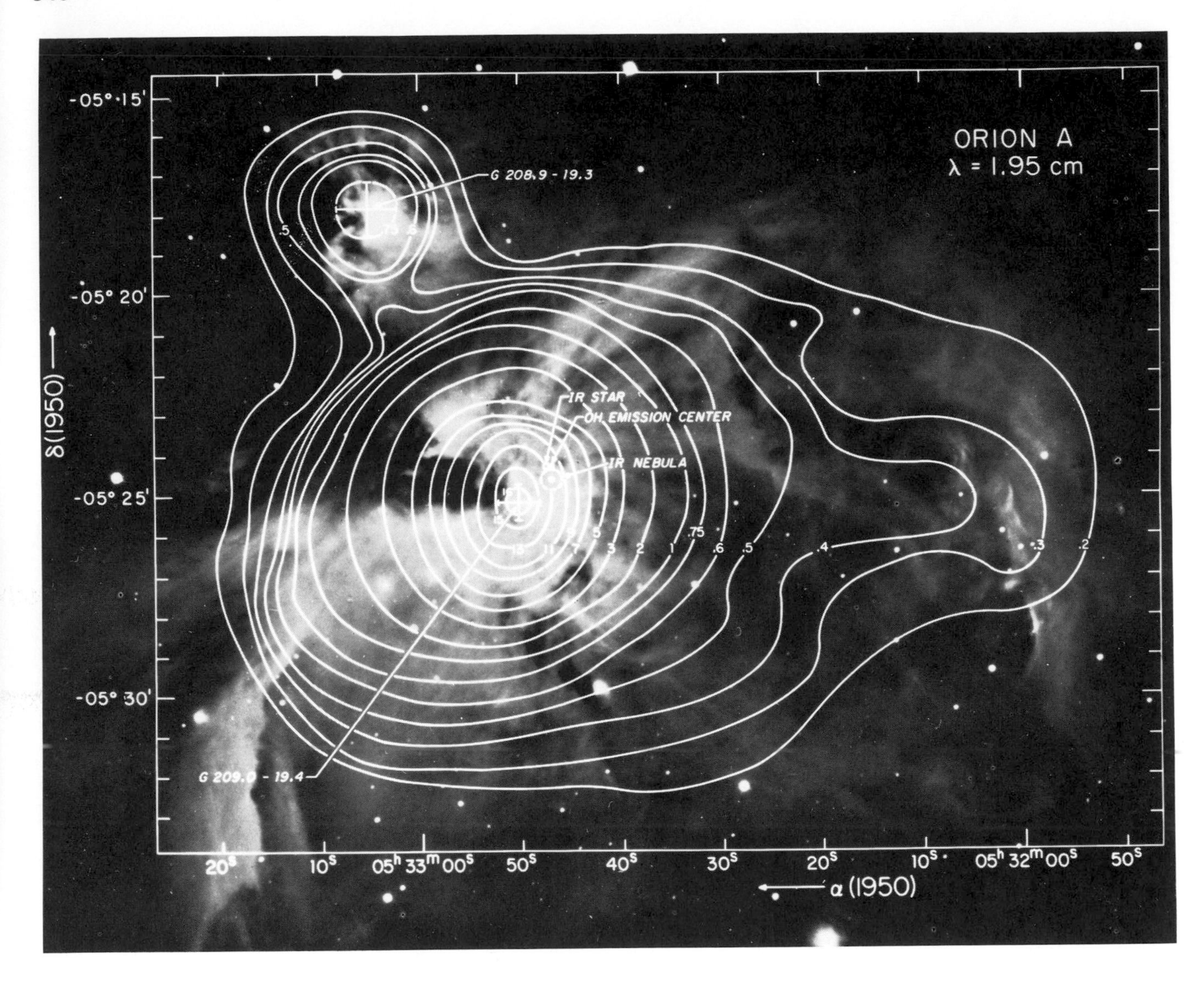

Figure 11–4. The Orion Nebula contains several strong sources of infrared and radio-wave photons. This picture of the Orion Nebula has superimposed on it a radio map of the intensity of photons with wavelengths of 1.95 centimeters. The contour lines that map the radio emission show that the peak number of photons come from a location near, but not precisely at, the source of the most intense infrared emission. By studying the photons of microwave and radio wavelengths from the Orion Nebula, astronomers have found large numbers of photons emitted or absorbed at the energies and wavelengths characteristic of ammonia, methane, hydroxyl (OH), water vapor, methyl alcohol, and other molecular species.

of water above give sufficient shielding to organic molecules from the life-making, yet life-destroying, ultraviolet photons.

So we can imagine a sort of primitive ferment of life in the oceans that led to the formation of the first cells capable of reproduction. The biggest mystery in the process now seems to be how organisms ever started to manufacture DNA molecules: Once you have DNA, the rest is duck soup. The increasing complexity of life has arisen from the occurrence of slight changes in already-existing species, coupled with the pressure from competing life forms. Slightly different organisms may do far better at reproduction, and the effects of this differential reproduction multiply. In the simple broth of the early oceans, organic molecules were present for the taking by any larger conglomeration that could use them. Such conglomerations could encounter new supplies of organic molecules purely at random. But as various new kinds of molecules formed through the impact of ultraviolet photons or cosmic rays, the first molecules that were capable of duplicating themselves presumably burst on the scene after enough molecular combinations had been formed. The advantages of being able to replicate must have been so great that these organisms with DNA, or something like it, soon became the dominant sort of molecular chains. Since then, evolution has led to further refinements of the replicating DNA molecules, which were the biggest breakthrough in life on Earth about three or four billion years ago. Organisms containing more than a single cell first appeared on Earth more than two billion years ago.

The evolutionary process in life relies on the existence of variation between individuals, part of which arises from spontaneous changes called "mutations" that appear from one generation to the next. We call these changes "spontaneous" to indicate that we don't know what causes them, perhaps cosmic rays (Page 245). Mutations allow for trials among alternative possibilities for an organism to deal with life's problems. Most mutations yield results that are unsuccessful from a reproductive point of view, and quickly disappear. Others give the organism a better chance for survival and reproduction, and if these changes can be passed on to the organism's descendants, they will tend to become more abundant because the organism is more likely to have plenty of descendants. Geographical isolation from other members of the same species can cause the variation within a species to produce a new species of organisms that no longer will mate with the original species members should they later have a chance to do so. Thus the mutations cause variation within a species, and the spread of organisms throughout the world leads to the diversity of species. Today many million species of plants and animals inhabit the Earth, and they owe their diversity originally to the mutations within a lesser number of species. If all mutations ceased tomorrow, we would realize this had happened from laboratory observations of bacteria, but millions of years could pass before we noticed the changes in the larger world around us from this lack, because the *effects* of mutations on many plant and animal species would take this long to appear.

Early organisms, tiny bacteria that began in the ocean and later spread to the air and to the land, developed when oxygen molecules were extremely rare. These organisms were "anaerobic," meaning "without oxygen," and in fact oxygen molecules would be deadly poison to anaerobic bacteria. As ultraviolet photons from the sun continued to bombard the Earth's atmosphere, they kept splitting water molecules into hydrogen and oxygen atoms. The hydrogen atoms, which are the lightest of all atoms, apparently have escaped from the Earth's gravitational forces during a period of the past few billion years. Oxygen atoms tend to pair into oxygen molecules, and they also can form a small amount of ozone molecules (three oxygen atoms) that, together with the oxygen molecules, shields the water against further destruction by ultraviolet photons. Water vapor diffused through the atmosphere helps to trap the infrared photons radiated from the Earth's surface, thereby providing most of the "greenhouse effect" that raises the surface temperature (Page 289). But the proportion of oxygen in the Earth's atmosphere made by the dissociation of water molecules would never have reached the 20% that we have today: Most of the oxygen that we breathe comes from photosynthesis in plants. The action of the ultraviolet photons billions of years ago sufficed to start the oxygen and ozone shield against further destruction of water molecules; it also broke down the ammonia into nitrogen and hydrogen atoms, and the methane into carbon and hydrogen atoms. The hydrogen tended to escape, while the nitrogen atoms linked into pairs (nitrogen molecules) that now constitute 78% of our atmosphere. The carbon atoms could join with oxygen atoms to form carbon dioxide, or they could form larger molecules that settled to the Earth's surface.

As some oxygen molecules entered the Earth's atmosphere—and later, as a great many more came from plants—living organisms had to adapt or die. Some anaerobic bacteria survive today, living far inside protective environments that shield them from oxygen. For example, our digestion employs anaerobic bacteria that live happily in our intestines. Most organisms, including everything that we call "higher" forms of animal life, are "aerobic" creatures that require oxygen for survival; that is, they adapted completely to the flood of oxygen that plants released into the atmosphere about two billion years ago.

Photosynthesis gave plants (and still does give them) the chance to use the photon energy from the sun directly instead of waiting for some organic molecules to float by. The process of photosynthesis takes carbon dioxide (CO_2) and water (H_2O) and some energy from visible-light photons to make sugars (types of carbohydrates, which contain groups of

CH_2O) and oxygen molecules (O_2). These carbohydrates supply the plant with material to grow larger, and their chemical bonds can release energy to the plant as the carbohydrates are assimilated. Some plants, such as the "snow plant" of the Sierra mountains, are incapable of photosynthesis and live on the decaying matter of other (dead) plants. Almost all of the plants familiar to us, however, use photosynthesis, thus bearing witness to the usefulness of the process, and it is the plants capable of photosynthesis that have conquered the Earth and changed its atmosphere. Two billion years ago, simple plants floating in the oceans began to release oxygen into the atmosphere (the oxygen diffused out of the surface layers of the ocean) as they developed photosynthesis. Geological testimony of this historical occurrence comes from the fact that rocks older than two billion years show deposits of iron compounds made without oxygen, whereas rocks made more recently show the characteristic red color of iron-oxygen combinations, such as the marvelous rock layers in the Grand Canyon. Having oxygen in the atmosphere is a tremendous thing for us humans, but most of it got there as the by-product of photosynthetic activity in plants.

Animals eat plants, or they eat other animals that eat plants, or so on down the "food chain." All animals therefore function as scavengers, letting plants do the basic work of making carbohydrates out of carbon dioxide, water, and sunlight. The oxygen that animals breathe can react with other molecules to "oxidize" or burn them, thereby releasing energy stored in the chemical bonds of the molecules eaten for food. The chemical combination of oxygen and carbohydrates liberates energy of motion stored in the chemical bonds, just as it took energy of motion (in photons) to make the carbohydrates in photosynthesis. The unwanted carbon dioxide released in the oxidation must be flushed out of the animal, and new oxygen molecules must be breathed in constantly to continue the oxidation. Many animals have well-developed circulatory and ventilation systems for both these functions. These animals strive to support a high metabolic rate (rate of functioning through life processes such as oxidation). In human beings, and to a lesser extent in other mammals, about one third of metabolic activity when at rest goes toward keeping the brain running, proof that thinking is indeed hard work.

Since animals are basically scavengers, they compete with one another for the fruits of photosynthesis, plants or plant-eating animals. This competition has bred specialized abilities in some animals to find food more efficiently than other animals. It is not so long since human beings struggled directly with other animal species, either to eat them or to avoid being eaten by them, or (as with hyenas, lions, hawks, snakes) sometimes merely to eat the same food.

What we consider to be the last big hurdle in the history of humanity, the development of human intelligence, came relatively late. Hard as it is to define life precisely, it is harder still to define what we mean by "intelligence." If we mean an ability to solve problems, then most vertebrate animals—fish, amphibians, reptiles, birds, and mammals, and even many invertebrates such as squid—show evidence of intelligence. If we use "intelligence" to denote the ability to speculate, to dream, to see beyond the present, then intelligent life began on Earth quite recently (some say not yet). Human beings became a separate species less than a few million years ago, and early humans did not possess much more imagination than the animals with whom they competed. But somehow our brains evolved into more than just an organ to interpret information from our sense organs and to run our bodily functions.

So here we are, alive by the billions, and by our own self-defined standards, intelligent. Perhaps self-awareness could be used to define intelligence—but is it necessary? At any rate, as humanity has grown progressively more aware of where we are, the urge has grown in certain quarters to see whether we can find other intelligent species to communicate with.

A number of questions arise at once. First, can we find other "intelligent" species? Second, should we find them? Third, who will do the talking?

The Search for Extraterrestrial Life

The first of these questions draws the most attention. The likelihood of finding other intelligent species depends on how closely distributed these species are. Most scientists today think that life forms should have developed naturally on many other planets around many other stars. This line of reasoning proceeds from our sun's "mediocrity" among the stars. The Earth and the other planets of the solar system apparently formed as a natural part of the protosun's contraction. The composition of the sun and the Earth includes a mixture of elements typical of material in younger stars, which we think all has been processed through supernova explosions that made the heavier elements. The Earth's gravity has proven insufficient to hold most of the original hydrogen and helium, but otherwise our mixture of elements seems "normal." If many of the hundred billion stars in the Milky Way galaxy have planets, say five or ten per star, then if even one planet in a thousand has the right conditions for life to develop, there should be a billion life-sustaining environments in our galaxy alone.

What are the right conditions? Our ignorance is deep but not infinite. We have established some *sufficient* conditions by experiment: Take water, ammonia, methane, and hydrogen, add photon energy or electrical discharges, and you get amino acids, the basic units that make proteins. But what conditions are definitely *necessary* to make living organisms? Although we should remember that we might yet find life forms stranger than our wildest dreams, still the conclusions we have today seem reasonable. To form anything like our kind of life, you need the right temperatures, the right "solvent" (water), and the right chain-making atom (carbon).

Number one, the right temperatures. Atoms can link together to form molecules, but at high temperatures the molecules will destroy each other by collisions. At low temperatures, all the atoms and molecules move about so little that they are unlikely to bind together into complex structures. Although we shall consider later the far-out possibility of life forms made only of giant nuclei (Chapter 12), life made of ordinary molecules seems likely to arise only at temperatures near 300 degrees absolute ("room temperature"). The possible range perhaps extends from 200 degrees to 370 degrees absolute, but does not include 100 degrees or 1000 degrees. Thus planetary surfaces that are located the "proper" distances from their stars to have temperatures of a few hundred degrees absolute appear to be natural sites for life to develop.

Number two, the right "solvent," water. We all have heard that our bodies are about 65% water,[1] and we carry this reservoir around for a reason. In a typical cell in our bodies, the complex molecules (such as the DNA in our chromosomes) carry the "information." Thus, for example, our chromosomes determine how the next generation will look—a lot like ourselves. The larger molecules move in a bath made primarily of water molecules, the heritage of the original environment of all living things, the oceans. Water provides the "solution" or bath in which chemical reactions can occur easily. The water molecules, which are constantly in motion, buffet the other molecules gently (the larger these other molecules are, the less they feel the impact of the water molecules), and thus the different molecules in a cell stand a chance of eventually meeting one another. The chemical reactions that occur among the larger molecules, such as the formation of new protein molecules (or cellulose in plants), comprise the "life processes" that we rely on, and they occur in a watery solution that represents the best available medium.

Why is water so wonderful? The answer lies in an organism's need to *regulate* its "metabolism," or life functioning. Because temperatures vary in time on almost any conceivable planet, a living organism must survive temperature changes. (The exceptions are found deep in the seas.) Most kinds of life on Earth spend their efforts in obtaining the raw materials for further growth and for reproduction, and in keeping up with the changes in temperature. Plants and animals use the water inside them to cool off. When the temperature increases, they

---

[1]Some jellyfish are 95% water, some plants only 40% water.

"transpire" or breathe out water vapor, as we know from our experiences on a hot day. Water has a marvelously *high* resistance to being vaporized: It takes 540 times more energy to vaporize a gram of water than to heat it by one degree absolute. Thus water molecules provide an efficient way to be prepared for a change in temperature; they are a cushion against increasing temperatures as well as the solvent for chemical reactions. No organism can "live" without water, although some can dry out for years and revive when wet again.

Ammonia is water's closest competitor as a means of temperature regulation among common molecules, but ammonia is only about half as efficient as water, since it takes only 300 times more energy to vaporize a gram of ammonia than to heat it by one degree absolute. Since ammonia has about the same temperature of vaporization and of freezing that water does, ammonia could suffice as a solvent, although not as well as water does. In addition, water has the almost unique property of expanding, not contracting, as it freezes. Thus water ice floats in liquid water, and soon will remelt into liquid, instead of sinking to the ocean bottoms where it might accumulate until the seas were frozen solid. Ammonia oceans would be in more danger of freezing completely during prolonged cold periods.

Thus water, which is quite common on Earth, has a definite value to us. Our atmosphere protects water molecules from destruction by ultraviolet photons with the same ozone layer that protects the larger, carbon-chain molecules in our bodies. Water droplets high in our atmosphere find additional protection by forming ice crystals that give important shielding to the molecules inside them. Planets without protective layers of ozone and oxygen molecules are not likely to have large amounts of water on their surfaces, and without water such planets are less likely to have life.

Number three, the third requirement for life is the right chain-forming molecule, carbon. A carbon atom can combine with as many as four other atoms at one time, something that neither hydrogen, nitrogen, oxygen, nor helium atoms can do. Carbon's ability to combine with so many other atoms allows it to form the backbone of the long-chain molecules in living cells. These molecules can encode vast amounts of information to be passed on to the next generation by molecular self-replication.

Sometimes scientists speculate that silicon atoms, which also can form bonds with four other atoms at once, could serve as well as carbon atoms, and thus scientists imagine silicon-based "organic" material. However, silicon atoms cannot match carbon atoms in all respects. Silicon atoms hold on to other atoms very weakly, compared to carbon-carbon and carbon-hydrogen bonds. Thus large molecules containing silicon tend to fall apart quite readily. Also, silicon atoms are far less abundant than carbon atoms. In particular, oxygen atoms are far more numerous than silicon atoms (about ten times more abundant in the sun and other stars), whereas carbon atoms are not so rare in comparison to oxygen atoms (oxygen is only three times more abundant than carbon in stars). Silicon atoms tend to link together with oxygen atoms to form networks of silicates such as quartz (silicon dioxide). The silicon-oxygen bond is the strongest bond that silicon atoms can form. We in fact are living on the result of this bond: Most of the Earth's crust consists of silicon-based rocks, seen in their simplest form as quartz crystals. *All* of the silicon has gone into forming such oxygen-linked structures, because oxygen atoms outnumber silicon atoms by ten to one. Sometimes we find calcium atoms combining with silicon and oxygen to form calcium silicate ($CaSiO_3$). Calcium silicate combines easily with carbon dioxide to form calcium carbonate ($CaCO_3$) and more silicon dioxide (SiO_2). This process tends to remove carbon dioxide from the atmosphere and produce calcium carbonate rocks such as limestone. Seashells also are calcium carbonates made by animals from the carbon dioxide dissolved in sea water. All of the processes mentioned above lock up silicon atoms on Earth in silicon compounds that do not bind easily to other molecules. Such "inert" compounds are unlikely to form long chains such as those in carbon-based organic molecules.

This brief, one-two-three summary cannot cover all the possibilities for forming complex molecules, but we can report that many indications point to life being carbon-based, at the proper temperatures (near 300 degrees absolute), and with water used as a solvent. An additional piece of evidence for this conclusion comes from a meteorite that fell near Murchison, Australia, in 1969. The material in this meteorite presumably had been orbiting around the solar system, exposed to the sun's ultraviolet photons, for billions of years. As part of the meteorite, astronomers discovered various amino acids, the relatively large carbon-based molecules that form the proteins we find in animal life on Earth. However, these meteoritic amino acids did not include just the specialized subgroup found in living beings, but embraced a wider range of amino acids, some of which have been left out of terrestrial organisms.[2] The discovery of amino acids in the Murchison meteorite serves to reaffirm that the normal mix of elements in the solar system (or in other stars) will yield amino acids under photon bombardment. Amino acids,

[2]When we reflect that evolution has tried and rejected all sorts—almost every sort, probably—of molecular combinations made of common atoms, we can see why *any* nonorganic substance is likely to be toxic to us. If molecules helped organisms function, then they probably were included; if not, they have been deliberately excluded by evolutionary pressures. Therefore any "new" (that is, previously rejected) pollutant already has proven harmful once before—before we began to reintroduce it through human activity.

of course, are not alive; even a protein molecule made of a thousand different atoms is not alive. But the first step toward life does seem to follow naturally near the sun, and this first step may well be the step that takes the longest time.

In regions of relatively low density, such as interstellar clouds, even amino acids may never form because it takes too long for the smaller molecules to encounter one another. Thus, although some interstellar clouds have the right temperatures to make complex molecules, only in the densest clouds could the molecules even begin to form in a few billion years.

When we examine the other planets in our own solar system, we see that they lack one or more of the keys to success in life. All of the planets have a good supply of carbon atoms, but they have the wrong temperatures and little water. Only on the surface of Mars (barely) and perhaps deep in the atmospheres of the giant planets (especially Jupiter) will we find the proper temperature range. Venus and Mercury are far too hot for any large molecules to exist on their surfaces, although high in Venus' atmosphere we find temperatures near 300 degrees. Furthermore, the scarcity of water on Mars, Mercury, and Venus seems to rule out any advanced life forms that need a solvent for temperature regulation. The giant planets have lots of ammonia, so one might speculate that far beneath their cloudtops, ammonia-solvent life forms exist at temperatures near 200 degrees absolute.

Pleasant as speculation may be, proof is much better and more expensive. American scientists, engineers, and taxpayers have sent the Viking spacecraft to land on Mars, scoop up a sample of surface, and test it to see what, if any, sort of organic molecules the surface contains. Viking should provide the first direct test for life on another world, and even negative results from this search will tell us more about the conditions needed for life to develop.

When we look at the host of stars around us, our speculation flows more freely. Maybe among these stars, on the average, each planetary system in existence has only one planet with the right temperature. Maybe only one planet in a thousand has protected enough water molecules, should they indeed be essential for life. But with so many planets, we easily can imagine that all sorts of life forms have developed. Since what we call "intelligence" arose on Earth from the pressures of evolutionary selection, we can expect to find it elsewhere. Naturally, other life forms hardly would be just like ourselves. Still, we expect that the basic process of photosynthesis might have developed on many other planets. If there are "animals" that eat the results of photosynthesis, they probably will have some locomotive system and some "eyes" in front to use their star's light to see where the plants are, or where their enemies are. On a smaller, low surface-gravity planet, birds might be the dominant life form. On Earth, birds barely can fly (aerodynamically speaking) and must sacrifice size and strength to take to the air. In contrast, on a high surface-gravity planet there might be no birds, nor any animals as large as elephants and rhinoceroses.

The number of intelligent planetary civilizations in our galaxy depends on the number of planets with life, on the fraction of these that develop intelligent beings, and on the average length of time that intelligent civilizations last before destroying themselves.[3] The life span of an intelligent civilization is hard to estimate, as are the other probabilities in this calculation. A poll of speculators to predict the future life span of our own civilization might produce answers ranging from five to five billion years. We humans have been self-conscious for about 50,000 years, say one hundred thousandth of the Earth's history. If we last another 50,000 years, we shall span two hundred thousandths of this history. The life span of an average intelligent civilization turns out to be the chief determining factor in estimating the number of other intelligent civilizations in our galaxy. When we multiply the different factors together, we find that the number of other intelligent civilizations in the Milky Way roughly equals this average life span, measured in years. Thus if we are a "typical" intelligent species and last for a million years, there should be a million (to within a factor of ten or a hundred) other civilizations in our galaxy at any moment.

Communication With Other Civilizations

Intelligence is one thing, communication is another. Until recently we lacked the ability to send messages far from Earth, supposing that we had the imagination and the desire to do so, and we could barely have distinguished between an extraterrestrial visitor and a conquistador from overseas. Only during the past fifty years have humans created the technological capacity to communicate effectively with other civilizations in our galaxy. This onrush of technological achievement covers less than a thousandth of the span of human civilization, and just over one hundred millionth of the age of the Earth. If we think about what might happen to another civilization on another planet, it seems quite probable that any developing civilization will pass from no communications ability to great ability within a tiny fraction of the planet's age. Once discoveries start, they spark more discoveries, and since the basic principles of electrons and photons are the same everywhere, most civilizations should develop radio and television. The advantage of using the photons with radio and television wavelength for communication

[3]We are not pausing here to consider life forms that may be called intelligent though not technological (e.g., whales and dolphins). Our limited imaginations push us toward the conclusion that technologically oriented civilizations will be those with which we might establish communications.

purposes resides in the photons' ability to penetrate clouds and walls, which are too thin to stop long-wavelength photons (Page 18). Radio waves of certain wavelengths also can bounce off the Earth's ionospheric layer, so the ionosphere can reflect them part way round the world.[4] Higher-frequency (shorter wavelength) television and radio waves do not reflect, so we need a direct "line of sight" to the transmitter to receive them.

If we seek to communicate with other civilizations, radio photons offer the same superiority over visible or infrared photons for this purpose as they do for terrestrial communications: They penetrate our atmosphere, and they also should penetrate most other planetary atmospheres. Furthermore, radio and television messages are inexpensive, despite the high cost of Western Union. The Mariner space probes to Mars and Venus, hundreds of millions of kilometers away, sent back thousands of pictures (by radio encoding) using the same amount of electrical power as a household light bulb. Of course, other stars are much farther away, but the cost of sending them a short television program would not exceed the price of a new automobile. If we know where to point our radio message, we could use the Arecibo radio telescope (Figure 11–5) as a transmitter to focus the beam. This type of telescope also makes the best available detector of radio signals, because it gathers radio waves over a full 300-meter diameter. If a civilization *anywhere else in our galaxy* had an instrument equal to the Arecibo telescope, we could detect its messages and they could detect ours, *provided that each of us knew where and when* to point the telescope. (Of course, the photons might take 10,000 years or more to travel in each direction, but we've got time.)

Some radio-astronomy observatories have spent some time "listening" to likely regions of the sky, but no sounds of other civilizations have been found during these brief experiments. Such searches are difficult because we don't know exactly where to point the telescope, nor do we know at what frequency we should be listening. The situation is like trying to tune a television far out in the country by turning the antenna and changing the channels simultaneously, except that there are millions of possible channels and there are also millions of directions, because the antenna beam of a radio telescope can be pointed with great accuracy.

Nonetheless, astronomers are ready to try. First of all, we consider the frequency problem. If we ask ourselves what frequency another civ-

---

[4]The interference by planetary ionospheres, plus a general increase in background radio emission around the galaxy at longer wavelengths, makes 21 centimeters about as long a wavelength as we can use for attempts at interstellar communication.

Figure 11–5. The radio antenna at Arecibo, Puerto Rico, which is three hundred meters across, is the largest radio telescope in existence today. Thousands of mesh panels, supported by a network of cables below them, form a reflecting surface whose position can be held to within a few centimeters at all times. (Notice the special shoes worn by the workmen to avoid damage to the reflecting panels.) The telescope can be used either to receive radio signals—by focusing them to the detectors two hundred meters above the dish—or to send them. To use the antenna as a pinpoint transmitter, we emit radio waves from the focus (high above the dish) and bounce them off the reflecting surface to form an almost perfectly aligned beam of photons.

ilization might broadcast on, our minds turn to the most abundant element in the universe, hydrogen, which constantly produces radio-wave photons at a frequency of 1420 MHz (Page 36). Any other civilization would know about these radio waves, and if they were seriously broadcasting to other civilizations, they probably would transmit signals at a frequency near 1420 MHz.[5] To speculate still further, if another civilization had water as the basic solvent for its organisms, they well might choose to broadcast at a frequency between the basic hydrogen frequency, 1420 MHz, and the frequencies near 1700 MHz at which OH

[5]No one would broadcast at *precisely* 1420 MHz (21 centimeters) because of the huge amount of natural emission of radio photons by hydrogen atoms at this frequency and wavelength.

molecules emit radio photons, since H plus OH makes H_2O, water. Bernard Oliver, a man who has thought much about these matters, therefore called the frequency band between 1420 MHz and 1700 MHz the "water hole" and, in his ear-catching way, has suggested that galactic civilizations meet at the water hole.

It is interesting to speculate that if our Earth has an "average" civilization, many other civilizations must be much older (and many others much younger, still in the precommunications stage). If our reasoning is correct, and it's not hard to establish contact among civilizations provided that all parties are willing, then already many civilizations have been communicating for a long time. If other planets reached communications ability in 3½ billion years instead of our 4½ billion, they may have been talking for a billion years now. Since photons take fifty thousand years to cross half the Milky Way, even widely separated civilizations could have exchanged thousands of messages in a billion years. We end up thinking that the older civilizations that cared to get in touch with one another may well have established a complete galactic two-way television system. As other civilizations reach the stage of communications ability, the ability to receive pictures and to send them, they may be encouraged to add their own two photons' worth, or perhaps they face a rigorous screening process before receiving a broadcast license—or even a work permit—from some galactic regulatory agency.

All this speculation may seem far-fetched as well as unimaginative, since we have assumed that other civilizations must "think" as we do. However, some of these conclusions follow from logic and science. If certain civilizations don't care to build antennas and transmitters, they simply won't enter the communications arena. But if they do try to communicate, the rules of science make television the natural way to proceed. Television is cheap (witness its success on Earth), and television pictures are a straightforward means of sending information. From simple pictures accompanied by markings, one quickly can learn a written "language" where symbols are shorthand ways to express pictures of increasing complexity. With enough pictures, language becomes less important for understanding another civilization. The need for economy in sending messages arises from the fact that planets' surfaces are finite, and even a star's entire resources are finite. Why waste effort doing things the hard way? This rule must prevail on many other planets.

Our first effort in the field of galactic communications should be to *listen*. We might think that an ironic situation would arise if everyone listens and no one transmits. However, we in fact have been transmitting messages—the radio and television photons used for global communication—since 1900 or so. Some of these radio and television photons constantly leak out into space, and travel at the speed of light in all directions.[6] The wave of photons announcing that humans have developed radio now forms a sphere *seventy light years* in radius around the Earth and the sun. The stray television photons form another sphere forty light years in radius. Both of these spheres are expanding constantly at the speed of light, regardless of what we do on Earth. The sphere of leaked television photons now includes the hundred nearest stars; soon it will blanket a thousand, and in a hundred thousand years it will contain the entire galaxy. Of course, these signals grow fainter and fainter as they embrace larger and larger volumes of space, and could never compare in strength to a deliberately aimed signal from, say, the Arecibo telescope. Yet even these weak, leaked photons could be detected by a technologically advanced civilization, and the time has come to ask whether we should try to eavesdrop on the stray signals from *other* civilizations.

According to knowledgeable scientists and engineers, humanity now could build an array of giant radio antennas that could tune across a large band of frequencies to detect the stray radio and television noise from a civilization such as our own, anywhere among the million nearest stars. The construction of such an array of antennas has been proposed by the imaginative physicist Bernard Oliver, who calls the array "Project Cyclops" after the one-eyed giant in Homer's *Odyssey*. To build the antennas and connecting cables for Project Cyclops would cost about ten billion dollars, the price of a small war. Project Cyclops would consist of a thousand antennas, each a hundred meters in diameter, set in a circular pattern ten kilometers across (Figure 11–6). The antennas all would have to point in precisely the same direction and listen at the same frequencies at any one time, but this can be achieved with our present capabilities. The key question is whether human beings want such a system enough to spend about three dollars per person (fifty dollars per person if the United States builds it alone) for its construction.

Project Cyclops represents a far more efficient way to seek other civilizations than making attempts to visit them in person. Just to land a man on the moon cost more than Project Cyclops, and the star nearest the sun is a hundred million times farther from us than the moon. To travel in a rocket takes huge amounts of energy, since we constantly must push material out behind to make the rocket accelerate. In contrast, photons travel for free, and at the speed of light, once they are produced. No civilization ever will be totally rich, and the successful

[6]Notice that if we used only cable television and radio, and turned off all radar antennas, we would reduce our output of stray photons by a factor of several hundred. It may be that not every civilization of "advanced" capabilities still leaks as many radio-wave photons as we do.

Figure 11–6. An artist's conception of how Project Cyclops would look from the air (left) and from the ground (right). Each of the thousand antennas would be a hundred meters across, which is one third the diameter of the Arecibo telescope. If we could arrange for all the antennas to point to the same location in space, and could interpret all the information they received simultaneously, we would have a system capable of detecting a civilization such as our own (in radio power) anywhere within three hundred parsecs, *without* having them send a message directed at ourselves. About a million stars lie within three hundred parsecs of the sun.

ones probably learn how to limit their instinctive desires to waste natural resources. If we can send a message to the galactic center for a thousand dollars, but it costs a trillion dollars to go there (and at a speed of even 30,000 kilometers per second, this would take ten times longer than photons do), which course of action seems preferable? And which would another civilization choose? Furthermore, if we could eavesdrop on another civilization's radio and television, we could decide whether we want to try to contact it directly—a key political decision that our present leaders have little training in resolving. Similarly, any other civilization listening to our television could form its own opinion of civilization on Earth. This could explain why we have received no direct messages even if a galactic network of civilizations already exists.

Despite the fact that space travel costs enormously more than photon-based communication systems—and takes longer, too—our apparently innate desires to believe that the galaxy loves us have led to a multitude of reports claiming that extraterrestrial visitors have been seen. This book is not the place to review the "UFO question,"[7] but we can note

[7]For a good review of this question, see *UFO's: A Scientific Debate*, edited by Carl Sagan and Thornton Page (Cornell University Press, Ithaca, N.Y., 1969).

that *if* extraterrestrial civilizations are choosing to visit us, for whatever reasons, they seem to be trying to remain undetected, since reports of their appearance always involve only a few people. Any visitors from space could, if they chose, always appear on national television or at sports events (which they might take to be our terrestrial religion). If, in fact, a more advanced civilization visits us and chooses to remain undetected, its superior technology almost surely will allow it to do so. Then the question of other civilizations returns to its original form of how to establish contact with the other civilizations that, on a probability basis, almost surely exist throughout our galaxy. The reports of contact with other-worldly creatures touch our continuing, intuitive feeling that *we* are the center of the universe. The common term "spaceman" shows how we project our interpretation of reality onto the question of life in the galaxy. How much easier to have other civilizations visit us than to have to seek them out—perhaps even to prove that we are worthy of attention! Many a science-fiction novel or movie turns on the opposite idea, that something about *us*—our good looks, our natural resources, our bad karma—is drawing the attention of other civilizations to planet Earth. Much likelier, however, would be the tendency to ignore our solar system as long as it poses no direct threat to the billions of other stellar systems in the galaxy.

Speculations that run opposite to these views include the idea that life on Earth itself originated not in the "natural" way described earlier in this chapter, but rather as the result of a visit billions of years ago by space creatures who left behind simple living creatures or just organic molecules. This "garbage" theory of the origin of terrestrial life—sometimes modified into the "zoo" theory that we are but a scientific experiment of some extraterrestrial Frankensteins—still assigns the Earth a special role as worthy of the expense, for one reason or another, of the long journey here. (It also begs the question of where the "real" form of life originated, and how.) The ancient Sumerians had a legend that Oannes, a fish with a man's head and feet, taught people writing, science, architecture, and agriculture before disappearing into the sea. Such legends attempt to encapsulate the long history of human progress in a coherent way—but who knows for sure? Perhaps Oannes came to Sumeria, and perhaps our zoo keepers soon will appear to move in a fresh babel of creatures. The logic of economy suggests that these possibilities are most unlikely, but if Oannes returns, I for one will not insist on logic.

As the result of our technological advances, we live in a generation capable of communicating with other civilizations in our galaxy. Among a hundred thousand human generations, ours may be the first to experience such communication. Surely this event, even a television picture from a thousand light years away, would have an enormous impact on the way humans think of themselves and of our parent Earth. Let us hope that this impact—whenever it arrives—will be one of increased respect for the universe that brought us here, rather than the fear, divisiveness, and resentment that humans often have shown in times of stress. Yet we have survived, and may yet communicate.

Summary

Life, defined as a system capable of reproducing itself at a small expenditure in energy, appears to have arisen on Earth several billion years ago as a natural consequence of the interactions among atoms and molecules. However, as far as we can tell now, certain conditions must be met for living organisms to occur in this manner; namely, the right temperature, the right "solvent," and the right atom, carbon, to form the molecular bonds. Although we can argue that these conditions may be too restrictive, if they are generally valid, then the most likely sites for life should be the surfaces of planets, the thick atmospheres of planets like Jupiter, and the interiors of dense interstellar clouds, provided that the proper temperatures and materials exist at these locations.

The chief types of molecules employed in the functioning of living systems on Earth are "enzymes" (made of special kinds of protein molecules) and "nucleic acids," both of which are made up of simple atoms that exist throughout the universe. It is hard to estimate the chances of such atoms forming these more complex molecules elsewhere, but if we consider ourselves to inhabit an "average" planet, then we easily can believe that many other sites in our galaxy and outside it must support living systems, some of which should be enmeshed in civilizations at least as highly organized as our own. Many of these civilizations already may have established mutual intercommunication, and whether membership in this network will come easily or only with difficulty for human beings stands as a key question, perhaps to be answered during our lives.

Questions

1. Are you satisfied with the definition of life given on Page 334? Can you propose a more satisfactory definition?
2. Should we limit the definition of life to something that can produce near-identical copies of itself? How many intermediate stages might we allow? (Remember the old saying: "To an egg, a hen is merely a way to make another egg.")
3. In what ways was the Earth different at the time when life appeared (three or four billion years ago) from the way it is now?

4. How does a DNA molecule encode the directions for forming amino acids? How does the molecule transmit this coding to other DNA molecules?
5. Why are mutations important in the process of evolution?
6. Which stars are most likely to have some form of life on planets around them? What conditions on these planets appear to be necessary for life to emerge? Do you think that these conditions are indeed necessary ones?
7. Where on Jupiter might we expect to find some form of life? Why?
8. Why should we expect that other civilizations are trying to communicate with one another?
9. If we represent the history of the Earth since life began—say 3.2 billion years—by a single year (32 million seconds), what amount of time in the year is occupied by the time since humans developed the effective use of radio (about 64 years ago)?

Further Reading

I. Asimov, *The Genetic Code* (Signet Science Library, New York, 1962).

G. Hardin, *Exploring New Ethics for Survival: The Voyage of the Spaceship Beagle* (Penguin Books, Baltimore, Md., 1973).

C. Sagan and J. Shklovsky, *Intelligent Life in the Universe* (Dell Books, New York, 1966).

R. Bracewell, *The Galactic Club* (W. H. Freeman and Co., San Francisco, 1974).

C. Sagan (Editor), *Communication with Extraterrestrial Intelligence* (CETI) (Massachusetts Institute of Technology Press, Cambridge, Mass., 1973).

Chapter **12**

Black Holes, Pulsars, and the Future of the Universe

"But thought's the slave of life, and life time's fool,
And time, that takes survey of all the world,
Must have a stop."

William Shakespeare, *King Henry IV*

Black holes are the point of no return in the universe, the last sink of stellar failure. Even though we have no clear proof that black holes exist, we may yet discover that *most* of the matter in the universe has fallen into black holes. A "black hole," as astronomers use the word, is an object whose gravitational forces completely prevent any light or any other kind of particle or radiation from leaving the object. A black hole therefore vanishes, leaving behind only its gravity to interact with the rest of the universe. Since no light, no radio waves, no particles can escape from a black hole, we never can have any two-way communication with such an object. Our knowledge of black holes therefore must remain circumstantial, based on the physics that we know and on the few astronomical observations that relate to the possible existence of black holes.

What does it mean to say that gravitational forces "prevent" any light or other radiation from leaving a black hole? The resolution of this question lies in gravity's ability to attract *everything*, including photons. True enough, photons have no mass, so that the usual way of expressing the law of gravity would imply that they do not feel gravitational forces. For particles *with mass*, the gravitational force between any two particles varies as the product of the two particles' masses divided by the square of the distance between the particles' centers. But particles with no mass still obey the gravitational attraction from a massive object such as a star. Photons, the most important massless particles, actually will deviate from straight-line trajectories as they pass close to the sun (Figure 12–1). The sun's gravitational force attracts the photons and causes the light rays to bend toward the sun. This prediction can be verified during a total solar eclipse, by photographing the positions of stars that appear close to the sun while it is eclipsed. If we photograph the same stars six months later, when the Earth's motion around the sun has caused the sun to move to another part of our sky, we can measure the change in the stars' apparent positions produced by the sun's bending of their light rays from the stars behind it. Astronomers first made these observations in 1919, to determine whether Einstein's prediction of the amount of this bending was correct. These observations, and

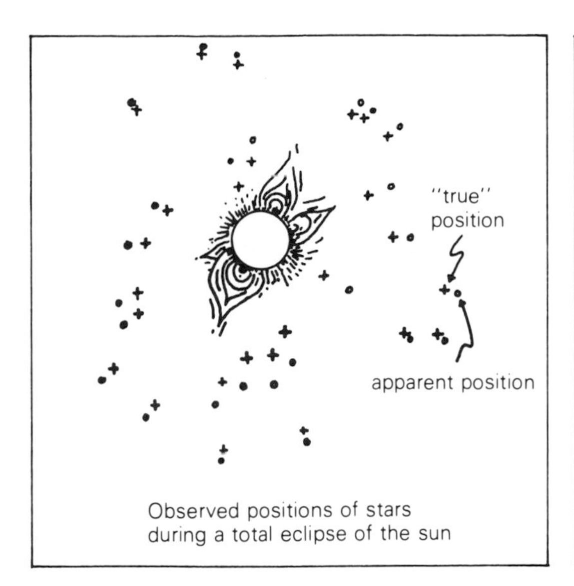

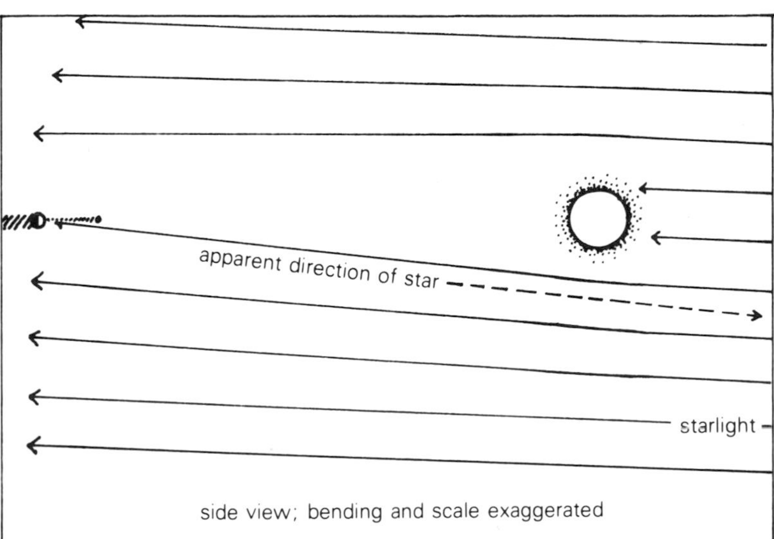

Figure 12–1. We can verify that the sun bends light rays toward it by photographing the stars that appear close to the sun during a total solar eclipse. Six months or so later, the Earth's motion will cause the sun to appear in a completely different part of the sky, and we can rephotograph the same stars to see if their positions appear to be different when the sun's presence does not affect the tendency of photons to move along straight lines.

many since then (more recently, made by using radio photons produced by distant radio sources) have upheld Einstein's calculations of the amount that photons will deviate from straight lines as they pass close by massive objects.

Because everything feels the effects of gravitational forces, sometimes physicists find it more convenient to talk in terms of *space* being curved in the presence of massive objects. Using this approach, we can say that light rays passing close by a massive object such as the sun will continue to travel in straight lines, but the sun bends space itself, so that the straight line looks curved to us (Figure 12–2). In this view, gravitational forces are equivalent to low spots in curved space, toward which objects tend to fall as the natural result of the curvature. Since the gravitational effect remains the same, we can use either the concept of the sun bending light rays out of straight lines, or of the sun bending all of space so that "straight" lines are bent. Calculations based on either concept give the same result.

The special property of gravitational forces that lets us think of them as bending space is that *gravitational forces always attract.* If we

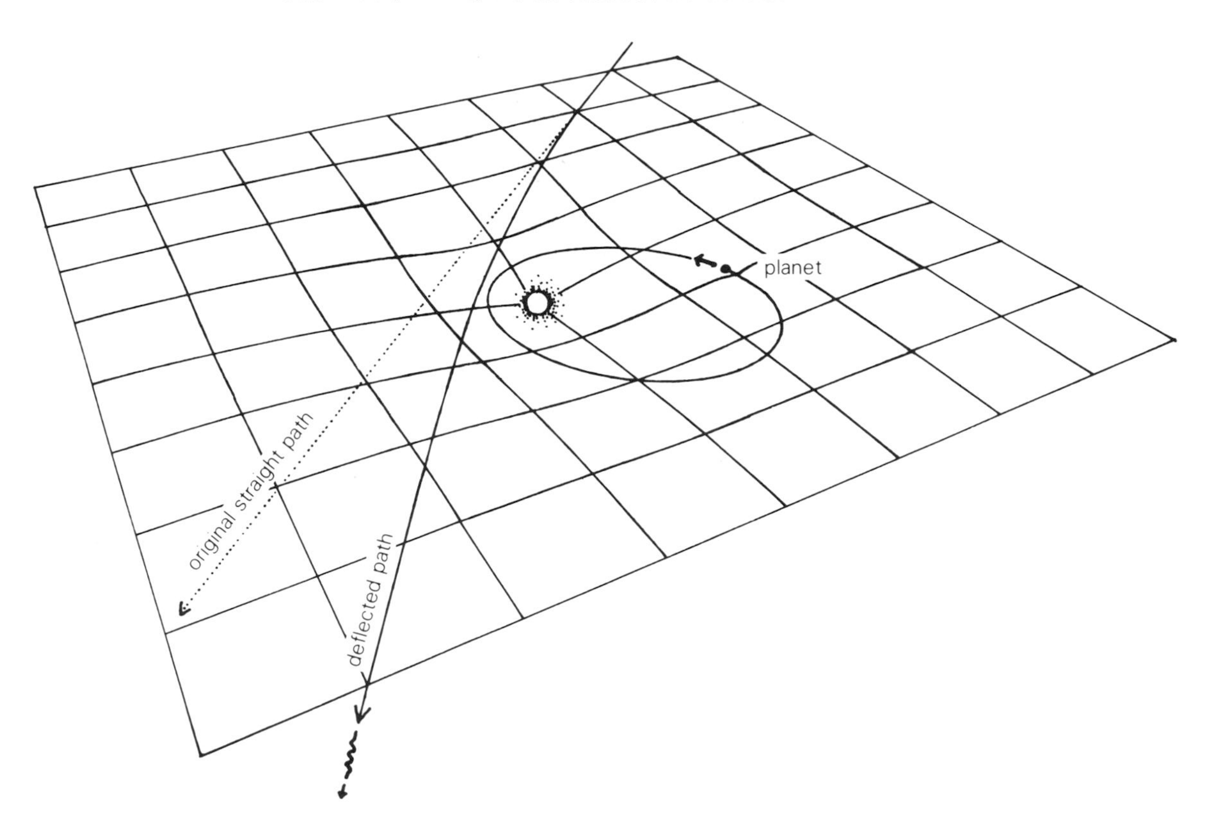

Figure 12–2. Physicists sometimes find it convenient to picture massive bodies as affecting the curvature of space, so space becomes bent in the presence of stars. The paths of light rays and of massive objects (such as planets) near the star then follow the trajectories of least effort in curved space.

wanted to think of electromagnetic forces as bending space, then space would have to bend one way for the positive charges and another way for the negative charges, at the same place and the same time. Since this is logically impossible, we cannot use the "curved space" concept for electromagnetic forces (or for strong or weak forces, because of their different effects on different types of particles), only for gravitational forces.

Gravitational forces always attract, so for larger and larger objects, these forces overwhelm all other types of forces. In particular, the existence of attractive *and* repulsive electromagnetic forces means that most electromagnetic forces are almost canceled by a similar but opposite force from the opposite charge. As we discussed in Chapter 2, the electromagnetic force between the particles in our bodies and the particles in the Earth exceed by far the gravitational force between these particles, but *all* the gravitational forces attract, whereas half of the electromagnetic forces attract and half repel, giving a total electromagnetic force between ourselves and the Earth that stays close to zero.

Photons feel gravitational forces, as shown by observations of stars close to the sun on the sky. If a photon passes close to a star, the photon will feel the star's gravitational attraction. What about photons escaping from the sun? They too feel the effect of the gravitational force from the sun. This force does not decrease the photons' speed (still the speed of light), but it does reduce the photons' energies. Every photon that leaves the sun loses a tiny fraction of its energy in escaping. The sun's gravitational forces try to hold the photon; they cannot do so, but they do rob each photon that leaves the sun's surface of about one two-thousandth of a percent of its energy of motion. This modest reduction in energy has little effect on us, but it exists for the sun and for every other star. In Chapter 5 we discussed, but rejected, the possibility that quasars' large Doppler red shifts result from this sort of gravitational red shift.

The fraction of their energies that a star's photons must lose in escaping from its surface depends on the strength of the gravitational force at the star's surface. This force will increase if the star has more mass, *or* if the star grows smaller. If we halve the star's radius, while keeping its mass constant, the gravitational force at the star's surface will increase four times; if we make the radius one tenth of its original value, the gravitational force at the surface will increase one hundred times. If we continue to compress a star such as the sun to a smaller and smaller radius, we make it harder and harder for a photon to escape from its surface. Calculations show that the fraction of the energy that an escaping photon loses will increase as one over the star's radius, if the star's mass stays constant. The amount of energy it takes to escape from the force of gravity at any object's surface is proportional to the object's mass divided by its radius. Therefore, if the sun's radius suddenly shrank two thousand times, photons would lose one percent of their energy, rather than one two-thousandth of a percent, in escaping from the sun's surface. We then would receive about one percent less energy from the sun (supposing that the sun managed to keep on producing the same amount of energy at its surface) because of the gravitational red shift.

If the sun's radius were to decrease not just two thousand times but two hundred thousand times, escaping photons would lose one hundred percent of their energy—that is, they would not escape at all. The sun thus would become a *black hole*, because no photons could escape. If a photon tried to get away from the sun, it would lose all its energy of motion, which amounts to saying that the photon would cease to exist. Among all the particles in the universe, photons are the lightest and the

$$F = \text{constant} \times \frac{M_1 \times M_2}{D^2}$$

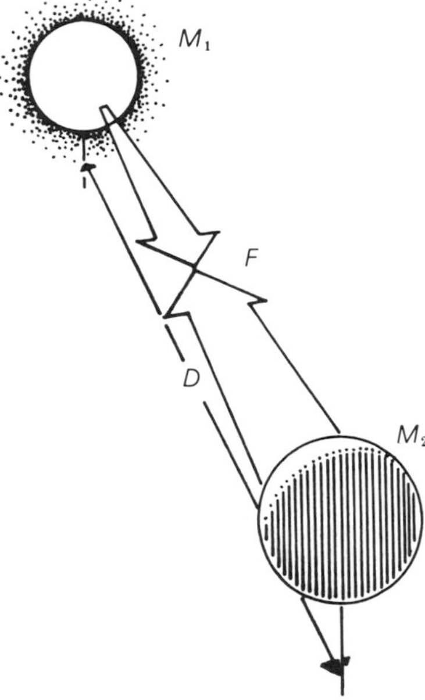

Figure 12–3. The gravitational force of attraction between two objects depends on the product of the objects' masses divided by the square of the distance between their centers. In algebraic terms, the amount of force equals a constant number times the term $(M_1 \times M_2)/D^2$.

liveliest: They have no mass and always travel at the speed of light. If photons can't escape from a star, no particle can, so a star whose photon's can't get away from it has a gravitational force strong enough to hold anything back. Such black holes are truly black, and they are holes too, because particles can continue to fall in even if nothing can get out.

The sun would become a black hole if its radius decreased 200,000 times (actually close to 227,000 times), from 700,000 kilometers to three kilometers. If the sun shrank like this, without losing any of its mass, we would see it grow redder and fainter, slowly at first as the effect started, then ever more rapidly. With a radius of one hundred kilometers, the sun still would extract only 3% of the escaping photons' energy; at a radius of ten kilometers, it would take 30%. Because of the nuclear fusion reactions in the sun's interior (Page 200), the sun will not collapse from its self-gravitational forces, at least for a while. But if it ever does, it will become a black hole when its radius shrinks to three kilometers, provided that its mass remains constant.

Any object can become a black hole if it shrinks far enough. As the object's radius decreases, if the mass in the object remains the same, then the density inside the object will increase steadily. Another object far from the first one will *not* notice any change in the gravitational force from the first object, because the gravitational force between two objects with mass depends only on their masses and the distance between their centers (Figure 12–3). But any particle on the *surface* of the shrinking object will notice a tremendous increase in the strength of the gravitational force. Every time we halve the object's radius, we quadruple the gravitational force at its surface, since the surface now lies twice as close as before to the object's center (Figure 12–4). Even the Earth would become a black hole if it shrank to a radius of about one centimeter. A person with a mass of seventy kilograms would become a black hole if that person shrank to a size of about a trillionth of a trillionth of a centimeter.

Unlikely as such shrinkings may be, every object with a given mass will become a black hole if it shrinks to a small enough size. The shrinking will increase the gravitational force at the object's surface and make it impossible for anything to escape. The radius at which photons no longer can escape, called the "black hole radius," varies in direct proportion to the object's mass. The sun has 300,000 times the mass of the Earth, so the sun will become a black hole if its radius decreases to three kilometers, whereas the Earth's radius would have to be 300,000 times smaller than three kilometers, or one centimeter, for it to become a black hole.

Inside a black hole that has the sun's or the Earth's mass, the density of matter must be enormous. If the sun were to shrink from its present radius of 700,000 kilometers to its "black hole radius" of three kilo-

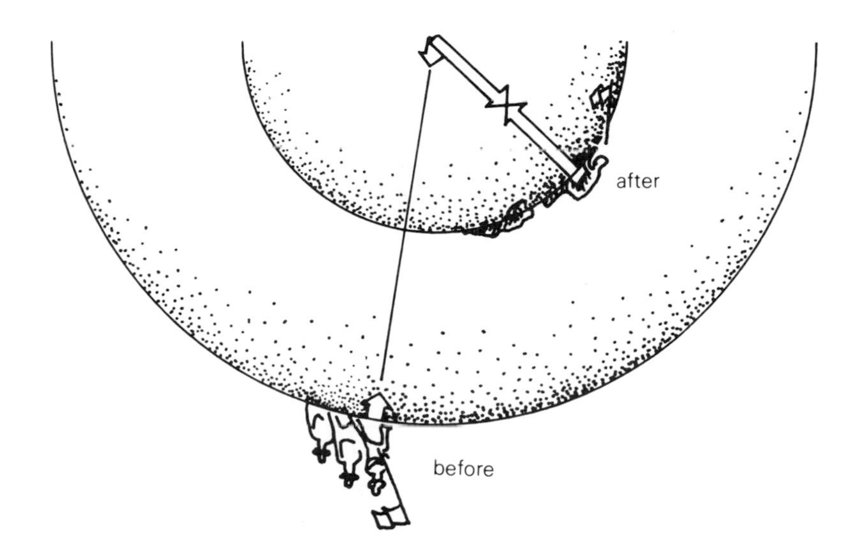

Figure 12–4. If the Earth suddenly were to become half as large as its present radius without losing any of its mass, we would find the Earth's surface twice as close to the Earth's center. As a result, the gravitational force at the Earth's surface would grow four times larger.

meters, the density of particles inside the sun would increase ten thousand trillion (10^{16}) times. If the mass stays constant, the density varies inversely as the *cube* of the radius, because an object's volume increases as its radius or diameter raised to the third power. No one knows how any matter ten thousand trillion times denser than the sun would behave, or what elementary particles start to appear when such tremendous densities exist. If anyone volunteered to go into a black hole to find out, this effort would be in vain as far as we are concerned, because even if such a person could survive inside a black hole, no message from inside could ever reach us.

What objects are likely to become black holes? The obvious answer is collapsed objects, objects with tremendous densities and tiny sizes. We have seen that stars form the basic units in the universe. Energy released at the stars' centers balances each star's self-gravitational forces. This balance works for objects with starlike masses, but not for objects much larger or smaller than stars. Smaller bodies such as the Earth do not liberate energy of motion, and they support themselves against their relatively weak self-gravity by electromagnetic forces among the molecules that form them. Objects more massive than stars, such as galaxies and galaxy clusters, contain millions or billions of stars in orbits around one another. These orbits provide another example of energy of motion (in the stars' orbits) balancing gravitational forces (among the stars themselves).

Stars that run out of nuclear fuel and thus lose their ability to liberate energy in their interiors are in danger of collapsing from their

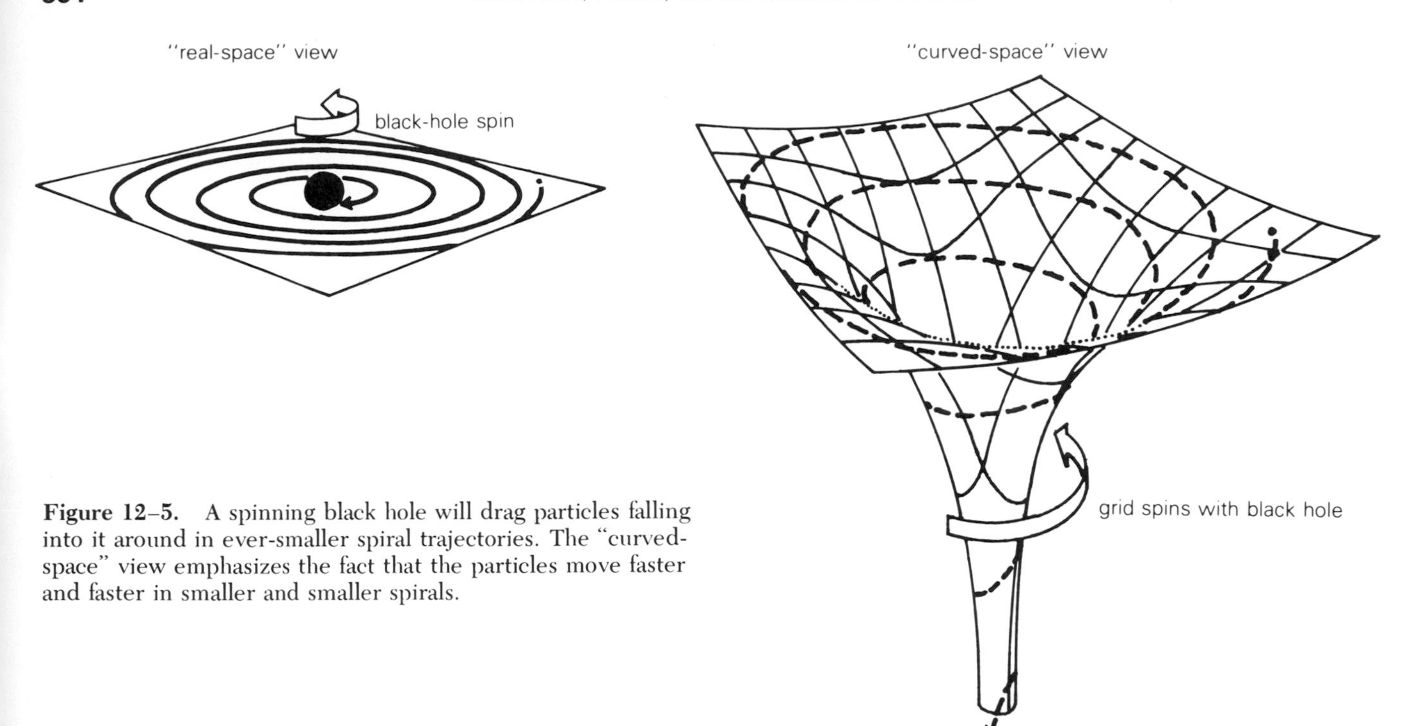

Figure 12–5. A spinning black hole will drag particles falling into it around in ever-smaller spiral trajectories. The "curved-space" view emphasizes the fact that the particles move faster and faster in smaller and smaller spirals.

self-gravity. We have seen (Page 235) that stars with less than 1.4 times the sun's mass can avoid such collapse through the support that the exclusion principle gives to their "degenerate" matter. Such stars become white dwarfs forever, as far as we know. As we shall see later in this chapter, other stars may contract to become "neutron stars," which are smaller than white dwarfs, but larger than the black hole radius for these stars. But some stars may collapse finally, completely, and utterly, to a size less than the black hole radius. Then the star will *be* a black hole and we shall never see it again.

Consider a star that collapses when it exhausts its nuclear fuel. If the star has been rotating, it will rotate much faster once it collapses. This increase in the rotation rate reflects the conservation of angular momentum (Page 93), which we know from our experience in acrobatic diving. For a star, the conservation of angular momentum means that if the star contracts 200,000 times in size without losing any mass, then the star will spin faster by 200,000 times *squared*, or forty billion times. The collapsed star, which previously rotated once a month, could form a black hole spinning forty billion times more rapidly, or forty thousand times a second. This fantastic rate of spin does affect the particles falling into a black hole, dragging them around in circles as they spiral in toward oblivion (Figure 12–5). A black hole can let the rest of the universe know about only three of its properties: First, its mass, shown by its ability to attract photons and any other kind of particles; second, its total angular momentum, which drags the particles around in spirals as they fall in; and third, its total electric charge, which will continue to attract particles of the opposite sign and to repel particles with the same sign of electric charge. We never can hope to know anything about a black hole except its total mass, total angular momentum, and total electric charge. Furthermore, any black hole would be likely to have a total electric charge of zero, since the hole would quickly attract enough particles of the opposite sign of charge to balance any excess of positive or negative charges.

Our typical black hole, formed from a collapsing star, thus should be a rapidly rotating invisible object, surrounded by a disk of particles spiraling inward toward the black hole radius (Figure 12–5). Every particle that falls into the black hole will slightly increase the mass of the black hole, thereby making the object's "black hole radius" a trifle larger, but the black hole remains a black hole inevitably. If two black holes should collide, they could form a single black hole with their combined mass, but such a collision never could destroy a black hole or pull the particles out of one and into the other.

Can We Detect Black Holes?

What would we see near a black hole? Could we expect to see anything? Luckily for our curiosity, we *can* hope to detect black holes, both by observing matter falling into them and by deducing their existence from their gravitational pull on other stars. A rotating black hole should have around it a disk of infalling matter that it has attracted from nearby interstellar gas and dust. As these particles swirl around in spirals, they move faster and faster as they come closer and closer to crossing over the critical black hole radius to oblivion. Although the final stages of the particles' infall must remain unobserved from outside, the earlier stages could be seen, and we would expect to "see" a rotating disk of extremely hot particles, each with lots of energy of motion. Collisions among the particles would strip any atoms down to individual electrons and nuclei, and the high speeds then would cause the particles to radiate photons as they pass by each other. This emission process, called "bremsstrahlung,"[1] transfers some of the energy of motion of charged particles to newly created photons, if the particles pass close to one another. To create x-ray photons, the particles must pass by one another at nearly the speed of light. Calculations of the energy of motion of the infalling particles and of the photons that they produce show that most

[1]"Bremsstrahlung" is not the name of a famous German scientist, but is the German term for "braking radiation," which physicists in this country never have translated for fear of losing the elegant simplicity of physics. The charged particles decelerate each other a bit as they pass by, and the energy of this deceleration appears as "bremsstrahlung" photons.

of the photons will be x rays. Thus we would not expect to "see" the disk of particles spiraling into black holes, but we could hope to observe them with x-ray detectors on satellites. X-ray astronomy must be carried out above our atmosphere, which shields us from harmful ultraviolet, x-ray, and gamma-ray photons, so astronomers have developed satellites that point their detectors upon command from the ground to look for sources of x-ray photons, or to examine known sources of x rays in more detail. At the present stage of x-ray astronomy, no source can be located on the sky with the same precision that the best visible-light telescopes on Earth can achieve, so x-ray astronomers often have problems in figuring out exactly where a given source lies in relation to the sources of visible-light photons. Even with much better resolving power, we could not hope to detect the infalling disk around a black hole as anything but a point of x-ray emission. The entire disk of material spiraling into the black hole would be far smaller than a star, and stars appear as simple points of light even in our largest telescopes.

To deduce that we have found a black hole, we must combine the direct observations of x-ray emission (which do show various pointlike sources) with conclusions made from what we know about orbital motions under gravitational forces. If we observe that x-ray photons from a particular source vary in number over a regular cycle, we can conclude that the x-ray emitting object may be orbiting around another object in such a way that the second object eclipses the x-ray source every time it comes in front of it. Following this line of reasoning, astronomers have looked in the direction of x-ray emitting sources to see if any stars appear at the same position as the x-ray sources. And indeed, in several cases they have found stars that vary cyclically in brightness with the same pattern as the x-ray sources that lie in the same direction. Astronomers interpret these observations as indicating that a (relatively) ordinary star and a condensed object are in orbit around each other, each one eclipsing the other in turn over a single orbit. The condensed object attracts particles from the star, and as the particles fall into the condensed object, they emit x rays by the bremsstrahlung process. For at least one object, called Hercules X-1, observations show that both the x-ray photons and the photons from the star undergo cyclical but opposite changes in their energies. If we ascribe these changes to the Doppler effect, then we would expect to see the x-ray photons shifted to higher energies when the visible-light photons shift to lower energies, and vice versa. Exactly this behavior appears in the observational results, so astronomers have become convinced that Hercules X-1 consists of a star and a condensed object in orbit around their common center of mass.

But is the condensed object a black hole? The best result from observing the Doppler shift of both the x rays from the condensed object (or rather, from the radiation produced close to the object's surface as matter falls on it) and of the visible light from the star, comes when we find that we can calculate the *masses* of both the star and the condensed object. We already know the orbital period from observing the cyclical variations in the light from the star or in the x-ray emission from the condensed object. If we also can measure the object's velocity in orbit, as determined from the Doppler effect, we can calculate the masses of the two objects. The reason that we can do this is as follows. Knowing the velocity in orbit and the time to complete an orbit (one complete cyclical change in brightness) gives us the size of the orbit, since distance equals velocity times time. Once we know the size of the orbit and the velocity in orbit, we can use the fact that more massive bodies must move more rapidly in a given-sized orbit to keep from falling into each other. That is, if two stars, each with twice the sun's mass, orbit around their common center of mass, they must move more rapidly in orbit than two stars with a mass equal to the sun's would move in the same-sized orbit, because the more massive stars exert more gravitational force on one another.[2]

By determining the masses of the star and of the collapsed object that orbit around their common center of mass in Hercules X-1, we can find whether the collapsed object could be a white dwarf or a neutron star. Neither white dwarfs nor neutron stars can exist for long if their masses exceed a few (perhaps three) times the sun's mass. (These calculations are extremely difficult, so their results are not as exact as we would like.) Thus if we determine the mass of a collapsed object to be more than a few solar masses, we probably have found a black hole. This line of reasoning is not airtight, but in our present state of knowledge, by far the most likely way for a condensed object (smaller than a star) with many times the sun's mass to exist at all is to be a black hole.

Hercules X-1 turns out not to be a black hole. The condensed object in orbit there has a mass about equal to the sun's, and most probably is a neutron star. The current best candidate for the first black hole found by humanity is Cygnus X-1, which is not an eclipsing x-ray source but is quite interesting nonetheless. At the same location as the source of x-ray photons in Cygnus X-1, we find a blue supergiant star, high up on the main sequence, which we know from our experience with similar stars should have a mass at least fifteen times the sun's mass. When we observe the spectrum of this young, hot star, we find that the lines in the star's spectrum shift back and forth in energy (or wavelength) over

[2]To use this method well, we need two objects of comparable mass in orbit around each other. If one object is far more massive than the other, as is the case for a planet orbiting the sun, we cannot determine the mass of the smaller body (the planet) by studying its orbit, although we *can* determine the mass of the sun in this manner.

a time cycle of 5½ days (Figure 12–6). If we interpret this shift as arising from the Doppler effect, the size of the shift in energy corresponds to a velocity in orbit of about 70 kilometers per second. The trouble is that we don't know how the orbit is tilted with respect to our line of sight, and we only see the projection of the velocity in orbit along our line of sight (Figure 12–7). When we see an eclipsing pair of stars, we know that we have our line of sight right along the plane of the orbit, so this projection effect does not interfere with our conclusions. In any case, the velocity in orbit of the young, hot star must be *at least* 70 kilometers per second, since the part of the velocity that we can see measures this much. If the companion to this star can make it move at this speed in orbit, we can calculate that the mass of the compact companion must be at least four times the sun's mass, and is more likely to be eight times the sun's mass. The x rays that this compact object emits suggest the infall of particles that we discussed previously, and the object's mass implies a black hole for lack of anything else.

Although the nature of black holes will prevent us from ever detecting them straightforwardly, astronomers daily grow more comfortable with the idea that in Cygnus X-1 we have found a real black hole. The apparent brightness of the young, bright, B0-type star in mutual orbit with the x-ray source suggests that the distance to this pair is about 2500 parsecs, and they lie in the plane of the Milky Way galaxy, as the sun does, but somewhat farther from the galactic center than we are.

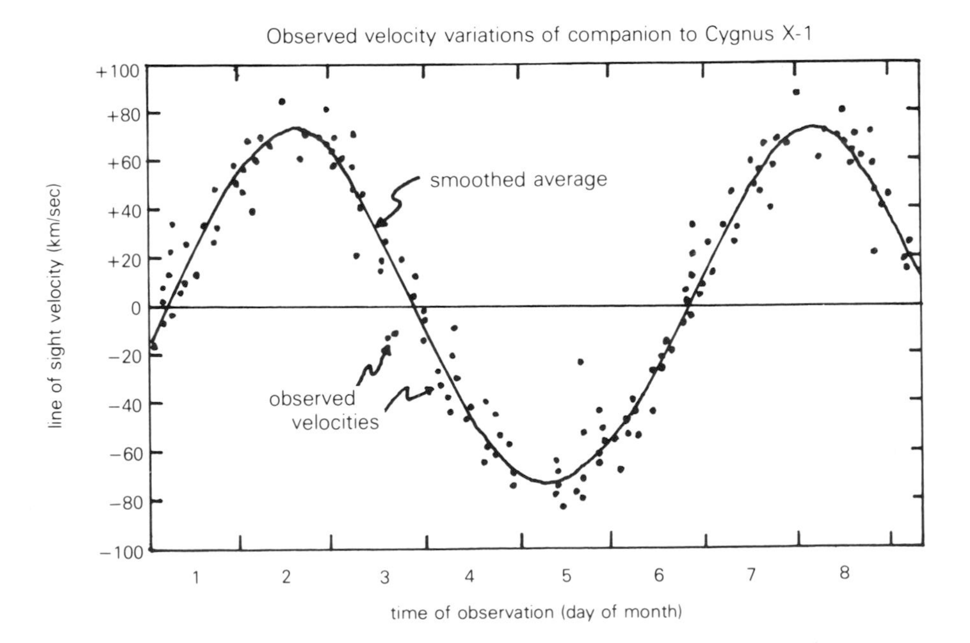

Figure 12–6. The star that is located at the same position on the sky as the x-ray source Cygnus X-1 shows a change in its velocity along the line of sight with respect to ourselves. This velocity change is cyclical, with a period of 5½ days, and the velocity changes by about 70 kilometers per second in each direction away from the average velocity along our line of sight to the star. The graph of the line-of-sight velocity at different times shows the actual observations (dots) and their smoothed average (solid line).

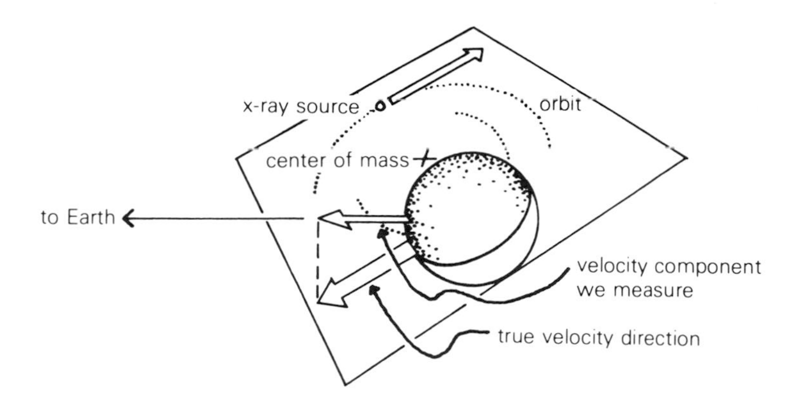

Figure 12–7. If the orbit of Cygnus X-1 and its companion star around their common center of mass occurs in a plane that is tilted with respect to our line of sight, then we see only the part of the velocity that lies along our line of sight. This projection effect means that the true maximum velocity must be greater than or equal to the 70 kilometers per second that we observe for the companion star to Cygnus X-1.

If one black hole exists, why not thousands? Or billions? Although we should hesitate to leap too far on scanty evidence, we may note that the companion star to Cygnus X-1 seems to be a fairly typical, young, hot supergiant. Could Cygnus X-1 itself be simply a more rapidly evolving star that burned itself out and collapsed? (Then Cygnus X-1 should have more than fifteen times the mass of the sun, at least during its main-sequence lifetime.) And if this is true, might there be many black holes in our galaxy that didn't happen to be part of double or multiple-star systems? Such black holes would not be easily detectable: They would not have such an easy time attracting particles as they would in mutual orbit with a star, and even if they emit x-ray photons, we cannot determine their masses to "prove" that they must be black holes. The fraction of our galaxy's mass that now resides in stars that have become black holes could be anything under 85%. If more than 85% of the galaxy's mass is in black holes, we would notice their cumulative effect on stars' orbits around the galactic center.

What about black holes that do not form from stars? Could there be black holes with much greater masses, or much smaller masses? Remember that we do not know how a black hole could form except from a star that collapses, but if we think about it we can see no reason why, for example, much more massive black holes should not exist. Of course, this does not prove that such black holes *do* exist. If they do exist, where are they? One imaginative suggestion puts a black hole at the center of every spiral galaxy, or perhaps inside every galaxy. A black hole with a hundred million solar masses at the center of our galaxy could explain how the galaxy formed in the first place. (Of course, this does not explain how the *black hole* formed.) Recall from Chapter 4 that we need significant initial perturbations in the density of particles to have them clump together to form galaxies. A massive black hole would attract other particles toward it that could settle into orbits around the galactic center. Some particles would fall into the central black hole, and these might be responsible for the intense photon emission, especially of infrared photons, that we observe from the nucleus of our galaxy. Finally, it is tempting to expand the speculation to include the possibility that Seyfert galaxies (Page 133) and quasi-stellar objects also contain massive black holes, and that matter falling into them radiates the vast amounts of energy that these galaxies and quasars emit.

Such speculations are fascinating although unverified.[3] Also, they do not explain everything, rather they push the problems back to making the black holes instead of the galaxies or quasars. When we examine the photons radiated from our galaxy's center, or the photons radiated by quasars, we find details not easily explained by assuming that particles falling into black holes are producing the photons. In view of our uncertain knowledge of what happens near black holes, we cannot state definitely whether galaxies and quasars (or even stars) each have a black hole at their center. We therefore shall content ourselves with the mild speculations mentioned above. But we can, if we like, calculate some characteristics of a black hole with a hundred million times the sun's mass. The black hole radius of such an object would be a hundred million times the sun's black hole radius, hence three hundred million kilometers. That is, an object a hundred million times the sun's mass will be a black hole if its radius is less than the diameter of the Earth's orbit around the sun.

What would the density be like inside such a black hole? Not so incredibly large as we might think at first. If we put a hundred million solar masses into a region with a radius of three hundred million kilometers, the average density of this matter will be only twice the density of matter in the sun. The reason for this low density inside a black hole comes from the enormous mass in the black hole. An object's critical black hole radius increases in direct proportion to the object's mass, but the volume inside an object varies as the *cube* of the radius. Thus for any object the average density inside the black hole radius, which varies as the mass divided by the volume, will be proportional to the mass divided by the cube of the black hole radius. Since the black hole radius varies directly as the mass, this density is proportional to one over the square of the black hole radius, or to one over the square of the mass of

---

[3]Some theoreticians have suggested that inside the *sun* (and by implication, inside other stars) there is a black hole with a small fraction of the sun's mass and a diameter of a few millimeters or centimeters. The innermost parts of the sun are falling into the black hole, but the rest of the sun is all right for now. The motivation for this suggestion is, first, that the black hole might have helped to form the sun (and other stars), and, second, that this might explain why the sun emits far fewer neutrinos each second (see footnote, Page 238) than calculations would predict. Hence the theoreticians call this "the black hole that ate the solar neutrinos."

the object involved. A black hole with a hundred million (10^8) times the sun's mass will have an average density of matter inside it that is ten million billion (10^{16}) times *less* than the density inside a black hole with the sun's mass.

Extremely massive objects therefore could become black holes at densities that are quite ordinary by our standards. To make a black hole with a hundred million times the sun's mass, we simply assemble a hundred million suns side by side, filling a sphere whose radius equals twice the radius of the Earth's orbit (Figure 12–8). This black hole would be minuscule on a galactic scale, but it would have enough mass within a small enough radius to make a black hole, even though its density is only twice the density of matter in the sun. The essential feature of a black hole, the fact that a photon must surrender all its energy in trying to leave the surface, depends on the ratio of the object's mass to its radius, while the density of the object varies as the mass divided by the cube of the radius.

What if we think about still more massive black holes, where the average density of matter could be smaller and smaller? Such thinking is eminently allowable, and it leads us toward the greatest black hole possibility—the universe. Acute readers already will have noticed that our definition of a black hole resembles our definition of the universe in one significant feature—nothing can escape. If we choose to regard the universe as a black hole, we are led to ask what mass and size a black hole would have if the average density inside it equalled the average density of matter in the universe. Figure 12–9 shows a graph of mass against radius for various objects in the universe. The critical black hole radius appears as a straight line in this diagram, because the black hole radius varies in direct proportion to the mass. Anything above and to the left of this straight line is a black hole, because its size is smaller than the black hole radius for an object with that particular mass. Objects that we can see, such as stars, galaxies, and galaxy clusters, all lie below and to the right of the dividing line between black holes and the other half of the diagram.

The average density in the universe, to within a factor of ten or so, is 10^{31} times less than the average density of matter within the sun. A black hole whose mass and black hole radius gave it this density must have a mass equal *at least* to 10^{23} times the sun's mass.[4] The black hole radius

Figure 12–8. A black hole could have a hundred million times the sun's mass and a density of matter about the same as the density of matter in the sun. To imagine such a black hole, suppose that we put a hundred million stars like the sun next to one another. These hundred million stars would fill a volume equal to the size of the solar system past the orbit of Mars.

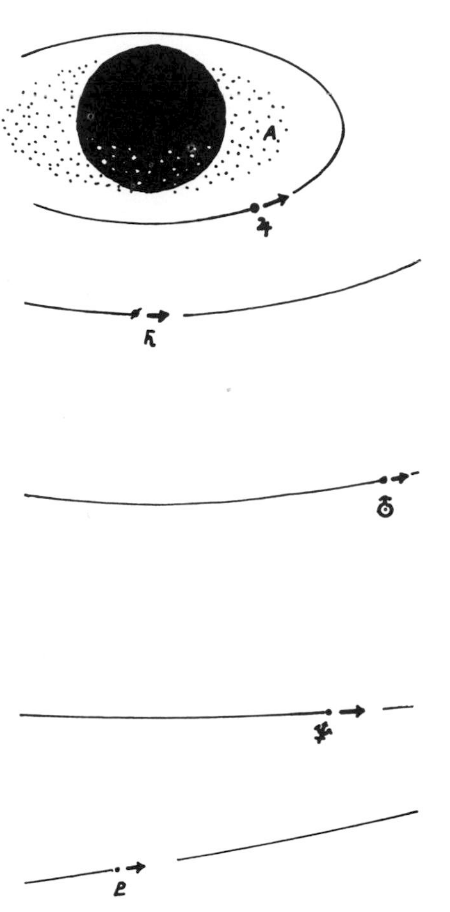

[4]The density inside a black hole is proportional to one over the square of the black hole's mass, so the mass is proportional to one over the square root of the density. Inside a black hole with the sun's mass, the density is at least 10^{16} grams per cubic centimeter. A black hole where the density is 10^{-31} grams per cubic centimeter has a density of matter that is 10^{47} times less than 10^{16} grams per cubic centimeter, so the black hole must be at least $10^{23.5}$ times more massive than a black hole with one solar mass.

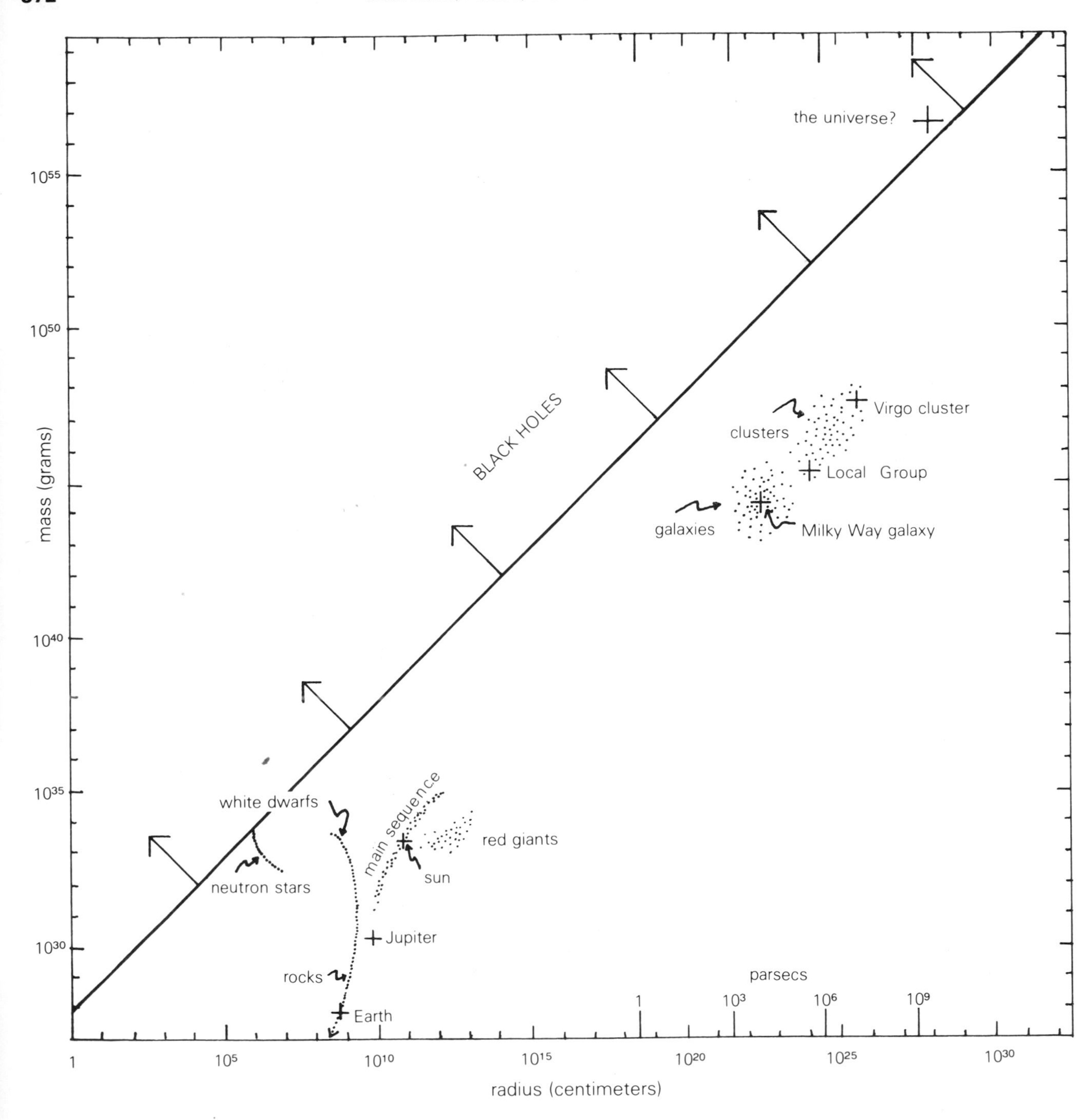

the universe?
BLACK HOLES
Virgo cluster
clusters
Local Group
galaxies
Milky Way galaxy
white dwarfs
main sequence
red giants
sun
neutron stars
Jupiter
rocks
Earth
parsecs
1
10^{3}
10^{6}
10^{9}
mass (grams)
10^{55}
10^{50}
10^{40}
10^{35}
10^{30}
1
10^{5}
10^{10}
10^{15}
10^{20}
10^{25}
10^{30}
radius (centimeters)

◀ **Figure 12–9.** A graph of mass against radius for various objects in the universe shows the critical line that corresponds to the "black-hole radius" for an object with a given mass. Any object whose mass and radius place it in the diagram above and to the left of this critical line will be a black hole.

for such an object would be 10^{23} times the sun's black hole radius, or 3×10^{23} kilometers. This radius equals ten billion parsecs, which is about 1¼ times the distance to the farthest known quasars (if quasars are really the farthest objects) and three times the distance to the farthest known galaxy. Given our uncertainty in establishing the average density of matter in the universe, this result for the size of the "universal black hole" is not exact, but it should tickle our fancies to notice that if we view the universe as a black hole, its radius does not far exceed the most distant objects that we think we can see.

Are we but one black hole among many in a larger metauniverse? We cannot rule out this possibility, but it is not as likely as it might appear. If we are inside a black hole, we can never get a signal out, but we could receive signals from outside. So far we have not detected such signals. More significant is the fact that inside a black hole we hardly would expect to find matter spread out as uniformly as it is in our universe. The immense gravitational forces should make a black hole denser in the middle than near the edges. The universe, in contrast, shows a remarkably constant distribution of matter, averaged over the largest distances we can see. The uniformity of the microwave background (Page 80) from all directions gives another indication of the regularity of density for the matter and photons in the universe. Still, we might yet find that what we call the universe is but a tiny part of the inside of a gigantic black hole.

Pulsars and Neutron Stars

Although black holes have not yet been definitely observed to everyone's satisfaction, neutron stars have (probably). Neutron stars make pulsars go (we think), and pulsars are real (we believe).

Neutron stars represent one final stage in the lives of stars, as white dwarfs represent another, and black holes a third. No star, as far as we can calculate, will become both a white dwarf and a neutron star in its lifetime, unless a white dwarf somehow could accrete enough mass to become a neutron star. Stars with masses less than 1.4 times the sun's mass almost surely will become white dwarfs, whereas stars with masses between 1.4 and "a few" solar masses should end as neutron stars. We say "a few" solar masses for the upper limit on neutron-star masses because calculations of this upper limit remain a bit too complex to be solved completely. Stars with more than "a few"—say three for a good round number—times the sun's mass can become neither white dwarfs nor neutron stars, thus these massive stars give us the most likely candidates for black holes.

A neutron star forms when a star's insides collapse so fast that the degeneracy of the electrons cannot stop the collapse, or if the star is too massive to be supported by electron degeneracy. As we saw from our study of white dwarfs (Page 230), the "exclusion principle" that describes the behavior of closely packed electrons keeps the electrons apart. The exclusion principle keeps the electrons from all being in almost the same place with almost the same velocity, and this exclusion principle acts among particles of the same type, in particular among groups of electrons, of protons, or of neutrons—but not among helium-4 or carbon-12 nuclei. Inside a white dwarf, the electrons feel the exclusion principle's effect first because they tend to move at much larger velocities than the nuclei and thus feel constrained in their tendency to have large velocities before the much heavier nuclei do.[5]

In white dwarfs, the exclusion principle keeps the electrons apart, and they in turn prevent the nuclei from all rushing together by electromagnetic forces. This works fine if you have lots of electrons. Most stars in fact do have plenty of electrons, but neutron stars do not. A neutron star forms from the collapse of the central regions of a star that has exhausted all possible energy-liberating processes. In the collapse, the nuclei collide with each other so violently that they break apart into protons and neutrons. To break nuclei apart, a collision must be a million times more violent than that needed to strip electrons from their orbits around a nucleus, as occurs in the outer layers of hot, ordinary stars. In a collapsing star, the protons all capture electrons, not into orbits, as in hydrogen atoms, but into a fusion reaction that yields a neutron and a neutrino. The neutrinos escape from the scene of the fusion reaction (blowing off the outer layers of the star as they do so), but the neutrons remain behind. All of this happens before the electrons have time to stop the collapse through the action of the exclusion principle. As the nuclei and electrons in the collapsing star's insides rush together, in less than one second the nuclei come apart and the electrons all fuse with protons before they can establish a counter-pressure to oppose the collapse.

In the twinkling of an eye, then, the star becomes all neutrons, either broken off from the nuclei or made by protons fusing with electrons. Notice that since electrons do not feel strong forces, they do not fuse with protons easily. The fusion of a proton and an electron to produce a

---

[5]Recall that for mysterious reasons, the exclusion principle does not act among nuclei with an *even* number of nucleons (protons plus neutrons), but only among nuclei with an *odd* number. Thus nuclei such as helium-4 and carbon-12 never feel the effect of the exclusion principle directly.

neutron and a neutrino is a weak reaction, which is the opposite of the decay of an isolated neutron into a proton, an electron, and an antineutrino. We might think that because electrons and protons attract each other by electromagnetic forces, they would love to fuse together, but this is not so. The electromagnetic forces always get the electron into an orbit around the proton, thereby making a hydrogen atom, and never make the two fuse together. For such fusions we need weak reactions, and since weak forces act only over distances less than 10^{-13} centimeter, we need to make the protons and electrons collide violently and get this close together before the electromagnetic forces make the electron orbit around the proton.

Such violent collisions occur inside a star that collapses within a few seconds because it has too much mass to become a white dwarf. The electrons all fuse with protons before the exclusion principle works on them, and the neutrons they form collapse further to a still greater density. Eventually the exclusion principle among the *neutrons* keeps them from collapsing any farther, but to reach this point the star must shrink to a tiny size and a huge density. Because neutrons are 1839 times more massive than electrons, they move far more slowly (about $\sqrt{1839}$, or 43, times more slowly) than electrons with the same energy of motion. Thus neutrons can be packed more closely before they feel the effect of the exclusion principle (Page 231). When the neutrons do feel this effect, the collapse stops, and we have a neutron star, which is a star made almost entirely of "degenerate" neutrons that are supported against collapse by the exclusion principle. The distance between neighboring neutrons inside a neutron star averages about five hundred times less than the distance between neighboring electrons in a degenerate white dwarf star. Therefore the density of matter inside a neutron star exceeds that inside a white dwarf star by the cube of five hundred or about a billion times. Dense as white dwarf stars are, they are like the air compared to neutron stars. A typical white-dwarf star may contain the mass of the sun in a space the size of the Earth, thereby giving the star an average density a million times the density of water, or of the average density of matter in the sun. A neutron star may contain twice the sun's mass in a sphere just ten kilometers in radius, thereby producing an average density a million billion (10^{15}) times the density of water. Table 12–1 shows the radius and density of a main-sequence star, a red-giant star, a white-dwarf star, a neutron star, and a black hole with a starlike mass. The ratio of the density of the matter inside a neutron star to the density inside a white dwarf is a thousand times greater than the ratio of the density inside a white dwarf to the density inside the sun.

A neutron star ten kilometers in radius consists almost entirely of neutrons (about 10^{57} of them), and we can think of the entire star (if we like) as one gigantic "nucleus." As we saw in Chapter 2, an isolated

Table 12–1
Sizes and Densities of Various Kinds of Stars

Type of star	Approximate mass (in solar masses)	Radius (in kilometers)	Average density (in grams per cubic centimeter)	Fraction of original energy of a photon lost in escaping from star's surface
Main sequence (sun)	1	700,000	1	0.000005
Red giant	3	10,000,000	10^{-3}	0.00000125
Red supergiant	10	150,000,000	10^{-6}	0.00000025
White dwarf	1	7,000	10^{6}	0.0005
Neutron star	2	10	10^{15}	0.65
Black hole	1	less than 3	more than 10^{16}	1.00

neutron will decay into a proton, an electron, and an antineutrino. Some of these decays occur inside a neutron star, but the protons and electrons that they produce soon fuse again into neutrons and neutrinos. Thus the star remains mostly neutrons, with a few protons and electrons present at any time.

Neutron stars are almost as small and as dense as black holes, but not quite. A star with twice the sun's mass will become a black hole if it collapses to a radius less than six kilometers. If it instead becomes a neutron star with a radius of (say) twelve kilometers, it will have twice the radius, and only one eighth of the density, of a black hole with the same mass. These differences can be crucial: A neutron star can last indefinitely, supported against gravity by the exclusion principle, with a radius twice the black-hole radius that would lead to the star's complete disappearance.

Neutron stars are great lumps of neutrons with real surfaces and a real interior. On the surface of a neutron star, the force of gravity will be enormously greater than the force of gravity at the surface of the Earth, of the sun, or even of a white dwarf star. Photons trying to leave a neutron star's surface would lose half their energy in doing so, but they still can escape.

It is tempting to think about possible life forms that could exist on a neutron star's surface. The sorts of life that we discussed in the previous chapter all were based on reactions among atoms and molecules, which depend on electromagnetic forces. Atoms and molecules stay together at temperatures less than a few hundred degrees, or a few thousand degrees, above absolute zero. The temperature at the surface of a neutron star exceeds a million degrees absolute, so atoms and molecules certainly do not exist there. But we can speculate that strong forces lead to the formation of large chains of neutrons, protons, and other, "exotic" particles briefly produced at the star's surface. Such chains of elementary particles would interact with each other on time scales of a billion-billionth of a second, and we can imagine creatures made of them with sizes measured in trillionths of a centimeter and lifetimes in billionths of a second, so that whole civilizations rise and fall in the time it takes to read this sentence. Such creatures might communicate with photons of their favorite wavelength, presumably something like their own sizes, so they would be using extremely high-energy gamma ray photons. Their "giant radio telescopes" deployed in "Project Cyclops" to detect long-wavelength photons might be a millionth of a centimeter across, capable of detecting ultraviolet photons.

Speculation is fun, but observation gives results. During the early 1940's, when neutron stars and black holes first were imagined, few astronomers thought that such strange collapsed objects would be detected soon. A generation later, astronomers may have detected black holes, and they almost certainly have found neutron stars. Neutron stars produce "pulsars," which are radio sources that emit photons in regularly spaced pulses. These pulses of photons apparently arise near the surfaces of rapidly spinning neutron stars, and their rhythmical beat reflects the neutron stars' rotation cycles.

Neutron stars all should be rotating quite rapidly, for the same reason that black holes also rotate rapidly. As a star collapses, if it keeps about the same mass it must spin more quickly to conserve its angular momentum, as a high diver spins more rapidly when she contracts her body. The star's rate of spin will increase in proportion to one over the square of its radius, so if the star grows a thousand times smaller it will spin a million times more rapidly. A neutron star that forms as the interior of a star collapses might begin its collapse with a radius of 120,000 kilometers. If it forms a neutron star by collapsing to a radius of 12 kilometers, the neutron star will be ten thousand times smaller than the original collapsing star. Thus if the original star was rotating once each month, the neutron star will rotate a hundred million times more rapidly, or a hundred times per second.

Along with this great increase in the star's rate of spin goes a similar increase in the star's magnetic field strength. Most stars have a magnetic field associated with them, and as they rotate the field spins around too. Since electrically charged particles feel the electromagnetic forces produced by the star's rotating magnetic field, these forces push the particles around as the star and its magnetic field keep rotating. For a star such as the sun, the magnetic field strength is relatively low, and most particles near the sun feel stronger forces than the electromagnetic forces from the sun's magnetic field. The magnetic field does tend to make the charged particles in the sun's outer atmosphere follow along the direction of the field, but the magnetic field's rotation does not make these particles keep spinning along with the field. However,

Figure 12–10. Inside particle accelerators called "synchrotrons," electromagnets are used to make charged particles follow circular paths instead of the straight-line paths in which their momentum would carry them.

if the magnetic field should grow much stronger, its electromagnetic forces could control the motions of nearby charged particles in the same way that magnets in a particle accelerator make charged particles trace a curved path instead of the straight lines they want to follow (Figure 12–10). The strength of the magnetic field at a star's surface, like the star's rotation rate, increases as one over the square of the star's radius. The magnetic field in a star that collapses to one ten-thousandth of its original size will increase in strength by a hundred million times. This enormous increase will make the neutron star into a giant spinning magnet, as seen by a charged particle close to its surface. The star's rotating magnetic field will make any charged particles nearby sweep around along with the star, although the particles can slowly spiral outward away from the star (Figure 12–11). Such a spinning magnet, carrying along all the nearby charged particles, should produce what we call a "pulsar."

Pulsars seem to be rotating neutron stars left behind by supernova explosions. More precisely, pulsars produce photons in regularly spaced pulses close to the surface of rotating neutron stars. The term "pulsar" originally meant "pulsating radio source," and indeed the hallmark of all known pulsars is the cyclical burst of radio photons, repeated over and over again. Some pulsars also have been found to emit pulses of x-ray photons and of visible-light photons, so pulsars are more than radio sources that regularly pulse on and off. The reason why a rotating neutron star emits photons in pulses, at regular intervals equal to the period of rotation, remains a subject of debate. It seems likely that some part of the pulsar emits more photons than other parts, and as the more highly emitting part sweeps past us, we see a pulse like the flash from a rotating ambulance beacon (Figure 12–12).

Consider what happens to a neutron star that remains after a star collapses its core, blows off its outer layers (Page 239), and leaves behind a neutron star spinning around at something like a hundred times per second. The star's rotating magnetic field pushes charged particles around, and as these particles spiral away from the star, they emit photons by the process of synchrotron radiation described on Page 39. This process works when charged particles moving at almost the speed of light accelerate or decelerate in a magnetic field. The charged particles with the greatest energy of motion emit gamma-ray and x-ray photons, particles with less energy of motion emit visible-light photons,

Figure 12–11. A collapsed, rapidly rotating neutron star with an intense magnetic field will use its electromagnetic forces to push charged particles around with it as it spins. The charged particles accelerated in this manner will slowly spiral farther and farther away from the rotating star. As the particles move, they produce photons of synchrotron radiation. ▶

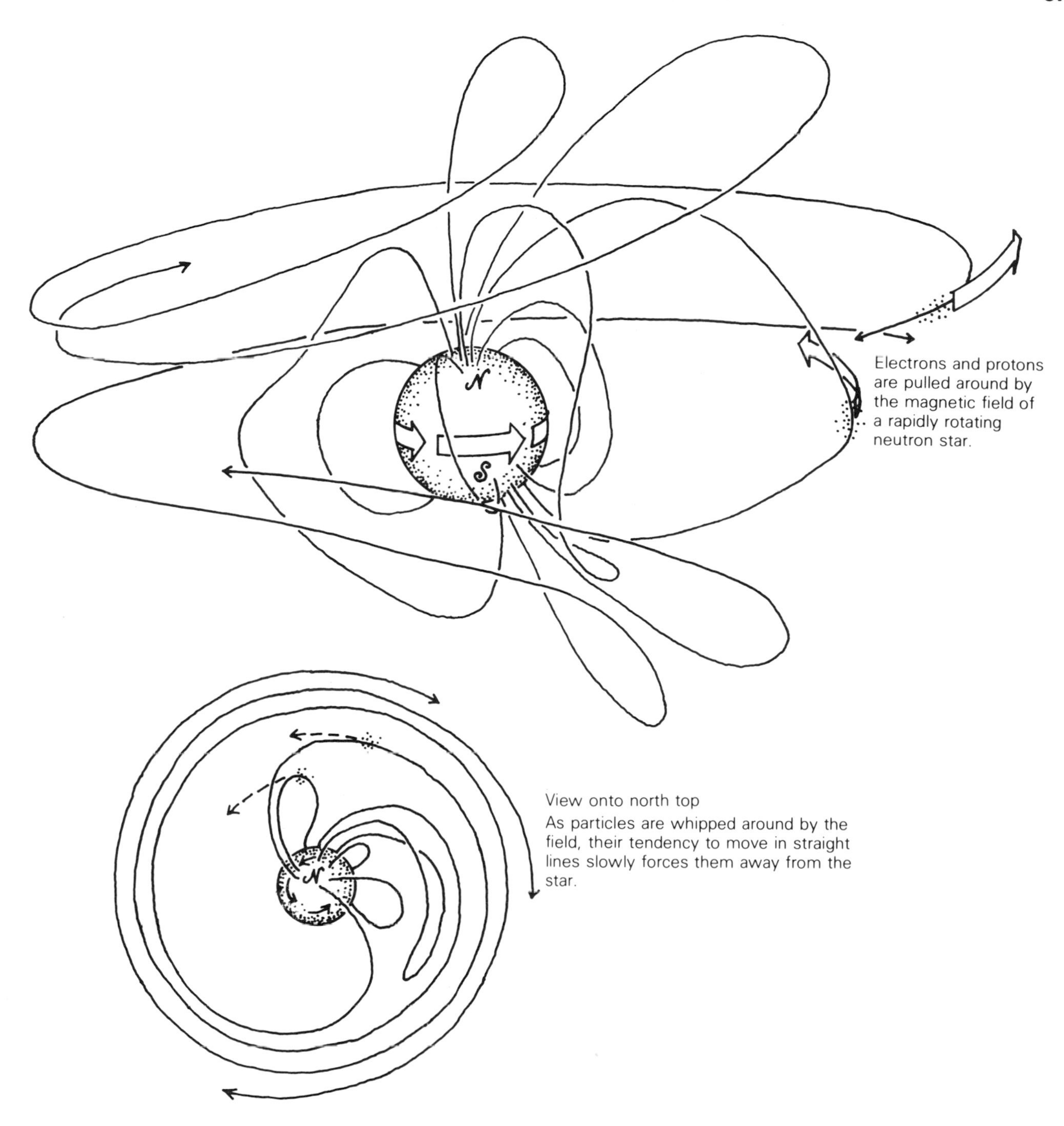

Electrons and protons are pulled around by the magnetic field of a rapidly rotating neutron star.

View onto north top

As particles are whipped around by the field, their tendency to move in straight lines slowly forces them away from the star.

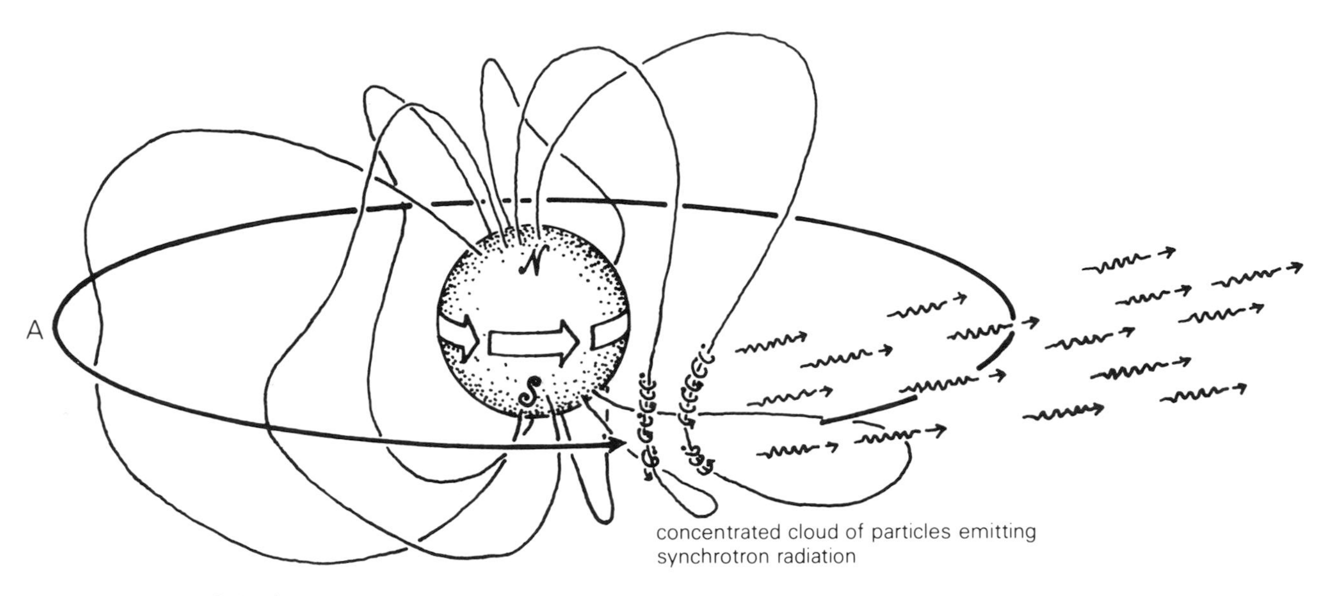

Figure 12–12. Pulsars presumably contain some regions that emit more photons per second than other regions do. As the regions of greater emission sweep past us, we see a flash of photons. Some regions might produce more photons because they contain larger numbers of charged particles than other regions do.

and the particles with the least energy of motion (but still enough to be moving at almost the speed of light) emit radio-wave photons by the synchrotron process. As these charged particles emit synchrotron photons, they lose some of their energy of motion. This lost energy appears as the energy of the photons produced by the particle's changing motion through the magnetic field. The star's rotating magnetic field will accelerate the charged particles to almost the speed of light, but it can't do this forever because each such acceleration takes some energy of motion away from the rotating star and gives it to the particles, which in turn pass on some energy to the photons they produce. Thus the rotating neutron star, left behind by the supernova explosion, gradually should decrease its rate of spin as it radiates away pulses of photon energy.

Some of the charged particles accelerated by the rapidly rotating magnetic field escape into space with extremely large energies, thereby forming part of the background of "cosmic rays" that we discussed on Page 245. In particular, the cosmic-ray protons and electrons with the highest energies of all are believed to come from the vicinity of the rotating neutron stars, whereas the majority of cosmic-ray particles, which have somewhat less energy per particle, come from the exploding outer layers of the supernova that preceded the neutron star. The highest-energy protons can have as much as ten billion ergs of energy, which is equal to the energy of a well-hit tennis ball, in a particle whose mass is just 1.67×10^{-24} gram. Hence we think that supernovae make, both during and after their explosion, all the cosmic rays that we have detected—protons, electrons, neutrons, helium nuclei, and smaller amounts of nuclei heavier than helium, all moving at speeds close to the speed of light.

The slowing down of pulsars that we can predict from the way that they produce photons shows up when we study the pulsars carefully. Of the hundred or so pulsars that have been detected so far, the best example of a pulsar with a decreasing rate of spin occurs in the Crab Nebula. As we discussed in Chapter 8, this web of hot gas filaments was left behind by the supernova that Chinese astronomers observed in 1054 A.D. At the center of the nebula, a radio source pulsating thirty times per second was discovered in the 1960's. The source also turned out to be blinking on and off at the same rate in x-ray and visible-light photons (Figure 12–13). During the past few years, astronomers have

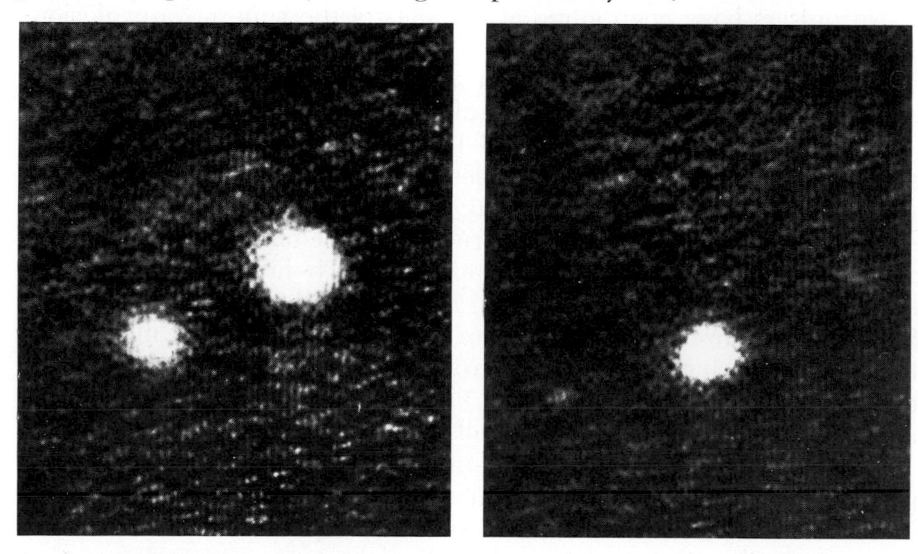

Figure 12–13. The central star in the Crab Nebula, which exploded as a supernova, turns out to be a pulsar that blinks radio-wave, visible-light, and x-ray photons on and off about thirty times each second. The two pictures each represent a composite of many exposures, one timed to have the shutter open at the peak of the pulsar's radio flashes, the other to have the shutter open at the low point in the flash cycle. We can see that the pulsar appears brighter in the photograph timed for the peak of the pulse cycle. Can you tell which photograph shows the peak of the pulsar's cycle?

timed the interval between pulses precisely enough to determine that the pulsar is slowing down by one tenth of a percent each year. In addition, astronomers also have measured sudden deviations from the gradually slowing down rotation rate. Some of these deviations indicate small, temporary *increases* in the pulsar's spin rate. Quite possibly, "starquakes" in the outer crust of the neutron star can explain these temporary deviations from the overall, steady decrease in the neutron star's rate of spin.

The pulsar in the Crab Nebula spins more rapidly than any of the other known pulsars. The second-fastest pulsar appears in the "Vela Nebula" (in the constellation Vela, seen only from southern latitudes), and flashes radio-wave and x-ray photons on and off eleven times per second. All the other pulsars flash radio-wave photons on and off anywhere from five times per second to once every four seconds. We can estimate the age of a pulsar (since it began pulsing) from its rate of pulsation: The youngest pulsars should spin the fastest, and thus flash on and off with the shortest time cycles. Using this idea, together with additional information from observing the slowing down of the pulsars' spin rates, astronomers have estimated the pulsars' ages to vary from one thousand to one million years.

No pulsars have been found at the sites of the supernova explosions observed in 1572 A.D. and in 1604 A.D. The logical conclusion to draw from this fact seems to be that not all supernovae produce pulsars, or perhaps not all the narrow beams of photons from pulsars sweep past us. If the supernovae of 1572 and 1604 had left pulsars behind, they should have the shortest pulse cycles of any pulsars that we know. We should hardly be surprised that not all supernovae produce rapidly rotating neutron stars. Some supernovae might blow their parent star apart completely, whereas others might leave behind a collapsed core too massive to form a neutron star, and thus will produce black holes. Such massive cores become neutrons for a moment as they collapse, but the exclusion principle among neutrons cannot halt the collapse through the core's black hole radius and into the limbo of the lost, where unknown states of matter could well exist.

The legacy of supernovae thus embraces several possibilities—complete destruction of stars, neutron stars, and black holes. The outer layers of a supernova burst into space and provide us with heavy elements and cosmic-ray particles. The rotating neutron stars that supernovae leave behind continue to accelerate particles to cosmic-ray energies. Supernovae thus turn out to be the biggest thing since the big bang for making the universe as we know it into a cosmic paradise. In this cosmic Elysium we orbit our own, life-giving sun, one of the majority of stars that contain the products of supernova explosions but that will never themselves become a supernova, a neutron star, or a black hole—just the degenerate dwarfs that provide a quiet end to a star's efforts.

The Future of the Universe

We live in an expanding universe made of stars and permeated with the photons left over from the earliest years of the expansion. Through gravitational forces the stars cluster in galaxies, which themselves form larger groups. By the interplay of gravitational, electromagnetic, strong, and weak forces, stars liberate energy of motion from energy of mass, fusing protons into helium nuclei and helium into heavier nuclei. The energy of motion escapes in photons from the stars' surfaces and in neutrinos from the star's centers. Small lumps of heavy-element debris close to stars may be situated favorably to receive the stars' photons, which can induce chemical reactions such as photosynthesis, which deposit some of the photons' energy of motion in living organisms called plants. This energy stored in molecular bonds can become energy of motion once again when "animals" consume the plants and break the molecular bonds apart in their digestive systems.

When stars exhaust their ability to liberate energy of motion through nuclear fusion, they usually fade away as degenerate white dwarfs. Since the minority of stars with more than 1.4 solar masses cannot exist as white dwarfs, they must lose enough mass during their red-giant phase to get under the limit, or their interiors will collapse when the stars no longer can liberate energy of motion to oppose their own self-gravitational forces. Such a collapse will produce either a rapidly spinning neutron star (from the less massive stars) or a black hole (from stars whose cores contain more than a few solar masses).

So much for a two-paragraph summary of the universe. Is the universe really this straightforward? Even leaving out exotic objects such as quasars, which may contain important physical messages that we don't understand yet, there are a few more points to consider. The future of the universal expansion depends on the strength of the universe's self-gravitational forces. Thus the determination of the density of matter in the universe ranks high among our inquiries into the future (Page 77). We also should pause to ask whether advanced civilizations could ever alter the evolution of stars, of entire galaxies, or even of the universe itself.

As we discussed in Chapter 3, the hard part about determining the density of matter in the universe is finding the matter that does *not* shine. We can count galaxies and galaxy clusters well enough, but we cannot easily measure the density of matter spread out among the galaxies and galaxy clusters, or the amount of matter within galaxies that does not emit noticeable amounts of photons. The hardest problem of all is to guess how much of the universe's mass lies inside black holes, or in white dwarfs and neutron stars that have grown too old to be visible.

If the proportion of stars with various masses always remained constant, we could be relatively sure that most of the mass once in stars has not fallen into black holes. Stars with less than the sun's mass now contain more than 90% of the mass in our galaxy that appears in stars, and we believe that all of these lower-mass stars will end up as white dwarfs, not as black holes. But what if during the first few billion years of our galaxy, large numbers of high-mass stars were born, grew old, and collapsed into black holes? Could most of our own galaxy's mass—and thus, by implication, most of other galaxies' masses—now reside in black holes? At the present time we can only say that the total mass in black holes inside our galaxy could not be more than about five times the mass in stars that shine, provided that the mass in black holes is distributed through the galaxy in about the same way as the mass in stars. Five times! Quite a lot, considering that the mass in stars is 10^{11} times the sun's mass. Furthermore, this possibility comes tantalizingly close to providing the extra mass that would explain the observations concerning the "deceleration" of the universal expansion (Page 78). If galaxies have, say, one black hole with 25 solar masses for every ten stars with an average mass one half of the sun's, then the extra mass might explain why the "deceleration parameter" observations seem to indicate that the universe may cease to expand and begin to contract. Also, additional black holes might be located outside galaxies; perhaps entire galaxies with billions and billions of stars have become this many black holes by now, because all the stars had masses too great to become neutron stars or white dwarfs.

When we try to determine the amount of mass in a galaxy cluster from observing the motions of the galaxies in the cluster, we find that an average cluster should have five or ten times more mass than the galaxies (counting by their visible light) or we would observe the galaxies all escaping from the cluster. Whether the additional mass exists as diffuse intergalactic gas or as invisible black holes, or whether the observations are somehow incorrect, remains a puzzle for the next generation of astronomers to solve. But we would feel much more comfortable about the entire black hole situation if we could just see even one black hole as it formed, to confirm that black holes really exist.

Gravity Waves

The best hope for "seeing" a black hole as it forms seems to be to detect the "gravity waves" from a collapsing star. Gravity waves are similar to electromagnetic waves, but they consist of "gravitons" instead of photons. Gravitons, like photons, should have no mass and always travel at the speed of light. We can produce a gravity wave by wiggling some mass back and forth, just as we can produce electromagnetic waves (photons) by wiggling an electric charge back and forth. When gravity waves strike an object with mass, they should make the mass wiggle back and forth, as electromagnetic waves will make an electric charge oscillate when they hit it. The difficulty is that gravity waves are so incredibly weak. On the elementary-particle level, gravitational forces are about 10^{39} times weaker than electromagnetic forces (Page 47). In large bodies, gravitational forces overcome this weakness by combining and combining, so that 10^{57} particles in the sun all add their attractive forces, whereas the electromagnetic forces cancel. We need a *huge* object to overcome the inherent weakness of gravity: Even if we spin a steel I-beam twenty meters long a hundred times per second, the gravity waves that the beam emits carry an energy less than 10^{-20} erg, which is less than one trillionth of the energy released in a *single* proton-proton fusion reaction!

Thus it might seem that we never could hope to observe gravity waves, but again the enormous mass of stars allows them to overcome this deficiency. If an entire star collapses in a few seconds or less, the gravity waves from this rapid motion of a large object can carry away as much energy as the photons can. Since gravity waves come from rapid changes in the positions of large masses, we need the star to collapse as swiftly as possible to make a detectable gravity wave. We can try to observe gravity waves from collapsing stars by building a gravity-wave detector (Figure 12–14). We suspend a large mass in perfect balance, and wait for it to oscillate when gravity waves strike it. These gravity-wave detectors now exist at the University of Maryland and elsewhere. They are huge aluminum cylinders carefully isolated from outside influences, so precisely balanced that they can record a change of less than 10^{-14} centimeter in the distance between the two ends of the cylinder!

The deployment of these gravity-wave detectors has begun only recently, but already there seem to be preliminary indications that they have detected gravity waves from outside the Earth. To determine the *direction* from which these gravity waves arrive is not easy, because the detectors are only slightly more sensitive in one direction than another. However, the preliminary studies suggest that the greatest source of gravity waves, as measured from the Earth, lies in the direction of the center of the Milky Way galaxy.[6] If these results are confirmed by further experiments, they lead to the conclusion that a black hole—or a group of black holes—near the center of our galaxy is producing enough energy in gravity waves each year to equal the energy of mass of a thousand solar masses! That is, each year near the galactic center, one thousand times the sun's mass is being turned into the energy of motion of

---

[6]As a matter of uneasy fact, other experiments designed to test the original results obtained by Professor Joseph Weber have failed to confirm Professor Weber's results. This disagreement has provoked considerable acrimony and debate among gravity-wave scientists.

Figure 12–14. A gravity-wave detector constructed at the University of Maryland by Professor Joseph Weber consists of an aluminum cylinder hung with extreme care and monitored with incredible precision. The cylinder weighs about four tons (four thousand kilograms).

the gravitons that form gravity waves. This energy, and the mass that produced it, soon would vanish from the galaxy, just as photons eventually take their energy far from their point of origin. If the gravity waves are the last gasp of matter falling into black holes, then many times one thousand solar masses per year might be falling in, with a fraction of this mass (equal to one thousand solar masses) radiated away in gravity waves. Some calculations suggest that as matter falls into black holes, about one percent of its energy of mass is converted into gravity waves, while the other 99% either ends up inside the black hole or escapes as photon energy as the matter falls in.

The figure of one thousand solar masses—or much more—being lost from the galaxy each year doesn't seem right. Our galaxy has only a hundred billion, or at most a trillion (10^{12}) solar masses. If it loses a thousand solar masses or more each year, the galaxy would last at most only a billion years, yet it has been here ten billion years or more. Something must be wrong either with the observations, or with the conclusions we derive from them. Still, gravity waves indeed may already have been detected; if the experimental results are anywhere near correct, they indicate a tremendous outpouring of energy in gravity waves from the direction of the center of our galaxy.

What we would like to do is to detect the burst of gravity waves from the collapse of an individual star. Then we could wait a while to see whether we observe a pulsar (hence a neutron star) at this location; if we don't, we would imagine that we saw the formation of a black hole. Given the steady improvement in the technology of gravity-wave detectors, we may be able during the next decade to find the true explanation of the apparent gravity waves from the galactic center, and to detect individual bursts of gravity waves as stars in our own galaxy—or even in other galaxies—collapse. Just as radio astronomy (in the 1950's) and x-ray astronomy (in the 1960's) have opened up new prospects in electromagnetic radiation (photons), so too we can hope to investigate gravitational radiation through gravity-wave astronomy. Already, as with many a new observational technique, gravity-wave astronomy apparently has produced a puzzle—the gravitational radiation from the galactic center—that may rank with the photons from quasars in demanding new insights into the laws of physics for a successful solution.

Modification of Galaxies by Advanced Civilizations

During the past few generations, humanity has become able to alter its home planet; we now can change the weather, the composition of the atmosphere, the level of the oceans, and the patterns of vegetation. Most of these changes, if they were to occur rapidly on a biological time scale, would harm the prospects of life on Earth, so we make them only inadvertently. Through such unplanned changes we may eliminate all life on Earth, or at least all human life, or we may go on to discover how to reach harmony with our environment even as we evolve.

It is no coincidence that humanity faces the problem of eliminating itself just as we become capable of communicating with other civilizations. The technological advances that have given us the capacity to build radio telescopes, and to send radio and television messages, also threaten our continued existence, if we underestimate our capacity for error. If we, or any other civilization, should manage to pass through this initial flush of power over our ability to alter the home planet, we may do so by realizing that with the right attitude, the human race might last not just millions of years, but hundreds of millions or even billions. Such a civilization could take the long view of affairs, unafraid that self-generated catastrophes would destroy it.

Such a long-living civilization might not have much interest in advanced technology, or it might decide that the price of such an interest would be self-destruction. However, if we are any guide, civilizations will strive toward those technological changes that they think will increase their well-being. What sort of changes could we expect to see?

Well, first of all, genetic ones. Even today, scientists are close to figuring out in detail how organisms duplicate themselves. Before long, we probably shall be able to create the life forms we want. This might not happen for several generations, but if we survive far longer than that such results seem extremely likely. If we could manipulate genetic material at will, we could effectively assure our own immortality, either by halting the process of aging, or by providing our brains with new bodies. Other civilizations might have acquired this ability long ago, and will have set out to restructure their planetary systems and the stars around them for the long haul.

A growing civilization with time on its hands might decide to use all the power from its parent star. On Earth we now seek to use solar power cheaply, as we deplete fossil fuels at an ever-increasing rate. If we could turn only one hundredth of a percent of the energy of motion of the sun's photons that reach the Earth into useful energy, we would have an energy supply equal to all the energy of motion we obtain by burning coal and oil. Yet the Earth intercepts less than one billionth of the photon energy that the sun emits. Hence by capturing all of the sun's photons, we could have an energy source ten trillion times greater than our fossil fuels can provide for only a few centuries at best. Some civilizations may have done just this with their own stars: They might have colonized an entire sphere around the stars to support a population far greater than any planet could hold.

How would such colonization begin? Gerard O'Neill has pointed out that even today we could construct rotating cylinders in orbit around the sun that run on solar energy. Each of these cylinders, 20 kilometers long and five kilometers across, could have a population of ten thousand to a hundred thousand people (Figure 12–15). The cylinders' rotation would provide an analogue to the Earth's gravity, although the reflection of sun-

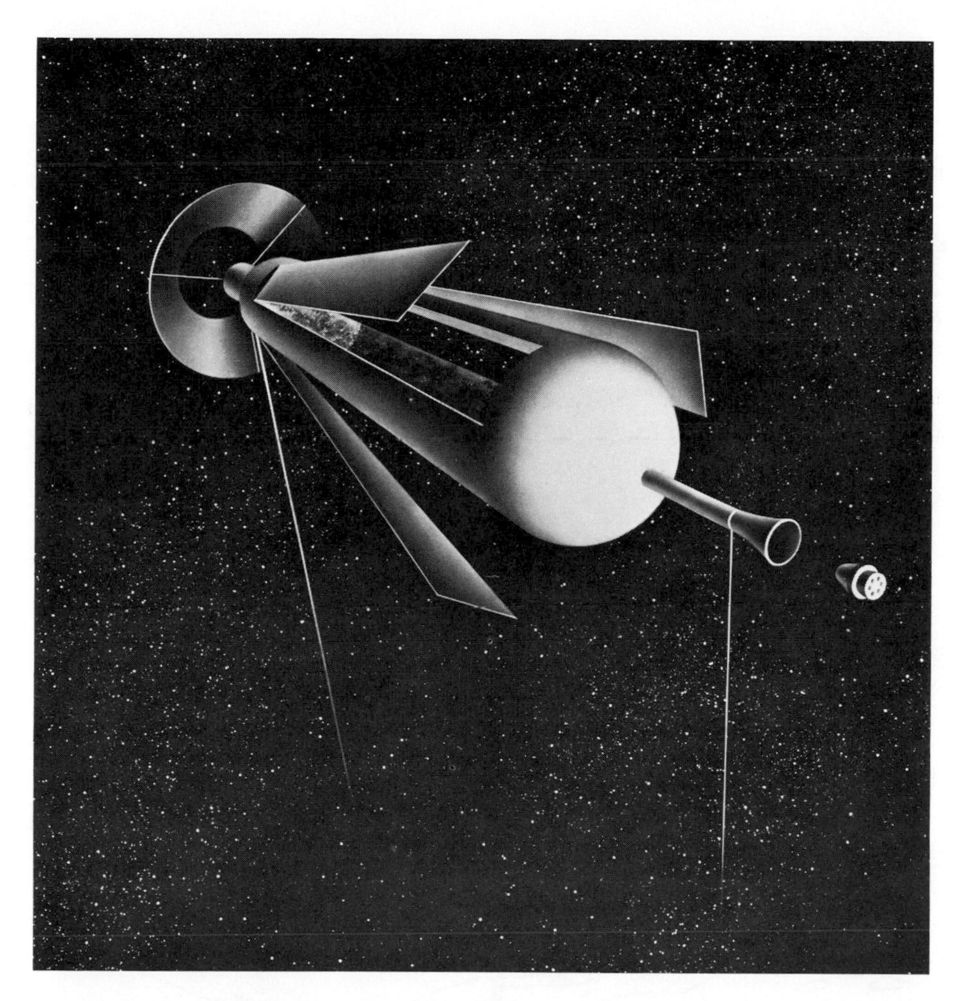

Figure 12–15. The physicist Gerard O'Neill has suggested that we now could begin the colonization of space by constructing cylindrical environments such as the one drawn here. The mirror at the upper left helps reflect sunlight onto the three planar mirrors and in through windows between the "land." If each cylinder were five kilometers across and 25 kilometers long, and rotated once every two minutes, people living on the inside would feel an acceptable substitute for Earth gravity on the floor of the cylindrical living space.

light that O'Neill plans would make for weird horizons (Figure 12–16). Initially we would have to make such cylindrical colonies from materials sent from Earth, but eventually the space colonies could obtain their basic raw materials from, say, the asteroid belt. The asteroid Ceres alone contains about one ten-thousandth of the Earth's mass, which could make a layer two hundred meters deep all over the Earth. Thus Ceres contains far more matter than we are likely to excavate here, enough matter in fact to make a hundred million space colonies. Each

◀ **Figure 12–16.** Inside the rotating cylinder, each of the three "land" strips would see the other two strips high in the sky. The apparent "gravity" would decrease as one sailed toward the center line of the cylinder, so that sports such as skiing and hang-gliding could be practiced in novel ways.

colony might have a different sort of population and a different political system, or they might all form one giant totalitarian network.

When we look from the human viewpoint at the idea of constructing a hundred million space colonies, each with a hundred thousand people, our minds may boggle. Still this sort of thing may happen someday, and it may well already have happened around other stars in our galaxy. If a civilization has arranged itself all around its parent star to use the star's entire energy output, we would not be able to see the star's visible-light photons. Such civilizations would emit infrared photons, as we on Earth emit photons from the heat liberated in our cities and factories. Thus the likeliest candidates for the sites of these civilizations are the "infrared stars" that emit low-energy photons and no visible-light photons. However, we think that most of the sources of infrared photons found so far in fact are extremely young stars still condensing out of interstellar gas clouds (Page 176). The best way to find an advanced civilization for sure would be to listen to its radio and television, an effort that we have not yet seriously undertaken (Page 353).

If some civilizations support themselves by capturing all of their star's photons, they might start to colonize nearby stars, and might spread all through their galaxy. Such colonization could take millions of years for the travel alone, but the civilization will see itself as having tens of millions of years to work in. The centers of spiral galaxies, which emit especially large amounts of energy (Page 370), might attract the attention of these civilizations, and we can envision a galaxy where one or more civilizations has taken over the entire photon output from the center or even from the complete galaxy. Of course, there may be many a civilization with no expansionist tendencies, but if any of the colonizing type do exist, they are the ones that will try to take over their home galaxies. We can say for sure that this corner of the Milky Way remains free, a New World waiting for its explorers and exploiters. Perhaps this concept of increasing, overwhelming mechanisms of colonization remains merely the sort of paranoid fantasies that lead nations to war. Since stars are rather efficient liberators of energy, the easy method of controlled thermonuclear fusion consists of capturing the energy from stars rather than trying to construct thermonuclear fusion reactors on planetary surfaces. The extent to which other civilizations have made this logical possibility an everyday reality remains unresolved.

When we turn to the realm of galaxy clusters and the universe at large, we reach the comforting thought that no civilization is likely to affect matters over such large distance scales. A galaxy cluster spans perhaps thirty million parsecs, thus some super-civilization that could travel at nearly the speed of light would be able to colonize it in, say, less than three hundred million years.[7] But as we consider the distances between galaxy clusters, we see that if it would take many hundred millions, or even billions of years, to travel between them, there has simply not been enough time for any civilization to take over thousands of galaxy clusters billions of parsecs apart. As for altering the universal expansion, by our current understanding of the problem, there's not a chance. All of space somehow started to expand about fifteen billion years ago, and only gravity can make it stop. Without producing new mass, everywhere, no one can slow the universal rate of expansion.

But it won't do to act overwise. Just as this chapter would have been dismissed as untested moonshine a generation ago, so too another generation of scientists may open new areas of knowledge and speculation undreamt of before. It is traditional to end textbooks on a happy note to reward the reader who can finish, but I believe that it doesn't much matter what we think about the universe, whose joys and sorrows far outreach our capacity to wonder.

Summary

Black holes arise from objects that shrink so much that the increasing gravitational force at their *surfaces* prevents anything—even light or radio waves—from escaping. Gravity affects all particles, including massless photons and neutrinos. If the gravitational force grows strong enough at an object's surface, nothing can leave and the object can send no message to the outside universe, save for its continuing gravitational force and its amount of angular momentum. Matter can fall into a black hole, but nothing that passes inside the critical radius around the center of a black hole will ever emerge.

The objects most likely to become black holes are the collapsing cores of some stars that explode as supernovae. In a double-star system, such a collapsed core could continue in orbit with the other, "normal" star,

---

[7]The colonizers would be in for some rude surprises if they reached a galaxy made of antimatter (or, if the colonizers were made of antimatter, if they reached a galaxy made of matter).

and astronomers think that they may have found some such star–black hole pairs by means of the photons produced as matter falls toward the black hole. Any object, in theory, can become a black hole, and we someday might find black holes with millions of solar masses (as have been suggested to lie at the centers of galaxies) or even greater masses. The universe itself could turn out to be the inside of a fantastic black hole, although this does not "prove" that an outside to the black hole necessarily exists.

Some collapsing stellar cores do not form black holes, but become neutron stars. Neutron stars have sizes slightly larger than the critical size at which they inevitably would become black holes. A neutron star twice the sun's mass, for example, would have a radius of about ten kilometers rather than the six kilometers that would make it a black hole. Stars that collapse to form neutron stars should end up rotating quite rapidly, and astronomers believe that "pulsars" contain such neutron stars, which drag an intense magnetic field around with them as they rotate. The magnetic field's rapid motion accelerates charged particles away from the neutron star's surface, thereby causing (we believe) the regular pulses of radio-wave, visible-light, and x-ray photons that pulsars emit.

Questions

1. What is a black hole? What makes it work?
2. If the sun decreased its radius about 200,000 times, it would become a black hole. How many times stronger would the gravitational force at the sun's surface become as a result of such a contraction?
3. Is the sun likely to become a black hole? What types of stars are the most likely to collapse so far that they become black holes?
4. How can we hope to discover whether a particular object really is a black hole?
5. How massive can a black hole be?
6. Can the entire "universe" be the inside of a black hole? What arguments against this view exist?
7. What produces neutron stars? What do neutron stars produce?
8. Can a neutron star expand to become a white-dwarf star? Can a white-dwarf star shrink so much that it becomes a neutron star? Will neutron stars eventually become black holes?

9. Why do pulsars blink on and off with an almost constant interval between photon pulses?
10. Why does the interval between pulses slowly change? Which is likely to be younger, a pulsar that pulses ten times per second, or a pulsar that pulses once every two seconds?
11. Why do we have such great difficulty in determining the average density of matter in the universe?
12. What are gravity waves? How are they related to collapsing stars?

Further Reading

J. Taylor, *Black Holes* (Avon Books, New York, 1974).

J. Ostriker, "Pulsars," *New Frontiers in Astronomy* (W. H. Freeman and Co., San Francisco, 1975).

K. Thorne, "The Search for Black Holes," *New Frontiers in Astronomy* (W. H. Freeman and Co., San Francisco, 1975).

H. Gursky and E. van den Heuvel, "X-Ray Emitting Double Stars," *New Frontiers in Astronomy* (W. H. Freeman and Co., San Francisco, 1975).

H. Shipman, *Black Holes, Quasars, and the Universe* (Houghton Mifflin, Boston, 1976).

Appendix

Glossary

Absolute brightness: The true, or intrinsic, brightness of an object, as contrasted with its apparent brightness.

Absorption: The removal of photons of a particular energy, frequency, and wavelength from a beam of light, usually the result of the photons' interaction with atoms or molecules.

Absorption line: A dark feature in the spectrum of photon energies of a source of photons, caused by the absorption of photons of a particular energy or wavelength.

Amino acid: Simple molecules that contain an amino group (NH_2) along with carbon, oxygen, and other hydrogen atoms, and form the building blocks of proteins.

Angular momentum: A measure of the amount of spin an object possesses, determined by multiplying the mass of the object times its rate of spin times the square of the size of the object in the direction perpendicular to its spin axis.

Angular size: The fraction of a circle (360 degrees) over which an object appears to extend as we see it, measured in degrees of arc, minutes of arc (one sixtieth of a degree), or seconds of arc (one sixtieth of a minute).

Antimatter: The complementary form of "ordinary" matter, which possesses the same mass but opposite electric charge as the corresponding particles of matter.

Apparent brightness: The brightness or luminosity that an object appears to have as we see it.

Astronomical unit: The distance from the Earth to the sun, equal to 149,500,000 kilometers.

Atom: The smallest unit of an element, consisting of a nucleus of protons and neutrons, surrounded by one or more electrons in orbits around the nucleus.

Atomic number: The number of protons in the nucleus of an atom.

Atomic weight: The number of protons plus the number of neutrons in the nucleus of an atom.

Balmer series: The absorption lines produced by hydrogen atoms when the atoms absorb photons that can make the atom's electron jump from the second-smallest to a larger orbit; conversely, the photons that are emitted at the same energies as the Balmer absorption lines when a hydrogen atom's electron jumps from a larger orbit into the second-smallest orbit.

Big bang: The original explosion that began the expanding universe some fifteen to twenty billion years ago.

Black hole: An object so compacted that the gravitational force at its surface prevents any particle or photon from escaping from the surface.

Bremsstrahlung: The emission of photons caused by the close passage of charged particles, such as electrons.

Center of mass: The point within an object, or within a group of objects, that makes the quantity of mass times distance the same on either side of that point in any direction.

Cepheid variable: A member of a class of stars that vary periodically in light output, with large absolute brightnesses that are related to the stars' periods of light fluctuation.

Closed universe: A model of the universe that has no boundaries but nonetheless contains only a finite volume; the three-dimensional analogue of the two-dimensional surface of a sphere.

Comet: A fragment of primordial solar-system matter, orbiting the sun as a lump of gases frozen around dust and rocks, which can produce a tail several million kilometers long when close to the sun as the result of the vaporization of some of its material.

Corona: The outermost atmosphere of the sun, heated from below to several million degrees absolute, and extending past the Earth as an expanding solar wind.

DNA: Deoxyribonucleic acid, the part of the chromosomal material that determines inherited characteristics, made of two long helical chains of molecules joined by cross-linking pairs of molecules.

Deceleration parameter: A measure of how the rate of expansion of the universe differed in the past from its present value.

De-excitation: The passage of an electron in an atom from a larger to a smaller orbit, following naturally after the passage from a smaller to a larger orbit, called atomic excitation.

Degenerate matter: Matter in which the exclusion principle plays an important role, not just in the structure of the atoms that form the matter, but in the ability of the particles in the matter to assume otherwise natural positions and velocities. (Also see exclusion principle.)

Density wave: A pattern of alternately greater and lesser densities thought to rotate around the centers of spiral galaxies.

Doppler effect: The change in energy, frequency, and wavelength observed for photons arriving from a source that moves with respect to an observer, along the observer's line of sight.

Eccentricity: A measure of the elongation of an orbit, ranging from zero (for a circular orbit) to almost unity (for a greatly elongated orbit); eccentricities of unity or greater imply that the orbit does not form a closed curve.

Eclipse: The partial or total obscuration of one object by another, usually of the sun by the moon or of the moon by the sun.

Electromagnetic forces: Forces that act between electrically charged particles, either of attraction (for opposite charges) or of repulsion (for like charges), decreasing in strength as one over the square of the distance between the particles.

Electromagnetic radiation: Streams of photons carrying energy of motion outward from a source of radiation.

Electron: An elementary particle with one unit of negative charge and a mass of 9.1×10^{-28} gram. Electrons form the outermost particles of atoms in orbit around the nucleus.

Element: The set of atoms that all have the same number of protons in the nucleus.

Elementary particle: One of the basic units of nature (although perhaps divisible into still other particles), such as electrons, protons, and neutrons.

Elliptical galaxy: A galaxy whose shape, seen in a photograph, is that of an ellipse or of a circle; in fact, the galaxies have an "ellipsoidal" shape.

Emission: The production of photons of a particular energy or of a range of energies.

Energy of mass: The energy equivalent to a given mass, equal to the amount of mass multiplied by the square of the speed of light; the amount of energy (potential for doing work) that would be obtained if all of the mass of an object were converted into energy of motion.

Energy of motion: The energy associated with the motion of a particle or of a group of particles.

Enzymes: Certain proteins that catalyze and regulate processes within living organisms.

Excitation: The passage of an electron in an atom from a smaller to a larger orbit, usually caused either by photon impact or by collisions among atoms.

Exclusion principle: A summary of the observational fact that for certain kinds of elementary particles within a collection of identical particles (in particular, for electrons, protons, neutrons, and nuclei with an odd number of nucleons), no two particles can be at almost the same location with almost the same velocity.

Exploding galaxy: A galaxy in which, so far as we can tell, a violent outburst has occurred within the past few million years.

Frequency: The number of times that a given photon oscillates per second, measured in cycles per second (also called "Hertz").

Galaxy: A large group of stars, containing anywhere from one million (10^6) to one trillion (10^{12}) stars, and having a diameter from one thousand to one hundred thousand parsecs.

Galaxy cluster: A concentration of galaxies, containing between ten and a few thousand member galaxies, moving in orbit around their common center of mass under the influence of their mutual gravitational attractive forces.

Gamma rays: Photons with the largest energies, usually defined as having energies in excess of one millionth (10^{-6}) of an erg.

Giant planets: The four planets Jupiter, Saturn, Uranus, and Neptune within the solar system, each more than a dozen times more massive than the Earth and far larger in diameter.

Globular cluster: A spherical or ellipsoidal agglomeration of ten thousand or more stars, with a total size of a few parsecs, representing some of the first concentrations of matter to form within a contracting protogalaxy.

Gravitational forces: Forces of attraction that act between all particles. These forces are proportional to the product of the masses of the particles and to one over the square of the distances between them.

Gravitational redshift: A decrease in the energy of photons leaving the surface of an object. The shift is caused by the loss of energy in opposing the force of gravity and is proportional to the mass of the object divided by the radius of its surface.

H II region: A sphere of ionized hydrogen and other gases surrounding a bright, hot star (or stars) that emits photons capable of ionizing the neighboring gas, thereby causing the gas to produce photons when the ionized atoms recombine into hydrogen atoms (H I region).

Helium flash: The sudden increase in absolute brightness that occurs within an aging star as the star's central regions become hot enough to make helium nuclei fuse.

Hertz: Scientific notation for a frequency of one cycle per second.

Hubble's law: The summary of Hubble's discovery of the expanding universe, that the recession velocities of distant galaxies equal a constant, H (the "Hubble constant"), times the galaxies' distances from us.

Ideal radiator: An object in complete equilibrium with its surroundings that emits and absorbs the same number of photons every second, thereby producing a characteristic shape in the spectrum of its photon emission.

Inert atoms: Atoms that participate in chemical reactions most unwillingly or not at all, as the result of having the number of their outer electrons equal to the maximum number of electrons allowed in that orbit.

Infrared radiation: Photons with slightly smaller energies and frequencies, and slightly greater wavelengths, than the photons that form visible light.

Instability strip: The region of the temperature–luminosity diagram that contains stars that are late in their evolutionary history and are likely to pulsate in size and absolute luminosity.

Interstellar matter: Material spread among the stars in spiral and irregular galaxies, consisting mainly of hydrogen atoms and hydrogen molecules, with a few percent (by mass) of dust particles that absorb photons over a wide range of photon energies.

Ion: An atom from which one or more electrons has been removed or added.

Ionization: The formation of ions by the removal of electrons from atoms, either by photon impact ("photo-ionization") or by collisions among atoms.

Irregular galaxy: A galaxy that is classified as neither spiral nor elliptical.

Isotopes: Atoms that have the same number of *protons* in the nucleus. Since the number of *neutrons* can vary among isotopes of the same element, isotopes all have the same atomic number but have different atomic weights.

Kinetic energy: See energy of motion.

Light: Certain types of electromagnetic radiation, usually defined as either visible light or visible light plus infrared and ultraviolet radiation.

Light year: The distance light travels in a year, equal to about ten million kilometers or 0.3 parsecs.

Local Group: The small cluster of about seventeen galaxies to which the Milky Way belongs.

Magellanic Clouds: Two irregular galaxies that are satellites of our Milky Way; the closest galaxies to our own but much smaller than the Milky Way galaxy.

Magnitude: A measure of relative brightness, most often used to compare apparent brightnesses, in which each additional magnitude implies an additional faintness of 2.512 times, so five additional magnitudes imply an additional faintness of $100 = (2.512)^5$ times.

Mantle: The layer of the Earth (or of one of the other terrestrial planets) between the crust and the inner core.

Maria: The Latin name for the lunar "seas," which are actually plains of former lava flows.

Mass: A measure of the amount of material in an object, as determined either from the object's resistance to an accelerating force or from the object's ability to exert gravitational force.

Median plane: The plane of symmetry in a spiral galaxy that bisects such a galaxy into a nearly identical top half and bottom half.

Megaparsec: One million parsecs.

Meteor: One of numerous fragments, ranging in size from specks of dust to large boulders, that orbit in swarms around the sun and appear as glowing streaks in the sky from their friction in the Earth's atmosphere if they traverse it;

more technically, the bodies are "meteoroids" and the streaks are the meteors.

Meteorite: The part of a meteor that survives its passage through the Earth's atmosphere and collides with the Earth's surface.

Microwaves: Electromagnetic radiation consisting of photons whose energies and wavelengths place them in the high-energy, or short-wavelength, part of the radio spectrum, usually defined as those photons with wavelengths between one millimeter and a few centimeters.

Microwave background: The photon emission left over from the first few minutes after the big bang, which has not interacted significantly with matter since a time 300,000 years after the big bang and carries an energy spectrum characteristic of an ideal radiator at a temperature of three degrees above absolute zero.

Milky Way: The galaxy of which the sun and its planets are a part, whose central regions appear as a band of light on the sky made up of the photon emission from millions of stars closer to the galactic center than the sun.

Minute of arc: A measure of angular distance, equal to one sixtieth of a degree, or 1/21,600 of the circumference of a complete circle.

Molecule: A stable grouping of two or more atoms, bound together by the electromagnetic forces among the electrons and nuclei in the atoms.

Neutrino: An elementary particle with no mass and no electric charge (but quite different from a photon), characteristically emitted or absorbed in reactions among elementary particles that involve weak forces.

Neutron: An elementary particle with a mass of 1.6747×10^{-24} gram and no electric charge, which is stable when bound by strong forces into an atomic nucleus, but if left by itself will decay after about a thousand seconds into a proton, an electron, and an antineutrino.

Neutron star: A highly compacted object, formed from a collapsed star, in which most of the protons and electrons have combined to form neutrons under the immense force of the star's self-gravitation.

Nucleic acid: One of a class of complex molecules, of which DNA is the best example, involved in the functioning and reproduction of living cells.

Nucleon: A proton or a neutron.

Nucleus: The central region of an atom, composed of one or more positively charged protons and none or more neutrons, which have no electric charge.

Open cluster: A loose grouping of stars, numbering several dozen to several hundred or a thousand, several parsecs in diameter and typically younger than a globular cluster.

Open universe: A model for the universe in which the extent of space is infinite.

Organic: Strictly speaking, compounds containing carbon atoms; more loosely, having properties associated with life on Earth.

Ozone: Molecules made of three oxygen atoms (O_3), as contrasted with ordinary oxygen molecules (O_2), which shield the Earth's surface from ultraviolet radiation.

Parallax: The apparent change in position of an object caused by the motion of the observer; for stars, parallax usually can be detected only as relative motion with respect to a background of more distant stars.

Parsec: A unit of distance, equal to 3.258 light years or about thirty trillion kilometers, and equal to the distance to an object whose apparent shift back and forth as a result of the Earth's orbiting the sun is one second of arc in each direction from the object's average position.

Photo-ionization: Ionization produced by photon impact.

Photon: The elementary particle that forms electromagnetic radiation, having no mass and no electric charge, and always traveling (in a vacuum) at the speed of light, c = 300,000 kilometers per second.

Planet: The nine largest bodies in orbit around the sun, ranging in mass from one thousandth of the sun's mass to one ten-millionth of the sun's mass.

Planetary nebula: An envelope of gas surrounding an aging star. The star emits enough high-energy photons to excite the gas around it and to make the gas emit photons upon de-excitation.

Population I: The younger stars, formed after an earlier group of stars had seeded the protostar material with elements heavier than helium; the sun is a typical older Population I star, whereas the young Population I stars outline the spiral arms of galaxies like our own.

Population II: The older stars, typified by the members of a globular cluster, far less concentrated toward the median plane in our own and other galaxies than Population I stars.

Positron: An antielectron.

Proper motion: The apparent motion of a star through space that remains after we allow for the parallax effect produced by the Earth's motion around the sun; the amount of proper motion depends on the star's velocity with respect to ourselves and to its distance from us.

Protogalaxy: A galaxy in the process of formation.

Proton: An elementary particle with a mass of 1.6724×10^{-24} gram and one unit of positive electric charge; one of the basic constituents of atomic nuclei.

Proton-proton cycle: The series of three nuclear fusion reactions through which most stars liberate the bulk of their energy of motion from energy of mass.

Protostar: A star in the formation process that has not yet begun to liberate energy of motion through nuclear fusion reactions.

Quasi-stellar radio source (quasar): An object with a starlike appearance that emits large quantities of optical, infrared, and radio photons, and with an optical spectrum that indicates a sizable Doppler shift to lower energies.

RR Lyrae variable: A member of a class of stars that vary periodically in light output, having a slightly smaller true brightness and a slightly faster cycle of light variation than Cepheid variable stars.

Radio galaxy: A galaxy that emits as much, or more, energy per second in radio photons as in visible-light photons, in contrast to "normal" galaxies, which emit perhaps one millionth as much energy in radio photons as in visible-light photons each second.

Radio waves: Electromagnetic radiation made up of photons with energies less than 10^{-20} erg; the form of electromagnetic radiation with the least energy per photon.

Recombination: The re-formation of atoms following their ionization.

Red giant: A star that has evolved past the main-sequence phase and has expanded to a much greater size than it had as a main-sequence star.

Red shift: A shift in the spectrum of photons emitted by a photon source to lower energies, lower frequencies, and lower wavelengths as the result of the relative motion of the source away from the observer (see Doppler effect).

Second of arc: A measure of angular size, equal to one sixtieth of a minute of arc or 1/3600 of a degree.

Seyfert galaxy: A member of a class of spiral galaxies characterized by intense visible-light, infrared, and radio emission from a small nucleus (a few tens or hundreds of parsecs in diameter) at the center of the galaxy.

Solar system: The sun plus its family of planets, asteroids, meteoroids, and comets, all of which orbit around the sun.

Solar wind: The outermost parts of the solar corona, which expand past the Earth and the other planets as a stream of electrons and ions.

Spectrum: The distribution of photons emitted by a particular source, shown as the number of photons with a given energy plotted at each energy, or the number of photons with a given wavelength plotted at each wavelength; the photons themselves can be used to furnish this plot in the visible-light regions of the spectrum when we spread the light out into its various energies with a prism and photograph the result.

Spiral galaxy: A galaxy characterized by a flattened distribution of its stars and interstellar matter, showing spiral "arms" outlined by the peak density of hydrogen gas and by the young, hot stars that recently have condensed from the gas.

Star: A mass of gas bound together by its own self-gravitational forces and hot enough at the center to liberate energy of motion from energy of mass.

Steady state: A model of the universe in which the density of matter remains constant as the universe expands, with new matter appearing from nothing to offset the density decrease that arises from the universal expansion.

Each particle will, on the average, have twice as much energy of motion at a temperature of 400 degrees absolute as at 200 degrees absolute. Water freezes at 273.16 degrees above absolute zero, and boils at 373.16 degrees on the absolute scale. (Notice that there are no *negative* temperatures when we measure temperatures on the absolute scale.) The average surface temperature of the Earth is close to 290 degrees above absolute zero; the sun's surface has a temperature of 6000 degrees absolute, and its center is about 13 million degrees on the absolute scale (23 million degrees Fahrenheit).

Answers to Selected Questions

Chapter 1

3. One sixtieth ($1/60$) of a parsec, or about half a trillion kilometers.
4. Light takes 32½ years to travel the 10 parsecs from Arcturus, and 32½ million years to travel the ten million parsecs from M 87.
5. The record of human history would span 3200 seconds, or a little less than an hour.
6. A first-magnitude star is ten thousand (10^4) times brighter than an eleventh-magnitude star. A third-magnitude star is one hundred million (10^8) times brighter than a twenty-third-magnitude star.

Chapter 2

2. A photon with a wavelength of one centimeter has ten times the energy of a photon with a wavelength of ten centimeters.
4. The nearer star (Polaris) is $25/16 = 1.56$ times brighter than the farther star, as we see them.
5. The *increase* in energy for the photons from Eltanir is half as much as the *decrease* in energy for the photons from Aldebaran.
9. An atom of carbon-13 has six electrons.

Chapter 3

3. The Corona Borealis cluster is three times farther away.

Chapter 4

9. M 81 is twice as close to us as M 83 is.
10. M 87 has 16 times the true brightness of NGC 7793.

Chapter 5

9. The frequencies are reduced to $1/1.55 = 64.5\%$ and to $1/1.87 = 53.5\%$ of their original values. The photon energies are reduced by the same amounts as the frequencies. The farther quasar, 3C 196, is about $0.87/0.55 = 1.58$ times as far as the nearer one.

Chapter 7

14. The hotter stars are sixteen times as bright and eighty-one times as bright as the cooler stars.

Chapter 8

8. A star of ten solar masses with an absolute luminosity 4000 times the sun's will last about $4000/10 = 400$ times *less* than the sun on the main sequence, or about 25 million years.

Chapter 9

1. Jupiter would have a mass of six kilograms (6000 grams) in the model, and the Earth's mass would be twenty grams, about equal to the weight of four nickels.
2. The Earth would be 280 meters from the sun in the model, and Jupiter would be 1400 meters (1.4 kilometers) away.
5. A planet nine times the Earth's distance from the sun would orbit around the sun once in 27 earth-years. A planet 25 times farther from the sun than the Earth is would take 125 years to make one orbit.

Chapter 11

9. If one year represents 3.2 billion years, then in this model 64 years will be represented by 0.64 second.

Chapter 12

2. The gravitational force would increase forty billion (4×10^{10}) times.

Index